FOOD ADDITIVES TO EXTEND SHELF LIFE

WITHDRAWN
UTSA LIBRARIES

WITHDRAWN
UTSA LIBRARIES

FOOD ADDITIVES
TO EXTEND SHELF LIFE

Nicholas D. Pintauro

NOYES DATA CORPORATION

Park Ridge, New Jersey   London, England

1974

Copyright © 1974 by Nicholas D. Pintauro
    No part of this book may be reproduced in any form
    without permission in writing from the Publisher.
Library of Congress Catalog Card Number: 74-75902
ISBN: 0-8155-0548-5
Printed in the United States

Published in the United States of America by
Noyes Data Corporation
Noyes Building, Park Ridge, New Jersey 07656

LIBRARY
University of Texas
At San Antonio

FOREWORD

The detailed, descriptive information found in this book is based on U.S. patents relating to chemical additives which can be used in foods to prolong their shelf life. This book serves a double purpose in that it supplies detailed technical information and can be used as a guide to the U.S. patent literature in this field. By indicating all the information that is significant, and eliminating legal jargon and juristic phraseology, this book presents an advanced, commercially oriented review of how to prolong the shelf life of foods by means of chemical additives.

The U.S. patent literature is the largest and most comprehensive collection of technical information in the world. There is more practical, commercial, timely process information assembled here than is available from any other source. The technical information obtained from a patent is extremely reliable and comprehensive; sufficient information must be included to avoid rejection for "insufficient disclosure."

The patent literature covers a substantial amount of information not available in the journal literature. The patent literature is a prime source of basic commercially useful information. This information is overlooked by those who rely primarily on the periodical journal literature. It is realized that there is a lag between a patent application on a new process development and the granting of a patent, but it is felt that this may roughly parallel or even anticipate the lag in putting that development into commercial practice.

Many of these patents are being utilized commercially. Whether used or not, they offer opportunities for technological transfer. Also, a major purpose of this book is to describe the number of technical possibilities available, which may open up profitable areas of research and development. The information contained in this book will allow you to establish a sound background before launching into research in this field.

Advanced composition and production methods developed by Noyes Data are employed to bring our new durably bound books to you in a minimum of time. Special techniques are used to close the gap between "manuscript" and "completed book." Industrial technology is progressing so rapidly that time-honored, conventional typesetting, binding and shipping methods are no longer suitable. We have bypassed the delays in the conventional book publishing cycle and provide the user with an effective and convenient means of reviewing up-to-date information in depth.

There is an index of chemical additives. The Table of Contents is organized in such a way as to serve as a subject index by food category. Other indexes by company, inventor, and patent number help in providing easy access to the information contained in this book.

15 Reasons Why the U.S. Patent Office Literature Is Important to You —

1. The U.S. patent literature is the largest and most comprehensive collection of technical information in the world. There is more practical commercial process information assembled here than is available from any other source.

2. The technical information obtained from the patent literature is extremely comprehensive; sufficient information must be included to avoid rejection for "insufficient disclosure."

3. The patent literature is a prime source of basic commercially utilizable information. This information is overlooked by those who rely primarily on the periodical journal literature.

4. An important feature of the patent literature is that it can serve to avoid duplication of research and development.

5. Patents, unlike periodical literature, are bound by definition to contain new information, data and ideas.

6. It can serve as a source of new ideas in a different but related field, and may be outside the patent protection offered the original invention.

7. Since claims are narrowly defined, much valuable information is included that may be outside the legal protection afforded by the claims.

8. Patents discuss the difficulties associated with previous research, development or production techniques, and offer a specific method of overcoming problems. This gives clues to current process information that has not been published in periodicals or books.

9. Can aid in process design by providing a selection of alternate techniques. A powerful research and engineering tool.

10. Obtain licenses — many U.S. chemical patents have not been developed commercially.

11. Patents provide an excellent starting point for the next investigator.

12. Frequently, innovations derived from research are first disclosed in the patent literature, prior to coverage in the periodical literature.

13. Patents offer a most valuable method of keeping abreast of latest technologies, serving an individual's own "current awareness" program.

14. Copies of U.S. patents are easily obtained from the U.S. Patent Office at 50¢ a copy.

15. It is a creative source of ideas for those with imagination.

CONTENTS AND SUBJECT INDEX

INTRODUCTION

Primitive man was restricted to the food supply immediately at hand. This supply was determined in both quality and quanity by the soil and climate. When there was sufficient quantity, man devised methods to protect and extend his food supply beyond the immediate future and sometimes to the next season by sun-drying, salting, fermenting, smoking and other simple methods. Modern food manufacturers use canning, freezing, refrigeration, dehydration, and modern packaging as methods of preservation. Another effective modern approach for protection against spoilage is the use of additives.

Food additives, as defined by the National Academy of Sciences, are those chemicals that may be incorporated in foodstuffs, either directly or indirectly, during growing, storage or processing of foods. Every chemical should serve one or more of these general purposes: improve or maintain nutritional value, enhance quality, enhance consumer acceptability, improve keeping quality, and facilitate preparation. In modern applications, food additives are combined with classical methods of food preservation to maximize stability for extended shelf life.

One of the most important factors producing food deterioration is oxidation. Food oxidation occurs when the oxygen present unites with other chemical elements or groups of elements in foods to form oxides which cause the breakdown of fatty or other carbohydrate food components resulting in the ultimate deterioration of protein matter. The reactions arising out of and accompanying oxidation are believed to cause various chemical and biological changes resulting in such undesired results as rancidification of fats, molding of carbohydrates and putrefaction of proteins. Basically, the undesirable effects of food oxidation appear to stem primarily from microorganism activity and chemical changes and/or reactions.

All organic foods also contain microscopic forms of parasitic plant and vegetable life called microorganisms which live on larger forms of life. Among the most common microorganisms in food, useful to man and animal, are bacterial molds and yeasts. Not all such microorganisms, however, are necessarily harmful or undesirable. For example, lactic acid bacteria are usefully employed in making sauerkraut from fresh cabbage and silage from fresh grass. In most foods, however, the excessive growth of yeast, bacteria or molds indicates and accompanies spoilage.

In addition to microorganisms, raw organic foods contain substances called enzymes which appear to act somewhat as organic catalysts. A common example of such activity is demonstrated in the fruit ripening process. Even after the fruit is picked, enzyme activity

1

continues, and if not halted, eventually will cause the fruit to spoil and rot. Generally speaking, enzymes act upon foods in two ways: either to add to or reduce water content or take up oxygen. There are three generally recognized types of food enzyme activities: lipolytic, amylolytic and proteolytic, and each type produces marked effects which can contribute either to the spoilage or the preservation of foods.

For instance, lipoxidase, which is a member of the lipolytic enzyme family, acts on fats and may cause butter, for example, to turn rancid. Cellulase and diastase, contrastingly assist in the preservation of foods by converting cellulose and starches into sugars, alcohol and esters, the latter giving the characteristic flavors to foods. Cellulase is especially effective in hydrolyzing cellulose, while diastase, being high in alpha-amylase with some beta-amylase, is particularly active on starches. The proteolytic enzyme family includes, by way of example, papain, ficin, bromelin and others which generally act to hydrolyze and reduce proteins.

It is generally acknowledged that the processes and evolutions involved in food spoilage and deterioration are not fully understood, but it is known that they are greatly accelerated by the presence of air, moisture and heat, particularly in excessive amounts. Since these factors are naturally present, it is extremely difficult to preserve and store foods without taking abnormal precautions to avoid these factors. For example, the surface exposure of foods to air greatly accelerates rancidification of fatty components, which action weakens the natural antioxidants present in foods and thereby helps to bring about complex adverse effects on enzyme splitting systems which are interlocked in the protein, carbohydrate and fatty food constituents. In this way, food becomes susceptible to destructive attack by bacteria, molds and yeasts.

In the tremendous expansion of the food industry in the past twenty years there have been great strides to formulate, process and package foods to provide greater stability and longer shelf life. Consequently, there has been a great demand for additives to prevent or retard food deterioration. These additives include antioxidants, antibacterial agents, mold inhibitors, color stabilizers, anticaking agents, antibrowning agents, cloud stabilizers, metal scavengers, enzyme inhibitors, etc.

Food additives classified as preservatives totaled 50 million pounds in the United States in 1965. It is expected that this volume will reach 75 million pounds by 1975 for a value of $50 million at the chemical manufacturers' level. Processes for use of additives to extend shelf life of foods, as given in this volume, are organized according to food category or food group. It is believed that this approach will be better for the understanding of the reader than if the additives are discussed according to their function in several food groups. In addition, there is an index of additives at the end of the book.

The three Appendices include general information on food additives as published by the National Academy of Sciences, selected abstracts from the Federal Food, Drug, and Cosmetic Act, and several recent pronouncements on policy in the *Federal Register* by the Food and Drug Administration.

ANTIOXIDANTS FOR FATS AND OILS

BHA AND METHYL ANTHRANILATE

Rancidity is one of the major contributors to deterioration of fatty materials and is greatly promoted by the influence of air, heat, heavy metal ions, moisture and similar factors. It is well known that antioxidant materials may be effectively added to fats and oils to preserve them, principally by opposing oxidation and inhibiting related reactions promoted by oxygen and/or peroxides which are thought to be largely responsible for rancidification. Among the more common antioxidants commercially used for this purpose are such materials as butylated hydroxytoluene (BHT), butylated hydroxyanisole (BHA), nordihydroguaiaretic acid, ethoxyquin, propyl gallate, ascorbic acid and tocopherol.

Certain of these antioxidants appear to be more efficient in combination with one another and therefore various blends and compositions of antioxidant materials are offered commercially as fat preservatives according to the manufacturers' specifications. Rarely, however, do such compositions contain more than approximately 60% active ingredients.

A typical commercial formulation for an antioxidant fat preservative, for instance, consists of 20% BHT, 20% BHA, 10% propyl gallate, 10% citric acid and 40% carrier. While citric acid is not an antioxidant by itself, it appears to increase the activity of the commonly used antioxidants, promoting their general effectiveness and capabilities. Citric acid therefore, is referred to commonly as a synergist because of its apparent ability to increase the activity of the antioxidants over and above their normal capabilities.

While such known antioxidant compositions and formulations have proven generally successful in their application as preservatives for fats, oils and fatty substances, nevertheless they are limited in their application and effectiveness. Among such limitations is their limited capability of preserving fatty substances over prolonged periods. Additionally, most such commercially available antioxidant compositions readily break down and are ineffective at temperatures substantially in excess of 180°F. This means that these known preservative materials have little or no active ability to restrict and control unfavorable oxidative reactions which take place during normal cooking and rendering processes when reducing raw fatty substances to commercial grade products.

As a result, practice dictates the use of these materials by adding them to the fats after the latter have been rendered and cooled to temperatures below 180°F. This means that normal autooxidative reactions which are accelerated by the extremes of atmospheric oxygen, moisture and heat, present in the rendering processes, remain unchecked during the

3

rendering stages so that the burning of carbonaceous materials and oxidative rancidity take place freely in the rendering stages. This leads to lower quality of the produced fats and cracklings, particularly as to color and free fatty acid values which greatly affect their commercial value.

In the area of edible fats, particularly those used in animal foods, for example, such known synthetic antioxidant preservatives are further subject to certain limitations and restrictions of use due to their known toxic effects at higher percentages. Consequently, tallows and greases which are preserved by known antioxidant preservatives, applied within their permissible nontoxic limits, cannot, as a rule, be sufficiently stabilized against rancidity thereby, to prevent their eventual contamination and loss of quality and palatability.

With respect to stabilization periods, one of the more accepted methods for determining the degree of stabilization of fats treated with antioxidants is based on the number of hours' protection afforded to the fat. This is determined by the Accelerated Oxygen Method of testing stability and is commonly referred to as Hours AOM Protection. Recognized authorities on fat stabilization, more particularly The American Meat Institute, recommend 20 hours AOM Protection as the minimum standard requirements for preservation of fats. This period is said to be equivalent to substantially 6 months' protection against rancidity under refrigerated storage conditions.

An effective combination of antioxidant materials is described by *T.B. Tribble and E.L. Rondenet; U.S. Patent 3,318,819; May 9, 1967; assigned to Flavor Corporation of America.* The combination consists of known antioxidant materials such as BHT and BHA, and ortho-aminobenzoic acid (anthranilic acid) or its esters, such as methyl anthranilate and ethyl anthranilate.

Free fatty acid content of fats is a determinative indication of the rate of oxidative potential. In general, free fatty acid (FFA) results from the breakdown of fat with the dissipation of glycerine, leaving fatty acid. The FFA number of fat analysis refers to the percentage of free fatty acid present, and generally speaking, fats having a low FFA number are preferred and considered to have a greater commercial value and quality than fats with higher FFA numbers. It has been found that these fat preservatives are very effective in lowering the FFA or free fatty acid content of the treated product.

In addition to the free fatty acid content, the color rating of fat is also commercially important. Color rating is referred to as the FAC color number in fat analysis; the same being based on a color reading matched against samples of fat having standardized colors as recommended by the Fat Analysis Committee of the American Oil Chemists' Society. Generally speaking, the FAC color number usually assays substantially one or two times the FFA acid number and since this ratio appears to be fairly consistent, the two numbers are deemed to have an interrelationship. This relationship is demonstrated by the lowering of the FFA or fatty acid number of fat being accompanied by a lowering of the FAC color number as well.

In theoretical consideration, the higher the FFA number of fat, the greater amount of antioxidant required to stabilize the same against oxidative rancidity. Fats having high acid numbers are usually classified as inedible and unfit for human consumption, although they are acceptable for use in animal foods. Since this preservative demonstrates a definite ability to lower the free fatty acid content of fats treated therewith, it likewise has the ability to lower the FAC color number, thereby resulting in finished greases and tallows having low FAC color ratings.

Still another characterizing feature of these preservatives resides in the apparent continuing operative activity of the antioxidant material in the treated fats after rendering. This continuing antioxidant activity brings about a progressive decline in both the acid and color number values of fats even after the same are placed in storage after rendering. This feature therefore causes fats treated with these antioxidant preservatives to show progressive improvement in FFA acid number and FAC color number while in storage, resulting in improvement of their commercial value. Some of the results produced by the antioxidant treatments are illustrated in the following Table 1.

TABLE 1: NORMA-HOFFMAN STABILITY

	Sample Treated at Temperatures Below 180°F	Time to 5-lb Pressure Drop, hr*
1	Untreated choice-grade tallow	2.0
2	Tallow plus 0.05% anthranilic acid	2.0
3	Tallow plus 0.05% methyl anthranilate	3.0
4	Tallow plus 0.05% citric acid	2.5
5	Tallow plus 0.025% citric acid plus 0.025% anthranilic acid	3.0
6	Tallow plus 0.05% BHA	3.0
7	Tallow plus 0.05% BHA plus 0.025% anthranilic acid	3.5
8	Tallow plus 0.05% BHT	5.0
9	Tallow plus 0.05% BHT plus 0.025% anthranilic acid	5.5
10	Tallow plus 0.009875% BHT plus 0.0046875% methyl anthranilate	3.75
11	Tallow plus 0.046875% plus 0.00234375% methyl anthranilate	2.5

*Norma-Hoffman Bomb Test

The information of Table 1 demonstrates the stability of choice-grade tallow, both in an untreated state, that is, without synthetic preservatives added and the comparative effect of certain recognized antioxidant preservative materials along with the improved results obtained by the special antioxidants. The indicated results were obtained by treating the rendered tallow at temperatures below 180°F according to the Norma-Hoffman bomb test as specified by the American Oil Chemists' Society. In this respect, the Norma-Hoffman test determines the time required for the fat sample to oxidize under certain accelerated conditions. This is brought about by placing the sample in an enclosed chamber into which excessive oxygen is introduced until an atmosphere having a constant pressure, substantially 15 psi is obtained.

Thereafter heat is applied to the chamber or bomb and the time required for the bomb's atmospheric pressure to drop 5 psi is then measured as determining the time required for the sample to oxidize. Viewed in another manner, this time is indicative of the oxygen interception ability of the stabilizing additive. Thus, the hour figures set out in Table 1 indicate the time required for oxidation of the test sample under the accelerated test conditions, with the higher time values indicating a greater resistance to oxidation.

As noted, Sample 1, the untreated choice-grade tallow, had a bomb test time of 2 hours. It will be noted that the addition of 0.05% methyl anthranilate to the tallow (Sample 3) resulted in a bomb test time of 3 hours, showing a marked increase in the treated tallow's ability to resist oxidation. The addition of citric acid to the tallow, as demonstrated by Sample 4, served to increase the bomb test time for the tallow by $5/10$ hours, while in Sample 5, the addition of one-half the amount of citric acid employed in Sample 4 plus an equal amount of anthranilic acid resulted in a bomb test time greater than that obtained by Sample 4, treated with the citric acid.

The results obtained from Samples 6 and 7 in Table 1 illustrate the ability of anthranilic acid to increase the antioxidant effect of BHA, with similar relationship as to BHT being demonstrated by the observed results of Samples 8 and 9. The results of utilizing comparatively minor nontoxic amounts of BHT and methyl anthranilate in combination are found in the results obtained from Samples 10 and 11 which compare favorably with results obtained from Sample 7, for example, containing toxic percentages of the antioxidant. In this respect, the recognized total toxic limit for the antioxidant additives, such as BHT or BHA, is 0.02% for edible fats prepared for human consumption and 0.05% for fats used in animal feeds. The most effective results obtained appear in Sample 9 wherein BHT and anthranilic acid are combined in a ratio of substantially 2:1 and a total additive percentage of 0.075%, giving an extended bomb test time of 5.5 hours, compared to the untreated tallow bomb test time of the 2 hours, as seen in Sample 1. Generally speaking, it is apparent that the utilization of methyl anthranilate or anthranilic acid, particularly in combination with known antioxidants such as BHA or BHT, effectively extends the stability of the tallow. The following Table 2 further serves to illustrate some of the results, based on AOM stability findings.

TABLE 2: HOURS AOM STABILITY

	Sample Treated at Temperatures Below 180°F	AOM Stability, hr
1	Untreated choice-grade tallow	5
2	Tallow plus 0.05% citric acid	5.5
3	Tallow plus 0.05% anthranilic acid	7.0
4	Tallow plus 0.05% methyl anthranilate	5.5
5	Tallow plus 0.01875% BHT plus 0.009875% methyl anthranilate (0.028625% total)	81
6	Tallow plus 0.01875% BHT plus 0.009875% methyl anthranilate plus 0.05% citric acid (0.07625% total)	113
7	Tallow plus 0.0375% BHT plus 0.01875% methyl anthranilate (0.056% total)	113
8	Tallow plus 0.05% BHT plus 0.025% anthranilic acid (0.075% total)	121

The untreated choice-grade tallow, Sample 1, demonstrates an AOM stability of 5 hours. By adding 0.05% citric acid to the tallow (Sample 2) at temperatures below 180°F, the AOM stability was increased by $5/10$ hour. By comparison, in Sample 3, the addition of 0.05% anthranilic acid to the tallow increased its AOM stability to 7 hours, while the substitution of the same percentage of methyl anthranilate in Sample 4 produced results equivalent to that produced by the citric acid additive utilized in Sample 2.

Sample 5 shows the effect of using nontoxic percentages of the antioxidant. It also will be noted that the hours AOM obtained from Samples 6 and 7 are identical, although the total percentage of the additive ingredients employed in Sample 7 is substantially less than that employed in Sample 6. In this respect, the percentages of BHT and methyl anthranilate in Sample 7 are twice the percentages of those ingredients in Sample 6, but the citric acid has been eliminated from Sample 7. In each instance, however, the BHT and methyl anthranilate bear a ratio of substantially 2:1.

Sample 8 of Table 2 shows the greatest number of hours AOM stability obtained utilizing BHT and anthranilic acid in a ratio of 2:1 and a total ingredient percentage just slightly less than that utilized in Sample 6, where BHT and citric acid combination was used. The following Table 3 serves to illustrate some of the results produced in achieving stability in acid and color.

TABLE 3: COMPARATIVE ACID AND COLOR VALUES

	Sample	FAC Color No.	FFA Acid No.
1	Choice-grade tallow (untreated)	9	1.9
2	Tallow plus 0.0125% BHT plus 0.00625% citric acid treated at 225°F	9	1.9
3	Tallow plus 0.0125% BHT plus 0.00625% methyl anthranilate treated at 225°F	Less 3	1.2

As shown in Table 3, the untreated choice-grade tallow of Sample 1 showed an FAC color number of 9 and an FFA acid number of 1.9. After treating the tallow of Sample 1 with a combination BHT-citric acid preservative, as in Sample 2, there were no observed changes in either FAC color number or FFA acid number over the Sample 1 findings. By way of contrast, the tallow of Sample 3, treated with BHT and methyl anthranilate, showed a marked depression in color number to less 3 scale value with a corresponding lowering of the free fatty acid number to 1.2. It will further be observed that both Samples 2 and 3 were treated at 225°F, substantiating the ability of the methyl anthranilate to promote antioxidant activity, particularly at elevated temperatures, while lowering the acid and color numbers of the resulting rendered product.

BUTYLATED HYDROXYTOLUENE (BHT)

The antioxidant of *L.A. Hall; U.S. Patent 2,981,628; April 25, 1961; assigned to The Griffith Laboratories, Inc.* is known as 2,6-di-tertiary-butyl-4-methylphenol, and known also as butylated hydroxy toluene. This material can be used in very small proportions, gives long shelf-life to fat and oil products, and carry-through to baked products such as crackers and cookies. It is also useful with fats or oils of either vegetable or animal origin. The product is not only antioxidant, itself, but forms synergistic mixtures with other well-known antioxidants. Examples of such synergistic antioxidant compositions are as follows. All the ingredients are shown in pounds.

	- - - - - - - - - - - - - - Examples - - - - - - - - - - - - - - - -						
	1	2	3	4	5	6	7
Di-tert-butyl-p-cresol	15.00	26.40	20.00	14.00	33.00	20.00	10.00
Propyl gallate	3.34	6.70	--	--	--	--	--
Citric acid	3.74	--	4.00	--	3.35	--	--
Citric acid (anhydrous)	--	3.35	--	--	--	4.00	4.00
Lecithin	22.68	20.10	--	--	20.10	--	--
Corn oil	55.24	43.45	--	--	43.55	60.00	60.00
Butylated hydroxyanisole	--	--	6.00	14.00	--	--	10.00
Propylene glycol	--	--	70.00	--	--	--	--
Lard oil	--	--	--	72.00	--	--	--
Fatty monoglyceride	--	--	--	--	--	16.00	16.00

The fatty monoglyceride is preferably cottonseed monoglyceride. These examples are used in oils or fats on the basis of from 0.005 to 0.02% of the di-tertiary-butyl-para-cresol. The amount of di-tertiary-butyl-para-cresol can be varied within these limits, maintaining the same proportions of propyl gallate, citric acid, and lecithin with respect to it which are shown in the above examples. The residue of the composition may then be made up with corn oil, assuming 8 ounces of the complete antioxidant composition for 1,000 pounds of lard or other fat or oil. Olive oil or cottonseed oil, lard oil, oleo, or other vegetable or animal oils may be used in place of the corn oil.

As an example of making the foregoing examples, anhydrous citric acid may be added to corn oil, previously heated to 265°F. The citric acid is added over a period of about 3 minutes and the mixture stirred continuously during this addition and for 45 minutes thereafter, until substantially complete reaction occurs. During this time, the temperature is maintained at about 265°F. The mixture is then cooled slowly to 120°F and agitation continued for about 20 minutes. The temperature is then raised to 160°F and the propyl gallate added with vigorous agitation for 15 minutes while maintaining the temperature at 160°F. After cooling to about 120°F, the di-tertiary-butyl-para-cresol is added, stirring slowly for about 20 minutes.

The citric acid and lecithin in Example 5 are incorporated as lecithin citrate in order to make the material oil and fat soluble. In Example 3 where propylene glycol is used as a solvent, the di-tertiary-butyl-para-cresol and butylated hydroxy anisole are dissolved in the propylene glycol with agitation at a temperature of approximately 120°F, together with the citric acid. In Example 4, the di-tertiary-butyl-para-cresol and butylated hydroxy anisole are dissolved in the lard oil with agitation at a temperature of 120° to 125°F. When tested with lard, the following improvement was obtained with respect to the AOM test and the Schaal Keeping Test.

	AOM, hours	Schaal Keeping Test, days
Control, no antioxidant	4.5	17
Di-tertiary-butyl-para-cresol, 0.01%	13	27
Di-tertiary-butyl-para-cresol, 0.02%	20	27
Control, no antioxidant	6.5	17
Di-tertiary-butyl-para-cresol, 0.005%	20.0	--
Di-tertiary-butyl-para-cresol, 0.01%	26.0	27
Butylated hydroxy anisole, 0.01%	27.0	31
Di-tertiary-butyl-para-cresol, 0.015%	34.0	--
Di-tertiary-butyl-para-cresol, 0.02%	38.0	40
Di-tertiary-butyl-para-cresol, 0.01% plus 0.003% citric acid	30.0	--
Example 1 (8 oz/1,000 lb lard)	41.0	--

BHA-BHT SYNERGISTIC COMBINATIONS

The process of *L.R. Dugan, Jr. and H.R. Kraybill; U.S. Patent 2,926,092; February 23, 1960* is based on use of BHT (2,6-di-tertiarybutyl-p-cresol) and BHA (butylated hydroxyanisole) in combination with each other to produce a synergistic effect which greatly increases stability of animal fats and oils and greatly reduces and retards rancidity in foods containing such fats and oils.

To achieve an efficient and complete dissolution of the antioxidant additives in the glycerides, fats or oils, any desired proportions of 2,6-di-tertiarybutyl-p-cresol and butylated hydroxyanisole may be first dissolved in suitable mutual solvents such as propylene glycol, ethanol, and mono- and/or diglycerides, and/or triglycerides including animal fats and oils or mixed animal and vegetable fats and oils. The effective amount of 2,6-di-tertiarybutyl-p-cresol and of butylated hydroxyanisole will vary within the range from about 0.0005% to about 0.02% on a weight percent based on the amount of material to be stabilized.

The stability of a lard (Kingan lard) treated by adding to, separately and concurrently, nominal amounts of 2,6-di-tertiarybutyl-p-cresol and BHA. The AOM test was followed in all cases and the experimental results are tabulated as follows.

AOM Stability Induced by 2,6-Di-tertiarybutyl-p-cresol and BHA in Lard

Antioxidant	Correlated AOM Stability (hr)	Difference from Control	Expected Stability Due to Components	Synergistic Effect (hr)
Control (Kingan lard)	11	--	--	--
0.005% 2,6-di-tertiarybutyl-p-cresol	37	26	--	--
0.010% 2,6-di-tertiarybutyl-p-cresol	53	42	--	--
0.020% 2,6-di-tertiarybutyl-p-cresol	64¾	53¾	--	--
0.010% BHA	45¾	34¾	--	--
0.020% BHA	54½	43½	--	--
0.005% 2,6-di-tertiarybutyl-p-cresol + 0.010% BHA	80	69	60¾	8¼
0.010% 2,6-di-tertiarybutyl-p-cresol + 0.010% BHA	101	90	76¾	13¼

The measured stability figures, listed in the second column, are obtained by a correlation of figures obtained from testing of several batches of the particular material. The Difference from Control column indicates the numerical difference in stability time directly attributable to the addition of the particular amounts of antioxidants. These figures are arrived at by subtracting the natural stability of the control (Kingan lard without additive) from the measured stability of the lard with the particular antioxidants incorporated.

The values of synergism listed in the last column for the combinations of antioxidants are values attributed to mutual synergism by virtue of the greater stability figures arrived at by AOM testing as compared to expected stability calculated by adding the stabilizing effects of the two antioxidants when used separately in the same amounts. In other words, the expected enhanced stability of using 0.005% 2,6-di-tertiarybutyl-p-cresol would be 26 hours which when added to that of 0.010% BHA, which is 34¾ hours, would result in an expected stability increase of 60¾ hours.

However, the use of the same amounts of the two components in combination produced an AOM stability test result of 69 hours. The difference between these two figures of 8¼ hours can only be attributed to a mutually synergistic effect created by the use of these two antioxidants in combination. It is also to be noted that an increase in the percentage of 2,6-di-tertiarybutyl-p-cresol used in conjunction with the same percent of BHA produced an even greater synergistic effect of 13¼ hours of enhanced stability. The stability differential would be the additive effect of 0.010% BHA, or 34¾ hours, which total would be 76¾ hours. This expected stability increase of 76¾ hours is less than the experimental results of 90 hrs and the difference of 13¼ hrs is likewise attributed to synergism.

To be completely satisfactory as an antioxidant, the substance so selected must also exhibit carry-through properties; that is, it must remain effective as an oxidation inhibitor in the material to be stabilized even after both that material and the selected additive have been subjected to high temperatures such as occur in cooking, baking, or frying operations. To demonstrate the fine carry-through properties of the combination of 2,6-di-tertiarybutyl-p-cresol and BHA the following data shows comparative stabilities of various food products prepared with untreated and with stabilized lard, the stability of the foods being determined by Schaal Oven Storage techniques where the foods are subjected to high temperatures so as to accelerate oxidative deterioration of the unsaturated fats and oils.

	Pastry
	Hours
Control (Kingan lard)	209
0.005% 2,6-di-tertiarybutyl-p-cresol + 0.010% BHA	1,533
	Crackers
Control (Kingan lard)	295
0.005% 2,6-di-tertiarybutyl-p-cresol + 0.010% BHA	981
	Potato Chips
Control (Kingan lard)	60
0.005% 2,6-di-tertiarybutyl-p-cresol + 0.010% BHA	602

Other AOM stability testing of stabilized lards which have been repeatedly used as a frying medium for potato chips gives results comparable in order of effectiveness to test results shown previously for freshly prepared lards. In the case of used lard where the untreated control lard had an AOM stability of ¾ hour, that with 0.005% 2,6-di-tertiarybutyl-p-cresol showed a stability of 8¼ hours and with 0.10% BHA of 17 hours. With an AOM test result of 41¾ hours for the lard with both the stated amounts of 2,6-di-tertiarybutyl-p-cresol and BHA incorporated therein, a synergistic effect for the combination was clearly evident in the used lard.

HYDROXY NAPHTHYLMETHYLAMINOBIS(PROPIONIC ACID)

M.F. Zienty; U.S. Patent 3,240,609; March 15, 1966; assigned to Miles Laboratories, Inc. states that certain derivatives of hydroxy naphthylmethylaminobis(propionic acid) display antioxidative properties when incorporated into edible fats and fatty oils and products containing these compositions. Specifically, the inhibitors are 1-hydroxy-2-naphthylamino-N-bis(propionitrile), 2-hydroxy-1-naphthylmethylamino-N-bis(propionitrile), and the corresponding acids derived therefrom. These compounds may be represented by the formulas shown on the following page.

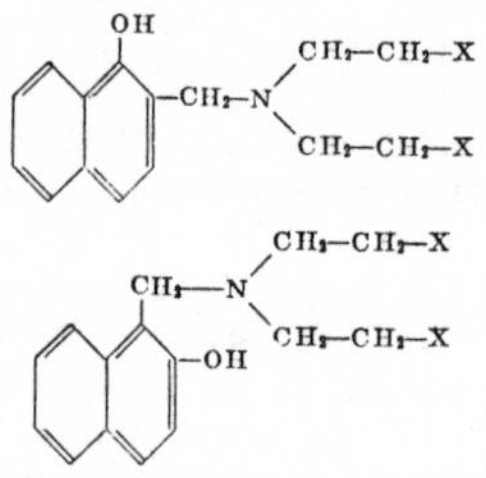

wherein X represents a functional group, such as –CN or –COOH. The nitrile compositions are conveniently prepared as reaction products of β,β'-iminodipropionitrile, formaldehyde and α-naphthol or β-naphthol. The corresponding acid products are obtained by hydrolyzing the nitrile compounds with a hydroxide base, such as potassium hydroxide or sodium hydroxide. The following equation, which illustrates the preparation of 1-hydroxy-2-naphthylmethylamino-N-bis(propionitrile) and the acid derivative, is representative of the overall reaction.

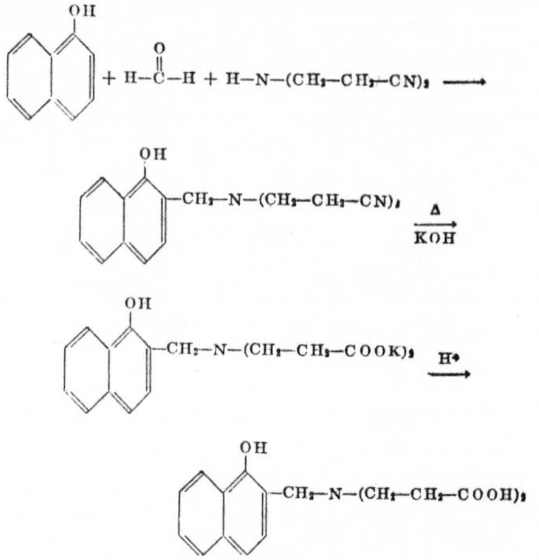

The 2-hydroxy-1-naphthylmethylamino-N-bis(propionitrile) compound and the corresponding acid derivative, are obtained via the same procedure illustrated above, by using β-naphthol in place of α-naphthol.

Specific properties of the aforementioned compounds which make them particularly useful as antioxidants are nontoxicity, solubility in fats and oils and absence of undesirable odors, flavors and colors when incorporated into fatty substances to be protected. The compounds are used to stabilize lard, castor oil and corn oil, but other materials which may likewise be protected include butter, beef tallow, linseed oil, rapeseed oil, olive oil, palm oil, coconut oil and peanut oil. To effectively inhibit oxidative deterioration in these materials the substituted propionic nitriles and acids are mixed in small amounts, for example, from about 0.025 to 0.1% by weight.

When tested by means of the Swift Stability Test, commonly referred to as the AOM method, the antioxidants were found to retard rancidity in fats and oils for substantially longer periods of time than butylated hydroxy toluene, as well-known lard antioxidant. Referring to the test data set out in the following table, the figures shown represent the time required (AOM time in hours) for rancidity to develop in lard, corn oil and castor oil when stabilized with these compounds as compared to stabilization with butylated hydroxy toluene.

Inhibitor	AOM Time in Hours		
	Lard	Castor Oil	Corn Oil
1-hydroxy-2-naphthylmethylamino-N-bis(propionic acid)	102	21	8
2-hydroxy-1-naphthylmethylamino-N-bis(propionic acid)	72	30	16
1-hydroxy-2-naphthylmethylamino-N-bis(propionitrile)	8	8	8
2-hydroxy-1-naphthylmethylamino-N-bis(propionitrile)	8	8	8
Butylated hydroxy toluene	40	8	4

Example 1: Preparation of 1-Hydroxy-2-Naphthylmethylamino-N-Bis(Propionitrile) —
12.3 grams (0.1 mol) of β,β'-iminodipropionitrile was dissolved in 125 ml of dioxane and 7.5 ml of formalin was slowly added dropwise. The resulting mixture was stirred for 1 hour at room temperature. To this mixture was added 14.4 grams (0.1 mol) of α-naphthol all at once, and the reaction mixture was heated under reflux for 4 hours. The solvent, dioxane, was stripped off under reduced pressure, to give the crude nitrile product in the form of an oily residue, which was shown to possess antioxidant properties.

Example 2: Preparation of 2-Hydroxy-1-Naphthylmethylamino-N-Bis(Propionitrile) —
12.3 grams (0.1 mol) of β,β'-iminodipropionitrile was dissolved in 125 ml of dioxane and 7.5 ml of formalin was slowly added dropwise. The resulting mixture was stirred for 1 hour at room temperature. To this mixture was added 14.4 grams (0.1 mol) of β-naphthol all at once and the reaction mixture was heated under reflux for 3 hours. The solvent, dioxane, was stripped off under reduced pressure, leaving an oily residue. To induce crystallization, the oily mixture was treated with ether. The crystalline product was collected by filtration, dissolved in hot ether and allowed to cool to room temperature. Upon cooling the individual crystals separated from the solution to give the final product as a white, amorphous powder. Yield: 27 grams; MP: 100° to 101°C. Analysis – calculated for $C_{17}H_{17}N_3O$: N, 15.05. Found: N, 14.84.

Example 3: Preparation of 2-Hydroxy-1-Naphthylmethylamino-N-Bis(Propionic Acid) —
61.5 grams (0.5 mol) of β,β'-iminodipropionitrile was dissolved, with stirring, in 500 ml of dioxane. 37.5 ml of formalin was slowly added dropwise and stirring was continued for 2 hours at room temperature. 72.08 grams (0.5 mol) of α-naphthol was then added all at once and the reaction mixture was heated under reflux for 6 hours. The solvent, dioxane, was stripped off under reduced pressure leaving an oily residue, which was allowed to cool to room temperature. After cooling, 300 ml of aqueous potassium hydroxide (65.0 grams potassium hydroxide in water) was added to the oily mixture and it was refluxed for 1 hour, during which time a considerable amount of ammonia gas evolved.

This solution was cooled to room temperature, acidified with glacial acetic acid and the resulting crude acid which precipitated was collected by filtration. The solid precipitate was dissolved in methanol, heated to 60°C, treated with activated charcoal, precipitated with water, filtered and dried. The final product was a fine, white powder, which melted at 298° to 300°C. Yield: 27.0 grams. Analysis – calculated for $C_{17}H_{19}NO_5$: N, 4.40. Found: N, 4.51.

Example 4: Preparation of 1-Hydroxy-2-Naphthylmethylamino-N-Bis(Propionic Acid) —
31.7 grams (0.1 mol) of 2-hydroxy-1-naphthylmethylamino-N-bis(propionitrile) was placed in a 3-necked flask equipped with a stirrer and reflux condenser. To this compound was added 150 ml of aqueous potassium hydroxide (37.5 grams potassium hydroxide in water) and the reaction mixture was heated under reflux for 3 hours. During the reflux period the potassium salt of the compound began to precipitate from solution. After reflux, the solution was cooled to room temperature, acidified with glacial acetic acid and the crude

acid which precipitated was collected by filtration. The precipitate was dried, dissolved in hot benzene, treated with activated charcoal and filtered. The filtrate was treated with petroleum ether until the first permanent opalescence was observed and then cooled to room temperature to precipitate the final product as a colorless powder, melting at 197° to 198.5°C. Yield: 28.5 grams. Calculated for $C_{17}H_{19}NO_5$: N, 4.40. Found: N, 4.56. Following are examples illustrating utilization of the antioxidants in stabilizing various edible fats and oils against oxidative deterioration.

Example 5: 5 mg of 1-hydroxy-2-naphthylmethylamino-N-bis(propionic acid) (0.025% by weight) was added to 20 grams of melted lard. A commercially available rendered lard product was used, which had an uninhibited induction period of approximately 6 to 8 hours. As shown by the test data in the previous table, stabilization can be increased to approximately 102 hours using the inhibiting compound of this example.

Example 6: 5 mg of 2-hydroxy-1-naphthylmethylamino-N-bis(propionic acid) (0.025% by weight) was added to 20 grams of castor oil. The castor oil used was a commercial product having an uninhibited induction period of approximately 20 hours. The data in the table indicate that the inhibitor of this example stabilizes castor oil for approximately 30 hours.

HIGHER ACYLACETONES

There is general agreement among chemists that the development of rancidity in fat involves chemical reactions which take place, in general, in a sequence of steps of which the first step is the conversion of portions of a representative fat molecule, which may be long chain aliphatic acyl groups characteristic of an undecomposed fat, into free radicals. In a succeeding step or steps, such free radical is converted into or converts other portions of fat into the substances characteristic of rancidity. In view of this theory, there have been employed, as rancidity-inhibiting substances numerous chemical compounds compatible with fats which are known to be free radical acceptors which, by accepting them, inactivate such radicals. By inactivating free radicals which may be formed, such substances, in even very small amounts, appreciably inhibit, but do not altogether eliminate the tendency of fats to become rancid.

When, eventually, sufficient free radicals have been formed that the capacity of such preservative substances to inactivate free radicals is exhausted, then the subsequent formation of free radicals in the fat gives rise to rancidity.

It has been found by *P.A. Wolf and A.K. Prince; U.S. Patent 3,443,970; May 13, 1969; assigned to The Dow Chemical Company* that a small amount of an acyclic higher aliphatic hydrocarbylcarbonylacetone very strongly suppresses the formation of free radicals, with the result that the initiation of rancidity is inhibited, and the further result that the capacity of a very small amount of free radical acceptor to inhibit development of rancidity is greatly enhanced. The acyclic higher aliphatic hydrocarbylcarbonylacetone compounds are such substances as oleoylacetone, stearoylacetone, and other acyclic higher aliphatic hydrocarbylcarbonylacetone compounds of which the acyclic higher aliphatic hydrocarbylcarbonyl groups are of from about 10 to 26 carbon atoms.

It is believed that the dicarbonyl structure

$$\text{acyclic hydrocarbon higher aliphatic } -\overset{\displaystyle O}{\overset{\displaystyle \|}{C}}-CH_2-\overset{\displaystyle O}{\overset{\displaystyle \|}{C}}-CH_3$$

in some way acts, in a fresh fat susceptible to degradation to form free radicals, to inhibit the formation of such free radicals before they form. Lower acylacetones appear to be of diminishing effectiveness or ineffective and, because they are more volatile they are more fugitive.

In the storage of fat-bearing food stuffs and in the manufacture of refined fats from animal or vegetable sources, the rancidity-inhibiting agents can be applied to fat-bearing tissue before such tissue is placed into storage or rendered to refined fat. The application is readily made by, for example, brushing, spraying or dipping with a solvent solution of the alkanoylacetone.

In treating bulk, rendered fat, relatively free from free radicals of the sorts which initiate rancidity, the fat is heated at a temperature whereat it is either melted or at least soft enough that a compatible liquid can be mixed with it. While the fat is in a soft or melted condition, the desired acylacetone compound or a mixture of such compounds is added. These compounds are characteristically liquids at room temperature or at temperatures not much above room temperature; they are miscible with fats over a wide range of proportions, and their addition in a liquid or solid form is readily carried out.

If desired, more quickly to effect uniform dispersion of the acylacetone compound in the fat, the acetone compound in a desired amount or proportion can first be dispersed in a fat and diluent, which may be a portion of the fat to be treated. This fat then containing the acylacetone in a concentration substantially higher than the desired concentration in the finished protected fat is added in a desired amount to the whole amount of fat to be protected and intimately mixed therewith so as to achieve relatively uniform dispersion of the acylacetone compound throughout the fat. The fat so treated may also at the same or at a different time, if desired, be protected by the inclusion therein of a free-radical acceptor. Among the free-radical accepting substances are butylated hydroxytoluene, butylated hydroxyanisole, and tocopherols. The combination with a free-radical acceptor provides unusually good protection to fats from becoming rancid under conditions favorable to the development of rancidity.

Example 1: About 100 pounds of freshly rendered lard, yet melted from the rendering operation, is divided into two portions; one portion being about 99½ pounds and the other portion about ½ pound. Into the ½ pound portion there is added with stirring 0.001 pound of n-decanoylacetone, and it is intimately mixed therewith by stirring. The resulting dispersion is added to the other portion of melted lard, the entire amount then being 100 pounds of lard with the indicated amount of decanoylacetone. The resulting product is thereafter routinely packaged and placed into storage.

Example 2: A lard is prepared as in Example 1 except that there is added to the half pound of fat, together with the decanoylacetone, a rancidity-inhibiting amount of a free-radical acceptor selected from the list published in the *Federal Register,* p. 847, Jan. 27, 1961. The resulting product is likewise packaged and stored.

Example 3: The efficacy of the acylacetones was demonstrated by testing a lard composition freshly prepared as in Example 2 for oxidation. The procedure employed and the results obtained were as follows. A substrate was prepared by combining $1/100$ of 1% by weight of butylated hydroxytoluene with freshly prepared lard. The resulting product corresponded closely to a commercially available lard product protected only by an antioxidant material of the free-radical acceptor type. This product was further modified by the addition of a small amount of a copper salt, 0.00156 weight percent of $CuSO_4 \cdot 5H_2O$ by weight of lard. Copper and its compounds are known to enhance the formation of free radicals in fats exposed to rancidity-developing conditions.

Of the substrate fat material thus prepared, samples were tested in unmodified form, and other samples were modified by the addition of various substances. The samples, in glass cups, were enclosed and tested one at a time within stainless steel bombs provided with constant temperature heating means, means for the introduction of oxygen under pressure, and means for continuous recording of gas pressure internal to the bomb. In carrying out the test, the sample was placed within the bomb, gaseous contents thereof were flushed out with oxygen and with an atmosphere of essentially pure oxygen, the bomb was closed and internal oxygen pressure raised to a pressure of 100 pounds per square inch by gauge with the bomb maintained at a temperature of 60°C.

In this situation, the bomb and contents were maintained over an extended period of time. An end point of rancidity inhibition was assumed to be reached when any sample began to take up appreciable amounts of oxygen occasioning thereby a distinctive decrease in internal bomb pressure as shown on a pressure recorder. The effectiveness of any system of rancidity inhibiting substances was then ascertained in terms of the duration of exposure of bomb contents to oxygen under the conditions before such end point was noted.

In one test, the substrate consisting essentially of lard, butylated hydroxytoluene, and copper, reached an end point after 108 hours exposure whereas the same composition with 0.02% by weight of stearoylacetone reached an end point after 125 hours; an increase of 15%. When the experiment was replicated, the substrate alone reached an end point after 180 hours whereas the substrate modified by the addition of 0.02% of stearoylacetone reached an end point only after 218 hours. This represents a gain of approximately 20% in stability.

A further experiment essentially the same except that the employed alkanoylacetone was oleoylacetone determined an end point for the substrate at 185 hours whereas after 240 hours when the experiment was discontinued, the fat substrate containing the oleoylacetone in the amount of 0.02% by weight of substrate had not yet reached an end point. Similar results are obtained when using decanoylacetone or hexacosanoylacetone. Among the higher acylacetones to be used are lauroylacetone, stearoylacetone, lignoceroylacetone, and similar higher alkanoylacetones.

N-p-HYDROXYPHENYL UREA

It is reported by *M.F. Zienty; U.S. Patent 2,967,775; January 10, 1961; assigned to Miles Laboratories, Inc.* that N-p-hydroxyphenyl urea and its isomer N-o-hydroxyphenyl urea are effective oxidation inhibitors. These antioxidants can conveniently and economically be obtained as reaction products of para- and ortho-aminophenols and potassium cyanate. These inhibitors possess the requisite qualities for food antioxidants, namely, nontoxicity, solubility in fats and oils and absence of bad odors, tastes or colors after incorporation in the material to be protected.

Stabilization of fatty materials is effected by adding from about 0.025 to 0.1% by weight of N-p- or o-hydroxyphenyl urea. When tested by means of the Swift Stability Test or, as it is also called, the Active Oxygen Method, the subject compounds were found to retard the development of rancidity for about 50 and 40 hours, respectively.

Example 1: 5 mg of N-p-hydroxyphenyl urea (0.025% by weight) was added to 20 grams of melted lard. The lard used was commercially available rendered lard which has an uninhibited induction period of approximately 6 to 8 hours. In accordance with the AOM referred to above, peroxide values were measured at various time intervals as follows:

AOM Time in Hours	H_2O_2 meq/kg Lard
24	6
31	8
44	16
50	16
56	31

Data of this table demonstrate that the lard tested did not become rancid until after 50 hours, i.e., after 20 meq of H_2O_2/kg of lard were present. This amount is commonly regarded as the yardstick for rancidity in the art of oxidation inhibitions.

Example 2: A sample of N-p-hydroxyphenyl urea (0.025% by weight) was added to castor oil which has an induction period of about 20 hours. The peroxide values at various time intervals are shown in the table on the following page.

AOM Time in Hours	H_2O_2 meq/kg Castor Oil
4	4
12	8
22	14
30	16
48	22

These data indicate that spoilage of the castor oil was not effected until approximately after 45 hours of exposure to air.

Example 3: In this example N-o-hydroxyphenyl urea was tested for its antioxigenic activity. Five mg were added to the melted lard. Peroxide values which show that rancidity set in only after approximately 40 hours were as follows.

AOM Time in Hours	H_2O_2 meq/kg Lard
11	4
17	6
26	9
35	17
43	23

Under the same conditions, butylated hydroxy toluene, the well-known lard antioxidant, protects lard against spoilage for only 40 hours, while butylated hydroxy anisole, commonly used for stabilizing vegetable oils, preserves castor oil from deterioration for only 30 hours. In contrast, the data contained in the above tables show that N-o-hydroxyphenyl urea considerably surpasses the abovementioned commercial inhibitors as regards their respective antioxigenic properties.

HYDROGEN CHLORIDE TREATMENT

It is theorized that reversion somehow is related to the presence of the phospholipids. Reversion manifests itself as development of flavors and odors which are characteristic of the source from which the oils are derived. For instance, in soybean oil, reversion is typified by a beany-grassy or even a fishy odor and flavor, in cottonseed oil a phenolic-like flavor and odor, and in fish oil a fishy flavor and odor. Different oils tend to revert at different rates. Soybean oil tends to revert more quickly than the other oils generally used for edible purposes.

F.H. Fryer and G.B. Crump; U.S. Patent 3,562,301; February 9, 1971; assigned to Standard Brands Incorporated found that if the oils are treated with one mol of hydrogen chloride for each mol of phospholipid present, the phospholipids being calculated as lecithin, the oils will not undergo reversion and oils which have reverted will have their original odorless and flavorless characteristics restored. Smaller amounts of hydrogen chloride may be used but will not retard or prevent reversion to the same degree; that is, oils treated with smaller amounts of hydrogen chloride will tend to revert more quickly than oils treated with larger amounts of hydrogen chloride. However, oils treated with the smaller amounts of hydrogen chloride have the tendency to be less prone to reversion than oils not treated at all with hydrogen chloride.

It is believed that the anhydrous hydrogen chloride reacts with the phospholipid present in the oils to form phospholipid hydrochlorides. This belief is based on the observation that when an oil which had reverted was treated with hydrogen chloride to provide a bland oil, and was subsequently subjected to alkali refining, the original reverted flavor and odor of the oil was observed. This indicates that the alkali disassociates the compound formed by the hydrogen chloride treatment, and thus frees the substances which impart the normal reverted flavor and odor characteristics.

This procedure does not take the place of the refining step of deodorization. The relative odoriferous and flavoring substances which are normally removed by deodorization do not seem to be affected by the hydrogen chloride treatment. Thus, to obtain an oil which is bland and not prone to reversion the oil must be subjected to the hydrogen chloride treatment of the process and a deodorization treatment.

Illustrative examples of alkali refined oils which may be treated are soybean oil, corn oil, cottonseed oil, safflower oil and fish oil. Since the phospholipid content of these oils varies and it generally is inconvenient to analyze each oil which is to be treated in order to determine the exact quantity of hydrogen chloride needed, it is advisable to treat the oils with an excess of hydrogen chloride. In this respect, it is preferred to treat soybean oil with from about 400 to 2,000 ppm hydrogen chloride; corn oil with about 200 to 1,000 ppm hydrogen chloride; cottonseed oil from about 100 to 1,000 ppm hydrogen chloride; safflower oil from about 100 to 900 ppm hydrogen chloride, and fish oil with about 1,200 ppm hydrogen chloride.

It is essential that the method be carried out under substantially anhydrous conditions. Substantially anhydrous conditions means that the mixture of oil and hydrogen chloride presents a single phase. Under normal conditions, the limit of solubility of water in oils varies between about 0.04 and 0.14% by weight. Contacting the oils with substantially anhydrous hydrogen chloride may be accomplished by a variety of methods, e.g., by passing the hydrogen chloride through the oils, by contacting a thin film of the oils with hydrogen chloride over the oils so that it diffuses into the oil. It is preferred, because of economic and other factors, to pass the hydrogen chloride through the oils.

The temperature of the oil during the treatment may vary over a relatively wide range, but it is preferred to carry out the treatment in a temperature range of from about ambient to about 150°C. Lower and higher temperatures may be used, for instance, from the congeal temperature of the oil, i.e., about 5°C, to its decomposition point, i.e., about 225°C. The higher temperatures are not preferred since the hydrogen chloride tends to attack or oxidize the equipment in which the treatment is being performed, and there is a tendency for the oils to oxidize more readily at such temperatures.

Example 1: A drum of commercial alkali refined, bleached and deodorized corn salad oil containing 0.1% phospholipids, calculated as lecithin, was held at ambient temperatures (60° to 85°F) for 30 days without access to air. The oil had a reverted flavor. Dry HCl gas (0.003 mol) (0.11 gram) was passed over a period of 3 minutes through a glass sparger immersed in 1,340 grams of the oil at ambient temperature. The oil was then steam deodorized in a glass laboratory deodorizer. A taste panel rated this oil at a level of 8.5, on the basis of a refined corn salad oil (not reverted) being rated at 10.

When the reverted oil was simply steam deodorized in a glass laboratory deodorizer, the oil was rated 8.5 by the taste panel, but the observation was made that the oil had a slightly reverted flavor which was not present in the oil treated by the method of this process. The untreated oil (simply deodorized) and the HCl treated oil were placed in an oven maintained at a temperature of 145°F. After 12 days of storage in the oven, the oils were cooled and rated by a taste panel. The taste panel rated the HCl treated oil 8, whereas the untreated oil was rated 7, on the basis of a refined corn salad oil (not reverted) being rated 10.

Example 2: A sample of 1,660 grams of commercial degummed, alkali refined, bleached and deodorized soybean salad oil at ambient temperature was treated with 0.33 gram of dry HCl gas over a period of about 30 seconds by passing the gas through the oil by the use of a sintered glass sparger. The treated oil was allowed to stand at ambient temperature for 40 minutes and was then deodorized in a glass laboratory deodorizer. A second sample of the same oil was treated with HCl gas in the manner described above but using 1,600 grams of oil and 1.7 grams of dry HCl gas. The untreated oil was allowed to stand at ambient temperature for 45 minutes, and was then deodorized in a glass laboratory deodorizer. Both treated oils were excellent in flavor, but the oil treated with 1.7 grams of HCl gas was more bland.

The treated oil samples were placed in an oven maintained at a temperature of 145°F. The oil sample treated with 0.33 gram of HCl gas reverted after a storage time of 18 days, whereas the oil sample treated with 1.7 grams of HCl gas did not revert after a storage time of 34 days.

Example 3: 400 pounds of commercial alkali refined, bleached and deodorized soybean oil were treated with 80 grams of dry HCl gas over a period of 20 minutes in the manner described in Example 2. The oil was deodorized in a pilot plant deodorizer. The oil was rated by a taste panel as being excellent. This treated oil and oil which was untreated were used to deep fry batches of towel dried, thin sliced, raw potatoes (potato chips). The oils were maintained at a temperature in the range of 365° to 375°F for frying. Two thousand milliliter samples of the oils and 50 gram batches of the potato chips were used for frying. After each batch of potato chips was fried they were evaluated by a taste panel. Some typical results of this evaluation are shown in the following table.

			Taste of potato chips	
Test No.	Total no. of batches of chips fried	Total time (min.) oil held at 365–375° F.	From treated oil	From untreated oil
A.....	1	30	Excellent..........	Good.
B.....	5	75	......do.............	Slightly Beany.
C.....	14	185	......do.............	Strong Beany.
D.....	35	800	......do.............	Bitter, Fishy.
E.....	60	1,410	Good, sl. rancid....	

From this table it is apparent that the soybean oil treated does not revert on heating for prolonged periods to the extent that untreated soybean oil does.

Example 4: A sample of 4,500 grams of commercial alkali refined, bleached and deodorized cottonseed oil at ambient temperature was treated with 4.5 grams dry HCl gas in the manner described in Example 2. The oil was held at room temperature for 1 hour and then steam deodorized in a glass laboratory deodorizer. This treated oil and the same oil which was untreated were used to deep fry batches of codfish sticks. Each batch was fried for 3 minutes and then evaluated. The first 10 batches of fish fried in the treated oil were rated as being excellent in flavor, whereas the first 10 batches fried in untreated oil were rated as having a pungent fishy flavor.

The treated and untreated oils were tested according to the procedure described in Example 3 and the first 7 batches of potato chips fried in the treated oil were rated as excellent while the first 7 batches of potato chips fried in the untreated oil had a musky, off-flavor.

PLICATIC ACID-THIODIPROPIONIC ACID

According to *A. Karchmar and K.L. McDonald; U.S. Patent 3,573,936; April 6, 1971; assigned to Rayonier Incorporated* a mixture of plicatic acid and thiodipropionic acid exhibits a synergistic antioxidant effect when added to animal fats and vegetable oils and foodstuffs containing these materials. This antioxidant combination is effective at both ambient temperature and at high temperature (190°C) which makes it particularly suitable in certain food processing operations.

The use of thiodipropionic acid (TDPA) as an antioxidant is known, but it is considered as being relatively ineffective as an antioxidant in animal fats and vegetable oils. With plicatic acid, however, it is a highly effective antioxidant in animal fats and vegetable oils. Plicatic acid (see structural formula below) is one of the components of the complex mixture of phenolic compounds occurring in the aqueous extract of western red cedar wood *(Thuja plicata)*. The identification of plicatic acid and a process for extracting it from red cedar wood with aqueous solvents are described in two articles entitled, "The Polyphenols of Western Red Cedar" by Gardner, Barton and MacLean, *Can. J. Chem.,* vol. 37, 1703-9 (1959) and "The Chemistry and Utilization of Western Red Cedar" by Dr. J.A.F. Gardner, Department of Forestry Publication No. 1023 (1963), Department of Forestry, Canada.

Plicatic Acid

Plicatic acid, in amounts as small as from about 50 to 150 parts per million, is a powerful antioxidant for use in animal fats and vegetable oils and foodstuffs containing these materials. When it is added to these materials in combination with a substantially equal amount of TDPA, its effectiveness is greatly increased.

Example: For purposes of the study, the samples of fats and oils selected were free of antioxidants; commercial lard and oils of the same type containing antioxidants were used as subcontrols. BHT and BHA combined and NDGA alone were introduced in antioxidant-free fats and oils for a straight comparison with plicatic acid.

A total of 54 samples of lard and oils was weighed into 150 ml beakers. A total of 54 samples of the various antioxidants was weighed, all in the same concentration of 0.01% based on the weight of lard or oil. Wherever a combination of two antioxidants was used (BHT + BHA) or an antioxidant and a synergist (plicatic acid + TDPA) respectively, each compound was weighed to represent 50 ppm or a total of 100. The antioxidants were weighed into microbeakers, dissolved in a small quantity of absolute ethanol, and transferred quantitatively into the lard and oils. The latter were then slightly heated to evaporate the ethanol and placed in the air-aerated oven at 57°C (± 2). All samples in beakers were mixed with glass rods twice a day to permit an even exposure of oils and fats to the oven temperature.

Table 1 represents the effect of various antioxidant additives in a sample of fresh lard as measured by its peroxide value (a measure of its oxidation) at 450 hours. It will be seen that the plicatic acid plus TDPA antioxidant additive was the most effective antioxidant tested. Generally, the plicatic acid is used in amounts of from 50 to 150 ppm. Concentrations of plicatic acid below 20 ppm are relatively ineffective; however, it is soluble in oils to a concentration of the order of 150 ppm. Preferably, the ratio of TDPA to plicatic acid will vary from about 1 to 10 to about 1 to 0.5.

TABLE 1: LARD[1]

Compound	Time (hrs.)	Peroxide test, me. per 1,000 g.
Plicatic acid, pure	450	32.5
Plicatic acid, crude	450	40.5
BHT plus BHA	450	47.5
NDGA	450	37.5
Plicatic acid plus TDPA	450	17.5
Commercial lard (containing BHT plus BHA plus PG)	450	25.0

[1] Lard (fresh, deodorized) (initial peroxide value, 0.5 me. per 1,000 g).

TABLE 2: SAFFLOWER OIL[1]

Compound	Time (hrs.)	Peroxide test, me. per 1,000 g.
Plicatic acid, pure	400	67.5
Plicatic acid, crude	400	68.5
BHT plus BHA	400	82.5
NDGA	400	73.0
Plicatic acid plus TDPA	400	57.5
Commercial oil (containing BHT plus BHA plus PG[2] plus PG[3] plus citric acid)	400	70.0

[1] Oil (safflower, fresh).
[2] Propyl gallate.
[3] Propylene glycol.

Table 2 represents the effect of various antioxidant additives in a sample of safflower oil at 400 hours as measured by its peroxide value. Here again, the plicatic acid plus TDPA was the best antioxidant of the group. The results of these experiments confirm that the plicatic acid is a more potent antioxidant than the combination of BHT and BHA and is about equivalent to NDGA in activity. The purified plicatic acid (nearly white in appearance), as compared to the crude type, shows only slightly higher antioxidant activity which suggests that, unless there is an objection to its brown color, the purification of the crude plicatic acid may not be necessary.

CEDAR POLYPHENOLS-THIODIPROPIONIC ACID

In another process, A. Karchmar; U.S. Patent 3,628,971; December 21, 1971; assigned to International Telephone and Telegraph Corp. describes the use of cedar polyphenols in mixtures with thiodipropionic acid. The combination of cedar polyphenols and TDPA exhibits a synergistic antioxidant effect with the result that the combination of the two compounds produces a far greater antioxidant effect than would normally be expected from either one of these compounds alone.

Polyphenols can be prepared from such sources as red cedar wood. Generally, red cedar wood (Thuja plicata) includes from 5 to 15% of a mixture of nonvolatile, water-soluble compounds. A major component of this mixture, varying in amount from about 1 to 5% of the weight of the wood, is plicatic acid. However, the aqueous extract of this wood includes not only plicatic acid, but also a substantial amount of less acidic phenolic compounds which are collectively known as polyphenols. The principal component of these polyphenols has been found to comprise lactones of plicatic acid having the same catechol-like grouping in the molecule as plicatic acid. Also present in the extract are various carbohydrates and some salts.

The results of investigations relating to the extraction of chemicals such as plicatic acid and the polyphenols from western red cedar wood, the identification of these chemical constituents, the yields obtained and the methods used for the separation of the extract into its various components have been in two articles referred to previously.

One method for the extraction of plicatic acid and the polyphenols from cedar wood and their purification comprises neutralizing an acidic aqueous extract of red cedar wood with a suitable base such as sodium hydroxide. The neutralized solution is then passed through an ion-exchange column filled with a suitable molecular adsorption-type phenol formaldehyde resin having its active exchange centers in the sodium state, a clean separation of the polyphenols and the plicatic acid is obtained. The polyphenols are adsorbed on the adsorption resin while the plicatic acid and carbohydrates pass through unchanged.

At the conclusion of the adsorption cycle, after the plicatic acid effluent has been removed, the polyphenols are eluted from the adsorption resin with a suitable organic solvent such as acetone, methanol or methyl ethyl ketone, leaving the resin in proper condition for the next adsorption cycle of the neutralized extract. The polyphenols are recovered from the eluate in pure form by simply evaporating off the solvent. The plicatic acid is recovered from the effluent by some suitable method, such as adding sodium chloride, acidifying the mixture with hydrochloric acid and then extracting the plicatic acid with a solvent such as methyl ethyl ketone. On evaporation of the solvent a relatively pure plicatic acid is obtained.

The major portion of the cedar polyphenols is believed to be open-chain substituted 2,3-dibenzyl butyrolactones having the structural formula shown on the following page, where R_1 is either H or OH, R_2 is either H or OH, and R_3 is either H or CH_3. More particularly it was found that mixtures of cedar polyphenols and thiodipropionic acid wherein the amount of polyphenols in the mixture is at least substantially equal to the amount of thiodipropionic acid, are synergistically more effective antioxidants for edible oils and fats than equivalent amounts of either component taken separately or in combination with

other known antioxidants. The ratio of polyphenols to thiodipropionic acid in forming
the antioxidant mixtures should be in a range of from about 1:1 to 2:1.

It has been found that only very small amounts of the antioxidant mixture of cedar polyphenols and thiodipropionic acid need be used to achieve antioxidant results. The actual
amount to be used will vary over a rather broad range depending upon such factors as
the particular product to which the mixture is added, the exact purpose for which it is
added and the like. Normally, it is preferred to use an amount of antioxidant mixture
ranging from about 50 to 400 ppm.

Example: For purposes of study, the samples of fats and oils selected were free of antioxidants; commercial lard and oils of the same type containing antioxidants were used as
subcontrols. BHT and BHA combined, cedar polyphenols and methionine combined,
cedar polyphenols and 1-proline combined, methionine alone and 1-proline alone were
introduced into antixoidant-free fat and oil samples for a straight comparison with poly-
phenyl-thiodipropionic acid antioxidant mixtures.

A total of 40 samples of lard and oils was weighed into 150 ml beakers. A total of 40
samples of the various antioxidants was weighed, all in the same concentration of 0.015%
based on the weight of lard or oil. Whenever a combination of two antioxidants was used
(BHT + BHA, cedar polyphenols + methionine, and cedar polyphenols + 1-proline) or an
antioxidant and a synergist (cedar polyphenyl + thiodipropionic acid) respectively, the
total mixture was weighed to represent 150 ppm.

The various antioxidants samples were weighed into microbeakers, dissolved in a small
amount of absolute ethanol and transferred quantitatively into the oil and lard. The latter
were then slightly heated to evaporate the ethanol and placed in an air-aerated oven at
58°C ($\pm$2). All samples in beakers were mixed with glass rods twice a day to permit an
even exposure of oils and fats to the oven temperatures. Table 1 represents the effect of
the various antioxidant additives in a sample of fresh lard as measured by its peroxide
value (a measure of its oxidation) after a given period of time. It will be seen that the
cedar polyphenols and thiodipropionic acid antioxidant additive was the most effective
antioxidant tested.

TABLE 1: LARD*

Antioxidant additive	Concentration of antioxidants (p.p.m.)	Time (hrs.)	Peroxide test (me./ 1,000 gr.)
Control	0	200	105.0
Cedar polyphenols	150	500	100.0
Thiodipropionic acid	150	300	102.5
Methionine	150	200	82.5
1-proline	150	200	95.0
Cedar polyphenols + Methionine	100 + 50	500	92.5
Cedar polyphenols + 1-proline	100 + 50	500	92.5
BHT + BHA	75 + 75	500	82.5
Cedar polyphenols + Thiodipropionic acid	100 + 50	500	50.0

*Fresh, deodorized, initial peroxide value = 0.5 me/1,000 g.

Table 2 represents the effect of various antioxidant additives in a sample of safflower oil after a given period of time as measured by its peroxide value. Here again, the mixture of cedar polyphenols and thiodipropionic acid was the best antioxidant of the group.

TABLE 2: FRESH SAFFLOWER OIL

Antioxidant additive	Concentration of antioxidants (p.p.m.)	Time (hrs.)	Peroxide test (me./1,000 gr.)
Control	0	350	97.5
Cedar polyphenols	150	450	87.5
Thiodipropionic acid	150	350	85.0
Methionine	150	350	90.0
l-proline	150	350	95.0
Cedar polyphenols + Methionine	100 50	450	85.0
Cedar polyphenols + l-proline	100 50	450	85.0
BHT + BHA	75 75	450	82.5
Cedar polyphenols + Thiodipropionic acid	100 50	450	67.5

CYSTINE STABILIZERS

H. Enei, S. Okumura, A. Mega and S. Ota; U.S. Patent 3,585,223; June 15, 1971; assigned to Ajinomoto Co., Inc. found that L-cystine or DL-cystine, when mixed with oils and fats in small amounts, provides good protection against oxidation, and that its stabilizing effect is not lost by heating.

Although most amino acids are known to have some antioxidant effect, this effect is weak and is further reduced by exposure to the elevated temperatures commonly encountered in the cooking of foods and in frying. The unique stabilizing effect of cystine among the amino acids is evident from the data of Table 1 which lists peroxide values determined by the active oxygen method on a blank and on 20 ml samples of soybean oil prepared by adding 5% by weight of the tested amino acid to 10 ml oil, heating the mixture to 260°C for 4 minutes, filtering the heated mixture, and diluting 5 ml of the filtrate with 15 ml untreated soybean oil. Each sample was subjected to three heating steps and the peroxide value was determined after each step. Each step involved heating at 140°C for 3 hours. In the second step, the high-temperature treatment was followed by heating at 97.8°C for 1 hour, and in the third step by heating at 97.8°C for 4 hours.

TABLE 1

Amino acid	Peroxide value after— First step	Second step	Third step
None	109.1	189.5	440.0
Cystine	7.9	11.1	22.2
Cysteine	32.3	64.7	247.2
Methionine	61.8	97.0	231.0
Glycine	127.9	198.6	403.0
Valine	114.6	193.5	428.0
Tryptophan	92.9	110.4	250.0
Serine	106.0	174.5	470.0
Arginine	116.7	126.6	261.2
Aspartic acid	86.4	112.0	271.0
Glutamine	122.2	166.8	302.3

When cystine-containing samples briefly heated to 260°C were stored at 32°C for 40 days, the peroxide value was still well below 20. Analogous results were obtained with other fats and oils of vegetable and animal origin, including the common edible and other oils and fats, such as cod-liver oil, lard, tallow, fish oil and chrysalis oil, linseed oil, cottonseed oil, safflower oil, rice oil, corn oil, palm oil, sesame oil, cacao oil, castor oil and peanut oil. Similar effects are produced on fats and oils derived from microorganisms such as yeast oil and Chlorella oil, on phospholipids, such as soybean lecithin and egg yolk lecithin and on fat-dissolved vitamins, such as β-carotene, vitamin A and vitamin E.

Foods containing a high percentage of fats or oils are equally stabilized against oxidation by cystine, and good results are obtained in cheese, butter, margarine, vegetable shortening, mayonnaise, ham sausages, prepared salad dressings, potato chips, fried rice crackers, and fried fish balls. L-cystine or DL-cystine may be added to the fat or oil in the form of the free acid, as a nontoxic salt, or as an addition compound with a physiologically tolerated acid, such as hydrochloric acid, and these salts and acid addition compounds will be understood to be included in the terms cystine, as used in this description.

Normally, at least 0.02% cystine, and equimolecular amounts of its salts and acid addition compounds are required for significant protection. Table 2 lists the peroxide values obtained on soybean oil samples prepared as described with reference to Table 1 to contain varying amounts of cystine and subjected to the aforementioned three-step heat treatment. The results for untreated soybean oil, and for soybean oil containing the legally permissible maximum amount of BHT are also listed for comparison.

TABLE 2

		Peroxide value after—		
Anti-oxidant	Percent	First step	Second step	Third step
None		159. 8	266. 2	464.0
Cystine	1. 0	4. 6	6. 8	14. 0
Do	0. 5	18. 2	32. 4	157. 4
Do	0. 1	93. 8	162. 0	322. 0
Do	0. 02	101. 5	203. 6	410. 0
BHT (0.02%)		162. 0	261. 2	456. 7

Under most conditions, the amount of cystine should be between 0.1% and 1.5% of the fat to be stabilized. The stabilizing property of cystine is greatly enhanced by heat treatment at temperatures above 140°C, and preferably at 180° to 300°C. The optimum time of heat treatment is inversely related to the temperature, and should be 3 to 10 minutes at about 180°C, and shorter, though not substantially less than 1 minute at 260° to 300°C. The lower limit of 140°C is critical, as will presently be shown, but the upper limit is determined mainly by the temperature of thermal decomposition of the fat or oil by chemical breakdown or partial volatilization.

The cystine is activated by heat treatment in solution in the fat or oil to be protected, and only a small portion of the fat and oil need be used for preparing a concentrated cystine solution which is then heated and thereafter admixed to the bulk of the oil or fat.

Example 1: 10 ml batches of soybean oil were mixed with 0.5 gram L-cystine, and the mixtures were heated respectively to 140° to 180°, 220°, 260°, and 300°C for 3 minutes or for 6 minutes. Each treated batch was added to 40 ml untreated soybean oil, and the mixture was heated at 140°C for 3 hours while air was blown through it at a rate of 310 milliliters per minute. The samples activated at temperatures of 220°C or more were additionally heated at 97.8°C for 3 hours. Peroxide values were determined after the heat treatment at 140°C, and after 1 and 3 hours at 97.8°C. These values are listed in Table 3.

For comparison, an oil sample free from antioxidant and samples containing 0.02% of BHT, BHA, and PG respectively were subjected to the same heat treatment and aeration. The corresponding peroxide values are listed in Table 4. The superiority of cystine over the conventional materials is obvious. It will be noted that L-cystine as a normal ingredient of food is not subject to the narrow legal limitations applied to the synthetic antioxidants

TABLE 3

	Activation time									
	Three minutes					Six minutes				
Activation temp., °C	140	180	220	260	300	140	180	220	260	300
Peroxide value after—										
3 hrs. at 140° C	99. 9	48. 1	10. 0	5. 8	5. 6	83. 9	46. 9	4. 7	12 9	9. 0
1 hr. at 97.8° C			24. 1	10. 9	13. 1			8. 5	17. 8	12. 6
3 hrs. at 97.8° C			18. 5	10. 0	25. 7			15. 8	24. 0	72. 5

TABLE 4

	Peroxide value after—		
Anti-oxidant	3 hrs. at 140° C.	+1 hr. at 97.8° C.	+3 hrs. at 97.8° C.
None	131.1	195.0	238.0
BHT, 0.02%	132.1	230.3	332.0
GHA, 0.02%	117.3	189.4	378.0
PG, 0.02%	115.7	180.6	342.0

Example 2: Samples of soybean oil, rice oil, palm oil, tallow, lard, and corn oil were mixed with 1% cystine by weight, and the mixtures were heated for 5 minutes at 260°C. They were then subjected to aeration at 140°C for 3 hours by bubbling 6.25 volumes of air through each volume of oil or molten fat per minute. The peroxide values of the several materials were determined after the aeration together with the peroxide values of correspondingly heated and aerated samples not stabilized by the addition of cystine. The results are listed in Table 5.

TABLE 5

	Peroxide value	
Oil or fat	With cystine	Without cystine
Soybean oil	6.0	140
Rice oil	4.6	140
Palm oil	2.6	35
Tallow	3.0	125
Lard	6.9	130
Corn oil	11.0	178

Example 3: Soybean samples were mixed with 1% L-cystine by weight, and were heated for 5 minutes respectively to temperatures ranging from slightly above room temperature (25°C) to 300°C. They were aerated for 8 hours at 97.8°C by bubbling air through them at a rate of 6.25 volumes/volume/minute. The results listed below indicate the favorable results obtained even without significant heating, and the very substantial further improvement obtained by activating the stabilizer at temperatures above 140°C.

Activation Temperature, °C	Peroxide Value
25	40.2
120	38.2
140	32.8
180	7.2
220	2.0
260	2.6
300	4.1

Example 4: Two batches of soybean oil containing 1% L-cystine were activated at 220°C and 260°C respectively for 3 min and filtered. One of each sample was then diluted with an equal volume of untreated oil. 20 ml samples of each type of pretreated soybean oil and corresponding samples of untreated soybean oil and of soybean oil containing 0.02% BHT and 0.02% BHA respectively were subjected to aeration with 125 ml air at 140°C for 3 hours, whereupon the peroxide values were determined. The results obtained are listed in Table 6.

TABLE 6

Antioxidant	Peroxide Value
Cystine 1%, 220°C	7.4
Cystine 0.5%, 220°C	19.7
Cystine 1%, 260°C	4.6
Cystine 0.5%, 260°C	18.2
None	159.8
BHT, 0.02%	162.0
BHA, 0.02%	159.0

3-METHYLTHIOPROPYLAMINE HYDROCHLORIDE

The preparation of 3-methylthiopropylamine hydrochloride and use in foodstuffs as an antioxidant are described by *H. Samejima, K. Nakayama and Y. Nagano; U.S. Patent 3,579,357; May 18, 1971; assigned to Kyowa Hakko Kogyo Kabushiki Kaisha, Japan.* The preparation of 3-methylthiopropylamine hydrochloride is as follows.

7.45 grams (0.05 mol) DL-methionine and 12 ml (0.01 mol) of acetophenone are introduced into a distillation flask and heated at 170°C for 30 minutes on an oil bath. During heating, carbon dioxide gas is generated. 50 ml of ether are added to the solution which is then extracted with 2 N hydrochloric acid. The aqueous layer is removed from the extract, concentrated and excess hydrochloric acid removed therefrom. The solution is then decolorized by treatment with active carbon followed by evaporation to dryness in vacuo. The substance obtained is dissolved in 30 ml of acetone and cooled to yield 6.5 grams of a crude product. The crude product is recrystallized using ethanol and acetone to give 4.9 grams of 3-methylthiopropylamine hydrochloride, in the form of leaf-like crystals.

The 3-methylthiopropylamine hydrochloride obtained is characterized as follows: Melting point, 144°C. Analysis — calculated as $C_4H_{12}S \cdot HCl$ (percent): C, 33.92; H, 8.48; N, 9.47. Found (percent): C, 33.90; H, 8.48; N, 9.89. The antioxidant compound is especially active when present in compositions in concentrations of 0.01 to 0.1% by weight. The property of this antioxidant is demonstrated and compared with the properties of known antioxidants in the following experiments. The substance subject to oxidation by atmospheric air used in these experiments was sodium linoleate. The substance was dissolved in an aqueous solution and was oxidized by heating to a predetermined temperature. The oxide obtained was periodically estimated as a measure of the antioxidant action of the compound under test.

1.4 grams of linoleic acid were dissolved in 5 ml of 1 N NaOH solution. The solution was diluted with 500 ml of 0.02 M boric acid buffer solution having a pH of 9.0 and was used as the substance subject to oxidation. 200 ml of the solution (0.2 mol of linoleic acid) was mixed in a test tube with 10 ml of 0.01 M antioxidant solution (0.1 mmol of the antioxidant to be tested). The antioxidants used were ascorbic acid, butyl hydroxy anisole, and 3-methylthiopropylamine hydrochloride. Each test solution was maintained at 37°C on a water bath and 2 ml samples of the solution were periodically taken and treated with 2 ml of 0.75% thiobarbituric acid followed by heating in boiling water for 15 minutes. The absorbance of the red color thereby developed was determined at 535 mμ, and this provided a measure of the degree of oxidation. Graphs plotted from the results obtained are shown in Figure 1.1

FIGURE 1.1: FOODSTUFFS CONTAINING AN ANTIOXIDANT

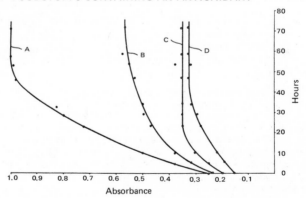

Source: H. Samejima, K. Nakayama and Y. Nagano; U.S. Patent 3,579,357; May 18, 1971

Each graph is a plot of the absorbance against time in hours. Graph A is of a control experiment with no antioxidant used. Graphs B, C and D are the various antioxidants ascorbic acid, 3-methylthiopropylamine hydrochloride and butyl hydroxy anisole respectively. The graphs show that 3-methylthiopropylamine hydrochloride has an antioxidant action higher than that of ascorbic acid, and almost equal to that of the widely used butyl hydroxyl anisole. Furthermore, it is advantageous that 3-methylthiopropylamine in its free base form or salt form is highly soluble not only in water but also in oil and fat.

Example 1: 100 grams of fried potato chips were prepared by frying sliced potato available upon the free market at about 140° to 160°C in natural butter (200 grams) to which was added 0.1 gram (0.05% by weight of butter) of 3-methylthiopropylamine hydrochloride and were sealed into a polyethylene envelope, which was kept at 40°C in order to determine at intervals the formation of peroxides (POV). The following POV values were obtained by extracting the sample with ether, removing ether from the extract and measuring the POV value in a similar manner to that according to the AOCS method. It is apparent from the results that the addition of 3-methylthiopropylamine HCl can well prevent the oxidation of fried potato chips.

POV Value (meq/kg)

Day	0	7	14	21	28	35	42
Control*	3.05	4.89	15.1	27.8	49.1	60.7	95.6
Sample**	2.41	2.35	4.61	7.26	7.90	12.3	11.8

*Without addition of 3-methylthiopropylamine HCl
**With addition of 3-methylthiopropylamine HCl (0.05% by weight)

Example 2: 0.13 g of 3-methylthiopropylamine HCl dissolved in 10 ml of water was added to a mixed nonhuman foodstuff composed of 100 g of exoleated soybeans and 50 g of fish meal with good stirring. The mixture was kept at 40°C in a sealed polyethylene vessel. POV values were determined as in Example 1. The results show that oxidation of the soybeans and fish meal can be inhibited by addition of 0.05 wt % of 3-methylthiopropylamine HCl.

POV Value (meq/kg)

Day	0	14	28	42
Control*	0.98	15.6	84.3	419.0
Sample **	0.93	14.0	21.0	33.6

See footnotes of Example 1

SOLVENTS FOR ANTIOXIDANTS

Acetin Solvents

Various solvents for fat and oil antioxidants are known and are currently in use. Perhaps the most widely used solvent is propylene glycol. This solvent is not completely satisfactory in certain cases; furthermore, a solution containing a high concentration of citric acid is impossible to achieve. Moreover, other antioxidants are only sparingly soluble in propylene glycol.

Although certain monoglycerides are good solvents for butylated hydroxytoluene, butylated hydroxyanisole, and propyl gallate, these monoglycerides are not useful in some applications requiring higher concentrations of an acid synergist, such as citric acid, because they are extremely poor solvents for citric acid. In these cases where a high concentration of acid synergist is desired, it is desirable to maintain the antioxidant concentration at as high a level as is possible and yet increase the acid synergist concentration. Since many antioxidant compositions are mixtures of several components which usually include citric acid or other acid synergist, it is highly desirable to have a solvent which will allow the preparation of such solutions.

The use of the acetins as solvents for antioxidants was found by *M.B. Knowles and*

H.S. Pridgen; U.S. Patent 2,944,908; July 12, 1960; assigned to Eastman Kodak Company.
Triacetin is a good solvent for all of the common fat and oil antioxidants. Mixtures of di- and triacetin are very good solvents for any of the combinations of the common fat and oil antioxidants, and especially for the antioxidant compositions containing acid synergists. Monoacetin is particularly valuable as a solvent for the various acid synergists which are employed in antioxidant compositions. It has been found to be very useful when incorporated with other solvents for antioxidants, including propylene glycol and the other acetins.

One or more of the acetins can thus be used as solvents for any of the well-known antioxidants or synergists. Among the antioxidants which may be dissolved in the acetins are the phenolic antioxidants, the acetins being particularly useful as solvents for phenolic antioxidants such as butylated hydroxyanisole, butylated hydroxytoluene, propyl gallate, nordihydroguaiaretic acid, gum guaiac, the tocopherols, hydroquinone, 2,4,5-trihydroxy-butyrophenone, 2,4,5-trihydroxyacetophenone, 5-acenaphthenol, etc. The acetins are also useful as solvents for other classes of fat and oil antioxidants such as glycine, lecithin, 3,3'-thiodipropionic acid and esters thereof, etc. as well as various mixtures of any of these or similar materials.

The acetins are valuable solvents for synergists such as citric acid, phosphoric acid, ascorbic acid, isopropyl citrate, and other similar materials. This is one of the most advantageous and unobvious features of this process since a common solvent for various solutes including synergists is quite unexpected. Synergists are sometimes called metal deactivators. The following table illustrates the solubility of typical antioxidants and synergists in the acetins when used singularly.

TABLE 1: SOLUBILITY (PERCENT BY WEIGHT)

	Citric acid	Propyl Gallate	Butylated Hydroxyanisole	Butylated Hydroxytoluene	2,4,5-Trihydroxybutyrophenone	Hydroquinone
Monoacetin	25	15	---------	---------	10	10
Diacetin	30	25	30	5	25	10
Triacetin	5	10	50	30	30	5

As is apparent from Table 1, individual acetins or mixed acetins can be used without any other solvent components to dissolve various antioxidant and synergistic components. These and other stabilizers mentioned above can be readily dissolved in acetin solvents (various acetic acid esters of glycerine). The process is particularly concerned with acetin solutions of mixtures of synergists and phenolic antioxidants (one or more of each type of component).

Any of the well-known solvents can be used in combination with the acetins and antioxidant compositions. Certain systems use glycerine, propylene glycol, acetic acid, naturally occurring fats and oils, and products derived from naturally occurring fats and oils such as higher fatty acid monoglycerides, diglycerides, etc. in various combinations with one or more of the acetins.

The following compositions were prepared and tested. These data are given in order to illustrate the type component that can be made and to indicate the AOM stability provided by the particular compositions. These data are based on tests conducted under equivalent conditions employing similar samples of lard in each instance. The AOM figures are in hours required for formation of 20 meq of peroxide/kg of fat, with air bubbling through the samples at 98°C. A peroxide value of 20 was taken, according to normal practice, as corresponding to a commercially objectionable rancidity level.

In Table 2, BHA represents a mixture of isomers consisting of 3-tertiarybutyl-4-hydroxy-anisole and 2-tertiarybutyl-4-hydroxyanisole; BHT represents 3,5-ditertiarybutyl-4-hydroxy-toluene. These antioxidants are frequently referred to in the art as butylated hydroxyanisole and butylated hydroxytoluene respectively. The latter compound is sometimes referred to as butylated p-cresol.

TABLE 2

Components of Antioxidant Composition	Concentration, Weight Percent	AOM Stability in Hours
(a)		

25% Solute composed of:
  40% BHA (10% of overall)
  40% BHT (10% of overall)
  20% Citric Acid (5% of overall)
75% Solvent composed of:
  4% Propyl gallate (3% of overall)
  48% Diacetin (36% of overall)
  48% Corn oil monoglyceride (36% of overall)
	0.05	63

(b)		

30% Solute composed of:
  66.7% BHA (20% of overall)
  20.0% Propyl gallate (6% of overall)
  13.3% Citric acid (4% of overall)
70% Solvent composed of: 100% Diacetin (70% of overall)
	0.05	93

(c)		
50% Solute composed of: 100% BHA (50% of overall) 50% Solvent composed of: 100% Triacetin (50% of overall)	0.04	31
(d)		
30% Solute composed of: 100% BHA (30% of overall) 70% Solvent composed of: 100% Diacetin (70% of overall)	0.067	31
(e)		
30% Solute composed of: 100% BHT (30% of overall) 70% Solvent composed of: 100% Triacetin (70% of overall)	0.067	52
(f)		
25% Solute composed of: 100% 2,4,5-Trihydroxybutyrophenone (25% of overall) 75% Solvent composed of: 100% Diacetin (75% of overall)	0.08	180
(g)		

30% Solute composed of:
  83.3% 2,4,5-Trihydroxybutyrophenone (25% of overall)
  16.7% Citric acid (5% of overall)
70% Solvent composed of: 100% Diacetin (70% of overall)
	0.08	190

(h)		
10% Solute composed of: 100% Hydroquinone (10% of overall) 90% Solvent composed of: 100% Diacetin (90% of overall)	0.20	150

Since most antioxidant compositions are mixtures of several components, it is impossible to assign specific limits to the concentration of acetins used. If the concentration of one component is varied, there will be an effect on the maximum concentration of the other components. However, as indicated in Table 1, certain maximum concentrations are possible to achieve and, generally, the maximum concentration is the desirable one.

Polyoxyalkylene Ether

A severe limitation in formulating antioxidant compositions arises from the antagonism between certain otherwise acceptable antioxidant solutes. For example, it has been impossible to formulate antioxidant compositions to exploit the beneficial properties of such solutes when this antagonism exists. This antagonism is usually manifested by the inability to maintain all of the antagonistic solutes in solution with each other. A classic example of antagonism between otherwise beneficial solutes is that which exists between butylated hydroxytoluene (BHT), and nordihydroguaiaretic acid (NDGA). Another example of such antagonism is that which exists between BHT and propyl gallate.

The problem is further magnified because BHT is substantially completely insoluble in propylene glycol, the generally accepted antioxidant solvent. A further problem arises from the need to have butylated hydroxyanisole (BHA) present to solubilize antioxidant formulations to be used in frying media because BHA releases an unpleasant odor when used in such media.

A special solvent system containing polyoxyalkylene ether is given by *L.V.W. Van Ness; U.S. Patent 3,390,098; June 25, 1968; assigned to Armour Pharmaceutical Company.* In general the system is prepared by first placing the solvent system into a steam heated glass lined tank preferably equipped with a mechanical agitator. The agitator is started and sufficient steam is applied to the tank to raise and maintain the inside temperature of the tank at about 250°F. To this tank containing the solvent, and prior to the addition of the solute, a small but effective amount of a suitable chelating agent is added, such, for example, as citric acid, phosphoric acid, ethylene diamine tetraacetate (EDTA), which functions to hold trace metallic ions which might otherwise catalyze oxidation. When the chelating agent is in solution, the tank is cooled to about 200° to 210°F as by the application of cooling water on the jacket of the tank.

At this point solutes are added either jointly or separately; first, for example, BHT, while the agitator is running. When this addition is completed, the inside tank tamperature is adjusted to and maintained at about 190° to 200°F. The agitation is continued until the added solute is substantially completely dissolved in the solvent as evidenced by the contents becoming clear. While the agitator is still running, the second solute for example, NDGA, is added. The temperature of the tank contents is still maintained at about 190° to 200°F. Again, the agitator is run until a clear solution is obtained. If desired, this solution can be now packaged or, when appropriate, a third solute is added, for example, propyl gallate. Subsequent solute additions will be accomplished in the same manner.

After all of the desired solutes have been introduced in the manner described, the final solution is withdrawn from the tank, preferably, while still warm, that is, at a temperature between 100° and 110°F, and put in suitable containers. In addition to the problem solutes specifically referred to above, viz, BHT, NDGA and propyl gallate, butylated hydroxyanisole (BHA) or 2,4,5-trihydroxy butyrophenone (THBP), or mixed tocopherols may be added. Tocopherol, includes so called alpha-, beta- and gamma-tocopherol. The common names for these tocopherols are, respectively 3,8-dimethyl tocol, 5,8-dimethyl tocol, and 7,8-dimethyl tocol. The full name using β-tocopherol as illustrative, is 2,5,8-trimethyl-2-(4',8',12'-trimethyl tridecyl)-6-chromanol.

An especially useful polyoxyalkylene ether solvent is polyoxyethylene sorbitan monooleate, a complex mixture of polyoxyethylene ethers of mixed partial oleic esters of sorbitol anhydrides (known as polysorbates). Fine results are obtained when these polyoxyalkylene ethers are used in combination with propylene glycol ester of soybean oil and, in certain instances, with propylene glycol.

Other polyoxyalkylene ethers suitable for use are the water-soluble polyoxyalkylene derivatives of hexitol anhydride partial long chain fatty acid esters such, for example, as the polyoxyalkylene derivatives of sorbitan monolaurate, sorbitan monopalmitate, sorbitan monostearate and sorbitan monooleate which are commercially available as Tween 20, Tween 40, Tween 60 and Tween 80, respectively.

Example 1: 2,400 pounds of solvent is placed into a 1,000 gallon glass lined tank equipped with an agitator serviced with steam and cooling water. The agitator is then started and steam is put on the jacket to heat the charge to 245° to 250°F. With the agitator running, 480 pounds of chelating agent is added to the charge while maintaining the batch temperature at 245° to 250°F. When the chelating agent is all in solution, 1,600 pounds of the first antioxidant is added and the temperature is allowed to drop to 190° to 200°F. When this antioxidant is added completely, the batch temperature is adjusted to 190° to 200°F by heating or cooling the batch, whatever is needed. A second and third antioxidant solute can be added in a similar manner, adjusting the temperature to 190° to 200°F after each addition and agitating until the solution is clear. Using this procedure, an antioxidant formulation was prepared having the following composition (percent by weight): butylated hydroxyanisole, 40%; nordihydroguaiaretic acid, 4%; citric acid, 6%; polyoxyalkylene ethers, 50%. The formulation was oil-soluble, water-miscible, has excellent antioxidant properties and remained homogeneous on the shelf during a shelf-life test extending in excess of 2 weeks at room temperature.

Example 2: A formulation was prepared according to Example 1 and contained (in weight percent): 25% BHA; 25% propyl gallate; 3% citric acid, and 47% polysorbate 80. This formulation was mixed into 5 hour prime lard in ratio of 1 part antioxidant to 2,500 parts of lard and 1 part antioxidant to 4,000 parts of lard and subjected to tests by the active oxygen method. The results are reported below.

Ratio	AOM, hr
1:2,500	96
1:4,000	72

Example 3: Using the procedure of Example 1, an antioxidant formulation was prepared containing (in weight percent): BHA, 13.3%; BHT, 13.3%; citric acid, 5%; polysorbate, 48.4%; propylene glycol ester of soybean oil, 20%. This formulation was added to a 5 hour prime steam lard at a ratio of 1 part antioxidant to 1,330 parts lard and 1 part antioxidant to 2,660 parts lard and the stability was measured by the active oxidation method. The results are reported below.

Ratio	AOM, hr
1:1,330	64
1:2,660	48

This formulation was completely uniform and stable and, when subjected to a descending temperature test, was found to remain stable at temperatures below 32°F.

METAL DEACTIVATORS

Citric Acid Esters

G.P. Touey and H.E. Davis; U.S. Patent 3,274,109; September 20, 1966; assigned to Eastman Kodak Company discuss the problem of the catalytic effect of trace amounts of metals on the oxidation of fats and oils. Such oxidation is undesirable and generally renders the material unfit for its intended use. Oxidation is accelerated by a rise in temperature, by the action of sunlight and the catalytic effect of certain metals such as copper, cobalt and manganese.

The abovementioned metals are harmful, either in the solid metallic form or as dissolved organic compounds such as acid salts. Only trace amounts (1 ppm) of a copper salt, such as cupric oleate, for example, can greatly reduce oxidation stability. Salts can occur naturally in the material or through the action of small amounts of organic or inorganic acids present in the material on copper-containing equipment used to process, store or utilize the material.

Avoidance of copper is thus often impractical or impossible. A widely used remedy is to employ a copper (or other metal) deactivator which will tie up the metal in an inactive form in which it is no longer able to function as an oxidation catalyst. Many types of organic molecules can act as such metal deactivators. For any specific application, however, the deactivator must be selected to meet the particular requirements of such an agent, such as efficiency, solubility, nontoxicity, compatibility, nonextractibility (such as by water) and reasonable cost. Generally, deactivators contain several groups such as phenolic OH, − COOH, −NH$_2$, − NHR, −SH, − SR, etc. A typical metal deactivator is N,N'-disalicylidene-1,2-propylenediamine. Its union with copper ion is shown on the next page.

$$\begin{array}{c} \text{H} \qquad\qquad\qquad \text{H} \\ -\text{C}=\text{N}-\text{CH}_2-\text{CH}-\text{N}=\text{C}- \\ \text{H}_3\text{C} \\ \text{OH}-\!\!-\!\!-\!\!-\!\!-\text{Cu}-\!\!-\text{HO} \end{array}$$

Citric acid itself is a well-known metal deactivator. However, it is also well-known that this acid is not soluble in edible fats and oils. As a result, when it is used as a metal deactivator for such systems it must be used in conjunction with certain polar-type organic solvents such as alcohols and glycols. Thus, for example, in using citric acid as a metal deactivator in lard a large proportion of propylene glycol is normally used as a cosolvent for the acid. Even with such cosolvents there is always the possibility that the citric acid will crystallize out of the final product and hence become ineffective to remove metallic ions.

One method for rendering citric acid more soluble is to esterify one of its three free carboxyl groups with a long chain aliphatic acid. Pertinent information on the citric acid ester type of metal deactivators has been reported in U.S. Patents 2,701,203 and 2,686,751 and in the literature, for example, *Food Technology*, 8, pp 6–9 (1954), and *J. Amer. Oil Chemist Soc.*, 32, pp 175–176 (1955). Although such methods for solubilizing citric acids are useful they have one obvious disadvantage. This is the blocking of one of the active carboxyl groups in the product by the formation of an ester group.

It is well-known that citric acid owes its metal deactivating power to the presence of its three carboxyl groups. Therefore, any derivative of the acid which would have less than three carboxyl groups available would have a reduced activity toward metal ions. In fact, it has been reported [*Food Technology*, 8, pp 6-9 (1954)] that di- and triesters are completely inactive as metal deactivators.

A citric acid-derived metal deactivator was developed based upon the broad concept of altering the citric acid molecule so as to provide a derivative which is soluble in those organic substances from which it is desired to remove metal ions and at the same time will have the three carboxyl groups of the citric acid molecule preserved intact and available to function to tie up the metal ions present in the substrate undergoing treatment as, for example, a gasoline or lubricating oil or an edible fat or oil. Specifically, the metal deactivators are polyoxyalkylene ethers of citric acid having the structural formula

$$\begin{array}{c} \text{CH}_2\text{COOH} \\ | \\ \text{H(OR)}_x\text{O}-\text{C}-\text{COOH} \\ | \\ \text{CH}_2\text{COOH} \end{array}$$

wherein R is propylene or butylene and x has an average value of from 2 to 10. The average molecular weight for these compounds ranges from approximately 300 to approximately 900. In general, however, the preferred average value of x will range from 3 to 6, which will give a molecular weight of from approximately 365 to 625.

Only the polyoxypropylene and polyoxybutylene ethers of citric acid have been found to be useful. This appears to be due to the fact that the oxyethylene groups as they are present in the structural formula above are not oil-soluble and are therefore not useful in making the citric acid derivative soluble in edible fats and oils. On the other hand, oxypropylene and oxybutylene groups are oil-soluble and water-insoluble, thus making them ideally suitable for making a citric acid derivative soluble in edible fats and oils.

The formation of such ethers is illustrated by the following equation for the preparation of an ether which may be identified as tripolyoxypropylene ether of citric acid.

$$\begin{array}{c} CH_2COOCH_2CH_3 \\ | \\ HO-C-COOCH_2CH_3 \\ | \\ CH_2COOCH_2CH_3 \end{array} + 3CH_2\!-\!\!CH-CH_3 \xrightarrow{BF_3}$$

(1)

$$\begin{array}{ccccc} CH_3 & CH_3 & CH_3 & CH_2COOCH_2CH_3 \\ | & | & | & | \\ HO-CH-CH_2O-CH-CH_2O-CH-CH_2OCCOOCH_2CH_3 \\ & & & | \\ & & & CH_2COOCH_2CH_3 \end{array}$$

(2)

$$(2) \xrightarrow{NaOH} \begin{array}{cccc} CH_3 & CH_3 & CH_3 & CH_2COONa \\ | & | & | & | \\ HO-CH-CH_2O-CH-CH_2O-CH-CH_2OCCOONa \\ & & & | \\ & & & CH_2COONa \end{array} + 3CH_3CH_2OH$$

(3)

$$(3) \xrightarrow{H^+} \begin{array}{cccc} CH_3 & CH_3 & CH_3 & CH_2COOH \\ | & | & | & | \\ HO-CH-CH_2O-CH-CH_2O-CH-CH_2OCOOH \\ & & & | \\ & & & CH_2COOH \end{array}$$

(Tripolyoxypropylene ether of citric acid)

The polyoxyalkylene ethers of citric acid will be soluble in corn oil to the extent of about 50%, while in lard the solubility, depending on the temperature, may be only 3%. However, even in the case of the lower solubilities it was found that these citric acid ether derivatives are valuable as metal deactivators.

In treating edible fats and oils, the compound is simply added to the substance undergoing treatment and thoroughly mixed therein in the desired amount, which, in any event, will be at least equivalent to the amount of metal ion content to be removed. This in most cases will range, for example, from 0.0001 to 0.5% by weight of the substance undergoing treatment. The actual amount used in any given instance will of course vary widely, depending, not only upon the specific nature of the material itself, but also upon the source from which it was derived. For example, an edible fat such as lard derived from one source may contain only a moderate amount of metal ions, while another sample of lard derived from another source may have a much higher concentration of contaminating metal present and will therefore require a proportionately larger amount to deactivate the metal ion component.

The metal contaminants which are generally met with in edible fats and oils are copper, cobalt, iron, nickel and manganese although other metal contaminants may also be present. Standard methods are used for determining the effectiveness of the use of the metal deactivators. For example, in testing an edible fat such as lard which has been treated with a citric acid ether derivative to deactivate the metal ion contaminant the so-called AOM may be employed. The citric acid ether derivative deactivators may be prepared as in the following typical example illustrating one embodiment of the process.

Example 1: Preparation of Polyoxypropylene Ether of Citric Acid – 8 mols (464 g) of propylene oxide was slowly added over a 3-hour period to a stirred mixture of 0.1 mol (6.8 g) of boron trifluoride in 0.5 mol (138.1 g) of triethyl citrate. The temperature was held at 35° to 45°C during this period by means of an ice bath. After all the propylene oxide had been added the solution was stirred for 1 additional hour. Ethyl ether (500 ml) was added and the solution neutralized with aqueous sodium bicarbonate and then washed with water. The ethereal solution was dried and the ether flashed off. The residue was heated to a pot temperature of 180°C at 1 mm pressure to remove any unreacted triethyl citrate as well as any propylene glycol. The viscous residue was dissolved in 1 liter of an ethanol (90%)-water (10%) solution containing 3 mols (168 g) of KOH. The solution was refluxed for 2 hours to completely saponify the ether ester. The ethanol and water were removed by distillation and the residue extracted with ethyl ether to remove any polypropylene glycol. The residue was made strongly acidic with hydrochloric acid and extracted continuously with ethyl ether for 48 hours. The ether extract was then concentrated under reduced pressure to give the polyoxypropylene ether of citric acid, a viscous yellow liquid with the following structure:

$$H-\left[OCH-CH_2-\atop CH_3\right]_x \begin{array}{c} CH_2COOH \\ | \\ O-C-COOH \\ | \\ CH_2COOH \end{array}$$

The neutral equivalent of the product was about 134 to 137, indicating that the product contained an average of 4 propylene oxide condensation units, i.e., the average value of x was 4. The product was soluble in such products as lard, corn oil, cottonseed oil and peanut oil, as well as in petroleum products. It was soluble in the common aliphatic and aromatic organic solvents, such as acetone, ethanol, dichloromethane, ethyl ether, propylene glycol, benzene, toluene, xylene. In water, however, the product was essentially insoluble.

Example 2: Preparation of Polyoxybutylene Ether of Citric Acid – 8 mols (576 g) of butylene oxide (mixture of isomers) was substituted for the propylene oxide of Example 1. The reaction was then carried out and the product processed as described in that example. The neutral equivalent of the product was about 130 to 134, indicating that the product contained an average of 3 butylene oxide condensation units, i.e., the average value of x was 3.

Example 3: Use of Polyoxyalkylene Ethers of Citric Acid as Metal Deactivating Agents For Iron in an Edible Fat – The polyoxypropylene ether of citric acid designated in the table below as PECA and the polyoxybutylene ether of citric acid designated as BECA were tested as metal deactivating or chelating agents when such agents were used to treat a commercial lard (stabilized with butylated hydroxyanisole (BHA) as an antioxidant) employing the AOM for the valuation. This test involves bubbling air through the lard sample at a temperature of 99°C whereby peroxides are formed and the oxidation is followed by a determination of milliequivalents of such peroxides per kilogram of substrate. Ordinarily, a peroxide value of 20 is the upper limit which can be tolerated in edible fats, for example. Above this value fats exhibit an objectionable degree of rancidity.

In carrying out the abovementioned test a stabilized sample of lard containing 0.01% BHA and 1 ppm iron was prepared as follows: a 0.01 gram sample of BHA was added to 99.99 grams of lard along with 1 ml of an ethanol solution of ferric oleate so that the resulting solution or mixture contained 1 ppm of iron as ferric ion. This mixture was placed on a steam bath heated to about 78°C and stirred continuously for 25 minutes during which time the ethanol was removed from the mixture by evaporation. 20 ml of the stabilized lard solution was placed in an AOM tube maintained at 99°C and air bubbled through at a rate of about 2.3 ml per second.

Periodically, a portion of the test solution was removed and the peroxide content quantitatively determined by iodometric titration, the results being expressed as meq/kg of sample. A control containing no additive was run simultaneously to determine the induction period of the unstabilized material. The effectiveness of the citric acid ether derivatives in controlling or eliminating the oxidative effect of metal ions in lard, a typical edible fat, is shown in Table 1.

The deleterious effect of iron is demonstrated in this table and the metal deactivating effectiveness of the polyoxypropylene ether of citric acid and of polyoxybutylene ether of citric acid is clearly shown. Thus, lard containing these agents and iron have substantially the same stability as lard stabilized with BHA in the absence of such metal contaminants.

TABLE 1: EFFECT OF PECA AND BECA AS DEACTIVATORS FOR IRON

Sample	AOM Value
Lard, control (no deactivator nor metal contaminant)	4
Lard + 1 ppm iron (added as ferric oleate)	0.5
Lard + 0.01% BHA	18
Lard + 0.01% BHA + 1 ppm iron	3
Lard + 0.01% BHA + 1 ppm iron + 0.05% PECA	20
Lard + 0.01% BHA + 1 ppm iron + 0.05% BECA	19

Example 4: Polyoxyalkylene Ethers of Citric Acid as Metal Deactivating Agents for Copper in an Edible Fat — Samples of lard were prepared as in Example 3 and tests were carried out using the same procedure as in that example, the BHA and metal deactivator being weighed out and mixed with the lard prior to test. The results obtained are indicated in Table 2 below.

TABLE 2: EFFECT OF PECA AND BECA AS DEACTIVATORS FOR COPPER

Sample	AOM Value
Lard, control	10
Lard + 1 ppm copper (added as cupric chloride)	1
Lard + 0.01% BHA	35
Lard + 0.01% BHA + 1 ppm copper	1
Lard + 0.01% BHA + 1 ppm copper + 0.05% PECA	40
Lard + 0.01% BHA + 1 ppm copper + 0.05% BECA	43

Ethylenediaminetetraacetic Acid Esters

The stability system proposed by *W.J. Lennon; U.S. Patent 3,497,535; February 24, 1970; assigned to Geigy Chemical Corporation* consists of ethylenediaminetetraacetic acid (EDTA) esters of the general structural formula

$$X$$
$$\backslash \qquad\qquad\qquad\qquad\qquad Y$$
$$N-(CH_2CH_2N)_n-CH_2CH_2-N$$
$$HOOCCH_2 \qquad\qquad CH_2COOH \qquad CH_2COOH$$

wherein n is 0 or 1; X represents $-CH_2COOR$ or $- (CH_2)_m OCOR$; Y stands for $-CH_2COOH$, $-CH_2COOR$ or $-(CH_2)_m OCOR$; wherein R represents straight- or branched-chain alkyl or alkenyl groups of up to 22 carbon atoms as for example, ethyl, propyl, butyl, pentyl, hexyl, heptyl, octyl, nonyl, decyl, dodecyl, tetradecyl, pentadecyl, heptadecyl, octadecyl, eicosyl, docosyl, etc., and the corresponding ethylenically unsaturated groups, and m is 1 or 2.

For example, it was found that simple addition of 10 ppm of didodecyl EDTA ester to lard containing 1 ppm of copper and 50 ppm of BHA prolonged the shelf life 5 times over lard containing 1 ppm of copper, 50 ppm of BHA and 25 ppm of citric acid. Significantly, these new metal deactivators possess besides solubility in fats and oils also other requisite qualities for food-stabilizing agents, namely, relative nontoxicity, absence of bad odors, tastes or colors after incorporation in the fats and oils to be protected. These chelating agents of the above formula, can conveniently and economically be obtained as reaction products of EDTA dianhydride and suitable alcohols.

For instance, the didodecyl ester of EDTA was prepared in accordance with the following procedure: a mixture of 5.1 grams of EDTA dianhydride and 7.44 grams of dodecyl alcohol in 50 ml of benzene was refluxed for 24¼ hours and was protected by a drying tube. The reaction mixture was filtered hot and washed with hot benzene. The benzene filtrate was cooled in an ice bath, filtered and the white crystalline solid was washed with cold benzene followed by washing with hexane and dried in vacuo over P_2O_5. Obtained was 6.9 grams (yield: 55%) of the desired material: MP, 98° to 99.5°C.

In an analogous manner, the diethyl and dioctadecyl esters of EDTA, MP, 107° to 110°C and 102.5° to 105°C, respectively, were prepared by using instead of dodecyl alcohol, ethanol and octadecanol. EDTA dianhydride is prepared, for instance, as follows: 364 grams of ethylenediamine-N,N,N'N'-tetraacetic acid, 519 grams of acetic acid anhydride, and 600 grams of pyridine are stirred at 65°C for 24 hours. The reaction mixture which turns orange but gives no clear solution, is then filtered at 24°C and the residue is washed first with acetic acid anhydride and then with diethyl ether. It is then dried at 60°C and 12 torr. 307 grams of EDTA dianhydride ⟨1,2-bis-[2,6-dioxo-morpholinyl(4)]-ethane⟩ is obtained as white crystals; MP, 195°C.

Analysis for $C_{10}H_{12}N_2O_6$ (MW 256.21): Calculated: C, 46.87%; H, 4.73%; N, 10.93%; O, 37.47%. Found: C, 46.71%; H, 4.65%; N, 10.93%; O, 37.29%.

EDTA esters stabilize fats, such as lard and tallow but other fats and oils which may likewise be stabilized include particularly margarine but also butter and shortening as well as castor oil, linseed oil, rapeseed oil, olive oil, corn oil, palm oil, coconut oil, peanut oil, safflower oil, soybean oil, and the like. Stabilization of these materials is effected by adding from about 0.5 to 200 weight parts per million of chelating agents of the above formulas in combination with effective amounts of antioxidants. When tested by means of the Swift Stability Test or, as it is also called, the Active Oxygen Method, the subject compounds were found to retard the development of rancidity for about 20 to 30 hours.

This process will be illustrated by the following procedure. EDTA diester samples were weighed and introduced directly into clean aeration tubes. 20 grams of molten uninhibited lard containing the appropriate amounts of copper and antioxidant (BHA or BHT) was then added to each tube without employing any mixing or dispersion procedure. The following stock solutions of BHT, BHA and copper were prepared in molten lard:

> BHT = 0.400 gram Tenox BHT/100 grams
> BHA = 0.500 gram Tenox BHA/100 grams
> Cu^{++} = 0.1250 gram Nuodex Copper Naphthenate (8% Cu)/100 grams

After thorough mixing, the stock solutions were frozen and retained in the freezer until use. Immediately prior to each run, the appropriate stock solutions were melted and were diluted with molten lard according to the following table:

> 10 grams BHT stock diluted to 100 grams = 400 ppm BHT
> 1 gram BHA stock diluted to 100 grams = 50 ppm BHA
> 1 gram Cu^{++} stock diluted to 100 grams = 1 ppm Cu^{++}
>
> No stock solution was used after two remelts

The samples were then fitted with air inlet tubes and placed in the aeration apparatus at 100°C. Lard samples were withdrawn periodically and checked for peroxide content by the following iodometric titration.

Using the air inlet tube as a pipette, 7 drops of the molten lard (0.200 g) was transferred to a clean 250 ml Erlenmeyer flask. The lard was dissolved in 20 ml of a 3:2 mixture of acetic acid and chloroform, followed by the addition of 1 ml of saturated KI solution. This mixture was gently agitated and placed in the dark for approximately 1 minute. The sample was diluted with 100 ml of distilled water, and titrated to a starch indicator (2 ml of 0.5% starch) endpoint with 0.002 N $Na_2S_2O_3$. If the sample solution was deep yellow or brown when removed from the dark, it was titrated to a pale yellow before addition of the indicator. The sodium thiosulfate and starch solutions were prepared as follows.

Sodium Thiosulfate: A 0.1 N solution of $Na_2S_2O_3$ was prepared by dissolving 25 grams of $Na_2S_2O_3 \cdot 5H_2O$ in 250 ml of water and diluting to 1 liter. After standing for 48 hours, the above stock solution was standardized by titration with 25 ml aliquots of a 0.1001 N solution of potassium dichromate (4.9034 grams dried, reagent grade $K_2Cr_2O_7$ in 1 liter). All 0.002 $Na_2S_2O_3$ used for titrations was prepared by a 1:49 dilution of the stock solution.

Starch Indicator (0.5%): 1 gram of starch was slurried with a small amount of water, followed by the addition of 200 ml of boiling water. The resulting solution was boiled for 2 minutes, cooled and bottled. Fresh solutions had to be prepared at least every 3 days due to bacterial growth. Since some lots of acetic acid contain large amounts of peroxide or develop peroxide upon standing, reagent blanks were run periodically. No corrections were required. In the following tables reported induction periods are averages of duplicate samples. The dependence of induction period on Na_2EDTA, citric acid and didodecyl EDTA ester concentrations in lard containing 50 ppm BHA and 1 ppm Cu is shown in Table 1.

TABLE 1

Chelating Agent	Concentration (ppm)	Induction Period (hr)
None	---	3.5
Na₂EDTA	10	3.5
Na₂EDTA	25	3.5
Na₂EDTA	50	3.5
Citric acid	10	4.0
Citric acid	25	4.0
Citric acid	50	5.5
Didodecyl EDTA ester	5	8.0
Didodecyl EDTA ester	10	20.5
Didodecyl EDTA ester	25	28.5
Didodecyl EDTA ester	50	29.75

From this table it is evident that under the test conditions used (no special mixing), even citric acid is almost completely ineffective at all concentrations, whereas the induction period approaches a maximum of approximately 30 hours as the concentration of didodecyl EDTA ester is increased. The effectiveness of other EDTA diesters upon copper deactivation in lard containing 50 ppm BHA and 1 ppm Cu was also determined. The results are given in Table 2.

TABLE 2

EDTA Ester	Concentration (ppm as EDTA)	Induction Period (hr)
None	---	2.5
Diethyl	5	21.5
Diethyl	10	29-45*
Didodecyl	5	18.75
Didodecyl	10	29-45*
Dioctadecyl	5	20.5
Dioctadecyl	10	29-45*

*Failed overnight

INACTIVATION OF COTTONSEED LIPASE

Fats useful for conventional purposes are ordinarily present as neutral glycerides and usually must be substantially free of significant quantities of free fatty acids. When these fats are to be used for edible purposes, they are normally refined to remove any free fatty acid present. The cost of this refining represents a distinct monetary loss in the preparation of fatty materials. These undesired fatty acids normally result from the hydrolysis of the fatty material. This hydrolysis is prevalent in fatty tissues derived from animals, and therefore, unless such tissue is maintained at a very low temperature or heated sufficiently to inactivate the lipases present, the free fatty acid content of tissue increases rapidly. For example, the following table shows that visceral tissue stored at 34°F increases rapidly in free fatty acid with increase in storage time.

TABLE 1: PERCENTAGE OF FREE FATTY ACIDS AT VARIOUS HOLDING TIMES

Type of Fat	20 Hours, Percent	60 Hours, Percent	100 Hours, Percent
Ruffle fat	0.9	2.9	5.0
Leaf fat	0.2	0.6	1.0
Ceul fat	0.1	0.3	0.7
Back fat	0.1	0.2	0.5

Storage at higher temperatures produces a great increase in the free fatty acids. Thus, the following table illustrates that at a temperature of 72° to 74°F a three to fourfold increase in free fatty acid results. At such temperatures, other enzymes may also be present, attacking protein and producing materials which have undesirable flavors and contaminate the oil.

TABLE 2: PERCENTAGE OF FREE FATTY ACIDS IN LEAF FAT STORED AT VARIOUS TEMPERATURES

Storage time	40° F., percent	72-74° F., percent
20 hours	0.05	0.17
60 hours	0.13	0.46
100 hours	0.22	0.85

In addition to animal fats, oils derived from oil bearing fruits also develop free fatty acids after harvesting. Palm oil is a good example of this. This oil generally contains more than 2 to 3% free fatty acid, due to the active lipase present in the fruit. Olives are another example, unrefined olive oil normally containing from 1 to 5% or higher free fatty acid content.

Although oilseeds are less sensitive to fatty acid buildup than are animal tissue and oil bearing fruits, the free fatty acid present in oilseeds generally presents a serious problem. In the case of such seeds as flaxseed, which are raised in a relatively cool climate, storage temperatures may be sufficiently cool to avoid fatty acid buildup. This is also true to a lesser extent with soybeans. However, cottonseeds which are grown in warmer climates are difficult to cool sufficiently to inhibit the free fatty acid buildup probably resulting from enzyme action. Much of the damage is done after summer harvesting and before the outside temperature is low enough to allow the seeds to be cooled to a sufficiently low temperature to avoid this enzyme action. There is some question as to whether enzymes in the seeds or microorganisms on the surface of the seeds are responsible for fat hydrolysis. But in either case, hydrolysis of the fatty oil to free fatty acid results.

The cottonseed industry may be taken as a typical example of the problem caused by free fatty acid buildup in oilseed materials. Cottonseed is normally purchased at harvest time and then stored until it can be processed. Since the harvest period may be as short as 2 to 4 weeks, and cottonseed mills prefer to operate year round, it is obvious that much of the raw material must be stored for several months before being processed. This storage presents no problem when good quality dry cottonseed is stored. However, in many sections of the country, harvested cottonseed is wet and contains appreciable amounts of free fatty acid (from about 1.5 to 10% or more). When such wet seeds are stored, there is a continued increase in the free fatty acid, and this increase accelerates with the passage of time. Thus, when the seeds are processed, they contain much more free fatty acid, and consequently less saleable neutral oil, than was present when the seeds were purchased.

Generally, in the storage of oil-bearing seeds such as cottonseed, the seeds are placed in a large building equipped with loading and unloading facilities and provided with air ducts and suitable fans so that outside air may be blown or sucked into the storage seed. In this way the seed can be gradually brought down to the temperature of the outside air and if the relative humidity of the air is low enough, the seed may also be dried to a limited extent. This method is not very satisfactory due to channeling which occurs in the air stream, the fact that temperature and humidity conditions of the air are often unsuitable for this treatment, and to the power cost involved. It is estimated that in the United States approximately one dollar is lost on each ton of cottonseeds processed due to deterioration during storage; and this deterioration is due primarily to the buildup of free fatty acids in the cottonseed.

A number of methods have been tried in the past to inhibit the hydrolysis of the fats to free fatty acids. Among these is the keeping of the fat containing raw material cool. This method is, however, generally too expensive to be feasible. Another method of inhibiting this hydrolysis has been heating the raw material to a temperature sufficient to inactivate the enzymes. This method can be effective, but it is normally expensive and the undesirable effects of the heat treatment may more than offset the beneficial effects of this treatment. The addition of chemicals has also been attempted as a mode of inhibiting this hydrolysis. Among the chemicals used have been propylene glycol dipropionate, 4,6-bis-chloromethyl benzene or xylene, and various halohydrins such as ethylene chlorohydrin and the like.

Other chemical inhibitors which have been used in the past include compounds containing the aldehyde group; cupric, mercuric, ferric and cobaltous ions; fluoride, bromide, iodide or chloride ions (chloride being the least effective); and chloro-, bromo- or iodoacetic acids. In general, these inhibitors have been toxic, expensive, or not effective on a commercial scale.

An active chlorine-containing material is used by *F.A. Norris and D.P Grettie; U.S. Patent 3,300,524; January 24, 1967; assigned to Swift & Company* to inactivate lipase enzymes to inhibit the hydrolysis of fatty material to free fatty acids. The fatty material is treated with a solution of sodium hypochlorite or another source of active chlorine, such as chlorine itself, hypochlorous acid, alkali metal and alkaline earth metal hypochlorites, such as calcium hypochlorite and sodium chlorite, or chlorine dioxide.

As pointed out above, there is disagreement as to whether the buildup of free fatty acids in oilseeds is due to the enzymes in the seed or to the action of microflora present on the outer surface of the seeds. However, the treatment has been found to inhibit the formation of free fatty acids on the cottonseeds in storage under adverse conditions as well as to inhibit glyceride hydrolysis in an isolated enzyme system where only pancreatic lipase is present. Thus, it is obvious that the instant process would inhibit hydrolysis whether it is caused by enzyme action or microflora.

In the treatment of oilseed materials, other oxidizing agents such as hydrogen peroxide, sodium peroxide, calcium peroxide, sodium perborate, potassium dichromate and potassium permanganate can be used, but these oxidizing agents are not preferred since in most cases they are difficult to handle, expensive and undesirably contaminate the raw material for future food or feed use. The solution of active chlorine-containing materials, such as sodium hypochlorite, may be brought into contact with the oilseed material by soaking the seeds in a solution of the active chlorine-containing material, by spraying the material on the seeds, or by similar methods. The active chlorine-containing materials are useful since they are oxidizing agents which, in general, inhibit enzymes; and also because they liberate free chlorine which diffuses through the seed mass and thus extends the activity generally through the mass. As a result of this treatment, the fat splitting enzymes of the oilseed are inhibited so that the seeds may be stored under relatively adverse conditions without appreciable buildup in free fatty acids.

Example 1: 950 grams of good quality cottonseeds were soaked for 5 minutes in a 0.5% aqueous solution of sodium hypochlorite (0.238% available chlorine), and then spread on a tray and allowed to dry at room temperature. An equal quantity of identical seeds was soaked in the same way in distilled water and allowed to dry similarly. At the end of the immersion, the moisture content of the control was about 30% and that of the treated sample 33%. By the next day, the control had dropped to a moisture content of 11.5%, but the treated sample was still at 23.6% moisture. The corresponding free fatty acid of the two samples was 1.0% in the control and 0.6% in the treated sample. These samples were placed in desiccators maintained at 95°F and equipped to maintain about 85% relative humidity. Periodic samples were withdrawn from the desiccators to test the change in free fatty acid content. The results of these tests are indicated by the following table.

Percentage of Free Fatty Acid at Various Storage Periods

Sample	At Start	4 weeks	7 weeks	12 weeks
Control sample:				
Free fatty acids	1.0	5.0	12.0	22.7
Moisture	11.5		13.8	14.5
Test sample:				
Free fatty acids	0.6	4.3	4.2	6.9
Moisture	23.6		15.6	15.3

These tests clearly show that although there was some initial buildup in free fatty acid content of the treated samples probably because of the very slow rate at which the samples dried on standing, the cottonseed treated with sodium hypochlorite was able to maintain a much lower free fatty acid content than did the control sample. It should be noted that these storage conditions were very severe since a combination of high humidity and high temperature was used. The following example was carried out to demonstrate the effectiveness of the instant process in inhibiting enzyme action which hydrolyzes the fats to free fatty acids. This example shows that the treatment is effective whether the hydrolysis results from microorganisms on the oilseeds or from enzymes present.

Example 2: 0.5 gram of cottonseed oil was placed in a 125 cc glass-stoppered bottle. 5 cc of bile in glycerol solution was then added, together with some glass beads. The bottle was placed in boiling water and shaken until the oil was completely dissolved. The bottle and its contents were then cooled to room temperature and 10 cc of a 0.05 molar ammonium chloride-ammonia buffer of pH 8 was added. 100 mg of calcium chloride in water and 0.25 cc of a 3% solution of phenolphthalein were then added. Subsequently, enough water was added to make the total volume 30 cc. The enzyme was then added as a water solution and mixed. An aliquot of 5 cc was withdrawn and pipetted into 75 cc of a mixture of 9 volumes of alcohol and 1 volume of ether. The solution was titrated with an alcoholic potassium hydroxide.

The digest was kept at a constant temperature and minute amounts of strong ammonia were added to it whenever the pH became more acid. At varying periods of time, 5 cc aliquots of the digest were withdrawn and titrated. The difference between the titration at any given time and the titration at the start represents the quantity of free fatty acids set free. In the titration, the ammonia that has been added has no influence. In this test, 20 mg of pure enzyme were added to each tube and the tube maintained at 104°F during the digestion period. A water solution of sodium hypochlorite was added as an inhibitor at active chlorine concentration of 0.315% to 0.00011%, corresponding to a range of 1 to 3,150 parts per million. The percent inhibition of lipase expressed by the formula

$$\frac{\text{titration of control} - \text{titration of test run}}{\text{titration of control}} \times 100$$

was found to be as follows:

Parts per Million of Active Chloride	Percent Inhibition of Lipase
3,150	96
1,630	82
810	82
390	71
110	54
11	6
1	0

It is apparent from this example that the instant active chlorine treatment of fatty material to prevent hydrolysis is effective in pure enzyme systems. The following example illustrates the effect of the instant treatment on fatty animal tissue.

Example 3: 2 liters of fatty tissue to which 100 ppm of active chlorine had been added was compared with 2 liters of fatty tissue containing 1,000 ppm of propyl para-hydroxy-benzoate, which is a system that prevents bacterial growth only. The same fatty pork tissue to which these additives were added, was used alone as a control. The samples were incubated at 98.6°F for 16 hours and then rendered to obtain sufficient fat for free fatty acid titration. Results were as follows:

Sample	Percent Free Fatty Acid
Pork fatty tissue-control	7
Control plus 100 ppm active chlorine	2.2
Control plus 1,000 ppm benzoate	3.08

The above table shows that the hypochlorite is effective in the inhibition of free fatty acid formations to a degree beyond that which could be attributed to the prevention of bacterial growth alone. It should be noted that in connection with the inhibition of free fatty acid formations in animal fats an oxidizing agent-like hydrogen peroxide is completely ineffective since it is rapidly destroyed by catalase enzyme present in the pork tissue.

STABLE ANTISPATTERING OIL

Monoesters of polyoxyethylene sorbitan and long-chain unsaturated fatty acid, and polyesters of polyglycerol and long-chain unsaturated fatty acids are used by *R.G. Cunningham, R.D. Dobson, L.H. Going and E.R. Purves; U.S. Patent 3,415,658; December 10, 1968; assigned to The Procter & Gamble Company* to prepare an improved cooking and salad oil with storage-stable antispattering properties.

Many foods which are prepared by frying in liquid oil have a high moisture content. In pan frying these foods (especially during the early stages of frying), the contact between the relatively cool water in the food and the hot liquid oil in the cooking pan vaporized the water into steam which in turn causes the hot oil to spatter. It is a common experience to have hot oil spatter from a frying pan when minute steaks, particularly frozen steaks, are placed in a frying pan containing hot oil.

A number of different substances have been suggested as antispattering agents to either inhibit or retard spatter during the frying of foods. Some of these substances are particularly useful in products which are oil-in-water emulsions, such as margarine, wherein they tend to reduce the oil spatter caused by the sudden breakdown of the emulsion when the emulsion is heated. Unless margarine contains an emulsifying agent, the application of heat causes the oil-in-water emulsion to break down and release large droplets of water. The sudden escape of steam from these droplets expels hot fat particles with sufficient explosive force to cause the fat to spatter. This phenomenon can be observed when margarine is heated in a pan before the addition of any food to the pan. The sodium sulfoacetate derivatives of mono- and diglycerides and other substances are typical margarine antispattering agents.

Spattering caused by contact between cold, moist food and hot frying fat presents a problem the solution to which is unrelated to the solution of the problem of spattering caused by the sudden breakdown of an emulsion and the accompanying release of water. The ordinary emulsifying agents which are useful in margarine and in various cooking and baking shortenings do not possess the antispattering properties desired for cooking and salad oils, particularly, clear oils.

The improved storage stability of the cooking and salad oils results from an improvement in the dispersion of the active antispattering agent, the monoesters of polyoxyethylene sorbitan, in the oil. In combination with certain polyglycerol esters, the antispattering agent remains uniformly dispersed in the oil for prolonged storage periods; for example, up to about 6 months. Without the polyesters of polyglycerol, the antispattering agent tends to settle from the oil composition during extended storage. This settling lessens the anti-spattering activity of the oil fraction from which the antispattering agent has settled.

The cooking and salad oil composition consists of a clear, liquid glyceride base oil containing an active antispattering agent from 0.125 to 0.3%, by weight of the total composition. The polyglycerol molecule has the following structure:

$$HO-(CH_2-CH-CH_2O)_{(n-1)}(CH_2-CH-CH_2)-OH$$
$$| \qquad\qquad\qquad |$$
$$OH \qquad\qquad\qquad OH$$

When this molecule is esterified the two terminal hydroxyl groups are not affected; however, esterification can take place at all other hydroxyls. The ratio of free hydroxyl groups to the fatty acid ester groups is determined by dividing the two terminal hydroxyl groups plus the total number of polymer units which can be esterified (n) minus those actually esterified by the number of fatty acid ester groups. For example, the polyglycerol ester, decaglycerol decaoleate, often referred to as 10-10-0, has two terminal hydroxyl groups and 10 fatty acid ester groups or a ratio of free hydroxy groups to fatty acid ester groups of 0.2.

Another polyglycerol ester, decaglycerol tetraoleate, often referred to as 10-4-0, has a ratio of free hydroxyl groups to fatty acid ester groups of 2.0 determined as follows: two terminal hydroxyl groups plus 10 polymer units minus 4 esterified units divided by the 4 fatty acid groups. As a further example, the ratio of free hydroxyl groups to fatty acid ester groups in hexaglycerol tetraoleate is 1.0.

It is known that certain general classes of materials such as the sorbitan partial esters and the polyoxyethylene ethers of sorbitan partial esters, respectively, have useful emulsifying properties for plastic and liquid shortenings. It was not previously known that these particular esters are useful in amounts of from 0.1 to 1.0%, by weight, of a clear, liquid glyceride base oil to impart antispattering properties to the oil.

It is essential that the antispattering agent be a monoester as distinguished from di-, tri-, or higher partial or complete ester. It is also essential that the fatty acid portion of the monoester contain from 14 to 18 carbon atoms and that it be derived from predominantly unsaturated fatty acids as distinguished from saturated fatty acids. Examples of suitable unsaturated fatty acids for this purpose are myristoleic, palmitoleic, oleic and linolenic acids. The saturated monoesters such as polyoxyethylene sorbitan monostearate and the more fully esterified products such as polyoxyethylene sorbitan tristearate destroy the clarity of the liquid base oil of the cooking and salad oil compositions at low storage temperatures.

As previously indicated, it is preferred that the level of the antispattering agent be from 0.125 to 0.3%, by weight, of the cooking and salad oil composition. At levels above about 0.3%, by weight, the composition develops what has been described by some taste experts as an off-flavor. For some persons, this off-flavor is not detectable at levels up to about 0.5%, by weight.

In order to insure good clarity, it is essential that the liquid glyceride base oil be substantially free of general purpose shortening emulsifiers such as mono- and diglyceride esters, lactylated glyceride esters, and any other materials which might tend to cloud the base oil or otherwise interfere with its clarity. The use of substances which are fluidizers, for example, aluminum tripalmitate, reduce the antispattering characteristics of the cooking and salad oil compositions; their use should also be avoided.

A wide variety of liquid glyceride base oils can be used in the cooking and salad oil. Included among suitable oils are the so-called natural salad oils such as, for example, olive oil, sunflower seed oil, safflower oil, and sesame seed oil. Other naturally-occurring liquid glyceride oils such as, for example, cottonseed oil and corn oil, are also useful; these oils are given a preliminary winterizing, dewaxing, or similar treatment to remove the higher melting stearins before being used as a base oil. Certain other oils such as, for example, soybean oil, can be partially hydrogenated before use to improve their resistance to oxidative deterioration during prolonged storage periods; the higher melting solids formed during the hydrogenation treatment are preferably removed by winterization.

Suitable clear liquid glyceride base oils also can be obtained by directed, low temperature interesterification or rearrangement of animal or vegetable fatty materials, followed by the removal of the higher melting solids formed during the reaction. Another group of oils suitable for use as the liquid glyceride base oil is that group of oils in which one or more short-chain fatty acids, such as acetic acid and propionic acid, replace, in part, the long-chain fatty acids present in natural triglyceride oils.

Example 1: Five cooking and salad oil compositions were prepared from refined, bleached and deodorized soybean oil partially hydrogenated to an iodine value of 107 and winterized after hydrogenation. Acid-treated Tween-80 and Drewpol 10-10-0 were dispersed in four of the five compositions in the amounts shown in the following table, by thoroughly mixing the Tween-80, the Drewpol 10-10-0 and the oil for 3 minutes in a Waring blender. The oil without these additives was used as a control.

Tween-80 is a mixture of polyoxyethylene sorbitan esters of predominantly monounsaturated oleic acid; it contains an average of about 20 oxyethylene units in the molecule [i.e., polyoxyethylene (20) sorbitan monooleate]. The commercial product has an apparent pH in the presence of a trace of water of about 8. The Tween-80 used in this example was acid-treated with phosphoric acid (85% solution) to reduce its apparent pH to about 6.5 in the presence of a trace of water when measured with a Beckman pH meter. The acid treatment was carried out by first heating the Tween-80 to about 145°F in a stainless steel bowl to reduce its viscosity before adding the amount of acid calculated to lower the pH of the Tween-80 to the desired level, and slowly stirring the mixture to insure that the acid became uniformly dispersed.

Drewpol 10-10-0 is a mixture of polyglycerol esters of predominantly monounsaturated oleic acid. It has individual polyglycerol units ranging from diglycerol to decaglycerol and it has an average degree of polymerization of the glycerol moiety of between 4 and 6 glycerol units per polyglycerol molecule. The average ratio of free hydroxyl groups to fatty acid ester groups is 0.2. The five liquid oil compositions were used in frying frozen minute steaks. In each instance, the oil and other material which spattered from the frying pan during frying were collected and weighed in order to determine the amount of spatter reduction achieved with the antispattering oil compared to the control which contained no additive. Frying was done in 10" square Sunbean electric frypans at 360° and 420°F.

The lower temperature is the customary frying temperature for frozen steaks. Frying was also conducted at the higher temperature in order to demonstrate the improved spatter reduction under very vigorous frying conditions. In carrying out each frying sequence, a frozen minute steak (75 grams) was placed in the center of a frying pan in which the oil (30 grams) had been heated to the test temperature. The steaks were fried for 2½ minutes on each side.

"Tween-80" weight percent concentration	"Drewpol 10-10-0" weight percent concentration	Frying at 360° F.		Frying at 420° F.	
		Grams of spatter	Percent spatter reduction	Grams of spatter	Percent spatter reduction
0	0	6. 12	-----------	7. 51	-----------
0. 125	0. 025	1. 86	69	3. 66	51
0. 1875	0. 0375	1. 55	75	2. 69	64
0. 250	0. 050	0. 93	85	2. 03	73
0. 50	0. 10	0. 48	92	1. 22	84

The oil and other material which spattered from the frying pan during frying were collected on a three foot by three foot square sheet of aluminum foil placed centrally under the frying pan. The amount of spattering was determined by the difference in the weight of the original foil and the foil with spattering. The table on the previous page indicates the grams of spatter and the percentage of spatter reduction obtained with the various samples versus the control. Substantially similar antispatter activity as that shown in the table is achieved when the amount of polyglycerol ester, Drewpol 10-10-0, is increased to about 100% by weight of the antispattering agent.

The cooking and salad oil compositions containing Tween-80 and Drewpol 10-10-0 were clear and remained clear even after prolonged storage, i.e., a period of time exceeding about 6 months. Similar compositions containing the same amounts of Tween-60 (polyoxyethylene sorbitan monostearate) and Drewpol 10-10-0 will become cloudy when stored at 40°F and 32°F for short periods of time. Compositions containing similar amounts of Tween-80 and Drewpol 10-3-M in place of Drewpol 10-10-0 also will become cloudy after prolonged storage. Drewpol 10-3-M is described by the manufacturer as decaglycerol trimyristate; it has an average ratio of free hydroxyl groups to fatty acid ester groups of about 3.

Example 2: The storage stability of the antispattering oils was demonstrated by composing two sets of antispattering compositions. In each instance, refined, bleached, deodorized and winterized soybean oil having an I.V. of 107 was used as the base oil. The first set of compositions contained 0.5% dispersed Tween-80 [polyoxyethylene (20) sorbitan monooleate], the second set contained an equivalent amount of Tween-80 and 0.1% dispersed Drewpol 10-10-0. The two sets of compositions were stored in pint bottles for 3 to 6 month periods at relatively constant temperatures of 50°, 70° and 90°F.

At the end of 3 months and 6 months, respectively, certain bottles in each set were randomly selected and divided into a top and bottom fraction of oil without intermingling the two fractions. Any active antispattering agent (i.e., Tween-80) which settled to the bottom of the composition during the storage period was isolated in the bottom fraction. The two fractions which were taken from the same bottle were used separately to fry frozen minute steaks at 420°F in the manner described in Example 1. The results of the frying with each fraction from the same bottles were compared to determine in which samples the bottom fraction had substantially more antispattering activity than the top fraction. The following table shows that the presence in the one set of compositions of the Drewpol 10-10-0 substantially retarded the deposition of the antispattering agent. The compositions containing both Tween-80 and Drewpol 10-10-0 were more storage stable after 3 and 6 months storage than the compositions containing only Tween-80.

Storage time and temperature	Grams of spatter with oil samples containing 0.5% by weight "Tween-80"—fraction of sample		Grams of spatter with oil samples containing 0.5% by weight "Tween-80" and 0.1% by weight "Drewpol 10-10-0"—fraction of sample	
	Top	Bottom	Top	Bottom
3 months at 50° F	3.97	0.85	0.92	0.71
3 months at 70° F	2.38	0.54	1.09	0.56
3 months at 90° F	1.97	0.59	0.79	0.50
6 months at 50° F	5.36	0.82	1.41	0.77
6 months at 70° F	5.11	0.92	1.95	0.75
6 months at 90° F	3.89	0.91	2.78	0.85

Substantially the same storage stabilizing effect of the polyglycerol ester is observed when the amount of Drewpol 10-10-0 is increased up to about 100% by weight of the active antispattering agent. By way of illustration two sets of antispattering compositions were composed using refined, bleached, and deodorized soybean oil having an I.V. of 107 which had been winterized. The samples contained 0.22% dispersed Tween-80 [polyoxyethylene (20) sorbitan monooleate] and 0.22% dispersed Drewpol 10-10-0. One set of compositions was stored in pint bottles for 1 month at a temperature of 50°F; the other set was stored at 90°F. In the manner previously described, two fractions were taken from a randomly selected bottle in each set and used separately to fry frozen minute steaks. The grams of spatter with the top fraction of the oil sample stored at 50°F was 0.99; the grams of spatter with the bottom fraction was 0.76. The results with the top and bottom fractions from the oil sample stored at 90°F was 1.02 and 0.75 grams of spatter respectively.

ANTIOXIDANT FROM OKRA PODS

The natural antioxidant of *L.R.B. Hervey; U.S. Patent 2,950,975; August 30, 1960; assigned to John A. Manning Paper Company, Inc.* is obtained from dried okra pods. The okra product (referred to as dried okra product) is a dry powder extracted from okra pods by any suitable process which permits removal of the water, subsequent drying and isolation of the mucilage-producing material contained in the pods. The dried okra product is prepared by a process which reduces the alcohol-soluble materials content to less than 0.8% based on the total weight of the dried okra product, and which as a final step gives a material at least 60% of the individual particles of which are in a size range from 45 to 75 microns.

One such process involves dehydrating and precipitating the mucilage-producing material of okra pods by means of an organic dehydrating agent such as one of the lower alcohols, and drying the dehydrated and precipitated material under conditions which will not degrade the final dried okra product. Another process by which dried okra product may be prepared achieves the extraction of the mucilage-producing material of the okra pod by a form of steam distillation using a water-immiscible hydrocarbon.

It is important that the temperatures and times of processing are such as not to render the final product unstable or degraded with respect to its mucilage-producing abilities. Although the reasons why dried okra product is or is not stable are not understood, it appears that the enzymatic activity associated with the naturally occurring pods must be substantially and rapidly arrested to prevent subsequent degradation of the final product.

Example 1: Samples of pure leaf-lard containing no antioxidant and containing dried okra product in several concentrations were evaluated to determine their stability in an oxygen atmosphere. The dried okra product used was prepared by dehydrating and precipitating the mucilage-producing portion of okra pods in isopropyl alcohol. It was found preferable to incorporate the dried okra product by ball-milling it for about 4 hours into the leaf lard. Since the effect of ball-milling leaf lard on its stability was not known, two controls without dried okra product (one ball-milled for 4 hours, the other not) were run. The results showing the stability of the controls and of leaf lard containing 1% by weight dried okra product are plotted in Figure 1.2a.

These evaluations were made by confining a known weight of the lard in an atmosphere of pure oxygen at 100°C. The entire system was kept in constant agitation to assure equilibrium between the lard and oxygen. At predetermined intervals, the volume of oxygen in the system was measured by means of a mercury manometer directly attached to the system. The measurement recorded directly the volume of oxygen consumed by the lard during the process of oxidative rancidification and these volume measurements were converted to volume of oxygen consumed per gram of lard. These converted measurements are plotted in Figure 1.2a. It will be noted that the curves break quite sharply and the lard then consumes oxygen at a very rapid rate. The intersection of the slopes of each leg of the curve is considered to be the end point for oxygen stability time, in hours, of the lard at 100°C.

From Figure 1.2a it will be seen that the stability time for the leaf lard which had been ball-milled is about 1 hour while that for the lard which had not been ball-milled is somewhat less than 2 hours. In contrast to these figures the sample of lard which contained 1% dried okra product had a stability time of between 5 and 6 hours. Experience has indicated that 1 hour of stability time measured in the above described method is roughly equivalent to 3 hours' stability time using the Active Oxygen Method (AOM) at an equivalent temperature.

Thus the lard which had been ball-milled would have approximately 3 hours AOM time; lard which had not been ball-milled would have about 5 hours AOM time; while lard containing 1% dried okra product would have something over 15 hours AOM time. It will be seen from these performance figures that 1% dried okra product has the effect of multiplying the stability time of leaf lard by a factor of about 5 if the leaf lard is ball-milled, and by a factor of 3 if the leaf lard is not ball-milled.

Similar tests were repeated using 0.1, 0.5 and 2% dried okra product in the leaf lard. The results of these tests show that the antioxidant properties of the dried okra product increase with increasing amounts of the additive up to about 1% by weight. Although there is some increase in stability time when 2% is used instead of 1%, this increase is slight and drops off significantly above about 4%. Quantities of dried okra product less than about 0.1%, although they show some antioxidant properties, are not considered feasible. A useable range is therefore between about 0.1 and 4% dried okra product based on the weight of the leaf lard; a practical range is between about 0.5 and 1% dried okra product.

Example 2: In a somewhat similar manner the effect of adding dried okra product to peppermint oil was measured. In this case two different samples of peppermint oil were used and dried okra product was added in a concentration of 2% by weight. The samples with and without the dried okra product were placed in a Warburg apparatus and subjected to an atmosphere of oxygen. The study was carried out over a period of 38 days and the results are plotted in Figure 1.2b where the mols of oxygen absorbed per 100 mols of peppermint oil are plotted against time.

Curve A represents one lot of peppermint oil without dried okra product and curve A' represents the same lot of peppermint oil containing 2% by weight of dried okra product. Likewise curve B represents an extremely fast-oxidizing peppermint oil without dried okra product while curve B' represents the same oil containing 2% dried okra product.

After 38 days oil A without dried okra product had absorbed 6.4 mols of oxygen per 100 mols of oil while with dried okra product this was reduced to 3.2 mols or one-half. Over a similar period of time oil B without dried okra product absorbed 25.8 mols of oxygen per 100 mols of oil while that containing the dried okra product, B', absorbed 6.4 mols of oxygen per 100 mols of oil. This means that the amount of oxygen absorbed in the presence of dried okra product was reduced to about one-fourth that absorbed when no dried okra product was added.

FIGURE 1.2: ANTIOXIDANT FROM OKRA PODS

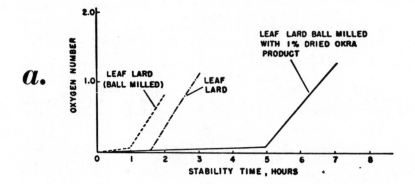

(continued)

FIGURE 1.2: (continued)

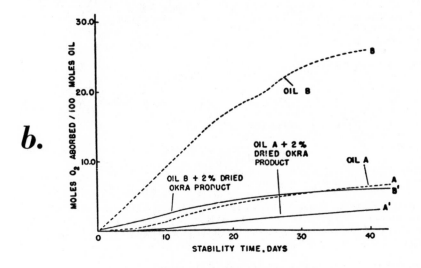

b.

Source: L.R.B. Hervey; U.S. Patent 2,950,975; August 30, 1960

Inasmuch as peppermint oil is known to be one of the most easily oxidized oils, it can be concluded that the okra product is an extremely effective antioxidant under even the most stringent conditions. The dried okra product may be added as a dry flour-like material to the substance to be protected or it may be thoroughly mixed into a small quantity of the material by any suitable technique, such as the ball-milling procedure used in the examples, and then added as a concentrated suspension or dispersion to the remaining portion of the material.

ANTIOXIDANT FROM MALTED SEEDS

The extraction and use of a natural antioxidant derived from malted seeds is described by *D.L. Baker and W.B. Dockstader; U.S. Patent 2,975,066; March 14, 1961; assigned to Basic Products Corporation.* The antioxidants produced are especially useful for delaying rancidification of edible fats and oils used for food and for feeds because they themselves are harmless for human animal consumption. The range of known and effective antioxidants, suitable for such purposes is quite restricted, and the more effective of those available have been limited in their utility because of their expense.

Suitable starting materials for the manufacture of the malt, from which the antioxidant materials are obtained, include commercial grains, for example, wheat, rice, maize or barley and, in some instances rye, although the seeds of nearly all seed propagated plants in some measure appear to provide a source of antioxidant substance. The starting seed, for example, barley, is first obtained in a mature condition after which it is cleaned and graded, and then introduced to the malting operation. The cleaned, graded and matured barley seeds are steeped in water for several days during which several changes of water occur, and the water content of the berries is raised to about 45%. Following the steeping operation , the soaked or steeped berries are spread in layers and held in a ventilated malting area to carry out the germination of the grain. The temperature during this time is carefully maintained in the range of from 55° to 80°F; generally for from 4 to 6 days. At the end of the usual growing or sprouting period rootlet development is evident and the acrospire will have extended to a substantial degree from its original location in the embryo at the base of the barley berry or kernel. Normal germination is concluded when the acrospire has grown from about three-quarters to the full length of the kernel in the majority of the kernels. This growth occurs beneath the hull of the barley and is not readily observable except by taking the hull off and examining a given barley kernel. When the acrospire has reached this point, growth is ordinarily concluded by kilning.

The kilning operation is a standard malting process during which time the temperature of the germinated berries is raised to not higher than about 200°F, and it is maintained at such temperature for a period of about 2 days. Ordinarily, the kilning operation occurs in two distinct phases, the first being approximately 1 day with the first 16 hours being a gradual increase in heat applied to the kernels. At the end of the 24-hour period the malt is ordinarily dropped to a lower kiln where it progresses through yet another drying period of about the same duration, concluding with a somewhat higher finishing temperature of 160° to 180°F or possibly somewhat higher. The moisture content after the first or initial drying of brewer's malt is reduced from 40 to 45% (which it had attained during the steeping operation and not measurably been reduced during germination) to about 8 to 14%. During the second period of kilning this moisture content is usually dropped to about 4%.

It is at the completion of the normal malting operation that the extraction process is ordinarily initiated. However, it has been found that the green malt, i.e., the undried, germinated seed, may also be treated to yield antioxidant. As shown in the accompanying flow diagram (Figure 1.3) there is a choice as to just how the antioxidant is to be concentrated.

FIGURE 1.3: ANTIOXIDANT PREPARATION PROCESS

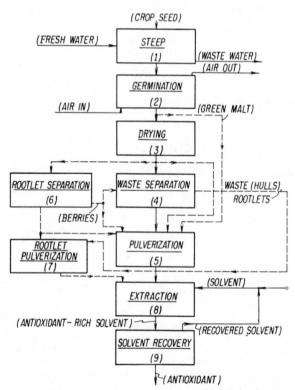

Source: D.L. Baker and W.B. Dockstader; U.S. Patent 2,975,066; March 14, 1961

At the conclusion of the usual kilning step the rootlets or sprouts which have extended a substantial distance from the kernel during germination are separated from the berries, following which the berries are dehulled and pulverized to a flour.

Since the hull of the malted barley kernel has no detectable antioxidant property itself, and particularly because it is not entirely desirable in any food product in which the antioxidant might be used, the usual and preferred technique is to remove the hull in a waste disposal step prior to extracting the antioxidant values from the malt flour. However, if desired, and certainly from an economy standpoint, if it is preferred to eliminate the dehulling operation it could be dispensed with. Thus, for example, where the antioxidant is to be used for the purpose of inhibiting autooxidation of an inedible organic material such as rubber or any other such material subject to oxidation, the entire product coming from the germination step may be pulverized and extracted. On the other hand, the rootlets have a definite antioxidant content and they alone may be separated and pulverized, followed by extraction as described.

The malt kernel separated from the sprouts or rootlets is pulverized to a flour. The phrase malt flour shall refer to that pulverized material obtained by pulverizing the dehulled barley berry which had been previously separated from the rootlets. The dehulled malt, prior to pulverization, is usually called pearled malt. The pulverized material, whether it be rootlets or the mixture of rootlets and berries, or even a combination of hulls, berries and rootlets, is subjected to an extraction with methanol, ethanol, isopropanol or isobutanol. The extraction is carried on by adding the pulverized malt derivative to refluxing solvent in a concentration of 20 to 25% by weight.

It is preferred to use relatively minor proportions of water and to thus have a mixture of water and alcohol as the solvent. The amount of water used with methanol or ethanol may vary from 5 to 25 or 30%, but it is preferred to employ from 10 to 20% water with such alcohols. It appears that a somewhat larger amount of water is desirable with isopropanol. As the molecular weight of the alcohol employed increases, the amount of water-soluble in such alcohol decreases. Thus, as a practical matter, butanol is miscible with only about 15% water and a butanol solution saturated with water has been found desirable.

Referring to Figure 1.3, the flow of materials along alternate paths to produce antioxidant from malted seed is shown. Thus, the seed, such as barley, is introduced to the normal steep stage **1** after which the steeped grain is passed to the germination zone **2**. Ordinarily the green malt leaving the germination step goes directly to the kiln **3** for drying. However, as shown in the drawing, it may be sent to the grinding or pulverization stage **5** or directly (not shown) to the extraction zone **8**. In the preferred method of operation, the green malt passes to the drying stage **3** where its water content is essentially completely removed. Upon leaving the kiln **3**, there is considerable choice as to just how the operation may be continued. Thus, all or part of the whole, dried, malted seed may be sent to the waste separation step **4**, where some of the hulls and all of the rootlets are removed from the dry, friable berry.

Alternatively, however, all or part of the material leaving drying zone **3** may be sent directly to the grinding step **5** or else it may be subjected first to rootlet separation in area **6** and then dehulling in zone **4**. Also, the rootlet-free berries (still with their hulls) after leaving zone **6** may be sent directly to pulverization zone **5**. The rootlets separated in zone **6** may then be pulverized individually in zone **7** or recombined with the pearled malt coming from waste separation step **4** and pulverized in zone **5**. In addition, when both rootlets and hulls are removed in zone **4** they may then be sent to the pulverization zone **7** and ground there with the rootlets from zone **6**.

After the respective parts or the whole malted seed have been ground to desired fineness, they are then extracted in zone **8**. After extraction is complete the solvent may, as desired, be recovered, purified and recycled to the extraction zone.

Example 1: 50 grams of Raymond hammermill ground rootlets (mill equipped with a

$1/64$" exit screen) were contacted with five successive 200 cc quantities of refluxing methanol. The composite antioxidant-rich solvent from the five refluxing operations was separated from the solid material by centrifuging and then evaporated to dryness, in a carbon dioxide atmosphere, to yield, based on original weight of rootlets, 17.7% or 8.85 grams of extract. A 5% by weight addition of this extract was then made to a sample of corn oil. That sample was then heated to 100°C for 30 minutes, left to stand overnight and then reheated to 100°C for 2 hours before centrifuging to remove the solids.

A comparison of the protection afforded this sample of corn oil with a control sample free of added antioxidant showed that the peroxide value of the treated sample was not significantly greater after heating such example for 10 hours at about 100°C with oxygen bubbling therethrough, than that of the fresh corn oil. On the other hand, the peroxide value of the control sample after heating for this same period at about 100°C, with the oxygen bubbling through, was about 1,000% greater than the thus treated sample.

Example 2: A portion of the extract obtained by evaporating the composite antioxidant-rich solvent was redissolved in methanol and the solution divided into two equal portions. These samples were then mixed with two volumes of chloroform and ten volumes of acetone, respectively, in a carbon dioxide atmosphere. A substantial quantity of solid was precipitated from each solution upon the addition of these materials. The precipitated solids were further evaporated to dryness in a CO_2 atmosphere and then added to equal volume samples of corn oil in the amount of 0.01% by weight.

After 8 hours of bubbling oxygen through the samples held at 100°C, the peroxide values of the two treated samples were about equal although the chloroform precipitate was somewhat lower than that containing the acetone precipitated extract. Both were significantly lower than the peroxide value of a control sample of corn oil subjected to the same conditions.

Example 3: 20 grams of finely ground malt flour (normal 23G brewer's malt flour, i.e., 95%, will pass through a 100 mesh screen, from which both rootlets and hulls had been removed before pulverization) were refluxed with 100 grams of methanol for a period of 15 minutes. The antioxidant-rich solvent was separated from the extract solution by filtration and the extract was then divided and part was dried at room temperature for 24 hours. The solid residue obtained was dissolved in propylene glycol so that 1 ml contained the solid content from 5 ml of the original extract. Propylene glycol was selected since it is a solvent used for marketable forms of known antioxidants.

1 ml of this propylene glycol solution was added to 20 grams of prime steam lard and tested in the standard AOM assay to determine the extent of rancidification. A 25 cc aliquot of the original undried extract was dried to a heavy syrup on a low heat hot plate (80°C) while a stream of CO_2 was continuously directed against the evaporating surface. The dried residue obtained was suspended in 5 cc of propylene glycol. 1 ml of this propylene glycol solution of CO_2 evaporated residue was then added to 20 grams of prime steam lard and tested in the standard AOM test to determine the extent of rancidification over varying periods.

One volume (100 cc) of the original undried extract was mixed with 2 volumes of acetone and the resulting precipitate sucked nearly dry on a Buchner funnel. The precipitate was then given 4 further washes with fresh acetone and 150 mg of a white to grey solid was obtained. This solid was readily dissolved in propylene glycol in the proportions of 1 mg per 10 ml of propylene glycol and tested by the standard, AOM procedure.

A comparison of the relative antioxidant activities in prime steam lard of the original undried extract as well as the residues from both atmospheric and CO_2-solid showed that the original undried extract gave about twice the number of hours of protection as either that prepared by evaporation in a CO_2 atmosphere or the acetone precipitation. The activity of the residue prepared by simply evaporating at room temperature without CO_2 protection was found to be much reduced, indicating the desirability, at least so far as methanol extraction is concerned, of blanketing in the evaporating solvent with CO_2. Other extractions similar to those described in the above examples were performed on both malt flour and ground rootlets with ethanol, propanol, isopropanol and butanol. These solvents were employed in both the anhydrous condition and with water added. Methanol, too, was tested with water added. It was found that the effect of water addition was neither predictable nor uniform. Thus, water addition to methanol up to a content of about 10% water (for the whole undried rootlet extract) seemed to give a very slight improvement in the activity of the extract but to fall slightly as a greater amount was added until 20% was reached.

With a solvent comprising 30% water content, the activity of the extract was found to be substantially zero. On the other hand, water addition to isopropanol showed a very significant increase in the activity of the whole undried extract produced from ground rootlets. Thus, samples of prime steam lard to which there had been added 1 cc of the original undried extract exhibited the following hours of protection/gram of original malt rootlets in the AOM test:

Solvent	Activity in AOM Test, Hr Protection/Gram of Malt Rootlets
Isopropanol, 0% H_2O	10
Isopropanol, 10% H_2O	30
Isopropanol, 20% H_2O	65 – 70
Isopropanol, 30% H_2O	30

The use of an atmosphere of CO_2 during the actual extraction has been found to give some striking, if not unexplainable results. Thus, comparable extractions of malt rootlets using anhydrous methanol as well as various aqueous dilutions thereof were carried out with and without a CO_2 blanket. The effectiveness of 1 cc of the undried original extracts from these experiments in inhibiting the rancidification of prime steam lard is shown below:

Solvent	Yield* (grams)	Activity in AOM Test, Hr Protection/Gram of Malt Rootlets	
		(No CO_2)	(With CO_2)
Methanol + 0% H_2O	1.85	2.0	2.5
Methanol + 10% H_2O	8.5	2.5	32
Methanol + 20% H_2O	9.6	1.5	40
Methanol + 30% H_2O	14.9	0	20

*Based on 100 grams of original rootlets extracted.

As may be seen, the final recovery or concentration of the active antioxidant which is derived from malted seeds, may be effected in various ways. Indeed, the solvent-wet extract itself (either partially evaporated to remove most of the solvent or an aliquot of the antioxidant-rich solvent) may be used effectively to inhibit autooxidation of either fats or oils. Generally speaking, it has been found that efforts to completely dry an extract, even in a CO_2 atmosphere, result in somewhat less overall antioxidant effectiveness. This seems to indicate that there is room for improvement in the methods of final isolation and protection of the active antioxidant material.

Nevertheless, it has been found possible to recover highly active antioxidants from malted grain and to effectively add the same to oleaginous materials, such as animal and vegetable fats and oils. While acetone has been specifically referred to above as a material capable of precipitating the active antioxidant substituents from the solvent, it should be understood that other organic solvents, such as chloroform, methyl ethyl ketone and ethyl ether may also be used if desired. However, for practical purposes acetone has been found most desirable.

ANTIOXIDANT FROM TEMPEH

Tempeh is a fermented soybean product. *P. Gyorgy; U.S. Patent 3,762,933; October 2, 1973* reports that oil of tempeh is useful as a stabilizer or antioxidant for edible fats and oils. Oil of tempeh, a component or extract of tempeh, is produced by extracting tempeh, with selective solvents. For example, oil of tempeh or tempeh oil is produced by extracting tempeh, preferably dry tempeh, obtained by fermenting soybeans with the fungus *Rhizopus oligosporus,* with a mixture of a normally liquid aliphatic hydrocarbon, such as an aliphatic hydrocarbon containing from 4 to 12 carbon atoms per molecule, e.g., hexane, and a normally liquid oxygen-containing, preferably aliphatic, polar organic compound containing from 1 to 10 carbon atoms per molecule, such as an alkanol, e.g., ethanol.

A solvent mixture made up of the aliphatic hydrocarbon, e.g., one volume of hexane and a polar oxygen-containing organic compound, such as an alkanol, e.g., 0.5 to 5.0 volumes ethanol, is used in an amount of about 5 volumes of this solvent mixture to one volume of tempeh. The resulting mixture of solvent and tempeh is maintained in contact, preferably at about room temperature, e.g., a temperature in the range of 10° to 40°C, for a sufficient period of time. Thereupon, the liquid solvent is separated from the undissolved or extracted tempeh and the separated liquid solvent and oil of tempeh-containing extract phase is then treated, such as by distillation, for the removal of the solvent.

This remaining material identified as oil of tempeh or tempeh oil exhibits remarkable stability and possesses antioxidative properties. For example, oil of tempeh when stored for 17 months in a freezer and then for 21 days at a temperature of 37°C exhibited a peroxide value (POV) of about 5.9 meq/kg. In contrast, commercial soy oil or soybean oil when maintained and tested under the same conditions exhibited a peroxide value of 81.4.

In these tests, the peroxide value was determined according to the method of D.H. Wheeler, *Oil & Soap,* 9, 89 (1932). This method is based upon the following procedure. A sample of the oil to be tested weighing 0.5 gram is placed in a 300 ml flask with a glass stopper and there are added thereto 10 ml of chloroform to dissolve the sample. Thereupon, 15 milliliters of glacial acetic acid are added along with about 1 gram of powdered potassium iodide KI. The resulting mixture is then refluxed on a water bath for 3 minutes and then cooled in an ice water bath. Thereupon 75 ml of distilled water are added and the resulting flask contents were vigorously shaken. After shaking there was added a drop of 1% starch solution and the flask contents then titrated with N/1000 sodium thiosulfate until the iodine I_2 color disappeared. The results of these tests are then calculated in meq/kg.

Another effective component is obtained by extracting oil of tempeh with a 50% aqueous ethanol solution and then evaporating the solvent from the resulting extract phase. After removal of substantially all of the solvent the residue is then extracted with petroleum ether to produce a liquid oily phase and an insoluble, substantially solid phase. The insoluble phase is separated and 50% aqueous ethanol added thereto to yield an aqueous phase containing the insoluble material dispersed or emulsified therein. Acetone is then added to precipitate the solid material which is then separated, such as by filtration, and dried.

Following the above procedure there is recovered on the basis of 1 gram of tempeh oil an antioxidatively active component in the amount of about 1 mg. This material is effective when added on the basis of about one part per thousand to an edible oil or fat, such as soybean oil, to protect the oil or fat for a substantial period of time against oxidation. Mixtures of commercial soybean oil and oil of tempeh were stored at 60°C for a number of days, 35 days and 43 days, and the peroxide values of these mixtures then determined. The results of these tests are set forth in the accompanying table.

Composition	POV, 35 Days	POV, 43 Days
100% soybean oil	118	270
90% soybean oil, 10% tempeh oil	53	82
80% soybean oil, 20% tempeh oil	33	30
70% soybean oil, 30% tempeh oil	25	20
60% soybean oil, 40% tempeh oil	10	13
50% soybean oil, 50% tempeh oil	7	8
100% tempeh oil	4	5

Further tests were carried out on mixtures of soybean oil and tempeh oil to determine stability of mixtures of these oils as measured by peroxide values, after storage for 35 days at 37°C. The results of these tests are set forth in the accompanying table.

Composition	POV
100% soybean oil	118
90% soybean oil, 10% tempeh oil	55
80% soybean oil, 20% tempeh oil	33
70% soybean oil, 30% tempeh oil	26
60% soybean oil, 40% tempeh oil	13
50% soybean oil, 50% tempeh oil	8
100% tempeh oil	7

Mixtures of commercial soybean oil and tempeh oil in the amount of 40 to 80% by weight tempeh oil and 60 to 20% by weight soybean oil after storage or incubation for 18 days at 37°C (1 gram in open scintillation counting vials) gave peroxide values in the range of 17 to 5 whereas a commercially available stabilized soybean oil gave a peroxide value of 23. This test clearly demonstrated the superiority of a soybean oil stabilized with tempeh oil over commercially available stabilized soybean oils. Further tests were carried out where mixtures of corn oil and tempeh oil and lard and tempeh oil were stored and incubated for 26 days at a temperature of 38°C (2 grams in open scintillation counting vials) and the peroxide values then measured. The results of these tests are set forth in the accompanying table.

Composition	POV
100% lard	220
100% corn oil	95
10% tempeh oil, 90% lard	10
10% tempeh oil, 90% corn oil	12
20% tempeh oil, 80% lard	5
20% tempeh oil, 80% corn oil	12
40% tempeh oil, 60% lard	6
40% tempeh oil, 60% corn oil	12

Further illustrative of the advantages of the practices of this process are peroxide values of various oil mixtures which were determined after storage at 60°C. The results of these tests are set forth in the accompanying table.

POV of Various Oils After Storage at 60°C

Oil Composition	Days of Storage				
	0	3	7	14	20
Cottonseed oil	1.9	101.7	202.8	350.6	–
Cottonseed oil + 10% tempeh oil	3.3	64.4	113.9	190.7	–
Cottonseed oil + 20% tempeh oil	5.4	34.8	89.2	112.4	–
Cottonseed oil + 30% tempeh oil	3.9	6.6	26.4	82.5	–
Cottonseed oil + 50% tempeh oil	10.8	5.7	5.3	48.3	–
Tempeh oil	16.0	3.9	1.7	5.5	–
Safflower oil	2.5	103.3	323.5	334.3	–
Safflower oil + 10% tempeh oil	3.7	70.1	133.1	258.4	–
Safflower oil + 20% tempeh oil	6.8	39.6	90.4	168.2	–
Safflower oil + 30% tempeh oil	8.6	11.6	61.1	100.2	–
Safflower oil + 50% tempeh oil	7.1	3.7	6.0	48.7	–
Tempeh oil	16.0	3.9	1.7	5.5	–
Lard	2.0	67.2	133.6	176.4	188.5
Lard + 10% tempeh oil	5.0	5.0	3.8	10.7	12.6
Lard + 20% tempeh oil	4.9	2.7	1.8	3.6	8.1
Lard + 30% tempeh oil	5.7	3.9	3.8	5.0	2.9
Lard + 50% tempeh oil	6.9	1.7	1.8	2.7	3.8
Tempeh oil	15.0	2.8	1.0	4.7	3.1

(continued)

Oil Composition	- - - - - - - - - - - Days of Storage - - - - - - - - - - -				
	0	3	7	14	20
Corn oil	1.0	17.3	93.3	314.0	–
Corn oil + 10% tempeh oil	2.0	5.7	17.2	89.0	–
Corn oil + 20% tempeh oil	2.8	4.8	6.3	32.3	–
Corn oil + 30% tempeh oil	2.8	3.4	1.7	9.0	–
Corn oil + 50% tempeh oil	5.5	1.9	1.9	3.5	–
Tempeh oil	15.0	2.8	1.0	4.7	–

Tests (iodine values) were also carried out on two samples of tempeh oil and two samples of commercial soybean oil. The results of these tests show that the two samples of tempeh oil gave iodine values of 132.6 and 123.5 and the samples of soybean gave iodine values of 139.6 and 133.2.

Substantial benefits are obtainable when only a minor amount of tempeh is incorporated in food compositions, particularly fat or oil-containing food compositions. Usually an amount upwards of about 0.5% of tempeh or oil of tempeh substantially improves the antioxidative properties of these compositions.

MEAT PRODUCTS

SALT-ANTIOXIDANT COMBINATIONS

A large proportion of pork sausage is commercially manufactured and then distributed and sold through retail outlets to consumers. In order to prolong the relatively short period of time in which the highly perishable fresh pork sausage may be stored before it is consumed, it is increasingly the practice to promptly freeze the product after it has been stuffed into casing and to hold it in the frozen condition until it is sold.

Port sausage deteriorates relatively rapidly due to both bacteriological spoilage and oxidative rancidity. The freezing of fresh pork sausage and holding in the frozen condition successfully protects the product against bacteriological spoilage but not against oxidative rancidity. As a result, an objectionable level of oxidative rancidity develops within approximately three weeks' time due to the catalytic activity of the salt which is present. The salt acts to greatly accelerate the development of rancidity in frozen pork, especially when the pork is held in a frozen condition.

It has been known to incorporate antioxidants into unfrozen pork sausage to protect the same against rancidity. In U.S. Patent 2,933,399 a procedure is given where small quantities of BHA (butylated hydroxyanisole and DMP (2,6-dimethoxyphenol) were added to the seasoning which was incorporated in the sausage batter. The resulting product was stored at a temperature above freezing and in the range of 40° to 43°F for up to 14 days for purposes of testing bacterial growth.

A salt formulation containing antioxidants is used by *D.L. Paul, R.H. Griesbach and J.F. Jaeger; U.S. Patent 3,366,495; January 30, 1968; assigned to Oscar Mayer & Co., Inc.* The antioxidants dispersed on the salts include citric acid and in addition at least one so-called hindered phenol antioxidant such as butylated hydroxyanisole (BHA), butylated hydroxytoluene (BHT) or propyl gallate. When all three hindered phenol antioxidants are used together, approximately 2 parts by weight of both BHA and BHT are present for each part by weight of propyl gallate.

Desirably a small amount of propylene glycol (i.e., a fraction of 1% by weight of the meat) is used to improve the distribution and dispersion onto the salt. The total amount of antioxidants that may be used will usually be dictated by governmental regulation. In general the total content should not exceed 0.02% of the fat content of the finished pork sausage product. The combined weight of the salt and the antioxidants dispersed equals about 2% by weight of the pork sausage batter. The following salt formulations having the antioxidants represent two preferred formulations.

Formulation 1

	Percent by Weight
Sodium chloride	97.090
Tricalcium phosphate	2.010
Butylated hydroxyanisole	0.266
Butylated hydroxytoluene	0.268
Citric acid	0.268
Propylene glycol	0.100

Formulation 2

	Percent by Weight
Sodium chloride (fine flake)	97.090
Tricalcium phosphate	2.010
Butylated hydroxyanisole	0.204
Propyl gallate	0.124
Propylene glycol	0.100
Citric acid	0.260
Butylated hydroxytoluene	0.204

In preparing Formulations 1 and 2 the propylene glycol may first be warmed to 125° to 135°F and then the citric acid dissolved therein, after which the propyl gallate (Formulation 2) is added until dissolved. The BHA and BHT may be added to the mass for stirring so that a completely uniform mixture is obtained. This mixture may then be added with stirring to the sodium chloride and tricalcium phosphate with mixing continued until a completely uniform blend is obtained.

Example: 300 lb of pork trimmings and 9 lb of ice and/or water are loaded into a chopper bowl. The spice in the amount of 29 oz/100 lb of meat, of either Formulation 1 or 2, together with such other ingredients as pepper, sage and dextrose, is spread evenly on the meat as the bowl rotates. The chopping may continue for such time as is required to obtain a uniform mixture. For example, chop for a total of 18 revolutions as follows: 5 rpm for 1 minute; 12 rpm for 1 minute; and then unload completely in 1 revolution.

The chopped pork sausage is promptly stuffed at a maximum temperature of approximately 36°F into casings and the casings are linked in known manner. The stuffed links are immediately frozen in a blast freezer or a chill tunnel to an internal temperature of 22° to 26°F. Thereupon the frozen product is packed in cartons and the cartons are transferred to a freezer at 0° to 5°F temperatures and frozen solid to a minimum temperature of 10°F. The product is retained in the hard frozen condition until such time as it is delivered to the retail store or other outlet.

When Formulation 1 comprising approximately equal parts of citric acid, butylated hydroxyanisole and butylated hydroxytoluene is used in preparing the sausage batter in accordance with the example, with 29 oz of the antioxidant ingredients dispersed on salt being used for each 100 lb of meat ingredients, it will be seen that the pork sausage product will have the following antioxidant content or composition:

	Percent
BHA	0.0049
BHT	0.0050
Citric acid	0.0050

It has been determined that when salt without any antioxidants dispersed thereon is used in producing fresh frozen pork sausage that oxidative rancidity develops in approximately three weeks' time due to the catalytic activity of the salt which is present. However, by incorporating salt which has the antioxidants predispersed thereon, the product may be held in the frozen condition for at least approximately six weeks before objectional oxidative rancidity develops. It has also been ascertained that if the antioxidants are not predispersed on at least about 90% of the salt content, but merely incorporated as another ingredient of the spice mix, they will not afford the full protection against rancidity.

COLOR STABILIZATION PROCESSES

Tetrazole for Fresh Meat

It is well-known that the desirable red color of fresh meats fades rapidly during storage. The same discoloration is also observed in meat products which are subjected to a preservation treatment, especially under light, vacuum refrigeration or frozen conditions.

This discoloration is due to chemical changes of the pigments present in the meat. By these pigments is meant compounds like myoglobin and hemoglobin comprising complexes of certain proteins with ferrous haem and ferric haem including both their oxygenated and deoxygenated forms. In their reduced state myoglobin and hemoglobin are purple-red, but in taking up oxygen they form oxymyoglobin and oxyhaemoglobin respectively, which are bright red. On oxidation the greyish brown metmyoglobin and methaemglobin are formed.

A.H.A. Van Den Oord and B. De Vries; U.S. Patent 3,615,691; October 26, 1971; assigned to Lever Brothers Company have found that by bringing about an interaction of meat products with a sufficient amount of tetrazole, meat products can be obtained having a desirable stable red color. Although the pH at which this interaction is brought about is not critical and may vary between rather wide limits, it is preferred to maintain the pH in the range of 4.5 to 6.5 which is the normal pH of meat products when in a consumable condition.

Tetrazole has several advantages over other products as stabilizers of the color of meat. In the first place it will form complexes with the indicated pigments, which are stable in the indicated pH range even without employing excessively high amounts. Furthermore, as opposed to the complex obtained with sodium nitrite and sodium nitrate, for example, the formed complex is insusceptible to the influence of light and oxygen. The complex is tasteless and it has no adverse effect on the taste and color of the meat product when fried for example.

The contact of the meat product with tetrazole can be achieved by several methods, which of course differ with the kinds of fresh and cured meat respectively. Thus it is possible to apply the tetrazole either in a solid form, for example, as a powder, or in the form of an aqueous solution, onto the surface of a body of meat or to soak or immerse the body of meat in such aqueous solution.

In the latter case, the final result will be dependent on the rate of penetration of tetrazole into the meat, so the time of treatment will of course vary with the form and dimension of the meat product used. Also, an aqueous solution of the tetrazole may be injected into the meat product. Finally in case of minced meat products tetrazole may be mixed intimately with the mince.

Since tetrazole is a weak acid, its aqueous solutions have a relatively low pH and a slightly acid taste. On contacting such solutions with meat products the buffering power of the latter is normally sufficient to maintain the final pH of the medium at the preferred level of 4.5 to 6.5. However, in particular when using concentrated solutions, it is preferred to adjust the pH of such solutions before use to about the same level, e.g., by neutralizing them with sodium hydroxide. This has no influence on the amount of tetrazole to be used, whereas the color of the treated meat products, especially when treated superficially, is slightly more stable even than that of products treated with acidic solutions.

In a case in which the process is applied to cured meat products, it can be combined with the normal curing process, by applying the tetrazole in combination with the normally used curing agents such as kitchen salt, sugar, sorbic acid and so on. Satisfactory results have been obtained by using tetrazole in the amounts specified below:

(a) When applied superficially in an amount of 0.2 to 0.6 mg tetrazole per 1 sq cm surface of the meat product.

(b) When applied by injection in pieces of meat or by mixing with minced meat an amount of 10 to 100 mg tetrazole per 100 grams meat product.

(c)   When applied by submersion and soaking an amount of 1 to 100 mg
       tetrazole per 100 grams meat product.

When applied to meat products which after the treatment are kept in a deep-frozen condition, e.g., at –20°C or below, the abovementioned preferred amounts of tetrazole can be reduced considerably, e.g., to about 20 to 30% of the given values or even less.

Example 1: Discolored beef was sprayed with a 20% aqueous solution of tetrazole at a temperature of 3° to 4°C, at a rate of 0.001 to 0.003 ml solution per square centimeter of the surface, corresponding to 0.2 to 0.6 mg tetrazole per square centimeter of the meat surface. A reddish color developed within 5 minutes after application and maximum color development was obtained after about 1 hour. This color was stable for the time that no bacteriological spoilage occurred in the controls, viz, 6 to 7 days at 2° to 4°C.

Example 2: On repeating the procedure described in Example 1, but using an aqueous solution of tetrazole which previously had been neutralized with sodium hydroxide to a pH of about 6.0, very similar results were obtained. The appearance of the beef was considered to be slightly better even than in Example 1.

Example 3: On repeating the procedure described in Example 1, but reducing the amount of tetrazole to 0.02 to 0.1 mg/cm^2 meat surface by decreasing the concentration of the spraying solution, also a marked improvement in color was produced, though the change in color was slower and less pronounced than with the more concentrated solution.

Example 4: On repeating the procedure described in Examples 1, 2 and 3, but using balls or similar shapes of discolored minced beef, the same results were obtained as with whole beef.

Example 5: Minced fresh lean beef was mixed with aqueous solutions of tetrazole containing 2.5, 5 and 10% tetrazole, at a rate of 1 ml solution per 100 grams beef (i.e., 25, 50 and 100 mg tetrazole per 100 grams beef), whereafter the beef was packed in a container sealed with an oxygen-permeable foil which was not in contact with meat. A bright red color developed which was stable for 10 to 12 days at 2°C. The lowest level still produced an appreciable improvement of the color although the concentration of 0.05% showed the best result.

Example 6: Minced fresh lean beef was mixed with aqueous solutions of tetrazole of the same concentrations as described in Example 4, whereafter the minces were packed in oxygen-impermeable pouches (Cryovac), vacuum-sealed and plate-frozen. Even after storage for 4 months at –28°C the minces had retained a fresh red color at all the concentrations used, whereas untreated controls had developed brown patches after less than 6 weeks.

p-Aminobenzoic Acid and Ascorbic Acid

Conventional meat color preserving agents, such as ascorbic acid and/or nicotinic acid, function by reacting with meat pigments either before or after they are oxidized. Nicotinic acid reacts with myoglobin and hemoglobin before being oxidized and forms a bright red compound that is relatively color stable and resistant to oxidation over a period of time. Ascorbic acid functions by reducing metmyoglobin and methemoglobin, which are brown or grey in color, to myoglobin and hemoglobin which react with oxygen upon exposure to air to form oxymyoglobin and oxyhemoglobin which are bright red in color. Continued exposure to the air, however, will cause an oxidation of the reaction product of nicotinic acid and myoglobin and hemoglobin, and of oxymyoglobin and oxyhemoglobin with a resultant change of color in the meat to brown or grey.

It will be noted that ascorbic acid performs its function by reacting with the meat pigments in their oxidized state and that the nicotinic acid performs its function by reacting with the meat pigments in their reduced state. In searching for new meat-color preservatives, the existing art has proceeded on the assumption that any new color preservative must also possess the ability to react with oxidized or reduced pigments.

A group of meat color preserving agents is reported by *E.W. Hopkins and K. Sato; U.S. Patent 3,597,236; August 3, 1971; assigned to Armour and Company.* This group is selected from the class of nitrogen bearing cyclic compounds substituted with a carboxyl group which consists of para-aminobenzoic acid, meta-aminobenzoic acid, isonicotinic acid, and N-ethylnicotinamide.

It is not known if these cyclic compounds enter into any reaction. The identity of the other reactants and of any reaction product is unknown, however, and no explanation for this phenomenon can be offered except that the color of meat is preserved when the meat is in contact with any one of these cyclic compounds and ascorbic acid. Some of the cyclic compounds have some color preserving ability when used alone but the combination of ascorbic acid with any one of the cyclic compounds preserves the color of meat for considerably longer periods of time than either ascorbic acid or any one of the cyclic compounds when used individually.

Treatment of whole meat, such as cuts of meat or whole carcasses, may be accomplished by dusting the exposed surfaces with one of the cyclic compounds and ascorbic acid or by spraying the surfaces with a solution. For carcasses, treatment may be accomplished by injecting a solution of the treating materials into the arteries and veins. For ground meat, the treating materials may be applied to the exposed surface of the ground meat or added to the meat before or during grinding to permit distribution thereof throughout the ground mass. It is preferred that the cyclic compound and ascorbic acid be uniformly mixed together before being applied to insure proper distribution.

When the treating substances are incorporated in ground meat, satisfactory results have been obtained by employing ascorbic acid in the amount of 100 mg/lb of meat and the cyclic compound in amounts of 50, 100, 200 and 400 mg/lb of meat. Good results are also obtained by employing ascorbic acid in amounts from 25 to 200 mg/lb of meat and the cyclic compound in amounts from 50 to 600 mg/lb of meat.

When the treating substances are dusted or sprayed on the surface of the meat, satisfactory results have been obtained by employing one part of ascorbic acid to one-half, one, two, or four parts of the cyclic compound but any ratio from 1 to 8 parts of ascorbic acid to 2 to 16 parts of the cyclic compound also give good results.

The time required to effect the treatment will also vary due to the different forms of meat which may be treated which results in different rates of diffusion. When the meat is ground and rapid diffusion is possible, the treatment may be effected within a few hours. For meat cuts and whole carcasses, several days may be required for effective treatment except when effective means of diffusion are provided.

Example 1: Five 2-lb batches of ground meat comprising 80% of beef chuck and 20% of kidney fat were made up and 4 of the batches were mixed with an additive having the following formulation for each 2-lb batch of meat: cyclic compound, 800 mg, ascorbic acid, 200 mg and dextrose (carrier), 4.5 grams. Each batch contained 1 of the 4 cyclic compounds disclosed by the process. The fifth batch contained no additives and was used as the control batch for purposes of comparison with the other four.

Dextrose is used as a carrier to provide a more uniform distribution in and on the meat of the meat preserving agents. Any other substance such as salt or starch which will serve this function and not have an adverse effect upon the meat or its color could be used as well. The concentration of the carrier used in all of the examples has been 4.5 grams per 2 lb of meat. Sodium bicarbonate was added to each of the batches having an acid cyclic compound and in an amount to neutralize the acid so the pH of the meat would not be changed.

Each of the batches were formed into 1-lb loaves and wrapped with a conventional fresh meat cellophane and held in a refrigerated room at 40°F. The color on the outside of the meat loaves was then observed at the time the loaves were initially wrapped and at various intervals thereafter. The color changes are tabulated as follows:

Additive	Initial	16 Hours	24 Hours	40 Hours	64 Hours
			Color Description After		
(1) p-Aminobenzoic acid and ascorbic acid	Red	Red	Red-bright red	Bright red	Bright red
(2) m-Aminobenzoic acid and ascorbic acid	Red	Red	Red-bright red	Bright red	Bright red
(3) Isonicotinic acid and ascorbic acid	Bright red	Red	Bright red	Bright red	Bright red
(4) N-ethylnicotinamide and ascorbic acid	Bright red	Bright red	Bright red	Bright red	Dark red
(5) Control	Red	Red	Dark red	Dark red	Brown

Example 2: Six 2-lb batches of meat were made up and additives were mixed in three of the batches except for the substitution of isonicotinic acid for meta-aminobenzoic acid. The fourth batch served as the control. The fifth batch was mixed with 200 mg of ascorbic acid and the sixth batch was mixed with 200 mg of isonicotinic acid. The batches were also prepared for the observation of color changes and the color changes are tabulated below:

Additive	18 Hours	42 Hours	68 Hours
	Color Description After		
(1) Isonicotinic acid (200 mg/lb) and ascorbic acid	Bright red	Bright red	Brown-red
(2) Isonicotinic acid (100 mg/lb) and ascorbic acid	Bright red	Bright red	Brown
(3) Isonicotinic acid (50 mg/lb) and ascorbic acid	Bright red	Dark red	Brown
(4) Control	Brown-red	Brown	Brown
(5) Ascorbic acid	Dark red	Brown-red	Brown
(6) Isonicotinic acid	Brown-red	Brown	Brown

Nicotinic Acid for Frozen Meats

In the production of frozen meats, the product is subjected to temperatures of around −10° to −30°F to set the structure of the meat product and guard against deterioration in flavor, color, odor and quality. Freezing is carried out at low temperatures so as to convert the product from the flaccid, yielding state to a hard, rigid, brittle block or chunk and the freezing step is usually carried out as rapidly as possible.

Rapid freezing is considered desirable since slow freezing tends to cause darkening of the meat. Since this rapid freezing is often carried out by blasting or blowing cold air over the surface of the product, and such procedure usually results in the evaporation of moisture from the meat, it has been the practice to carry out the freezing step after the meat has been enclosed in a covering or wrapping material which will inhibit dehydration of the meat and the resultant freezer burn.

Even in those cases where the wrapper or covering material does prevent freezer burn, there is often a loss of the desirable red meat color since many flexible packaging materials which are employed to prevent evaporation of moisture also inhibit the transmission of oxygen. It is important, if the bright red meat color is to be developed, that the meat be in contact with oxygen during or prior to the freezing step, in order to produce a desirable bloom.

More importantly, it has been known for quite some time that frozen meat tends to discolor much more rapidly than fresh meat when exposed to light. Generally speaking, the discoloration of fresh meat is due primarily to bacterial action and not lighting. On the other hand, frozen red meat in approximately 0°F freezer display cases under lighting of 50 to 150 foot-candles of intensity have been known to discolor in 4 or less days. This discoloration appears to be a light catalyzed oxidation of the meat pigments and it appears that the bright cherry red color which is desirable (oxymyoglobin) becomes oxidized to the objectionable, at least from a marketing standpoint, brown (metmyoglobin).

H.F. Bernholdt and H.L. Roschen; U.S. Patent 3,600,200; August 17, 1971; assigned to Swift & Company sprays a dilute solution of nicotinic acid on individual pieces of meat prior to freezing. The meat cut is then wrapped in a flexible transparent film having an oxygen permeability of at least 500 cc/100 in^2 in 24 hours at one atmosphere differential in order to assure good color during subsequent handling, freezing and marketing. The wrapped product is then sealed under vacuum in order to exclude entrapped air thereby eliminating objectionable frost pockets.

Normally, the wrapped product is frozen, placed on a tray and wrapped with a flexible netting and placed in a box for shipment. Alternatively, the wrapped product can be placed on a retail-type meat tray and overwrapped with netting or other high oxygen transmission film prior to freezing. Following wrapping, the product is then cryogenically frozen using a liquid nitrogen freezer tunnel, although other systems such as a plate or conventional freezers can be used.

In fact any type of freezing means may be used that possesses sufficient refrigeration capability to freeze the meat without formation of damaging ice crystals. During subsequent marketing of the product, the meat can be displayed in cases at temperatures below about 10°F, preferably below 0°F, under lighting of 50 to 150 foot-candles of intensity without showing discoloration for several weeks.

It has been found that concentrations of nicotinic acid within the ranges of 10 to 50 ppm give optimum results. Since the nicotinic acid is present on the surface of the meat, at least about 35 ppm of nicotinic acid should be present on a 1-inch thick steak. On larger cuts of meat, a 4-inch roast for example, the amount used, based on the total weight of the meat, will be approximately ¼ of 35 ppm or about 10 ppm.

In carrying out the process, no holding time is required, i.e., the nicotinic acid need not be worked into the meat as it is sufficient if the nicotinic acid merely coats the surface of meat. After contacting the meat with a small quantity of nicotinic acid, the product may be frozen prior to contact with the nicotinic acid but in such a procedure it is pointed out that adhesion of the acid to the meat is reduced somewhat.

Example 1: Fresh beef steaks having a temperature of 35°F and cut within 96 hours after slaughter were sprayed with an aqueous solution of 1% nicotinic acid with the resulting pickup of 0.5% (5,000 ppm). They were then placed in polyethylene pouches and vacuum sealed to exclude entrapped air thereby eliminating objectionable frost pockets. The product was placed on a retail-type meat tray and overwrapped using netting to hold the product in place. The product was then cryogenically frozen using a liquid nitrogen freezer tunnel. The product was displayed in 0°F freezer cases under lighting of about 150 foot-candles of intensity. The treated product retained its bright red color under these conditions for 40 days. Control samples, i.e., not treated with nicotinic acid were observed to go off color in 4 days resulting in a very unsightly product.

Example 2: Fresh beef steaks having a temperature of 40°F and cut within 96 hours after slaughter were sprayed with 1% nicotinic acid solution with a resulting gain of 0.35% (3,500 ppm basis meat) in weight. These steaks were placed, sprayed face up, on paperboard meat trays. Care was taken to have well filled packages and steaks were of sufficient thickness so that the nicotinic acid sprayed surfaces were higher than the package edges.

The packages were then wrapped in transparent film with high oxygen transmission characteristics (1,100 cc/100 in^2/24 hr at 1 atm pd) and were passed through a heat shrink tunnel to securely bring the film tight around the package and eliminate air pockets. The package units were then frozen in an air blast at 500 ft/min at a temperature of –20°F. Control packages were similarly prepared in identical fashion except that no nicotinic acid spray was employed. The frozen packages were exposed to merchandising conditions in a lighted frozen food display case under 110 foot-candle fluorescent illumination. Control steaks developed unacceptable color in 6 days, while treated samples were still satisfactory in appearance after at least 20 days.

Sulfites for Collagen-Containing Material

A sulfite agent is used by *V.L. Johnsen, E.V. Matern and R.S. Burnett; U.S. Patent 3,308,113; March 7, 1967; assigned to Wilson & Co., Inc.* to minimize color change in collagen-containing materials during heating and processing. Protein additives derived from collagen, to be useful for incorporation in food, cosmetic and pharmaceutical products, generally must be low in color, have a low ash content and be bland with regard to flavor and odor.

Collagen can be, and ordinarily is, hydrolyzed to gelatin under extremely mild conditions so that a bland, light colored gelatin is formed which is capable of producing water solutions which are viscous and have the power to form strong gels at relatively low concentrations. High viscosity and gel forming ability are properties having advantages for some uses and disadvantages for other uses, for example, use where the viscosity can interfere with the development of foam and result in products with undesirable texture and other physical characteristics, i.e., in the manufacture of aerated confectionery products, such as frappes, nougats, etc. In some cosmetic uses relatively high concentrations of protein are used in liquid preparations. If gelatin were used at these concentrations and at relatively low temperatures, gelation might occur which would be undesirable.

The gelling character of gelatin obtained from collagen can be reduced or largely destroyed by subjecting collagen or derived gelatin to high temperatures and pressures in the presence of steam and/or water. Treatment under these conditions is disadvantageous because the resulting solution of proteinaceous material has been rendered dark in color and possessed of an objectionable odor.

One conventional method of lightening the color of gelatin products of reduced gelling character has been to bleach the aqueous solution of product, but while this treatment removes color from the solution, the solid product obtained still retains an undesirable odor and flavor and tends to exhibit a marked darkening during drying.

Proteinaceous material of nongelling character means materials which are devoid of gelling character or are of reduced gelling ability. Processing of collagen directly to a nongelling material is less costly due to the elimination of ordinary gelatin preparation steps. Usually the nongelling proteinaceous derivative is prepared in one step by subjecting the raw material to high temperatures and pressures in an aqueous system. The disadvantages of the one-step system are that the resultant nongelling proteinaceous materials usually have objectionable odor and are too dark in color for many applications.

It was discovered that the treatment of collagenous material in the presence of sulfite ion or radical in solution during high temperature treatment of collagenous material to impart nongelling or reduced gelling character, inhibits development of appreciable color during the processing. The product of heat treatment in the presence of sulfite has color ratings which do not require bleaching and while the solutions have an objectionable odor similar to that resulting from the treatment in the conventional manner and in addition have an objectionable flavor including residual sulfite taste, the nature of the odor and flavor elements is such that they can now be eliminated by peroxide oxidation of the solution of the nongelling proteinaceous material.

Sulfite ion may be introduced into the aqueous solution of proteinaceous material by dissolving sulfur dioxide gas in the water to form sulfurous acid, by adding water soluble salts of sulfurous acid, and equivalent operations. Sufficient sulfite must be present so that at least 1,000 ppm, preferably 2,000 ppm (based on solids in solution), remain after the cooking operation to assure obtaining a light colored product. Other conditions being equal, the higher the temperature and the longer the time of heating, the larger the amount of sulfite required. Generally, an amount of sulfite agent is added which is capable of introducing into the solution between 0.1 and 5% of sulfite, i.e., SO_3 ion on a weight of the solids in solution basis.

Conditions of processing or the type of equipment used will govern the choice of the sulfite agent. Sulfur dioxide, which dissolves in an aqueous solution to form sulfurous acid may be used in jacketed pressure vessels that are not heated by direct steam because SO_2 introduction does not involve the introduction of ash forming cations. When vessels are used which are heated and pressurized by passing steam directly into the mixture, use of sulfur dioxide gas is uneconomic due to loss in the continuously vented gases.

When the solutions of proteinaceous matter are to be heated in vented vessels, the common salts of sulfurous acid such as sodium sulfite, potassium sulfite, sodium bisulfite, potassium bisulfite and mixtures thereof are utilized because of their greater stability under the conditions of the pressure cooking.

For some specific food uses, a low ash and especially a low sodium content proteinaceous material is required. If such product properties are to be met, use of sodium and potassium salts of sulfurous acid as a source of sulfite is not desirable because these salts or their soluble decomposition products would be contained in the finished product. If sulfur dioxide gas cannot be used, then salts such as calcium and magnesium bisulfite are used for the introduction of sulfite ion because these salts are converted to water-insoluble salts or to water-insoluble sulfites or sulfates during the processing, most of which materials can be removed by settling, filtration, etc.

The extent of the heat treatment required to eliminate the gelling characteristic of the proteinaceous material will vary with the type of collagen-containing raw material and with the type of end product desired. A product with virtually no gel strength can be obtained from most collagenous materials by cooking at 35 to 65 lb gauge steam pressure, i.e., at temperatures in the range between 275° and 310°F for from 2 to 5 hours.

Useful agents capable of freeing nascent oxygen in aqueous solution are hydrogen peroxide, barium peroxide, and sodium peroxide. Generally, use of between 0.05 and 0.5% of peroxide on the basis of solids in solution will be sufficient for treatment of any sulfite-containing, heat treated collagen solution. Addition of peroxide in the above specified amounts substantially eliminates objectionable sulfite flavor, and substantially eliminates the strongly objectionable odor generated during the treatment to reduce the gelling character of the solution.

If a product having improved properties such as low ash, reduced salt flavor, etc., is desired, treatment with various combinations of anionic and cationic resins may be used subsequent to the peroxide oxidation operation. Treatment of the extract with anionic exchange resin followed by treatment with a strong acid resin such as Amberlite IR-120H can reduce the ash content of a final product from a solution having 2.16% ash content to approximately 0.03% ash content.

Example 1: 20,000 lb of ground edible grade hind pigs' feet are cooked using steam at 40 lb pressure (288°F) with 2,500 gal of water and 60 lb anhydrous sodium bisulfite for 2 hours. The melted fat is drawn off and the liquid phase is decanted from the insoluble residue. The residue is again pressure cooked with 1,000 gal of water and 16 lb of sodium bisulfite, and the aqueous extract isolated. The aqueous extracts are combined and evaporated to 50% solids. Approximately 5,500 lb of the 50% solids material is obtained.

The concentrated protein solution is then heated to 200°F and passed through a precoated filter press. One-half of the clear solution was dried on a double drum dryer without further treatment. The remaining half was treated with 0.15% hydrogen peroxide and then dried on the drum dryer. When the water solutions of the untreated and peroxide-treated materials are tasted, the untreated sample has a predominant taste of sulfite whereas the peroxide treated sample has no sulfite flavor and a slight salty taste. The odor of the solution of the untreated sample is much stronger and undesirable than the peroxide treated sample.

Example 2: 18,000 lb of grounds pigs' feet are soaked in cold water for 2 hours. The water is drained off and the ground feet covered with water a second time and again drained. The feet are then covered with cold water which contains 25 lb of sulfur dioxide gas, and after steeping the ground feet overnight this solution is drained off. A calcium bisulfite solution is prepared by passing liquid SO_2 into a slurry of 27 lb of calcium carbonate until the solution becomes clear. This solution, plus enough hot water to cover the ground feet, is then added.

The material is then cooked with steam at 40 lb gauge pressure (288°F) for 2 hours. After cooking, the melted fat is drawn off, and the aqueous layer is removed. A second cook is carried out with added water at 5 lb pressure for one-half hour, the liquid extract removed and combined with the first cook-water extract, and evaporated to 50% solids. Approximately 4,800 lb of the concentrated material is obtained.

The material was filtered and split into portions A, B and C. The portion A was drum dried. The dried product had a sulfite content of 1,800 ppm and an objectionable taste and odor. The portion B was treated with 0.15% of hydrogen peroxide. The portion B was drum dried. The SO_2 content in parts per million of SO_2 on a solids basis was less than 125. The ash content of the dried material was 1%. Solutions of the dried material were almost odorless and had only a slightly bitter taste.

Example 3: Portion C of the liquid product of Example 2 was treated with 0.25% of hydrogen peroxide. The hydrogen peroxide treated liquid product may be passed through a tower containing approximately 2 cubic feet of anionic exchange resin (Rohm and Haas XE-168). After 800 lb of peroxide treated extract at 40% solids concentration is passed through the tower at a rate of a half gallon per minute, the pH of the composite sample of the effluent is 8.7. The SO_2 content in parts per million of SO_2 on a solids basis is less than 100. The pH of the treated extract is then adjusted to 5.5 with phosphoric acid and dried. The dried material has an ash content of approximately 0.4%.

In the event that it is desirable to avoid addition of an acid as the means of adjusting pH, the effluent from the anionic exchanger may be treated with a cationic exchange material such as Amberlite IR-120 H. A typical use for the proteinaceous materials which have limited gelling character is as whipping agent, in confectioneries such as nougat because it renders the mix low in viscosity at relatively low temperatures.

ANTIOXIDANTS IN COLLAGEN CASING

Over a period of many years, synthetic sausage casings have been prepared from animal collagen. Casings made of collagen have been prepared by processing animal hide to break the collagen into a fibrous structure and extrude the collagen fibers in the form of a doughy mass to produce tubular casings. The casings prepared in this manner have been hardened with formaldehyde and have been used as a removable casing for processing various sausages. These casings have not been edible even though collagen itself is an edible material.

More recently, edible sausage casings of collagen have been prepared and sold in commercial quantities. In the manufacture of edible collagen casings, considerable emphasis has been placed on the necessity for using collagen source materials which have not been subjected to a liming treatment. In fact, a number of recent processes describing the production of collagen have indicated that it is absolutely necessary to start with an unlimed collagen source material if an edible casing is to be obtained.

Casings which are packaged satisfactorily may be wholesome initially, but may often become rancid by oxidation upon storage. As a result, there has been a need for a treatment of collagen casings to improve its antioxidant properties. The importance of preserving the wholesomeness of collagen sausage casings is obvious. The importance of keeping the collagen sausage casing wholesome and palatable is especially important when storage is necessary. Stabilization of collagen casings to oxidation has been attempted by several means, such as choice of raw materials, proper packaging, deodorization, and exercising every precaution during processing to avoid unnecessary exposure to heat and light, or contamination with oxidation promoters.

All of these methods have their limitations in that their applications are not always feasible for collagen sausage casings; hence the stabilization of collagen casings may be improved by use of antioxidants. By the addition of traces of certain chemicals, it has been found that oxidative degradation can be inhibited and the age resistance of collagen casings considerably increased.

The critical conditions for use of antioxidants to preserve the freshness of collagen casings is given by *M.A. Cohly; U.S. Patent 3,567,467; March 2, 1971; assigned to Tee-Pak, Inc.* It is essential that the antioxidant should protect both the collagen itself and the glycerin plasticized in the casing during storage. Furthermore, the antioxidant must be capable of uniform distribution in the collagen casing. It is also noted that the fact that an antioxidant may be successful in protecting the casing from rancidity does not mean that it will also retard oxidative discoloration.

In procedures for preparing edible collagen casing, whether from limed or unlimed hides, the final step in the preparation of the casing prior to drying involves passing the extruded, coagulated, tanned, and washed casing through a plasticizing bath. The plasticizing bath is preferably an aqueous solution of glycerin or similar plasticizing material (e.g., sorbitol, dipropylene glycol or triethylene glycol) and preferably contains a softening agent of 0.5 to 5.0% weight of a fatty monoglyceride or an acetylated fatty monoglyceride. The softening agent is emulsified in the aqueous glycerin solution using a suitable emulsifying agent such as sodium dodecyl sulfate or polyoxyethylene (20) sorbitan monopalmitate, or the like.

The emulsifier is used at a concentration of about 10% by weight of the fatty monoglyceride or acetylated fatty monoglyceride. The monoglycerides which are used are fatty monoglycerides, such as glycerol monostearate, glycerol monooleate, glycerol monopalmitate, glycerol monolaurate, etc. (Myverol). Acetylated monoglycerides which may be used are the monoacetate and diacetate derivatives of fatty monoglycerides such as glycerol monostearate diacetate, glycerol monostearate monoacetate, glycerol monooleate diacetate, glycerol monopalmitate diacetate, etc. (Myvacet).

An edible, nontoxic Food & Drug approved antioxidant is incorporated in small quantities in the final plasticizing bath. Generally, a dilution of 1 to 1,400 or 1 to 2,800 (by weight) is sufficient to render the casing free from unpleasant odors due to microbial growth or glycerin oxidation over extended periods of storage.

During an experimental antioxidant test with edible collagen casings, it was noticed that antioxidant-treated collagen casing had much more sheen than the same casing not treated with antioxidant. This increased sheen or bloom is particularly noticeable when the casing was stuffed with pork sausage or frankfurter emulsion.

In Figure 2.1 of the drawings, the steps from extrusion through reeling or shirring are illustrated schematically in slightly more detail. The collagen slurry is introduced through inlet conduit **1** into die **2** having an annular die outlet **3** through which casing **4** is extruded. The die has an inner tube **5** which extends upwardly within the extruded casing to remove coagulating bath from within the casing. The die is located at the bottom of container **6** which contains a coagulating bath **7**. Coagulating bath **7** is circulated through conduit **8** from the tube for removal of the coagulating bath from inside the extruded casing. The casing which is coagulated in the coagulating bath passes over a series of rollers and is directed through a tanning bath **9**.

FIGURE 2.1: EDIBLE COLLAGEN CASING CONTAINING ANTIOXIDANT

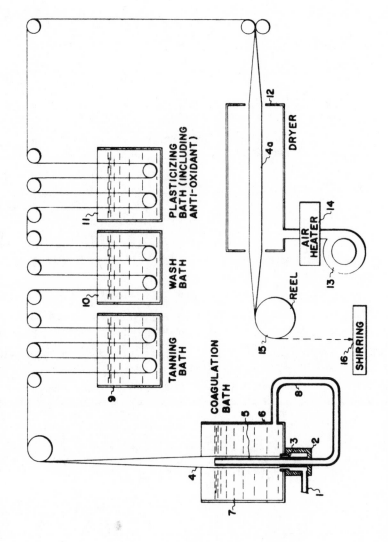

Source: M.A. Cohly; U.S. Patent 3,567,467; March 2, 1971

Tanning bath 9 consists of an aqueous solution of any suitable nontoxic tanning agent (e.g., vegetable tannins, nontoxic edible dialdehydes such as glutaraldehyde, and the aldehydes present in smoke condensates, and aluminum and ferric salts, preferably in olated form and rendered partially basic).

From tanning bath 9, the casing passes through a wash bath 10 where unreacted tanning agent is washed out of the casing. The casing is then passed through plasticizing and antioxidant bath 11 which introduces a small amount of a plasticizer such as glycerin, a softening agent such as fatty monoglyceride or an acetylated fatty monoglyceride, and an edible, nontoxic FDA approved antioxidant into the casing. From the plasticizing and antioxidant bath 11, the casing passes through dryer 12 where it is inflated as indicated at 4A and dried with air or other gas circulated by fan or blower 13 through air heater 14.

After leaving dryer 12, the casing may be collapsed and rolled up on reel 15 from which it is subsequently removed for shirring. In an alternate process, the casing may be passed directly to a shirring machine shown diagrammatically as 16. In either case, the shirring machine which is used for preparation of shirred strands of casing may be of any suitable type such as the types commonly used in the shirring of regenerated cellulose sausage casings. After the casing is shirred into individual short strands for convenience of handling, it is packaged for shipment to the meat packer. Sometimes it may be desired to cure the casing by heating at 60° to 80°C in an atmosphere of 20 to 50% relative humidity for several hours prior to shipment.

The links may be severed from each other and packaged in a suitable overwrap following conventional meat packaging techniques. When the sausage is cooked by the consumer, the casing is found to be quite strong and shrinks with the meat during cooking. The casing may be prestuck, if desired, to permit more rapid release of the fat during the cooking of the sausages. The casing which is prepared in this manner and treated with the plasticizer, antioxidant and (optionally) softening agent in the plasticizing and antioxidant bath is superior in antioxidative and appearance properties, particularly upon extended storage.

Example 1: In this example, the preparation of edible collagen casings from limed animal hides is illustrated using a plasticizing and antioxidant final treatment to produce a casing having improved antioxidative and appearance characteristics.

Selected cattle hides from carcasses certified fit for human consumption, weighing 65 to 75 lb each, are the starting material for this process. As soon as possible after flaying and inspection, the hides are washed in a large volume of circulating cool (10°C) water to remove adhering blood. After washing, the hides are fleshed fresh, without curing, to remove adhering fatty and muscular debris from the flaying operation.

The washed and fleshed hides are then treated in a liming bath containing 6% weight of fresh calcium hydroxide and 1.5% weight sodium sulfhydrate (the liming bath may contain up to 3% dimethylamine sulfate), as solution and/or slurry contained in about 450% weight of water at room temperature (15° to 20°C), all percentages being calculated on the weight of the hide treated. The treatment is carried out for a period less than about 6 hours, sufficient to remove most of the hair from the hide, and the hides are gently agitated from time to time to insure even penetration of the liming liquor.

After liming, the hides are removed from the liming bath and permitted to drain for a period of one-half hour while suspended. The limed hides are then gently squeezed, as between rubber rollers, to remove excess liming liquor. The hides which have been thus limed, drained, and squeezed are then cut or split in the plane of the hide into two approximately equal portions by weight.

The upper or outer hide surface contains all of the hair, hair follicles, and sebaceous and sudorific glands. The inner or corium layer consists essentially of collagen. The outer or hair-containing layer or split is discarded as unsuitable for use in the preparation of casing but may be used for the formation of leather laminates or coverings.

The corium layer or split is then placed in a tank or vat containing about 4.5 times the hide weight of water in a solution at less than about 15°C. Gentle agitation is used to insure even removal of debris and adhering lime solution and/or slurry. The hides are washed during a period of 20 to 30 minutes. The washings are removed and the corium splits resuspended in 4.5 times their weight of cool (15°C) water. Edible grade lactic acid, suitably diluted to 2 to 4 oz of 44% lactic acid per quart of cool water, is added in small portions at 15 min intervals, with gentle agitation for 5 min of each 15 min period.

The liquor is tested for pH before each addition, and the end point is regarded as the point when the pH is permanently depressed below 7.0. In general, this requires about 1.5% of the 44% lactic acid, based on the weight of the corium splits. This treatment is effective to neutralize the excess lime in the corium layer and to remove it as a soluble salt. The rate of addition of the lactic acid solution is carefully regulated so that the temperature of the bath is never permitted to rise above about 32°C.

The neutralized and delimed corium splits are then removed from the neutralization bath, drained, and rinsed in cool (15°C) water, packed into polyethylene bags to chill the prepared collagen and to maintain it below 5°C during storage and/or shipment prior to comminution and acid swelling operations. It should be noted, however, that the hides may, if desired, be cut into small pieces, or small pieces of scrap hide material may be used in the steps of liming, splitting and neutralization or deliming.

The delimed corium splits are cut into small square or rectangular sections, e.g., ¼ to 4 inches on a side, in preparation for grinding. The small pieces of treated hide are converted to a fine pulp by successive passes through a meat grinder. In this grinding operation, sufficient ice is mixed with the hide splits to maintain the temperature below about 20°C (and preferably below about 10°C). Successive passes through the meat grinder use successively small dies, the smallest being about ³⁄₆₄". At this point, the mixture is adjusted in water content by addition of sufficient water to bring the water content of the slurry to about 90%.

The collagen slurry or pulp is then treated with sufficient dilute lactic acid (other dilute or weak acids such as citric or acetic acid may be used) to produce a pH of 2.5 to 3.7. The acid is usually added as a dilute solution, e.g., about 0.8 to 2.0%. After thorough mixing, the pulp and acid are stored overnight at a temperature of about 3°C to swell. At the end of this time the collagen is swollen and has taken up all of the water in the slurry. The swollen collagen is then mixed with sufficient water and acid to maintain the pH of 2.5 to 3.7, thus producing a thin, homogeneous paste consisting of about 4% collagen and 1.2% lactic acid.

The swollen collagen slurry is passed through a homogenizer to further disperse the fibers and then is filtered to remove any undispersed fiber clumps or other solid contaminants. The paste is generally deaerated by storage under vacuum prior to extrusion. The process, from the washing of the limed hide through the acid swelling of the comminuted collagen, is preferably carried out in a period of about 6 to 12 hours, and generally no longer than 48 hours.

The homogenized and filtered collagen slurry is then pumped under pressure through the extrusion die, as previously described, into a coagulating bath consisting of about 40% ammonium sulfate (sodium sulfate can also be used) in water. When the collagen is extruded as a thin-walled tube into this concentration of ammonium sulfate, the collagen fibrils are dehydrated and collapse to form a film which is sufficiently coherent for further processing. As shown in Figure 2.1 of the drawings, the coagulation bath is circulated both inside and outside the tube to maintain the tube in an inflated condition and to insure proper coagulation of the casing both on the inside and the outside.

After the film is coagulated in the ammonium sulfate solution, it is necessary to tan the film to give it sufficient strength for further processing and for stuffing with sausage meat. If the film were taken from the ammonium sulfate coagulating bath and dried, it would be a film of moderate dry strength but would revert to a paste upon contact with water. It is therefore necessary for the casing to be tanned or hardened to provide the wet and dry strength required in an edible casing.

From the coagulation bath, the casing next passes into a first tanning bath which comprises a solution containing 10 to 20% of aluminum sulfate, $Al_2(SO_4)_3 \cdot 18H_2O$, 3 to 7% sodium citrate (or an equivalent amount of citric acid) and 3 to 7% sodium carbonate. The tanning bath is formulated so that the sodium citrate or citric acid forms a complex with the aluminum sulfate, and the sodium carbonate neutralizes a portion of the aluminum-citrate complex to render the same about ⅓ to ⅔ basic. This results in a tanning bath having a pH about 4.0 and permits the use of aluminum concentrations for tanning which are many times the concentrations available with other aluminum tanning baths, such as alum tanning baths. A suitable tanning bath may similarly be made from ferric salts by formation of a citrate complex and partially neutralizing the complex with sodium carbonate or other weak alkali to convert the complex to an olated form.

After the casing is thoroughly tanned it is passed through one or more wash baths to wash out any unreacted tanning or hardening reagent. The casing is then passed through a plasticizing and antioxidant bath consisting of a dispersion or solution of an edible, nontoxic antioxidant in an aqueous solution of glycerin (or equivalent plasticizer such as sorbitol, dipropylene glycol, triethylene glycol, etc.).

The plasticizing and antioxidant bath (which is preferably also a fat liquoring bath) has a small amount of antioxidant therein and introduces a substantial amount of the desired plasticizer, softening agent and antioxidant into the casing, which prevents the casing from becoming rancid and discolored after drying, particularly upon storage, and which gives the casing improved sheen characteristics. The plasticizing and antioxidant bath also contains 0.5 to 5.0% of Myvacet (type 9–40) which is a diacetylated derivative of glycerol monostearate, and is emulsified into the glycerin solution with about 10% sodium dodecyl sulfate based on the weight of Myvacet added.

The antioxidant is present in this final plasticizing bath in small amounts, generally in a dilution of 1 to 1,400 or 1 to 2,800 (by weight). This generally provides a finished casing having less than about 0.01% by weight of antioxidant (in the casing). If the casing has more of a concentration of antioxidant, the plasticizing bath is made more dilute in antioxidant to bring the concentration thereof in the final casing within the desired upper limit.

Satisfactory antioxidants are Aranox 7G, Aranox NE and Aranox 4E. Aranox 7G (Reheis Chemical Company) is a proprietary antioxidant having the following composition (by weight): butylated hydroxyanisole, 28.0%; propyl gallate, 12.0%; citric acid, 6.0%; and base (aqueous), quantity sufficient to neutralize.

Aranox NE is a proprietary antioxidant having the following composition: butylated hydroxyanisole, 40.0%; nordihydroguaiaretic acid, 4.0%; citric acid, 6.0%; and base (aqueous), quantity sufficient to neutralize.

Aranox 4E is a proprietary antioxidant having the following composition: butylated hydroxyanisole, 13.3%; butylated hydroxytoluene, 13.3%; citric acid, 5.0%; and base (aqueous), quantity sufficient to neutralize. In successive experiments each of the three Aranox compositions was used as the antioxidant (1 to 1,400 dilution).

The collagen casing is passed into the plasticizing and antioxidant bath prior to the drying and, by a system of multiple passes, maintained in the bath for a time ranging from 3 to 10 minutes. After the casing leaves the plasticizing and antioxidant bath it is dried, as described above, and shirred and packaged. The finished casing possesses a fine transparency and sheen, and passes rancidity and discoloration tests, particularly upon extended storage. In particular, it has no odor after one month of storage and no off-color such as brown or yellow. Furthermore, the casing has an excellent frying response, e.g., no splitting of the casing upon frying. In particular, Aranox 7G appeared to be the best antioxidant.

Example 2: This example illustrates the use of a combination of anionic detergents in preparing the plasticizing and antioxidant bath to produce a casing which is free from tackiness which is sometimes encountered in the casing during stuffing and linking. The casing is prepared as described in Example 1 or as in any of the alternate processes for casing manufacture.

In this example, however, the makeup of the plasticizing and antioxidant bath (of the experiment for each antioxidant) is modified slightly. The plasticizing and antioxidant bath contains sufficient glycerin to plasticize the casing and has 0.5 to 5.0% Myvacet (type 9–40) emulsified into the bath using 0.05 to 0.5% sodium dodecyl sulfate and 0.005 to 0.10% calcium stearate. The collagen casing is processed in this bath as described in Example 1 (using each of the three proprietary Aranox antioxidants in successive experiments) and the finished casing is found to be free from odors and discoloration often encountered upon storage, and gives excellent frying response. The casing is both shiny and transparent.

Example 3: In this example, the process described in Example 2 is repeated except that a nonionic detergent is substituted in emulsifying the Myvacet into the plasticizing and antioxidant bath (Aranox 7G, 1 to 2,800 dilution). Tween 20, sorbitan polyoxyethylene, (20) monopalmitate, is employed as the emulsifying agent at a concentration of 0.05 to 0.5% in preparing a bath containing 0.5 to 5.0% Myvacet. The casing is processed in a manner described in Examples 1 and 2, and after drying has a high sheen and transparency. The casing prepared using this plasticizing bath is free from odor and discolors upon storage, and has excellent response to frying.

SORBIC ACID AND HEAT TREATMENT

In the production of meat products such as sausage and ham or fish products such as fish sausage and boiled fish paste (kamaboko), the minced meat is preferably maintained at a pH of about neutral, since adverse effects are produced on the elasticity of the product if the acidity is increased. For example, in the case of sausage, the pH of the minced meat is suitably in a range 5.8 to 6.2. When the pH becomes lower than this, while good effects are produced on the development of the color of the meat, the water retainability of protein declines, with the consequence that adverse effects appear with respect to the structural formation of the sausage. Accordingly, it is obviously a disadvantage to add sorbic acid as a preservative to a sausage emulsion. With reference to use of sorbic acid as a food preservative, the following has been known:

(1) Sorbic acid can be used as preservative for meat products.

(2) The preservative effect of sorbic acid is higher as the pH of the food product is lower. Thus, in food preservatives of an organic acid type such as sorbic acid, the preservative action is associated with a nondissociating molecule. Accordingly, as the pH is lower, the number of nondissociating molecules increases and, therefore, the preservative effect is increased. Sorbates obtained by neutralization of sorbic acid with an alkaline substance, such as potassium sorbate and sodium sorbate, exhibit a lower preservative effect than sorbic acid per se, because such salts are alkaline or neutral while sorbic acid is acidic. In the production of meat products the following is known:

(3) The lowering of the pH of raw meat gives adverse effects to its water-retaining property, and the use of raw meat having a low water-retaining property results in a product of poor quality.

The above points (2) and (3) are contrary to each other and it is difficult to obtain products which are sufficient with reference to each of above matters (2) and (3). Accordingly, methods of preservation of meat products have been adopted with sorbic acid or salts using free sorbic acid with the idea of achieving a high preservative effect while sacrificing the water-retaining property of the product; using a sorbic acid salt with the idea of attaining a product excellent in water-retaining property while sacrificing the preservative effect of sorbic acid; or using a mixture of sorbic acid and its salt at a suitable ratio while sacrificing both water-retaining property and preservative effect to some extent. Accordingly, industry has long sought a method of preserving meat and fish which will eliminate the foregoing disadvantages associated with previous methods of preservation. Research has been conducted on the relationship between the pH of raw meat and the water-retaining property of the product.

R. Ueno, T. Miyazaki and S. Inamine; U.S. Patent 3,716,381; February 13, 1973; assigned to Ueno Pharmaceutical Co., Ltd., Japan found that although the lowering of the pH of raw meat results in a decrease in the water-retaining property of raw meat and accordingly a lowering of the quality of a product prepared from such raw meat, the lowering of the pH barely yields any adverse effects in the quality of the product if the lowering of the pH is effected after thermal denaturation of the raw meat.

Based on the above finding, a method has been developed which can provide meat or fish products satisfactory in both (2) and (3), and elution of sorbic acid is prevented in raw meat maintained at room temperature and sorbic acid is gradually eluted in the raw meat during a subsequent heating step.

The meat and fish preservative can be obtained, for example, by melting a hardened oil, adding and dispersing thoroughly a fine powder of sorbic acid, and spraying the melted mixture in a low temperature chamber, thereby obtaining the fine powder particles of sorbic acid as granules of 10 to 1,000 microns, the same being coated with the hardened oil. In producing the preservative in this manner, utmost care must be exercised in the management of the temperature for the following reasons.

If, in dispersing the sorbic acid in the melted hardened oil, the temperature of the melted mixture exceeds 90°C, the sorbic acid starts to dissolve in the hardened oil and upon reaching a temperature of 120°C, it becomes completely dissolved. When a melt in this state is sprayed, a part of the sorbic acid sublimes in the spraying chamber and a satisfactory product cannot be obtained.

Further, if the temperature of the melt exceeds 90°C, the composition of hardened oil-sorbic acid mixture changes as a result of the sublimation of the sorbic acid. Accordingly, the temperature of the melt must be maintained at below 90°C and it is especially preferred that it be adjusted to be within the range of 70° to 85°C. Again, if the temperature fluctuates, the sorbic acid particles dispersed in the hardened oil grow and a good product cannot be obtained. Therefore, fluctuation of the temperature must be avoided as much as possible after the sorbic acid has been dispersed.

The hardened oil used must be one whose melting temperature (MT) is within the range of 40° to 90°C. Since a hardened oil whose melting temperature is less than 40°C, melts at room temperature, the use of such an oil runs counter to the objects of this process. On the other hand, when a hardened oil whose melting temperatue exceeds 90°C is used, the noted inconveniences occur. Examples of the hardened oils suitable for use include hardened rape oil (MT 60° to 63°C), hardened castor oil (MT 80° to 85°C), hardened beef tallow (MT 54° to 60°C), hardened whale oil (MT 50° to 52°C), etc.

The hardened oil is used in an amount sufficient to coat the surface of the sorbic acid powder particles. The amount usually used is at least two times by weight based on the sorbic acid. There is no particular upper limit on the amount of the hardened oil used as long as no adverse effects are had on the food products by the use of excess hardened oil. However, it is customary to use an amount up to ten times that of the sorbic acid on a weight basis. A convenient amount is that ranging from 3 to 5 times by weight based on the sorbic acid.

Further, if necessary, a small amount (e.g., 2 to 10% by weight based on the hardened oil) of a surfactant for food use, for example, glycerol monostearate, acetylated monoglyceride, etc., can be used along with the hardened oil. As these surfactants promote the uniform dispersion of the hardened oil in the food materials, there is the advantage that their use produces a uniform distribution of the sorbic acid in the resulting food product.

Since the granular food preservative contains particles whose surface is composed of a hardened oil having a melting temperature 40° to 90°C, the hardened oil does not melt even though the preservative is added to the meat or fish emulsion during the process of producing meat and fish products, and accordingly the sorbic acid does not make contact with the minced meat, fish etc. It is only when in the final finishing step of the meat or fish products that they are heated at a higher temperature than the melting temperature of the hardened oil that the hardened oil coating of the particle surface melts to permit for the first time the contact of the sorbic acid of the particle interior with the meat or fish whereupon the action, as a preservative, of the sorbic acid begins to operate.

Accordingly, through the above procedure it has been discovered that it is possible to provide the preservative action of sorbic acid while eliminating any tendency toward a decrease in the water-retaining properties of the product.

Example 1: A homogenizer was used and 1 kg of sorbic acid powder comminuted to particle diameters of below 20 microns was added and thoroughly dispersed in a melt obtained by heating and melting 3.8 kg of hardened beef tallow (MT 60°C) and 0.2 kg of distilled glycerol monostearate. The temperature of the melt was maintained at 70°C during this time. This melted mixture was cooled by being sprayed into a chamber whose temperature was adjusted at 30° to 35°C, using a rotary dish type sprayer. As a result, a granular preservative of particle diameters 50 to 300 microns comprising sorbic acid powder particles coated with the aforesaid hardened oil was obtained.

Example 2: A homogenizer was used and 1 kg of sorbic acid powder comminuted to particle diameters below 20 microns was added and thoroughly dispersed in a melt obtained by heating and melting 3.8 kg of hardened rape oil (MT 63°C) and 0.2 kg of acetylated monoglyceride. The temperature of the melt was maintained at 70°C during this time. This melted mixture was cooled by being sprayed into a chamber whose temperature was adjusted at 30° to 35°C, using a rotary dish type sprayer. As a result, a granular preservative of particle diameters 30 to 400 microns and of the same constitution as in Example 1 was obtained.

Example 3: A homogenizer was used and 1 kg of sorbic acid powder comminuted to particle diameters below 20 microns was added and thoroughly dispersed in a melt obtained by heating and melting 4 kg of hardened rape oil (MT 63°C), the temperature of the melt being maintained at 70°C during this operation. When this melted mixture was cooled by being sprayed into a chamber whose temperature is adjusted at 25° to 30°C, using a rotary dish type sprayer, a granular preservative of particle diameters 30 to 250 microns and of the same constitution as in Example 1 was obtained.

Example 4: A homogenizer was used and 1 kg of sorbic acid powder comminuted to particle diameters below 20 microns was added and thoroughly dispersed in a melt obtained by heating and melting 2 kg of hardened rape oil (MT 63°C) and 1 kg of hardened beef tallow (MT 60°C), the temperature of the melt being maintained at 70°C during this operation. When this melted mixture was cooled by being sprayed into a chamber whose temperature was adjusted at 30° to 35°C, using a rotary dish type sprayer, a granular preservative of particle diameters 50 to 300 microns and of the same constitution as in Example 1 was obtained.

Example 5: After adding 24 grams of common salt to 660 grams of starting material fish flesh for preparing boiled fish paste, the material was ground for 30 minutes. This was followed by the addition of 21 grams of sugar, 60 grams of starch and a preservative indicated in the following table.

Preservative	pH before addition of preservative*	pH after addition of preservative*	pH after heating*	Quality of product**	Preservative effect***
Not added	6.84		7.05	Good	Spoiled after 36 hours.
Sorbic acid	6.85	5.90	6.28	Inelastic and easily broken	Spoiled after 72 hours.
Potassium sorbate	6.86	6.88	7.08	Good	Spoiled after 42 hours.
Preservative of Example 1	6.85	6.74	6.30	do	Spoiled after 72 hours.

*pH measurement: Ten grams of the specimen were suspended in 50 ml. of distilled water and measured with a glass electrode pH meter.
**Quality of product: The specimen was actually eaten and a sensory evaluation was made.
***Preservative test: The specimen was preserved in a constant temperature-constant humidity apparatus of 25° C. and RH 85%, and observations were made of the appearance of slime and mold, the time indicated being that at which the specimen became unfit to be eaten.

Example 6: Meat for Vienna sausage comprising 1,000 grams (40%) of pork, 700 grams (28%) of fat pork and 800 grams (32%) of lean beef was cured for 24 hours, then passed through a chopper and thereafter kneaded for a while using a silent cutter. To this mixture were then added such auxiliary materials as condiments and spices along with the preservative indicated in the following table, following which the mixture was kneaded for a further 6 minutes.

The kneaded meat was stuffed into a water-washed sheep intestine, dried for 60 minutes at 40°C, smoked for 30 minutes at 50°C, and then boiled for 10 minutes at a temperature of 70°C in the center. After the preparation of the sausage was completed, it is stored in a refrigerator for about 20 hours and thereafter allowed to stand in a constant temperature-constant humidity apparatus at a temperature 25°C and RH 85%. The results of the tests conducted with respect to the pH, product quality and preservative effect are shown in the table below. The tests were carried out as in Example 5.

Preservative	Amount added of preservative based on material meat (wt.%)	pH Before heating	pH After heating	Quality of product	Preservative test
preservative of Example 2	1.0 *	6.20	5.85	soft and elastic jelly(good)	spoiled after 96 hours
sorbic acid	0.2	5.82	5.90	hard and easily broken jelly (unsatisfactory)	spoiled after 96 hours
potassium sorbate	0.27 *	6.40	6.45	soft and elastic jelly(good)	spoiled after 96 hours
Not added		6.35	6.40	soft and elastic jelly(good)	spoiled after 48 hours

* 9.2 wt. % as calculated in terms of sorbic acid.

As can be seen from the results presented in the above table, the special preservative, as well as potassium sorbate, does not affect the quality of the product at all. The preservative of Example 2 also demonstrated a remarkable preservative effect which was comparable to that of sorbic acid. Namely, the shortcoming of sorbic acid, i.e., that it has adverse effects on the quality of the product, i.e., water-retaining property is eliminated.

ANTIMYCOTICS IN SAUSAGE CASINGS

In the handling and processing of cellulose sausage casings, one of the problems encountered is the control or inhibition of fungus growth. Spores of fungi cannot germinate without moisture and the development of fungus growth on cellulose can be controlled by keeping the moisture content below a predetermined level. In some cases, however, proper moisture control cannot be maintained and the formation of mold, yeast or other fungus growth may be an important but sporadic problem. In cases where fungus growth cannot be controlled by proper control of the moisture content, it is necessary to provide a chemical means to inhibit fungi. The problem of fungus growth on casings is encountered in the preparation of sausages of all types, e.g., cooked, smoked, dry and semidry sausages.

As meats will mold under almost any conditions, mold may appear on the sausage casing to some extent in the drying room. This mold may be held at a minimum by washing the sausage and removing the mold spores practically as fast as they develop. The removal of surface mold by washing the casing is at a substantial cost, nor is it entirely effective; thus, the mold has already had a chance to attack the cellulose and the printing, and removal of the mold tends to remove ink from the casing surface. Undesirable mold growth can also occur on the dry sausage after packaging for shipment.

The antimycotic treatment of cellulose sausage casings, whether for cooked, smoked, dry, or semidry sausages, has presented problems resulting from the composition of such casings and the method of use of the casings. The high water and glycerin composition in clear cellulose and fibrous (paper-reinforced) casings, both dyed and undyed, has made the antimycotic treatment of such casings somewhat difficult. Common antimycotics, both of the oil-soluble and water-soluble types, often do not adhere well and tend to leach out excessively in the presoaking of the casing in preparation for stuffing.

Attempts to incorporate antimycotics such as sorbic acid and its derivatives (e.g., potassium sorbate and calcium sorbate) as viscose additives have been unsuccessful because the sorbic acid or its derivatives are either water-soluble per se or are converted to water-soluble products by the viscose alkali and are washed out rather completely during subsequent processing steps.

Furthermore, simply adding such materials to the final plasticizing bath of the viscose process and drying will incorporate a degree of fungus resistance to large cellulose casings during storage, but such antimycotics will be removed during the soaking cycle which always precedes the stuffing operation and thus will not provide any functional antimycotic value on the stuffed product.

H.G. Rose; U.S. Patent 3,617,312; November 2, 1971; assigned to Tee-Pak, Inc. has discovered that cellulose sausage casings may be rendered resistant to fungus growth by application of a coating containing an antimycotic and which when fully cured is water-insoluble, partially water-swellable, and water- and gas-permeable. A suitable coating composition comprises a solution or emulsion of an antimycotic, such as sorbic or propionic acid or their sodium, potassium or calcium salts, or lower alkyl esters of para-hydroxybenzoic acid, together with a long-chain polyester and polyacrylic acid (alkali metal salt) in water.

Another suitable composition comprises an aqueous suspension or solution of a polyfunctional water-soluble polymer, a polyfunctional cross-linking agent, and an antimycotic. Either composition, when applied to a cellulose casing and dried, yields a coating which is adherent, water-insoluble, partially water-swellable, and permeable to gas and water vapor.

The coating composition used in treating clear or fibrous cellulosic casing is prepared by forming a solution or emulsion containing the antimycotic and the polymeric coating ingredients which produce the water-insoluble partially water-swellable permeable coating. The coating composition contains 0.1 to 5.0 percent weight (preferably about 1%) of the antimycotic. In normal application of the coating there is applied about 200 mg of antimycotic per square centimeter of casing treated. The minimum effective concentration of antimycotic in the casing is 100 to 200 ppm while the upper limit is determined mainly by economic considerations.

The coating composition includes a polymeric, at least partly organic, fluid, film-forming coating material which solidifies after having been spread out in a thin layer to form a thin, coherent, adherent, water-insoluble partially water-swellable, water- and gas-permeable, water-resistant layer of a satisfactory rigidity, porosity, and toughness. The preferred polymeric material includes a long-chain polyester and polyacrylic acid emulsion. The long-chain polyester is preferably a polymeric ester of polybasic acids and polyols.

Other polymers may be suitably employed together with the long-chain polyester and/or polyacrylic acid which, if used alone to coat cellulose sausage casings, may tend to hinder water vapor transmission which is required in sausage casings. Such polymers include the copolymers of vinylidene chloride with a comonomer such as acrylamide, vinyl acetate or vinyl chloride, polyvinyl ethylene, polyvinyl acetate based copolymers, various acetoxylated polyoxyethylene latex systems, certain natural resins, acrylic resins such as polyacrylates and polymethacrylates, polyethylenes, or urea-formaldehyde resins.

An alternate coating composition may include instead a water-soluble polymer having at least two reactive functional groups (e.g., gelatin, gum arabic, gum tragacanth, or egg albumin) and a difunctional or polyfunctional cross-linking agent, (e.g., diisocyanates, dialdehydes, aldehydes, polyepoxides, etc.) together with the desired antimycotic. Cure is effected by heating and drying the coating on the cellulose tubing to cross-link and insolubilize the same. The polymer and cross-linking agent in the coating composition must not react with water or with each other at ambient temperatures.

Water-soluble, polyfunctional, polymeric organic materials which are used in aqueous solution for coating cellulosic casings or other cellulosic substrates include a variety of materials such as gelatin, egg albumin, natural gums, such as gum tragacanth or gum arabic, starch derivatives, soluble cellulose ethers such as methyl cellulose, hydroxyethyl cellulose, carboxymethyl cellulose, etc. Gelatin, however, is the preferred polymer because of its low cost, high viscosity and adaptability to printing.

The cross-linking agents which are applied to insolubilize the polyfunctional, polymeric coating materials include a variety of cross-linking agents having two or more reactive functional groups per molecule, such as isocyanate, isothiocyanate, epoxy, chloroepoxy, acyl halide, acyl amide, ester, ketene, imino, halogen, or acid anhydride groups (including compounds having mixtures of such functional groups). Typical compounds which may be used as cross-linking agents include the following:

> Diisocyanates — the diisocyanate of dimer acid, 4,4'-methylene bis-
> (cyclohexyl isocyanate)
> Diisothiocyanates — the diisothiocyanate of dimer acid, 4,4'-methylene bis(cyclohexyl isothiocyanate)
> Polyepoxides — butadiene diepoxide
> Chloroepoxides — the reaction product of epichlorohydrin and a
> polyamide
> Acyl halides — adipyl chloride, sebacyl chloride
> Acyl amides — adipyl amide, sebacyl amide
> Polyesters — dimethyl sebacate, dimethyl suberate
> Polymeric ketenes — dimeric acid diketene
> Polyhalides — 1,6-dichlorohexane
> Diimides — 1,6-diimidohexane
> Polyfunctional nitriles — 1,6-dicyanohexane, tetracyanoethylene
> Polyfunctional acid anhydrides — pyromellitic acid anhydride
> Dialdehydes — glyoxal, malonaldehyde, glutaraldehyde, dialdehyde
> starch
> Aldehydes — formaldehyde

In applying the cross-linking agents to the various coating materials, the less reactive cross-linking agents are applied in admixture with the polymeric, polyfunctional coating material and antimycotic in aqueous solution. After the cellulosic substrate is coated with the solution, the drying of the product is effective to cause the cross-linking agent to react with and cross-link and insolubilize the water-soluble, polyfunctional material and, in some cases, cross-links the coating to the substrate. Where more reactive cross-linking agents are used, the cellulosic substrate or casing may first be passed through an aqueous solution of the water-soluble, polyfunctional, polymeric coating material and antimycotic and then cross-linked in a separate treating step.

The coating composition can be applied to the outer surface of the casing in any desired manner. Thus, application to conventional regenerated cellulose casing, fibrous casing, or any other kind of casing can be made by coating the outer surface of the tubular (inflated) casing by brushing, spraying or dipping. If desired, flattened casing (reel stock) can be coated.

At the time the coating composition is applied in the form of an emulsion or solution which contains the antimycotic, to be converted to a hard film, it must undergo a chemical or physical reaction (curing), which converts it from a mobile liquid into a hard, gell-like structure. The curing of the coating is preferably through evaporation and heating for a short time to form a water-insoluble film and is preferably accomplished during the drying of coated gel casing. In the curing step, the antimycotic is fixed within the casing and the coating.

Example: Heavy, large diameter, clear cellulose casing was coated with a coating composition consisting of an emulsion of 1.0% Kymene 557, 1.0% gum tragacanth (or gum arabic), 10% glycerol and 0.9% sodium sorbate in aqueous solution. The emulsion was applied by dip coating several pieces of the regenerated cellulose casing prior to drying to form a thin layer of film on the casing. The coated casing was then cured during drying in a conventional casing dryer.

In further experiments, the casings coated as described above were presoaked in hot water and stuffed with bologna emulsions of various types. In each experiment, a nontreated clear casing was run as a control. The casings coated in accordance with this method exhibited no fungus growth while the untreated controls exhibited some fungus growth thereon. Furthermore, coating adhesion to all of the stuffed treated casings after hot and cold water processing was highly satisfactory, as determined by paper-rub and Scotch tape removal of imprinted inks from the stuffed casings.

COATED ACID ADDITIVE

It is known that the effectiveness of organic acid type food preservatives, namely, sorbic acid, potassium sorbate, calcium propionate and sodium propionate are greatly influenced by the pH. The reason is that the preservative activities of these food preservatives are mainly due to nondissociated molecules in foods. In other words, as the pH is on a more acidic side, the number of nondissociated molecules increases and the preservative activities are enhanced.

It is desired that the pH of food be lowered as much as possible within such a range as will not impose bad influences to the quality of the food. The pH values for the activities of organic acid type food preservatives are generally in the range of 4.5 to 5.5, though they vary depending on the class of the food preservative to some extent. In the case of a food preservative of a high dissociation constant, the pH must be lowered as much as possible. In the case of a food preservative of a low dissociation constant, its preservative activity is expected at a relatively high pH.

In some food products, however, the lowering of the pH of the starting materials during the manufacturing adversely influences the qualities of the products. For instance, in the manufacture of fish products such as fish sausage and meat products such as meat sausage, the lowering of the pH of the starting meat emulsion to increase the activity of an organic acid food preservative results in decrease of the water retainability of the meat, with the result that adverse effects appear with respect to the elasticity of the end product prepared from such meat emulsion. Generally, the water retainability of meat is lowest at a pH of approximately 5.5, and when the pH becomes higher than this value, the water retainability increases. Accordingly, a higher pH is preferable with respect to the water retainability of meat.

Further, in the case of bread manufactured by yeast fermentation the lowering of the pH is not desired because the propagation of yeast is hindered by the increase of the activity of a food preservative and adverse effects are imposed on the fermentation of yeast, resulting in the formation of bread products with poor rising properties.

R. Ueno, T. Miyazaki and S. Inamine; U.S. Patent 3,692,534; September 19, 1972; assigned to Ueno Pharmaceutical Co., Ltd., Japan give a process to retard the lowering of pH during the manufacturing step when certain food preservatives are used. This is accomplished by adding to the starting materials of a food product granules comprising a pH lowering agent coated with a hardened oil in a manner such that the pH lowering agent will not be eluted during the manufacturing step conducted at room temperature but eluted during the heating step, together with a food preservative.

A pH lowering agent selected from the group consisting of fumaric acid, monosodium fumarate, tartaric acid, malic acid and citric anhydride is coated with a hardened oil having a melting point of 50° to 80°C to form granules having a particle size of less than 500μ in which the content of the powder is 25 to 40% by weight, is added to the starting materials of the food products together with an organic acid type food preservative. Any of known organic acid type food preservatives such as sorbic acid, potassium sorbate, calcium propionate and sodium propionate may be used.

The coating may be performed by various methods. For instance, a method comprising charging a molten hardened oil into a strong kneader, then adding a pH lowering agent such as fumaric acid, gradually cooling the system to room temperature while mixing to form granules of a diameter of 0.5 to 3 mm, and pulverizing them to less than 1 mm. Another method is to add the abovementioned substance to a coating pan, spraying it with a solution of a hardened oil in a volatile solvent and then removing the solvent by blowing in hot air. A suitable coating method is a spray granulation method which will be described below. With this method it is possible to easily obtain granules having a particle size of less than 500μ in which the content of powder uncoated with a hardened oil is extremely low.

A hardened oil having a melting point of 50° to 80°C is melted by heating, and a pH substance (namely, the abovementioned pH lowering agent) pulverized to less than 50μ is dispersed in the molten hardened oil. At this time the dispersion state is rendered better by addition of lecithin in an amount of about 0.1% based on the hardened oil. The so formed dispersion maintained at a temperature higher than the melting point of the hardened oil is sprayed into air of less than 40°C by means of a rotary disc sprayer, and thus cooled and solidified. The particle size of the granules can be easily and optionally controlled by adjusting the rotation rate of the rotary disc.

The content of the active substance in the coated granules is up to 40% (namely, the upper limit of the ratio of the active substance to the hardened oil is 40/60). The reason is that when the content of the active substance is too high, the dispersion becomes highly viscous and the spraying becomes difficult. A preferable content of the active agent is about 35%.

The coating state is evaluated based on the determination of the amount of the uncoated substance contained in the resulting granule. More specifically, 1 gram of the granules is added to 100 grams of water of room temperature and the mixture is agitated for 10 minutes by means of a magnetic stirrer, followed by filtration. The amount of the active substance in the filtrate is determined and expressed in terms of percent based on the active substance in the sample granules. When the active substance in the filtrate is an acid, the titration is conducted with the use of 0.1N NaOH, and when the active substance in the filtrate is a salt, it is converted to the corresponding free acid by an ion exchange resin and then the titration is conducted with the use of 0.1N NaOH. As a result of such measurements, it was confirmed that amount of uncoated active substance in the granules is less than about 5% when the active substance is fumaric acid, about 10% in the case of sorbic acid, about 20% in the case of the propionate, and 15 to 20% in the case of other pH lowering agents.

Example 1: Hardened beef tallow having a melting point of 59°C is melted by heating, and 0.1% of the lecithin is added thereto to obtain 6.5 kg of the mixture of hardened beef tallow and lecithin. While the mixture is maintained at 70°C, 3.5 kg of fumaric acid comminuted to an average particle diameter of less than 20μ is added thereto. Then, the mixture is sufficiently stirred by a homogenizer to disperse fumaric acid uniformly. While the dispersion is maintained at 70°C, it is sprayed into the air of 35°C by means of a rotary disc sprayer, and then cooled and solidified. As a result, granules having a particle size of 300 to 50μ are obtained.

Example 2: A starting material meat for preparing Vienna sausage comprising 1,400 grams (70% by weight) of pork and 600 grams (30% by weight) of beef is cured for 24 hours, then passed through a chopper and kneaded for a while by using a silent cutter. To this material mixture are then added such auxiliary materials as condiments and spices together with a preservative (and a pH lowering agent) indicated in the following table. Then, the mixture is kneaded for 5 minutes, and is stuffed into a sheep intestine casing.

The casing is used as it is washed with water and is not subjected to a special treatment. Then, the material mixture stuffed into the sheep intestine is dried at 40°C for 1 hour, smoked at 50°C for 30 minutes, then boiled at 70°C for 20 minutes and cooled. After the preparation of sausage is completed, it is stored in a refrigerator for about 20 hours, and then allowed to stand in a constant temperature-constant humidity apparatus at a temperature of 25°C and a relative humidity of 85%. The results of the tests with respect to the pH, product quality and preservative effect are shown below.

Preservative	Amount of preservative added based on material meat (wt. percent)	pH [1] Before heating	After heating	Quality of product [2]	Preservative test
Coated fumaric acid granule [3]	0.3	6.30	5.80	Soft and elastic jelly (good)	Spoiled after 96 hours.
Potassium sorbate	0.2				
Potassium sorbate	0.2	5.85	5.78	Hard and easily broken	Do.
Fumaric acid	0.09				
Potassium sorbate	0.2	6.40	6.45	Soft and elastic jelly (good)	Spoiled after 65 hours.
Not added		6.35	6.40	...do...	Spoiled after 48 hours.

[1] Fumaric acid content is 30%.
[2] The specimen is suspended in distilled water of a volume 5 times as large as that of the specimen and the pH is measured with a glass electrode pH meter.
[3] The specimen was actually eaten and a sensory evaluation is made.

As can be seen from the results shown in the above table, when the coated fumaric acid granules are used together with potassium sorbate in accordance with this process, the preservative effect is remarkably enhanced with no adverse effect on the quality of the product.

SURFACE TREATMENT WITH MONO- AND DIGLYCERIDES

Protection against chemical-microbiological changes that eventually will cause spoilage of fresh meat is described by *J. Scheide; U.S. Patent 3,667,970; June 6, 1972; assigned to Farbenfabriken Bayer AG, Germany.*

It is known that meat of warm-blooded animals remains palatable longer when the meat is stored and distributed at low temperatures. As an optimum temperature range, temperatures of from 0° to 6°C are maintained. Such meat is preservable for only a few days from the time of commercial sectioning and packaging under correct maintenance of the prescribed cooling temperatures. After that it is no longer saleable as fresh meat and must be otherwise utilized. Responsible for this low degree of preservability are the post-mortem changes setting in after slaughter of the animal. These post-mortem changes constitute biochemically irreversible alterations in the colloidal-chemical swelling state of the meat protein which results in a continuously increasing dehydration of the meat protein accompanied by changes in the permeability of the membranes. This dehydration manifests itself in loss, by seeping out, of the meat juice, in which are dissolved valuable meat proteins, amino acids, carbohydrates, and flavor components, such as the salts of ribonucleic acids.

As a result of the use of cutting and beating utensils in the slaughtering and cutting process, the loss of meat juice is accelerated by destruction of a relatively high number of the tubular muscle fibers whereby a rapid outflow of meat juice occurs at the points of injury. The loss of juice of the cut meat causes substantial technological difficulties in the distribution and storage of the packed cuts and at the same time represents a significant economic loss of protein.

In addition, meat juice is, because of its high water content, an especially ideal nutrient medium for microorganisms. Beginning with the rapid spoilage of the meat juice, the microbiological spoilage of the entire package proceeds quickly. To alleviate these difficulties, there have been developed packaging materials which are made of highly absorptive materials, such as, for example, treated cellulose, which is, however, not palatable itself. The juice seeping from the cut meat is absorbed to a substantial degree by the packing material. This form of packaging, however, does not extend the limited time of preservability of a few days, at temperatures from $0°$ to $6°C$, but only makes possible the distribution of commercially packed fresh meat.

Sufficient free meat juice remains to promote rapid microbial spoilage. Simultaneously, there is a substantial loss in meat juice absorbed by the packing material. It is also known to subject fresh meat immediately after cutting to a low-temperature freezing process at $-40°C$ and to store this deep frozen meat at a temperature of from $-16°$ to $-20°C$, especially at $-18°C$, to extend its preservability.

According to this process, a storageability of up to 12 months can be attained. However, even in this process considerable technical difficulties are caused by the above described post-mortem changes as a result of dehydration of the proteins and the loss of meat juice. The meat cuts are stacked prior to deep-freezing on conventional packaging materials and are then wrapped with a transparent packaging film. As a result of the high juice content of the meat surface and the high degree of water vapor saturation on the inside of the film wrapper containing the meat, a thick layer of ice crystals is formed, upon chilling, both on the meat surface and the inner film surface so that the meat cut itself is no longer visible after deep-freezing. To avoid these difficulties, several technical methods are used to make at least part of the meat cuts visible, but this involves a substantially higher cooling requirement.

It has been found that the fresh meat of warm-blooded animals can be given favorable properties by treating its surface with a coherent film of a mixture of:

(a) from 2 to 30% by weight of one or more monoglycerides; and
(b) from 98 to 70% by weight of one or more diglycerides and/or
triglycerides

of paraffinic and/or olefinic carboxylic acids containing from 8 to 22 carbon atoms and free of acetylenic unsaturation.

It is best to use the glycerides of naturally occurring unbranched carboxylic acids suitable for dietetic purposes. Examples of carboxylic acid glycerides of this kind include the uniform and/or mixed glycerin esters of caprylic acid, caproic acid, myristic acid, palmitic acid, stearic acid, arachic acid, oleic acid, linolic acid, linolenic acid and erucic acid. The following compounds are mentioned in particular: tricaproin, tricaprinin, trilaurin, trimyristin, tripalmitin, tristearin, triarachin, triolein, trilinolin, trilinolenin and trierucanin.

Triglyceride or mixtures of different triglycerides of the kind occurring in the form of liquid and solid vegetable and animal fats are preferred, in particular mixtures of triglycerides which are liquid at normal temperature or triglyceride mixtures with the monoglycerides and diglycerides of the carboxylic acids already mentioned above.

These mixtures show optimum handling properties when their ingredients have been mixed with one another in such ratios that at room temperature the completed mixtures are gel-like and pasty in consistency and show thixotropic properties. These particular properties can be attained by, for example, stirring the mixture to a homogeneous mixture, heating same to 65°C and immediately cooling it to 20°C. This has the advantage that, following their preparation, the mixtures change from liquids into gel-like pastes on completion of mechanical processing.

During their application, however, the mixtures revert to the easily handled liquid form due to the mechanical stresses to which they are subjected. After they have been used, the mixtures are converted back into gel-like pastes, thereby forming transparent, firmly adhering and coherent films over the surface of the meat.

It is necessary to balance the mixture in such a way as to guarantee adequate bond strength and film-forming capacity. It is also of particular advantage to produce this layer in such a limited thickness that it appears transparent while the piece of meat remains intact as regards its original appearance without any changes in the impervious seal over the pores of the meat. This result can be obtained, for example, by brush coating or by applying the glyceride mixture to the surface from a nozzle with the assistance of a compressed gas.

The usual compressed gases, such as compressed air, nitrogen, carbon dioxide, nitrous oxide, propane, butane and fluorinated hydrocarbons may be used as the propellent gases. Favorable results as regards the transparency of the film applied, the pore seal and the consumption of material are obtained for example by operating at pressures of from 5 to 6 atm and simultaneously using nozzles with a diameter of from 0.8 to 1.0 mm. When using such a technique, it is desirable to carry out the treatment with the mixtures at room temperature (20°C) or lower temperatures.

The mixtures may be applied either as such or following the addition of further digestible additives, such as for example, proteins, carbohydrates and natural aromatizing agents and flavorings. Suitable proteins include, for example, proteolytically active enzymes, such as papain, ficin and bromelin, while examples of suitable carbohydrates include monosaccharides, such as glucose, galactose and ribose.

The addition of certain proteolytic ferments promotes partial proteinolysis of the connective tissues of the meat, with the result that the food prepared for consumption is much more tender. The use of carbohydrates, pentoses in particular, promotes a rapid uniform browning of the surface of the treated meat, with the result that rapid crust formation and hence a further increase in juiciness are obtained, in addition to which the overall appeal is increased by the particularly appetizing appearance. Finally, it is possible by adding the natural aromatizing agents and flavorings to make even the inside of the meat particularly spicy, so that the appeal of the prepared food is still further improved.

Example 1: Two batches were made up, each of 100 grams, of the following compositions: (a) 5 grams of pure glycerin monostearate and 95 grams of liquid triglyceride mixture, and (b) 25 grams of pure glycerin monostearate and 75 grams of liquid triglyceride mixture.

In both cases, a mixture of olive oil and refined rape oil in a ratio of 1:2 was used as the liquid triglyceride mixture. The carboxylic acid component of the olive oil consisted essentially of 76% of oleic acid, 7% of palmitic acid, 2% of stearic acid and 5% of linolic and linolenic acids, while approximately 60% of the carboxylic acid component of the refined rape oil consisted of trierucain.

In order to obtain the requisite thixotropic properties, the mixture was initially heated with stirring to a temperature approximately 5°C higher than the melting point of the glycerin monostearate, that is, to 65°C. The mixture was left standing at this temperature until it was in the form of a homogeneous liquid. The liquid was then quickly cooled to room temperature (20°C) and the resulting mass was poured into suitable containers. After a while, the mass solidified into a gel-like paste which showed little or no flow properties. The mass could be converted into a free-flowing form at room temperature by intensive shaking or other mechanical working, and could be poured in such form into the supply container of a nozzle spray or into an aerosol can.

The following were treated with the mixture according to the process: 5 pieces of rump steak with a fresh weight of 850 grams, 5 pieces of pork cutlet with a fresh weight of 650 grams, and 5 pieces of veal steak with a fresh weight of 605 grams.

Immediately after it had been cut into portions, each individual piece of meat was sprayed all over with the mixture (a) or (b), under the following conditions: Two-component nozzle with a nozzle diameter of 1.4 mm — spraying pressure, 3.0 atm and propellent gas, nitrogen. The mixture was consumed in a quantity of 40 grams, based on the total amount of meat. The individual pieces of meat were coated with a transparent but, as a result of gloss, still visible film.

In a parallel test, 5 pieces of rump steak with a fresh weight of 790 grams, 5 pieces of pork cutlet with a fresh weight of 670 grams and 5 pieces of veal steak with a fresh weight of 600 grams were left untreated. After cooking salt had been added to the treated and untreated pieces of meat in a quantity of 1% by weight, 5 pieces each of the rump steak, pork cutlet and veal steak treated were successively fried until done on both sides in the absence of further additives in a kitchen frying pan which was kept at a constant frying temperature of 160°C by means of an electric cooking plate and a thermocouple. The frying time varied on average between 3.5 and 4 minutes, while the loss through frying varied between 12 and 15%. Pieces of meat with a juicy taste were obtained in every case.

Following the addition of a total of 90 grams of margarine, the samples which had merely been salted for the parallel test were fried until done on both sides in the frying pan under the same conditions. The frying varied on average between 6.5 and 7 minutes, while the loss through frying varied between 26 and 31%. In every case, pieces of meat which had a much drier taste as shown by a flavor test were obtained. When 1 and 2 gram amounts of the pentose xylose were respectively added to mixtures (a) and (b), the pieces of meat were much more uniformly browned and were given a more appetizing appearance.

Example 2: A mixture was prepared as described in Example 1 to contain the following constituents in a total of 1 kg: 885 grams of a liquid triglyceride mixture, 80 grams of a mixture of glycerin monostearate and glycerin distearate in a ratio of 1:1, and 35 grams of a mixture of natural aromatizing agents and flavorings.

Peanut oil whose carboxylic acid component consisted essentially of 60% of oleic acid, 20% of linolic acid, 4.5% of stearic acid, 4% of arachic acid and 4% of palmitic acid was used as the liquid triglyceride mixture. The mixture of glycerin monostearate and glycerin distearate was obtained by transesterifying vegetable oil with stearic acid and also contained 5% of monoglycerides and diglycerides of other carboxylic acids occurring in vegetable oils.

The mixture of natural aromatizing agents and flavorings contained extracts and distillates of pepper, nutmeg, mace, coriander, bay leaves and onion, monosodium glutamate and a mixture of the disodium salts of inosine-5'-monophosphoric acid and guanosine-5'-monophosphoric acid in a ratio of 1:1. Freshly cut pieces of meat were treated with this mixture as described in Example 1: 5 pieces of rump steak with a fresh weight of 810 grams and 7 pieces of veal cutlet with a fresh weight of 1,450 grams.

The working conditions were as follows: Two component nozzle with a nozzle diameter of 1.2 mm — spraying pressure, 4.0 atm and propellent gas, dinitrogen monoxide. A total of 25 grams of the mixture was consumed. The individual pieces of meat were covered with a transparent almost invisible film and were stored in a cooling chest at 4°C. At the end of the 7th storage day the samples were removed and examined. No weight loss was found. When the pieces of meat treated in accordance with the process were fried, they had a spicy, juicy and meaty taste. The loss through frying amounted on average to 18% for a frying time of 5 minutes.

Example 3: A mixture was prepared from the constituents listed below in the stated amounts to make up 1 kg of mixture: 860 grams of a liquid triglyceride mixture, 65 grams of a mixture of glycerin mono- and distearate in a ratio of 1:1, 45 grams of a mixture of natural aromatizing agents and flavorings, and 30 grams of carbohydrates.

Peanut oil whose carboxylic acid component consisted essentially of 60% of oleic acid, 20% of linoleic acid, 4.5% of stearic acid, 4% of arachic acid and 4% of palmitic acid was used as the liquid triglyceride mixture. The mixture of glycerin mono- and distearate was obtained by transesterifying vegetable oil with stearic acid, and also contained 5% of mono- and diglycerides of other carboxylic acids occurring in vegetable oils.

Extracts and distillates of pepper, celery, lovage, mace and coriander together with mono-sodium glutamate and the disodium salts of inosine-5'-monophosphoric acid and guanosine-5'-monophosphoric acid were used as the mixture of natural aromatizing agents and flavorings. A mixture of the hexose glucose with the pentose ribose in a ratio of 1:2 was used as the carbohydrate component.

In order to obtain the requisite thixotropic properties, the mixture is initially heated with stirring in the required ratio to a temperature about 5°C higher than the melting point of the mixture of glycerin mono- and distearates, i.e., to 65°C. The mixture is left standing at this temperature until it is in the form of a homogeneous liquid. This is followed by rapid cooling to room temperature (20°C), after which the carbohydrates are suspended in the mass which is poured into suitable containers.

After a while, the mass solidifies into a gel-like paste with little or no flow properties. The mass can be converted into a free-flowing form by intense shaking or other mechanical working, and can be poured in this form into the supply container of a nozzle spray or into an aerosol can.

Three freshly prepared cod fillets of a total weight of 552 grams were sprayed all over with mixture prepared by means of a 2-component nozzle, for which purpose a quantity of 5.1 grams was required. As a result, the cod fillets were surrounded all over by a continuous film of the mixture according to the process. Following the addition of a little (approximately 1%) common salt, the fish fillets were fried until cooked on both sides without further additives in a kitchen frying pan which was kept at a constant frying temperature of 160°C by means of an electric cooking plate and a thermocouple.

The frying time was 4 minutes, and the loss in weight, through frying amounted to 14.7%. The cod fillets were intensively browned all over and had a very juicy, spicy taste. In a parallel test, three cod fillets were left untreated. Their total weight was also 552 grams. Following the addition to the frying pan of a little cooking salt and 40 grams of margarine, the fish fillets were fried under the same conditions until cooked. The frying time was 6 minutes and the loss in weight through frying amounted to 29.3%. The cod fillets were only slightly browned in patches and had a distinctly drier and more insipid taste than the treated fish fillets.

ANTIBIOTICS

Chlortetracycline-Nystatin Combination

In view of the success of antibiotics in controlling the growth of many organisms, considerable attention has been directed toward the use of antibiotics for controlling meat spoilage. Various investigations have shown that when antibiotics, particularly the broad-spectrum antibiotics such as chlortetracycline, oxytetracycline and tetracycline, were applied to the surfaces of fresh meat, bacterial growth was inhibited. Further observations revealed that the shelf life of the meat so-treated was extended for several days.

Unfortunately, the shelf life of prepackaged cut meat is still not as long as desired and spoilage comes sooner than is desired. Also, there is a rapid deterioration of the bright red color of most cut meat which makes the meat undesirable to prospective consumers, even though the meat might still be sufficiently free from pathogenic microorganisms as to be safely edible. Sometimes the texture and odor of fresh cut meat deteriorates even though the meat may not be sufficiently spoiled to be inedible.

A.R. Kohler; W.H. Miller and H.M. Windlan; U.S. Patent 3,050,401; August 21, 1962; assigned to American Cyanamid Company use an antibiotic for preservation of meat in combination with nystatin or myprozine. The antibiotics that may be used should have as wide a bacterial spectrum as possible and be able to inhibit the growth of the bacteria that are generally encountered in spoiled meat. The antibiotic should be relatively stable under conditions found in meat, nontoxic and not adversely affect the color or taste. The antibiotic should also be water-soluble to the extent that the quantities which are necessary for treatment can be applied in the form of an aqueous solution if desired.

Preferred antibiotics for use are those of the tetracycline series, including chlortetracycline, bromotetracycline, oxytetracycline and tetracycline itself. Of these, chlortetracycline appears to have the most desirable properties. These antibiotics may be used in any of their water-soluble forms, usually as a mineral acid salt such as the hydrochloride. However, various other metallic salts such as the alkaline earth metal salts and the alkali metal salts may be used. Water-soluble complexes such as the borate or phosphate may also be used if desired.

An aqueous solution or substantially dry powder containing the antibiotic and either nystatin or myprozine is first prepared. Although no adjustment of the pH of the solution is necessary, it is possible, if desired, to stabilize the solution with suitable buffering agents. The solution or dry powder is then applied to the exposed surface of the cut meat at $40°$ to $50°F$. Although application at ordinary room temperature is possible, it is best that the treatment of the cut meat be carried out between $40°$ and $50°F$ in order to minimize the possibility of an onset of bacterial development which may occur at ordinary room temperature. Application of the solution may be done by dipping the exposed surfaces of the cut meat in the solution or, if desired, the solution can be sprayed on the exposed surfaces of the meat utilizing any mechanical sprayers or hand atomizers.

Application may also be made to the exposed surfaces of the meat either mechanically or by hand using a substantially dry powder mix containing the antibiotic and either nystatin or myprozine. However, in using the dry powder method of application, it is necessary that a bulk carrier be added in order to effectuate an even distribution of the dry powder over all the exposed surfaces of the meat.

The carrier may be common salt, flour salt, which is a finely divided form of common salt, monosodium glutamate, starch and glucose. The preferred carrier is common salt, although any edible powdered material which is inert to the meat and the antibiotic may be used. Following the treatment of the meat as described, the meat is then tray packed or placed on backing board, overwrapped in a suitable heat-sealable flexible film as, for example, polyethylene or coated cellophane and stored under refrigeration at $35°$ to $40°F$.

Nystatin is produced by cultivating culture A-5283 under submerged aerobic conditions in an aqueous nutrient medium containing an assimilable source of carbohydrate, nitrogen and inorganic salts.

Example: Fresh boneless chuck was cut into approximately one-inch cubes. The meat samples were divided into four groups having about 5 samples per group. Group 1 represents untreated meat and corresponds to the control. Group 2 represents meat treated with 3 parts per million chlortetracycline hydrochloride in substantially dry powder form homogeneously dispersed in flour salt. Group 3 represents meat treated with a substantially dry powder mixture containing 3 ppm chlortetracycline hydrochloride and 10 ppm nystatin homogeneously dispersed in flour salt. Group 4 represents meat treated with a substantially dry powder mixture containing 3 ppm chlortetracycline hydrochloride and 10 ppm myprozine homogeneously dispersed in flour salt. The powder used to dust the cut meat contained approximately 0.15% by weight of chlortetracycline and 0.5% by weight nystatin or 0.5% by weight myprozine.

The treatment was carried out by applying the dry powder mix containing the desired concentration to the exposed surfaces of the meat at the rate of 2 g/kg of meat using a conventional hand shaker. The portions of beef chuck were then double ground, prepared as patties, tray packed, overwrapped with water resistant coated cellophane and stored under refrigeration at about $36°F$ for the duration of the experiment.

Samples were removed periodically from each group and microorganism counts were made. Observations were made of the color and general appearance of the samples. Samples for bacterial counts were prepared by homogenizing the sample of meat in a sterile blender containing sterile distilled water in an amount equal to three times the weight of the meat for 3 to 5 minutes.

Based on the microorganism count and appearance, it was determined when the meat would be considered spoiled, that is to say unacceptable for consumer use and discolored when the change in color was sufficient to render the meat unacceptable for human use. The following table shows the number of days before spoilage and discoloration took place. In the case of spoilage, the test was not continued beyond the twenty-second day at which point the microorganism count was still within acceptable limits. This is indicated by the plus mark after the day number.

Group	Treatment Before Storage	Spoilage, days	Discoloration, days
1	Control (untreated)	7	5
2	Chlortetracycline hydrochloride 3 ppm	11	13
3	Chlortetracycline hydrochloride 3 ppm plus 10 ppm nystatin	22+	20
4	Chlortetracycline hydrochloride 3 ppm plus 10 ppm myprozine	22+	22

The results of the table show that when beef chuck was treated with combinations of chlortetracycline hydrochloride-nystatin and chlortetracycline hydrochloride myprozine, an increase in shelf life of at least 15 and 11 days was respectively obtained over untreated meat and meat treated with chlorotetracycline hydrochloride alone. It is also shown that good color appearance of the meat was maintained 15 to 17 days longer than the control and 7 to 9 days longer than the meat treated with chlortetracycline hydrochloride alone.

The process of *J.F. Pagano, A.E. Temple and H. Yacowitz; U.S. Patent 2,944,907; July 12, 1960; assigned to Olin Mathieson Chemical Corporation* also makes use of chlortetracycline-nystatin combinations to retard spoilage in meat and fish. A preferred composition contains 10 ppm chlortetracycline and 2.5 to 20 ppm nystatin with a potency of 3,500 units/mg.

Example 1: Laboratory-raised chickens, ranging in age from 6 to 10 weeks, were killed by bleeding and the legs were then removed, scalded and picked. The drumstick portion was then used in the dipping tests, each part being allowed to stand in the dip solutions for 1 to 2 hours at $5°C$. The drumsticks were then packaged and stored in a cold room at $5°C$. Samples were removed for initial bacterial, yeast and mold counts. Additional samples were removed periodically for odor tests and counts. Odor tests were conducted by members of a panel who had demonstrated ability to differentiate spoilage odors. Counts were made by dropping the drumstick in a flask containing sterile water and shaking 50 times. The water solution was then used for standard plate counts. Bacterial counts were done using Difco Yeast Beef Agar plus 100 micrograms of nystatin suspension per milliliter. Yeast and mold counts were done using Wort Agar containing 50 micrograms of chlortetracycline per milliliter. All plates were incubated at 37°C.

Bacteria (B), Yeast (Y) and Mold (M) Counts on Chicken Drumsticks Dipped in Chlortetracycline Solution

Dip Solution	Initial Count Before Dipping		Storage Time at 5° C.[1]					
			10 Days		18 Days		27 Days	
	B	Y and M	B	Y and M	B	Y and M	B	Y and M
	Count in millions per drumstick [2]							
Tap water	0.055	-------	3.9	4.3	156 (slight odor)	49	6,830 (putrid)	215
Tap water plus 23 p.p.m. chlortetracycline hydrochloride.	-------	-------	.006	1.7	.017 (yeasty)	233	0.13 (yeasty)	1,300

[1] Packaged in cellophane, three drumsticks per package.
[2] Three drumsticks pooled for each count.

Bacteria (B), Yeast (Y) and Mold (M) Counts on Chicken Drumsticks Dipped in Chlortetracycline and Chlortetracycline-Nystatin Solutions

	Count After 15 Days' Storage at 5° C.[1]		
	Bacteria	Yeast and Molds	Average Odor Score [2]
	Count in millions per drumstick [3]		
Tap water	23.0	1.3	1.2.
Tap water plus 5 p.p.m. chlortetracycline hydrochloride.	.05	4.3	sl. yeast.
Tap water plus 10 p.p.m. chlortetracycline hydrochloride.	.03	2.7	sl. musty.
Tap water plus 5 p.p.m. chlortetracycline hydrochloride plus 5 p.p.m. nystatin.	.3	.05	0.2.
Tap water plus 10 p.p.m. chlortetracycline hydrochloride plus 5 p.p.m. nystatin.	.035	.01	0.3.

[1] Packaged in Pliofilm, one drumstick per package.
[2] Odor score: 0=no odor; 0.5=very slight odor; 1.0=slight odor; 2=moderate to strong odor; 3=putrid. Three drumsticks tested per treatment.
[3] One drumstick per count.

As shown in the tables, it is to be noted that while chlortetracycline was effective in retarding spoilage as evidenced by the reduced bacterial count and odors, a pronounced increase in yeast and mold counts was observed, resulting in yeasty and musty odors. But, when chicken parts were dipped in an aqueous vehicle containing both chlortetracycline and nystatin, a marked decrease in yeast and mold counts and odors as well as a decrease in bacterial count was observed.

It has been further found that nystatin (as well as amphotericin) does not penetrate the outer surface of the meats treated, that is, is present only on the surface of the meats. Such inability to penetrate facilitates the inactivation of the antifungal agent during the cooking process.

Example 2: Experiments were conducted in which various nystatin solutions were placed in stainless steel cups, the bottoms of which were sealed with a piece of chicken skin or a

0.5 mm section of chicken breast muscle. The skin or muscle sections were fastened to the metal cups with thread. These cups, along with suitable controls, were placed on agar plates seeded with the test organism, *Saccharomyces cerevisiae.* Little or no diffusion of nystatin through chicken skin or breast muscle occurred. This indicated that nystatin was present only on the surface of the chicken parts. Since yeasts and molds normally grow on exposed surfaces, the localization of nystatin on the skin and muscle surfaces affords antifungal protection at the site most likely to be contaminated.

Example 3: It was found that the residues of nystatin remaining on the treated parts could be removed from the surface of the meat by repeated water washings using small volumes of water, and that the nystatin is inactivated during the cooking process; the nystatin being rapidly inactivated in boiling water.

Bacteria (B), Yeast (Y) and Mold (M) Counts and Odor Tests on Chicken Wings Dipped in Chlortetracycline and Chlortetracycline-Nystatin Solutions

Dip Solution	Initial Count Before Dipping		Storage Time at 5° C.[1]					
			7 Days		Average Odor Scores[2]	13 Days		Average Odor Score
	B	Y and M	B	Y and M		B	Y and M	
			Count in millions per wing[3]					
Tap water	7.3	0.014	1,419	0.158	3	2,300	.225	3.
Tap water plus 10 p.p.m. chlortetracycline hydrochloride.			8.8	32.0	sl. yeasty.	617	104.0	yeasty.
Tap water plus 10 p.p.m. chlortetracycline hydrochloride plus 1.25 p.p.m. nystatin.			36.0	7.8	.25	1,110	99.0	yeasty.
Tap water plus 10 p.p.m. chlortetracycline hydrochloride plus 2.5 p.p.m. nystatin.			8.3	1.7	0	2,512	84.5	yeasty.
Tap water plus 10 p.p.m. chlortetracycline hydrochloride plus 5.0 p.p.m. nystatin.			18.5	.465	0	967	52.0	0.
Tap water plus 10 p.p.m. chlortetracycline hydrochloride plus 10 p.p.m. nystatin.			13.0	.068	0	220	4.7	.25.

[1] Packaged in Pliofilm, two wings per package.
[2] See footnote of above table for scoring system. Two packages tested per treatment.
[3] Two wings per package, data represent average counts on two packages.
NOTE.—These tests were conducted using fresh chicken wings purchased from a commercial processing plant in the same manner as the tests in which the chicken legs were used. The chicken wings arrived packed in ice and had a relatively high initial count which resulted in rapid spoilage.

Residual Nystatin on Whole Uncooked Chicken

Portion of Chicken [1]	Weight of Portion in Grams	Nystatin Concentration	
		Micrograms per Gram	Micrograms per Portion
Skin		4.0	
Neck	121	0.5	60.5
Wing	124	0.5	62.0
Thigh	132	0.2	26.4
Drumstick	127	0.3	38.1

[1] Held for 4.5 hours in a nystatin solution containing 10.3 gamma (34 units) per ml. at 5° C.

Inactivation of Nystatin on Chicken Skin in Boiling Water

Initial Concentration of Nystatin, Micrograms per ml.	Concentration After 20 Minutes Boiling, Micrograms per ml.	Concentration After 40 Minutes Boiling, Micrograms per ml.
12.5	0.9	<0.6

Oxytetracycline Followed by Radiation

It is known that foods, such as meats and fish, can be helped to remain in a fresh condition for reasonably long periods of time by treatment with antibiotics. Such treatment has been effected by dipping the meat in brine, water or other conventional processing solutions, which contain the antibiotic dissolved or suspended therein.

Alternatively, the food may be sprayed with such a solution, or sprinkled with a dry antibiotic-containing powder, or packed in ice containing the antibiotic. Still another method of contact in the case of meats has been post-mortem infusion of the carcass after slaughter, pumping a solution of the antibiotic through the vascular system under pressure.

Despite their unquestioned value in food preservation, however, antibiotics per se are not a cure-all. They are most effective with foods which are of high bacteriological quality, and their preservative ability drops appreciably when they are used with meats and fish of poor quality, i.e., having initially high bacterial populations. The function of most antibiotics used in the food field, e.g., the generally preferred tetracyclines, is to retard the growth of spoilage organisms but not to kill them.

Among the surface contaminants that are able to grow even under the refrigerating conditions usually used for storing fresh meats and fish are a group of bacteria known as the pseudomonads. These, of which *Pseudomonas geniculata* is the most common species encountered on meats, are characterized by the ability to grow quite rapidly on the surface of meats even at temperatures slightly below 32°F. Extensive growth results in the usual browning, sliming, and musty odor so familiar to all. Any method which would prevent the contamination of meat surfaces with these bacteria, retard their growth, or kill them, would most certainly result in an increased holding time for fresh meats.

Work has been conducted in exploring the possibility of using ionizing radiations as food preservatives. These ionizing radiations are definitely lethal when applied at sufficient dosages, and pseudomonads are quite radiation sensitive. Unfortunately, however, sterilizing doses damage most foods with respect to organoleptic properties. Relatively low doses of ionizing radiations, in the order of 100,000 rep, result in a significant increase in the shelf life without seriously damaging the organoleptic qualities of the meat, but at such low radiation dose levels there are survivors of many bacterial species which still soon effect spoilage.

Two treatment steps are used by *R.C. Ottke and C.F. Niven, Jr.; U.S. Patent 3,057,735; October 9, 1962; assigned to Chas. Pfizer & Co., Inc.*, first the antibiotic treatment followed by radiation. When irradiation is used at pasteurization (i.e., 6.5×10^4 and preferably 10^5 rep) levels the effectiveness of the treatment is limited by the initial quality of the meat. With meat of better than median quality the shelf life limit obtained is about 15 days when the meat is stored at 2°C. If the meat is of bacteriologically poor quality, 10 days represents the shelf life limit.

Antibiotics alone at the level of 1 to 20 ppm, which seems to be about the preferred range, are even more limited by the level of initial contamination. In heavily contaminated meat they add little to the 5 to 10 day shelf life of untreated high quality meat. Combined irradiation and antibiosis, on the other hand, achieve much more than the expected effect. The process can procure a shelf life of two weeks even for very poor quality meats.

The tetracycline antibiotics (oxytetracycline, tetracycline itself and chlortetracycline) are preferred, because they are effective against many gram-positive and gram-negative bacteria, both aerobic and anaerobic, such as Salmonella, Micrococci and Clostridia. They are fairly stable at low and room temperature but decompose with relative ease at temperatures used in cooking. They may be used in any of their known, antibacterially active forms, such as the bases or acids per se or salts. For instance, the tetracyclines may be used as the free, amphoteric compounds, as the hydrochlorides, sulfates, phosphates, and other acid salts, or as the sodium, potassium and other metal salts, as well as in the form of various metallic complexes. If desired, other antimicrobial agents may be used in conjunction with the tetracycline antibiotics, but such other agents are not considered essential.

Gamma-irradiation is preferred for treatment of relatively large pieces, such as entire carcasses, since it has a high degree of penetrating power. On the other hand, beta-irradiation is extremely rapid, rendering it especially advantageous for surface treatment and for treatment of relatively thin cuts, such as steaks, and fillets. Suitable sources of gamma-radiation are radioactive isotopes, such as cobalt 60, spent fuel rods, fission wastes, and various gases employed in nuclear reactors. However, care should be taken to insure that the radiation, whatever the source, is completely free of neutrons. Suitable sources of beta-radiation include the linear accelerator, Van de Graaff generator, resonant transformer, capacitron, and the like.

The level of irradiation will vary somewhat, depending upon the particular foodstuff and its condition, among other factors. In most instances at least 65,000 rep should be employed, even with antibiotic treatment, and as much as 500,000 rep is not injurious with certain foodstuffs such as pork. However, this level is not usually desirable in the case of beef, which may undergo detectable color, odor or taste changes above 100,000 rep.

For best results, the foodstuff is subjected to treatment with a mixture of the antibiotic and a sorbic acid compound, i.e., sorbic acid, its esters and its salts, and then to irradiation. This combination provides a greatly enhanced effect, prolonging the shelf life of fresh meats, for instance, by at least another week. Generally, the sorbic acid is used in the proportions of 500 to 2,000 ppm; while 1,000 to 1,500 ppm, when 1 to 20 ppm of antibiotic is used, is a preferred range.

Example 1: Two hogs were injected intraperitoneally with 3 mg/lb oxytetracycline administered as the hydrochloride. Both animals were slaughtered 2 hours after injection. The injection was made through a 3½", 16 gauge needle that was directed inward through the body wall and inserted to the hub. The antibiotic, dissolved in sterile distilled water, was ready in the syringe that was then attached, and the calculated dose was delivered. The swine, which weighed approximately 200 lbs each, were satisfactorily held by an Iowa hog holder.

Following slaughter, the carcasses were hung at room temperature (80°F) for 2 days. Periodic observation showed spoilage at 24 hours in two untreated control carcasses, as evidenced by development of a putrid odor and a greenish hue to the meat; and by 48 hours gas bubbles were forming in the loose tissues. The treated animals, however, remained in excellent condition. After aging, the oxytetracycline-treated carcasses were chilled to 0°C, and carved into the usual cuts for marketing. These cuts were then loosely wrapped in aluminum foil and irradiated with a gamma source at 66,000 rep. Even after storing at 2°C for 30 days, the meat cuts remained of excellent bacteriological quality and good appearance.

Example 2: Ground beef of poor initial quality, i.e., having a high degree of bacterial contamination, was intimately contacted with oxytetracycline in the amount of 10 ppm by mechanically mixing the antibiotic thoroughly through the meat. Gamma radiation at 100,000 rep was then employed on the beef, following the procedure of Example 1. The beef was stored at 2°C and the bacterial flora determined and compared with controls, as shown in the following table. This table demonstrates clearly the significant increase in keeping time of poor quality meat when treated by the process, as compared with untreated meat or merely irradiated meat or merely antibiotic-treated meat.

Treatment	Bacteria per gram (×10³)	
	0 days	6 days
Control	100,000	1,000,000
Irradiated	300	10,000
Oxytetracycline	60,000	1,000,000
Irradiated + Oxytetracycline	100	100

The same process was repeated twice again, employing respectively tetracycline and chlortetracycline in lieu of the oxytetracycline. In each instance the same difference was noted in the irradiated and antibiotic-treated beef, as compared to the three controls.

Example 3: Ground lean beef was purchased at a chain supermarket, divided into 2 gram patties and wrapped in Visking frankfurter casing. The tetracycline antibiotics in 0.1 ml sterile distilled water, or 0.1 ml sterile distilled water only for the controls, were added to the patties. The casing was twisted closed and a loose overwrap of aluminum foil made. The test samples were then irradiated in a cobalt 60 furnace and stored at 2°C, or in some cases 10°, 30° and 45°C. Replicates were removed at intervals, disintegrated in a Waring blendor and plated in tryptone-glucose-yeast extract agar or potato dextrose agar with tartaric acid added. Control samples of three types were maintained; untreated, irradiated only, and treated only with the antibiotics.

FIGURE 2.2: PRESERVATION OF MEAT

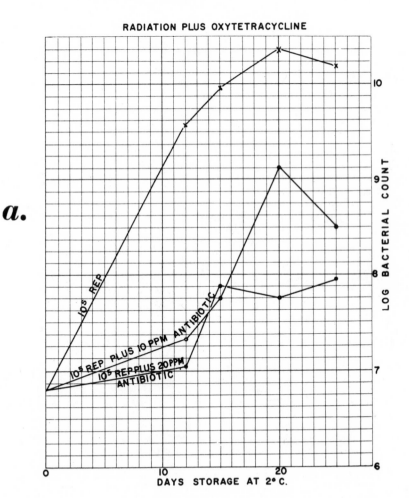

RADIATION PLUS OXYTETRACYCLINE

a.

(continued)

FIGURE 2.2: (continued)

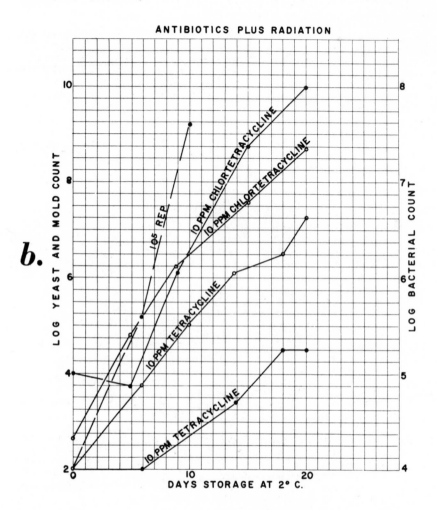

b.

Source: R.C. Ottke and C.F. Niven, Jr.; U.S. Patent 3,057,735; October 9, 1962

The meat procured from the chain supermarket ranged in initial bacterial count from
4×10^5 organisms per gram to over 10^8 per gram. The median and mode counts were both
approximately 4×10^6 organisms per gram. Meat at or above this median has a shelf life
of 2 to 5 days when stored at 2°C. Below the median count organoleptic spoilage may not
be apparent for from 4 to 10 days. Either the antibiotic alone at 1 to 20 ppm or irradiation
alone at 10^5 rep deferred spoilage until about the seventh day. (This irradiation level of
10^5 represents the maximum generally obtainable without detectable color, odor and taste
changes occurring in the beef.) The untreated meat had spoiled by the third day.

In Figure 2.2a the preservative action of 10^5 rep alone, and in combination with 10 and
20 ppm oxytetracycline is illustrated. In Figure 2.2b the effects of chlortetracycline and
tetracycline when combined with 10^5 rep are illustrated. Both bacterial and mycotic population changes are followed. The meat lot used in the tetracycline series had the lowest

initial contamination of any lot purchased, and this is reflected in a greatly increased bacteriostasis.

Example 4: Commercial grade steaks were obtained that averaged 3×10^4 microorganisms per gram. Sodium sorbate was taken up in a 1:4 propylene glycol-glycerol solution in combination with oxytetracycline. The antibiotic and the sorbate were in such concentration that addition of 0.1 ml of the solution to 5 grams of meat produced the desired final concentration of preservative per gram of meat (10 ppm of the antibiotic and 1,000 ppm sorbate). A 0.1 ml of the solution containing both preservatives, or oxytetracycline only, or a 0.1 ml of the propylene glycol-glycerol solution only, for controls, was spread over a 16 square inch section of Pliofilm.

A 5 gram sample of steak was then wrapped tightly and heat sealed in the treated wrapper. The samples were irradiated at 10^5 rep and held in humidified Pyrex casseroles at $2°C$. Samples treated with the sorbate plus antibiotic were microbiologically sound in the Pliofilm after 30 days. In both controls and test steaks the meat color had faded; however, in those samples with preservative added the color change was not as severe as in the controls.

Sample lots of aseptically prepared ground beef were treated in the same manner. Microorganisms were not detectable for the first two weeks of storage. In the samples treated with irradiation and antibiotic, spoilage levels of the microbial population were not reached for 35 days. Those samples treated with sorbate, irradiation and antibiotic had, after 47 days, only a small mycotic population.

ANTIOXIDANT FOR FREEZE DRIED MEAT

Freeze dried products and particularly fat containing freeze dried meat, as for example ham, become excessively rancid on exposure to atmospheric conditions within about 24 hours. Freeze dried meats should therefore be protected from such exposure in order to protect them from spoilage.

A potent antioxidant is used by *D.E. Mook and P.L. McRoberts; U.S. Patent 3,459,561; August 5, 1969; assigned to Borden, Inc.* to impart at least 30 days storage life to freeze dried meats. The freeze dried material may be any fat containing meat such as for example, beefsteak, ground meat, fowl and pork.

The antioxidants used are the hydroxy phenyl derivative class. Examples include butylated hydroxyanisole, butylated hydroxytoluene and propyl gallate. The antioxidants may be used separately or in combination and in further combination with chelating agents such as citric acid and salts of ethylenediaminetetraacetic acid. The preferred combination is butylated hydroxyanisole and citric acid. The process is carried out by spraying the antioxidant composition onto freeze dried meat in an inert atmosphere.

Warm ($90°$ to $120°F$) freeze dried ham is placed in a rotating and tumbling inducing apparatus such as for example what is conventionally called a "pill coater." The atmosphere prior to addition of the freeze dried meat may be purged with an inert gas such as nitrogen or carbon dioxide. An antioxidant composition is then added as by spraying in the form of a fine mist onto the surface of the rotating and tumbling freeze dried meat. The propellant generally used in the spraying operation is an inert gas, for example nitrogen or carbon dioxide. After treatment with the antioxidant the freeze dried meat may be packaged in an inert gas (e.g., nitrogen) for use in commerce.

Freeze dried products are especially sensitive to liquid, and particularly to liquid in bulk. Being essentially anhydrous, and also hydrophilic in the dried state, it is readily appreciated that merely dipping or otherwise coating the antioxidant onto the freeze dried product by methods whereby bulk liquids are used to coat the product, causes it to absorb liquid. Such soggy products are not useful for use and sale in commerce as originally intended. By spraying the antioxidant composition, via an inert atmosphere, onto a freeze dried product which is continually tumbling and rotating, the product is uniformly treated with the desired amount of antioxidant. As applied it does not absorb excessive liquid in local areas which would result in a soggy product. It is preferred to introduce the antioxidant composition (liquid) into the tumbling and rotating freeze dried product at a rate not to exceed about 2 fluid ounces per hour per pound of freeze dried product.

Generally the composition is sprayed at the rate of between 0.5 and 1.5 fluid ounces per hour per pound. The amount of inert gas carrier is not critical other than the economic limitation on excessive quantities and ineffective spray techniques with less than minimum amounts.

In general, about 150 to 500 parts by weight of antioxidant added, per million parts (ppm) by weight of fat content, results in a usable product. Generally, about 300 ppm will give a storage life period in excess of 34 days. The preferred range is 200 to 300 ppm. A chelating agent may be mixed with the antioxidant in proportion in the range of 15 to 25 parts by weight of antioxidant.

A preferred antioxidant composition consists essentially of a mixture of 100 parts by weight of butylated hydroxyanisole, 15 to 25 parts by weight of citric acid, and propylene glycol in an inert carrier atmosphere of nitrogen or carbon dioxide. Where the storage temperature is in excess of room temperature storage life is decreased for a given amount of active antioxidant agent.

The following table illustrates the proportions of antioxidant for freeze dried ham containing about 30% by weight of fat. Samples were stored in air at room temperature and at 100°F. Duplicate samples were kept in hermetically sealed containers in a nitrogen atmosphere at room temperature as reference samples.

		Storage Life, Days Until Development of Rancid Odor in Air	
ppm	Antioxidant	65°F	100°F
0	Control	1½	1
50	A*	6	5
	S-1**	6	6
100	A	9	8
	S-1	7	6
300	A	40	34
	S-1	8	7
500	A	76***	76***
	S-1	8	8

*5 parts of butylated hydroxyanisole for 1 part of citric acid in a propylene glycol carrier. (A)

**2 parts of propyl gallate for 1 part of citric acid in a propylene glycol carrier. (S-1)

***Sample did not develop typical rancid odor after 76 days and test was terminated.

The preferred antioxidant is a combination of butylated hydroxyanisole and citric acid in proportions of between 15 and 25 parts of citric acid for 100 parts of butylated hydroxyanisole.

Example: After freshly diced ham (¼" x ¼" x ⅛") has been freeze dried, antioxidant is added as follows. A solution was prepared of 40 parts BHA, 8 parts citric acid, and 52 parts propylene glycol. This was diluted to 5% in propylene glycol. A pill coater, which holds approximately 50 to 65 lbs of freeze dried diced ham was used to mix the ham during application of the antioxidant. As the pill coater rotated causing the freeze dried ham to tumble varying amounts of antioxidants were sprayed onto the freeze dried ham.

Spraying was accomplished with a spray gun, using nitrogen as a propellant. The spraying rate varied between 1 and 1.5 fluid ounces per hour per pound of ham. The following table tabulates the results obtained by varying the amount of antioxidants used.

Number	BHA Treatment Based on 25% Fat Content (ppm BHA)	Amount of 5% Solution Added to 50 lb of Ham (grams)	Storage Life (days) Until Development of Rancid Odor in Air	
			Room Temperature	100°F
1	0	0	1	1
2	100	28.3	15	10
3	200	56.5	34	23
4	300	84.8	52*	52*
5	400	113.0	52*	52*
6	500	141.3	52*	52*

*Typical rancid odor did not develop after 52 days and test was terminated.

The head space gas of each sample was checked by a Perkin-Elmer Gas Chromatography Unit. In all the oxidized samples, two definite peaks or compounds were present. These same compounds were not present in the unoxidized samples. The results were verified by organoleptic (odor) evaluations. The two peaks or compounds, found in the head space gas of the oxidized freeze dried ham samples were identified as pentanal and hexanal; the latter being an end product of oxidation of linoleic acid at the double bond. Oxygen Bomb tests also showed that the rate of oxygen absorption by the antioxidant treated freeze dried ham was notably lower than that of the untreated freeze dried ham.

PHOSPHORIC ACID

In the processing of protein-containing animal by-products it is often necessary to store the fatty, fleshy material for lengthy periods of time prior to the ultimate treatment such as by hydrolysis with an acid. Such prolonged storage at ambient temperatures produces partial degradation of any fat present in the by-product. As a result, fatty acids are formed and discoloration of the fat as well as emission of unpleasant odors occurs. Heretofore fat breakdown has been largely inhibited by storage under refrigeration. However, maintenance of such cold storage is uneconomical particularly in the storage of most animal by-products which are usually of little actual unit value.

W. Kuster; U.S. Patent 3,544,607; December 1, 1970 found that animal by-products can be stored for lengthy periods of time, at an ambient temperature such as room temperature or higher, in an aqueous solution of an inorganic acid without detrimentally affecting the fat content of the by-product. Furthermore, the acid used is advantageously that acid which is to be employed for subsequent acid hydrolysis. In this manner, the protein portion of the animal by-product can be conventionally hydrolyzed to the soluble polypeptide state merely by increasing the temperature of the aqueous storage solution and completing the processing.

The amount of inorganic acid required to inhibit fat deterioration during storage of untreated by-products is that amount which provides a pH for the resulting solution of about 1.5 to 2.5. It has been found that 2 to 4% by weight of inorganic acid based on the total weight of by-product will produce this desirable environment. Preferably the amount of inorganic acid used for inhibition of fat degradation during prolonged storage is that which corresponds to the amount necessary for subsequent hydrolysis. Although sulfuric acid, hydrochloric acid and the like may be employed, phosphoric acid is preferred.

The storage temperature can be varied, for example, with fluctuations in room temperature between from 60° to 80°F. In addition, higher or lower storage temperatures have not been found to adversely affect the conditions produced during prolonged storage. Storage at ambient temperatures is therefore possible even when there is a fluctuation. It is best that the animal by-product is comminuted so that the individual pieces do not exceed one inch in any dimension. In this manner, the by-product will be subjected to intimate contact with the acid solution. The size reduction can be accomplished in any conventional manner such as grinding or chopping, or prebreaking.

To further illustrate the process the following example is provided. Beef trimmings were obtained from a local butcher shop. The bone was removed. The remaining composition was approximately 85% fat by weight and 15% meat by weight. The raw materials were comminuted into pieces not exceeding about 1" in any dimension and placed in an open vessel. 3% by weight of commercial phosphoric acid (75 to 85% phosphoric acid) was added to an amount of water sufficient to cover the raw material in the vessel. A second batch of raw material was placed in a similar vessel filled with water. The materials were stored at room temperature for a period of 7 days. Portions of the materials were removed from the storage vessels at interim periods and tested for fatty acid concentration. The test results were as follows:

Treatment	Storage Time, days	Free Fatty Acid	Color	Odor
Under acid solution*	0	1.5	Light	None
Under acid solution	3	1.4	Light	None
Under acid solution	5	1.5	Light	None
Under acid solution	7	1.7	Light	None
Under water	0	1.5	Light	None
Under water	7	7.8	Dark	Strong

*pH adjusted to 2.2 with commercial phosphoric acid.

The raw materials stored under the phosphoric acid had a free fatty acid level of 1.7 at the end of 7 days. The raw material stored under a water solution had a free fatty acid of 7.8 after 7 days. The raw materials stored under the dilute phosphoric acid produced no off odors even at the end of the 7-day period. However, odors from the raw materials stored in a plain water solution were sufficiently strong as to be unbearable at the end of 3 days.

INTERMEDIATE MOISTURE FOODS

Higher Diols and Esters

During the past decade, considerable progress has been made in the development of intermediate moisture food compositions. These food compositions comprise a heterogenous group of foods which resemble dry goods in their resistance to microbacterial deterioration but which contain sufficient moisture that they cannot be considered as dry foods. Generally, the intermediate moisture foods are plastic and are easily masticated but do not produce an oral feeling of dryness. Notwithstanding their microbiological stability, intermediate moisture foods are subject to the same types of adverse chemical changes as observed with fully dehydrated foods. As a generalization, foods in the intermediate moisture range are more susceptible to the Maillard reaction than dry foods but less susceptible to fat oxidation.

The availability of water for spore germination and microbial growth is closely related to its relative vapor pressure commonly designated as water activity. Water activity (A_w) is defined as the ratio of the vapor pressure (P) of water in the food to the vapor pressure of pure water (P_o) at the same temperature. Within the range favorable to the growth of mesophilic microorganisms, A_w is practically independent of temperature.

In general, intermediate moisture foods have a total water content in the range from 20 to 60% by weight. However, this water is present in a form not readily available to microbes in that the water activity of the food is low. As pointed out, water activity is defined by the following equation:

$$A_w = P/P_o = ERH/100$$

A_w = water activity
P = partial pressure of water in food
P_o = saturation pressure of water at specified temperature
ERH = equilibrium relative humidity

Generally, a water activity below about 0.85 will prevent the growth of bacteria, while a water activity below about 0.8 will prevent the growth of yeast. Molds are most resistant to a lower water activity, some showing growth in a medium with a water activity as low as about 0.60. The intermediate moisture food compositions marketed are rendered stable against deterioration by incorporating into the prepared food one or more representatives of each of two types of ingredients, namely: (1) An osmotic agent such as salt, sugar and glycerol which has the effect of depressing the water activity of the food to levels at which most bacteria and yeasts will not grow; and (2) A mycostatic agent such as sorbic or propionic acids or their salts, which prevents the growth of certain molds and yeasts that are not readily inhibited by the osmotic agents.

Thus, unless these agents are added, most intermediate moisture foods are highly susceptible to spoilage due to the action of bacteria, yeast and molds. Prior methods of preserving foods by freezing and canning are very expensive. Frozen foods require freezer space for storage and usually cannot be refrozen, and are also less palatable than fresh or canned foods. Dried foods must be rehydrated before consumption and, in most cases, the rehydrated product is not so desirable organoleptically as the original fresh food. Thus, intermediate moisture foods provide an attractive alternative to prior conventional methods of food preservation. They are more convenient to use, require no special storage facilities and are potentially less expensive. When properly formulated, these foods have a high degree of palatability.

However, the development of intermediate moisture foods has been retarded because of the lack of suitable chemical additives. Desirable preservatives, in addition to being effective in controlling water activity and suppressing mold growth, must be safe to use, preferably nutritious, and must impart no undesirable flavor, texture or other organoleptic qualities to the finished product.

The preservatives commonly used are deficient in one or more of these requirements. For example, sugar, glycerol and propylene glycol are only moderately effective in controlling water activity and must be used at high levels in the food formulations. At these levels, these compounds add to the cost of the product and tend to impart an undesirable sweet flavor to the preparation. Propylene glycol, for example, is too toxic for use in the amounts required. A salt, such as sodium chloride, is effective but generally cannot be added at high enough levels in foods because of the salty taste and other undesirable qualities it imparts.

Sugars, such as sucrose, are not only too sweet for soft-moist food formulations but also present another difficulty due to their tendency to promote nonenzymatic browning. Nonenzymatic browning is caused by complex reactions between the amino groups of proteins and the keto groups of sugars (the Maillard Reaction). This leads to undesirable darkening of the food product as well as off-odors and flavors. Such interactions can also reduce the nutritional value of foods.

The above enumerated osmotic agents have very little mycostatic action and a separate mold inhibitor must be added. Commonly sorbic acid, propionic acid or their salts are used. These additives can only be used at low levels. In addition, being acids, they are less effective at neutral pH where many food products must be maintained.

J.W. Frankenfeld, M. Karel and T.P. Labuza; U.S. Patent 3,732,112; May 8, 1973; assigned to Esso Research and Engineering Company utilize linear aliphatic 1,3-diols to inhibit bacteria, yeast and mold in intermediate moisture foods.

When added to various food preparations, whether alone or in conjunction with one or more of the above named chemicals, these particular diols impart many very desirable qualities to the food product. These particular diols maintain such preparations in a bacteria-, yeast- and mold-free state, thereby increasing the shelf life of the product. In addition, they provide softness, or plasticity and enhance the palatability of food formulations. By replacing all or part of the salt, glycerol or sugar in current intermediate moisture foods, these diols obviate the difficulties caused by the strong flavors of such compounds and reduce the incidence of nonenzymatic browning. Finally, they are completely metabolized and actually improve the nutritive properties of the intermediate moisture foods in which they are incorporated.

The linear 1,3-diols or esters contain from 4 to 10 carbon atoms in the diol portion of the molecule. The ester portions of the molecule contain from 2 to 10 carbon atoms. The polyols contain hydroxy groups on at least the first and third carbon atoms of the molecule. It is this 1,3-dihydroxy configuration which renders these compounds very useful as food additives because of their inherent safety. Polyalcohols with hydroxyl groups in other positions on the carbon chain are more toxic and, therefore, are less useful as additives. In addition to being nontoxic and readily metabolized, the 1,3-diols and esters have certain other advantages, making them highly desirable as additives for intermediate-moisture or soft-moist foods; (1) they are stable, nonvolatile oils and have a long storage and shelf life; (2) they have an appreciable water-solubility and are readily emulsified, making them easy to formulate in various food preparations; (3) they are readily absorbed in the intestinal tract and they are completely metabolized.

A summary of the compounds along with their caloric densities and some of their physical properties is presented in Tables 1 and 2. The best esters are those with 5 to 8 carbon hydrocarbon tail in either the diol or ester portion of the compound combined with a concentration of polar groups in another part of the molecule, as for example, 1,3-octanediol-1-monopropionate or 1,3-butanediol-1-monooctanoate. Some especially valuable esters are shown in Table 2.

TABLE 1

Diol	B.P./mm. (° C.)	Taste, odor, etc.	Theoretical caloric density, kcal./gm.[1]
1,3-butanediol	202–203	Colorless, sweet odor, bitter taste.	6.7
1,3-pentanediol	78–81/0.5	___do___	7.4
1,3-hexanediol	81–82/0.2	Colorless, slight musty odor, bitter taste.	7.8
1,3-heptanediol	90/0.5	Colorless, slight musty odor, slight bitter taste.	8.2
1,3-octanediol	87–89/0.3	Colorless	8.5
1,3-nonanediol	126/1.1	___do___	8.7
1,3-decanediol	M.P.[2]=30–31	___do___	8.9
1,3-undecanediol	M.P.[2]=41–42	___do___	9.1

[1] Caloric density is the theoretically available energy in kilocalories per gram of the compound.
[2] M.P.=Melting point.

TABLE 2: PROPERTIES OF SOME 1,3-DIOL ESTERS

Compound	B.P.,° C. (mm.)	Rat feeding results: Caloric density (kcal./g.)		
		Observed	Calculated	Percent utilized
1,3-butanediol (parent diol):				
1-monopropionate				
1-monoctanoate	90–95 (0.3)			
1-monopalmitate	M.P.[1]=29–31			
Dipropionate	67–70 (0.4)			
1,3-hexanediol:				
1-monoacetate	59–62 (0.15)	6.7	7.0	95
1-monoctanoate	117–122 (0.3)	8.4	9.0	95
1-monopalmitate	124–126 (0.2)	7.3	9.3	78
Diacetate	81–82 (0.9)			
1,3-heptanediol:				
1-monooctanoate	85–90 (1 0)			
1-monopalmitate	M.P.[1]=38–39			
Dipropionate	90–92 (0.4)			
1,3-octanediol-1-monopropionate	83–86 (0.3)			

[1] M.P.=Melting point.

The diol and diol esters may be prepared by any suitable technique such as by the Reformatsky reaction followed by reduction, or by means of the Prins reaction of formaldehyde and the appropriate olefin. If the diol be a lower member such as 1,3-butanediol or 1,3-pentanediol, it is preferred to use 15 to 40%, such as 20 to 25% by weight based on the total food composition.

Also, the amount used will depend to some extent upon the pH of the intermediate moisture food composition. Thus, if the pH is in the range of 7.5 to 8.5, about 25% by weight of pentanediol or butanediol will be used. The preferred range for an ester of butanediol or pentanediol and for a higher diol or higher diol ester is 0.05 to 5.0% by weight based on the total food composition.

With respect to certain foods having a high pH, above about 5.0, it is desirable to utilize a mixture of diols, such as 1,3-butanediol and 1,3-heptanediol. This diol mixture is in the range from 75% butanediol to 99% butanediol, as compared to 1% heptanediol to 25% heptanediol. A very desirable diol mixture comprises about 95% butanediol and 5% heptanediol. Although all the compounds listed in Tables 1 and 2 are valuable as additives for intermediate moisture or soft-moist foods, some are more valuable for certain purposes than others. For example, the lower members of the diol series, 1,3-butanediol and 1,3-pentanediol are excellent humectants, plasticizers and are useful for controlling water activity. They can, therefore, be used to replace all or part of the salt, sugar, glycerol or propylene glycol which are conventionally used as osmotic agents in the soft-moist formulations.

These lower 1,3-diols also possess weak mycostatic action. In some formulations, they may be used without added mold inhibitors. This gives them a clear-cut advantage over salt, sugar, glycerol or propylene glycol systems where mycostats must always be added. In food formulations more conducive to mold growth, it is desirable to include a mold inhibitor in the 1,3-butanediol or 1,3-pentanediol systems. Any safe mycostat, for example, propionic acid or sorbic acid may be used. If so, a reduced quantity of such inhibitors can be used compared to the salt, sugar, glycerol or propylene glycol systems.

A desirable method to enhance the mycostatic action of 1,3-butane or 1,3-pentanediol formulations is by addition of 0.5 to 3% by weight, based on diol mixture, of a 1,3-diol of 6 or more carbon atoms as, for example, 1,3-heptanediol, or one of the diol esters as, for example, 1,3-octanediol-1-monopropionate. These compounds are superior mold-inhibitors. They have been shown to be significantly more effective than sorbic acid, propionic acid or their salts. These diols and esters are also effective at neutral or high pH where the mold-inhibiting action of organic acids is greatly reduced. An additional advantage is that these diols and esters are less toxic and more nutritious than sorbic or propionic acids.

Example 1: Nutrient broth was used as the basal nutrient medium for the growth of all microorganisms tested. 5 ml of nutrient broth medium (Difco Co.) were placed in 18 millimeter x 10 millimeter test tubes and the basal medium sterilized with steam at 15 psi for 15 minutes. After cooling, a sufficient amount of the various compounds were added to the basal medium to give the concentrations used. Normally a final concentration of 0.2, 1 and 2% were used.

After mixing the chemicals with nutrient broth, the tubes were inoculated with the various test microorganisms. The test microorganisms were grown 25 hours earlier in nutrient broth and 1 drop of the dense microbial suspension was added to the tubes. The tubes containing the chemicals and microorganisms were then incubated at the optimal growth temperature reported for each microorganism tested. Either 37° or 30°C was used. Growth in control tubes, as well as those containing chemicals, was observed visually. After a suitable incubation period, a small aliquot of the test solutions was streaked on an agar plate. This was done in order to confirm the visual readings of the presence of microbial growth.

The results are shown in Tables 3 and 4. The minimum effective concentration is the lowest concentration of additive which effectively prevented growth under the conditions of the test.

Food Additives to Extend Shelf Life

TABLE 3: PRESERVATIVE ACTION AGAINST BACTERIA

| | Minimum effective concentration against Salmonella | | |
	Staph. aureus, percent	Typhi- murium, percent	E. coli percent
Compound			
1,3-heptanediol	2	1	1
1,3-octanediol-monopropionate	+	0.2	2
1,3-butanediol-dipropionate	+	0.2	+
K-sorbate	+	2	+
Ca-propionate	+	+	+

NOTE: + = No effect at 2%.

It is apparent from the above that the effectiveness of these materials against a wide spectrum of bacteria is established by the typical data shown in Table 3. In these tests the test compounds are compared to the known commercial preservatives, potassium sorbate and calcium propionate, as to their ability to inhibit growth of various bacteria. It is apparent that several of the compounds are effective at lower concentrations than either of the current additives. Of especial interest is the result that some of the test compounds are active against Salmonella at levels as low as 0.2%. These tests were carried out under conditions conducive to prolific growth of the organisms.

Under conditions of normal food storage, the test compounds would be effective at even lower levels. Potassium sorbate was inhibitory in this test only at 2% or above under the test conditions and calcium propionate did not inhibit growth even at the 2% level. Salmonellae are important public health organisms frequently found in foods, especially meat, eggs, and dairy products. All members of the genus are considered as human pathogens. The additives are also very effective with respect to mold inhibition which is shown in the following Table 4.

TABLE 4: PRESERVATIVE ACTION AGAINST MOLDS

| | Minimum effective concentration against— | | | | | | | |
Compound	Tricho- derma 12688 [1], percent	Botrytis 9435 [1], percent	P. roque- fortii 6988 [1], percent	Fusarium 10911 [1], percent	B. fulva, percent	A. niger percent	A. flavus. percent	Bread mold, percent
1,3-pentanediol	+	+	2	+	+	+	+	0
1,3-heptanediol	1	1	0.2	1	0.2	0.2	1	0
1,3-pentanediol-monopropionate	0	0	1	0	0	1	0	0.2
1,3-octanediol-monopropionate	0.2	0.2	0.2	0.2	0.2	0.2	0.2	0
1,3-butanediol-dipropionate	0.2	0.2	0.2	1	0.2	0.2	0.2	0
1,3-butanediol-monooctanoate	0	0	0.2	0	0	0.2	0	0.2
1,3-pentanediol-monooctanoate	0	0	1	0	0	0.2	0	0.2
Potassium sorbate	2	0.2	0.2	2	2	2	2	0
Calcium propionate	+	1	0.2	+	+	+	+	0

[1] All microorganism numbers—American Type Culture Collection.

NOTE.— + = No effect at 2%; 0 = Not tested.

In Table 4 some selected diols and esters are compared with potassium sorbate and calcium propionate, commercial mold inhibitors, in effectiveness against various common molds. The lower the minimum effective concentration, the more effective the compound. It is apparent that several of the test compounds are significantly better than the currently used preservatives.

Example 2: In addition to these tests, some more definitive studies were conducted to determine the effectiveness of certain diols and esters in inhibiting the growth of two common molds under various culture conditions. These tests were carried out as described above except that the pH was varied by the use of suitable buffering agents and, in some experiments, either dextrose or glycerol were added to demonstrate the effectiveness of the diols and esters in different growth media. Molds were chosen because they were of especial importance in the spoilage of intermediate moisture foods.

For purposes of comparison, several commercial food preservatives were tested under the same conditions. Data showed that the diols and esters are significantly more effective than the commercial additives. Of special importance is the finding that 1,3-heptanediol and the esters are highly active inhibitors of molds, typified by *A. niger* and *P. roquefortii* at a pH of 6.8 (nearly neutral) where commercial additives are either only slightly effective or ineffective. This is very important for the preservation of many foods where the pH of the preparation must be near neutrality.

Glycerol Infusion

Fresh and cooked meat, vegetables and fruit at a moisture level in the neighborhood of 50% and higher, depending upon the character of the produce, are prone to undergo microorganic decomposition. Preservation of such produce by dehydration has been the subject of intensive and widespread investigation. But meats and vegetables rehydrate readily from a dried state of less than 20% moisture. It would be desirable to have such food items in a semi-moist form such that they can be eaten as is, simply warmed or if desired, moistened further for consumption and yet be shelf stable, i.e., require no refrigeration or even commercial sterilization preparatory to packaging. Unfortunately the edibility of meats and vegetables such as carrots and chicken in the 20 to 25% moisture range is unacceptable as the items are tough and dry to the taste and also subject to microbial decomposition under anaerobic packaging conditions.

In the case of meat, it would be desirable to avoid dehydration to a moisture content less than 15% due to the changes that take place during storage of the meat and as a result of dehydration. Dehydrated meat will be less rehydratable than might be desired in many food applications, particularly those where the meat is rehydrated with other material such as dehydrated vegetables. It would be desirable to have essentially a shelf stable meat preparation requiring no refrigerated storage which has a moist texture such that it can be eaten as is, warmed or further hydrated and which approaches conventionally cooked meat in eating quality. Workers have suggested that meat can be preserved by dehydration to say a moisture content of less than 40% and the curing of meat by such curing agents as sugar and salt is traditional. However, in many preservation treatments such curing is erratic and in some instances dependent upon the skill of the processor; in most such applications the food is characteristically tough and dry and excessively salty.

A process of dehydration of meat product followed by infusion of a stabilizing solution is described by *M. Kaplow and J. Halik; U.S. Patent 3,595,681; July 27, 1971; assigned to General Foods Corporation.*

Precooked meat and meat by-products are freeze dried under conditions where the moisture content of the meat or meat by-product will be reduced to less than 20% thereby leaving the meat tissue in a form whereby it will admit stabilizing solutes. After such desiccation the meat is immersed in a stabilizing aqueous solution having a substantial amount of polyhydric alcohol together with optional other stabilizing solutes such as sugars and/or salts where the material will be infused by such agents and will have a moisture content ranging anywhere from 20 to 40%.

In general the stabilizing solution will have a high concentration of solutes typically but not necessarily in excess of the ultimate weight of moisture intended to be present in the stabilized meaty material. The meat should be so dried that case hardening of the meat tissue is avoided. Usually this will be done after the meat has been at least pasteurized to an essentially pathogen-free condition and preferably after the meat is fully cooked either by roasting, boiling or otherwise; in this way the stabilized meat product can be consumed as is.

By subliming the water vapor present in the meat prior to infusion, its morphology is such that when a stabilizing solution containing the polyhydric alcohol is used, the solution will infuse the matrix of the meat effectively and uniformly in a comparatively short period of time. In this way, treatment in the stabilizing solution will not be unduly prolonged such that the meat will not unduly lose any of its desired character prematurely due to treatment by the stabilizing solute. Should the stabilized meat be further hydrated preparatory to consumption or incident to any further cooking, the cell structure of the meat is such that it will admit water readily and have a desired eating quality.

Various polyhydric alcohols of use in the intermediate moisture meat, vegetable and fruit products will be found effective not only as stabilizing solutes but importantly also in simulating product moistness and lubricity for enhancing eating quality, glycerol being the most preferred; other preferred polyhydric alcohols are sorbitol, mannitol and mixtures. For meat products, fat infused with the stabilizing solution has also been found to be an important additive in complementing the desired textural and flavor qualities provided by the polyhydric alcohol; salt and sugar, when compatible with flavor, are also useful additives as infusing solutes.

Antimycotics will generally be part of the stabilizing solution when the produce is to be aerobically packed, i.e., cold packed; propylene glycol has been useful both as a polyhydric alcohol and as an antimycotic infusion solute although propylene glycol will usually be preferred together with a sorbic acid compound such as potassium sorbate and at a comparatively low level in the stabilizing solution due to adverse taste.

Example: Fresh chicken (72% moisture) white meat without skin was cooked to an edible condition (62% moisture), cut in ½" to 1" pieces and freeze dried as follows. The pieces were quick frozen and inserted into a Stokes vacuum freeze dryer having a chamber temperature of $-8°F$ and a shelf heat temperature of $115°F$ for 20 hr. The product (1 to 2% moisture) was then immersed in an infusing solution having the following formulation:

	Percent	
	Solution	Final prod.
Freeze dried chicken solids		34. 3
Glycerol	49. 2	32. 4
Water	42. 5	27. 8
Chicken soup base [1]	6. 2	4. 1
Propylene glycol	1. 6	1. 1
Potassium sorbate	. 5	. 3
Total	100. 0	100. 0

[1] 6 parts salt, 11 parts sugar, 13 parts mono-sodium glutamate, and flavor and spices.

Weight percentage of solution infused was 65.7 which approaches closely the weight percentage of original water content of the chicken prior to freeze drying. Lesser amounts of liquid can be infused (partial infusion) by raising the soaking ratio of freeze dried chicken to solution and reducing the infusion time. The infused chicken product is essentially shelf stable. It can be packed without refrigeration and stored indefinitely in reopenable jars. When eaten as is the chicken is moist, tender and of acceptable flavor and appearance. The product can be further hydrated in water or a soup for other recipe applications.

Low pH Environment

In another process for manufacture of semimoist food products an acid system is described by *R.N. Du Puis; U.S. Patent 3,489,574; January 13, 1970; assigned to General Foods Corp.* The acid is effective to product hydrogenolyzed sugars as stabilizing solutes as well as to provide a lower pH environment. Amylaceous material is hydrolyzed to produce hexosic carbon chain lengths; then this hydrolyzate is subjected to hydrogenolysis whereby substituent aldehydic groups are converted to alcoholic groups and hexosic material is cleaved to subhexosic alcohols typically of three carbon atoms and 4 carbon reduction products and higher including nonhydrogenolyzed materials such as sorbitol.

Such hydrogenolysis processes call for use of a hydrogenolysis catalyst and hydrogen under pressure. The economic advantages are best realized when a crude amylaceous material is liquefied with an enzyme or an acid treatment; is simultaneously or subsequently saccharified to mono- and higher saccharides, depending upon conversion efficiency; and is thereafter hydrogenolyzed, incident to which the hexosic monomer will have its aldehydic groups reduced to methylol groups and will be cleaved to yield diols and polyols. However, instead of saccharifying a liquefied starch, the amylolytically treated material may be hydrogenolyzed directly, in which event oligosaccharides and sugars will be similarly reduced and molecularly cleaved to yield subhexosic polyols.

Example: Stabilization and browning effects of various hydrogenolyzed dextrose compounds relative to dextrose and sucrose in intermediate moisture meat- and vegetable-containing food products were evaluated in the following manner. A meat slurry containing 18 parts bovine tripe, 6 parts bovine udders, 4 parts beef cheek trimmings, 4 parts gullets and approximately 2 parts propylene glycol was charged together with 20 parts test solutes specified hereinbelow to a sigma blade Day mixer (jacketed for circulation of steam and/or water therearound) which had been preheated for 5 minutes by steam.

This meat slurry was mixed for 1½ minutes at the stated steam pressure, the Day mixer being covered and the product being maintained at a temperature of 210° to 212°F. Thereafter the steam supply to the Day mixer was shut off by opening a steam bypass and mixing was continued for 13 minutes. The thus pasteurized meat slurry was blended in the Day mixer with a preblended dry mix and liquid additives, the dry mix containing 32 parts soy flakes, 3.0 parts soy hulls, 3 parts dibasic calcium phosphate, 2.5 parts milk replacer, 1 part sodium chloride, 3.5 parts vitamins, coloring, minerals and spices, 0.3 part potassium sorbate and liquid constituents comprising 1 part bleachable fancy tallow and 1 part mono- and diglycerides. This mixture was then covered, the Day mixer was again subjected to steam as before until the product reached a temperature of 200°F. Thereafter the steam supply was disconnected and the cooked mixture was quenched by turning on cool water to cool the product to ambient conditions, i.e., 15 minutes, cooling the product to a temperature of 80°F.

The cooled mix was extruded by a meat grinding attachment into strands which were in turn subdivided and hand molded into patties which were then wrapped in a cellophane wrapping material and set aside for storage and color studies. Test solutes were included in three separate samples, formulated and processed as in the preceding example of semimoist products; one being formulated and processed to contain sucrose as the test solute for establishing an acceptance standard, coded A; a second being formulated like the first but containing dextrose as test solute, coded B; and a third sample identical to A and B but containing as test solute a hydrogenolyzed dextrose. This latter test solute consisted of a mixture of 51.2% propylene glycol, 35.3% of glycerin, and sorbitol 13.5%.

Listed in tubular form below are comparative Hunter Color readings for the samples coded A through C after 8 weeks storage, the lower a Hunter Color reading, the darker the product. It will be noted that whereas a dextrose sample included in an intermediate moisture formulation underwent a significant darkening in product as demonstrated by the reduced L and a values, the intermediate moisture product containing hydrogenolyzed dextrose was quite stable as evidenced by the substantial identity of the initial and 8-week color readings for sample C. At the end of 12 weeks, sample C was not only microbiologically stable, but also it displayed similar resistance to browning despite the derivation from monomeric saccharidal material which would normally undergo browning. It will be noted that sample C approximated the color acceptability standard, sample A, indicating the efficacy of such a hydrogenolyzate.

Hunter Color Readings

| | - - - - L Value- - - - | | - - - - a Value- - - - | |
Intermediate Moistures	Initial	8 Weeks	Initial	8 Weeks
(A) Sucrose	37.8	39.0	11.2	13.7
(B) Dextrose	39.3	33.0	11.2	7.6
(C) Hydrogenolyzed dextrose replicate	39.6	37.1	11.2	12.3

The hydrogenolyzate will permit formulation of higher moisture products than 30%, such as those having a moisture content as high as 40% and typically about 35%, in which formulations the level of total solutes including the hydrogenolyzate can advantageously be less than the moisture present; this advantage accrues particularly for formulations containing food grade acidulants such as phosphoric, malic or citric acids and their acid salts, which when used in a formulation will adjust its pH from a neutral one of, say 6 to 8 to one in the order of 3.5 to 5.9. For some animal food such as cat food it appears that felines enjoy a more moist product than one having a level less than 30% water. Also, such high moisture feline foods seem to be benefited palatabilitywise when an acidulant is employed, and so the formulation will be advantageously stabilized against putrefaction and other microorganic decomposition by the hydrogenolyzate as part of the total solute phase in combination with the hydrogen ion activity of the food acid.

Although the formulations described are of the animal or pet food variety, the benefits of microorganic stabilization will be likewise applicable to human foods where the hydrogenolyzate will similarly serve as a stabilizing solute. Among the human foods which will be so stabilized are sausage-like formulations, liquid preparations such as syrups, simulated cheeses containing vegetable and/or dairy solids and the like. All such products will be pasteurized by an elevation to a temperature where substantially all pathogenic organisms will be destroyed, say at a temperature of 180°F, and the pasteurized formulation may be either aseptically packed or cooled and aerobically packed in a nonhermetic wrapper such as cellophane. Such packaged products can be stored without refrigeration for periods of 3 to 6 months or longer.

Simultaneous Stabilization of Meat and Sauce

In another process *M. Kaplow and J.J. Halik; U.S. Patent 3,634,104; January 11, 1972; assigned to General Foods Corporation* give the steps to stabilize a food consisting of beef, vegetables and potatoes simultaneously with the sauce or gravy and using sorbate preservatives as part of the system.

Briefly stated the process involves infusion of a polyhydric alcohol within a food solid matrix in any one of a variety of degrees of subdivision and the formulation of distinct aqueous liquid also containing a polyhydric alcohol, the food solid and aqueous liquid having moisture contents usually less than 40% respectively and having a total concentration of water-soluble compounds respectively whereat the two dissimilar phases are anerobically or aerobically stable depending upon the packaging techniques intended. The respective water activities in the food solids and liquid phases are so related one to another that they equilibrate to one another during storage and retain their microorganic stability.

Thus, a beef stew is formulated to contain beef chunks and whole or subdivided carrots, peas and potatoes, each of which are respectively dehydrated to a moisture content less than 45% through the infusion therein of a stabilizing solute containing a polyhydric alcohol such as glycerol; a gravy is formulated as an aqueous liquid containing a suitably emulsified fat and thickening colloid together with edifying flavorants, colorings, spices and the like, which gravy is also of a moisture content less than 45% but has, as a plasticizing solute, a significant level of polyhydric alcohols like glycerol serving to provide a flowable plastic or at least semiplastic fluidity under the anticipated conditions of use.

The respective water activities of the dehydrated food solids phase and the liquid phase will be such that migration of aqueous fluids from one phase to the other will not result in a substantial change in the relative concentration of stabilizing solutes in the respective phases and commonly the level of soluble solids and moisture present in the solid phase and the liquid phase will substantially approximate one another such that any moisture migration that may occur incident to storage of the packaged foodstuff will be minimal and in any event will not adversely imbalance the concentration of stabilizing solutes in the food solid phase and the liquid phase.

The process will be found to be applicable to the formulation of such products as beef, lamb and meat and vegetable stews generally as well as casserole preparations having meat and/or fish with vegetables and grains such as rice and/or pasta foods such as noodles, and macaroni in various shapes. The liquid phase will be so formulated as to have the desired rheological characteristics compatible with food acceptability. In most applications, the gravy or sauce will be fluid but comparatively plastic or thickened in character. The liquid phase will, thus, be formulated to contain fat and/or colloids such as starches, dextrins, flour, as well as the condiments or spices and the water.

Customarily, it will be found practical to pasteurize the food solids phase by infusion at an elevated temperature, say in excess of 160°F, the pasteurization being carried out sufficiently to at least kill any pathogens or inactivate enzymes; a common range of immersion heating temperatures will be 180° to 210°F for 15 to 25 minutes depending upon desired product texture of the food solids phase.

In the case of the liquid phase prolonged heat treatment may be less critical and usually any heat treatment will be used that is sufficient to promote fluidity in the mixture of the fat and/or thickness with the aqueous and other liquids in the medium; thus, it may be desirable to elevate the temperature of the liquid to melt the fatty constituents and to promote sufficient emulsification to assure physical stability and minimize segregation of the aqueous and nonaqueous phases of the liquid.

As distinguished from canned stew, these products are intended to be packaged without commercial sterilization and advantageously may be cooled to subpasteurization temperatures prior to packaging, say to temperatures below 180°F thereby allowing greater manufacturing flexibility and a less expensive substantially nonhermetic package. On the other hand, it is not intended to foreclose the practice of hermetic packaging, or at least, combining the food solid and liquid phases at above pasteurization temperatures.

Example: Beef Stew with Gravy — A beef stew with gravy was formulated from the following individual constituents:

Liquid Phase Gravy Formulation

Ingredients	Percent
Corn syrup solids	34.95
Coconut oil (76°F congeal point)	30.00
Water	17.00
Glycerol	10.00
Seasonings and salt	4.00
Propylene glycol	2.00
Pregelatinized potato starch	1.00
Beef flavor	0.50
Potassium sorbate	0.30
Mono- and diglycerides	0.25
Total	100.00
Moisture content =	17.4 (vacuum-oven method)
A_W^* =	0.79

*The relative humidity of a headspace atmosphere in equilibrium with the food or liquid expressed as a decimal fraction of one.

The gravy phase was prepared simply by warming to above the congeal point of the fat and blending in a mixer.

Food Solid Phase — Fresh green peas were immersion cooked and then immersed in the infusing solution for 10 minutes, the infusion solution being at a temperature of 208°F, whereafter the peas were allowed to soak overnight and were withdrawn. The following was the weight constituency of the peas and infusion solution.

Peas, Fresh, Shelled	Percent	Grams
Solids	22.0	277.2
Moisture	78.0	982.8
Total	100.0	1,260.0

(continued)

Peas, Fresh, Shelled	Percent	Grams
Infusion solution:		
Glycerol	78.4	1,143.0
Water	11.0	160.5
Sodium chloride	6.2	90.0
Propylene glycol	4.1	60.0
Potassium sorbate	0.3	4.5
Total	100.0	1,458.0
Moisture content =	39.4 (vacuum-oven method)	
A_W =	0.80	

Fresh carrots, cubed beef and potatoes were infused by a like stabilizing solution using similar conditions, the cooking time for respective food solids was that at which optimal organic acceptability is afforded; the cubed beef was cooked for 10 to 15 minutes as were the potatoes and the carrots. The following are the formulas for infusing constituents for treating the carrots, beef and potatoes:

Fresh Raw Carrot Dices (One-Fourth to One-Half Inch)

Carrots, Peeled, Trimmed	Percent	Grams
Solids	11.0	275
Moisture	89.0	2,225
Total	100.0	2,500
Infusion solution:		
Glycerol	88.7	3,155
Water	5.5	197
Sodium chloride	3.7	130
Propylene glycol	1.6	56
Potassium sorbate	0.5	19
Total	100.0	3,557
Moisture content =	35.7 (vacuum-oven method)	
A_W =	0.71	

Beef, Cubed Chuck	Percent	Grams
Solids	35.0	1,617.0
Moisture	65.0	3,003.0
Total	100.0	4,620.0
Infusion solution:		
Glycerol	56.6	2,112.0
Seasonings	19.8	739.2
Water	8.0	300.0
Sodium chloride	7.4	277.2
Propylene glycol	7.1	264.0
Potassium sorbate	1.1	39.6
Total	100.0	3,732.0
Moisture content =	3.10 (vacuum-oven method)	
A_W =	0.77	

Fresh Raw Potato Pieces (One-Half to Three-Fourths Inch)

Potatoes, Peeled, Diced	Percent	Grams
Solids	20.0	1,000.0
Moisture	80.0	4,000.0
Total	100.0	5,000.0
Infusion solution:		
Glycerol	80.4	4,687.5
Water	10.4	610.0
Sodium chloride	4.3	250.0
Propylene glycol	4.3	250.0
Potassium sorbate	0.6	37.5
Total	100.0	5,835.0
Moisture content =	36.8 (vacuum-oven method)	
A_W =	0.79	

Composite Beef Stew with Gravy and Food Solids Phases

	Percent
Infused gravy	50.0
Infused potato pieces	20.0
Infused beef pieces	15.0
Infused carrot dices	7.5
Infused peas	7.5
Total	100.0
Moisture content =	26.2 (vacuum-oven method)
A_W =	0.78

The foregoing food solids and the gravy phases were simply mixed, mixing being carried out in a Hobart blender until a uniform distribution of solid and liquid phases was achieved. The stew is a spoonable plastic aggregation which can be stored as such under ambient room temperatures preparatory to packaging or may be packaged directly from the mixer using aerobic, subpasteurization temperatures. The preferred method of packaging the composite will be cold packed in a flexible pouch and heat sealed in a gaseous nitrogen atmosphere.

Levulinic Acid-Sorbate

It was found by *G.J. Haas; U.S. Patent 3,623,884; November 30, 1971; assigned to General Foods Corporation* that the proportions of potassium sorbate incorporated in animal foods may be reduced if levulinic acid is combined with potassium sorbate. This combination resolves: (1) the problems of the high cost of employing potassium sorbate in proportions sufficient to totally inhibit mold growth in meat or meat-product animal foods (0.3% by weight); (2) the reduced palatability to dogs of animal foods containing 0.3% by weight of potassium sorbate; and (3) is more effective than potassium sorbate or levulinic acid alone in preventing mold growth.

The types of antimycotics permitted in animal foods are few, and the characteristics and activities of these materials in moist systems have been adequately described in textbooks. However, little information is available on the activity of these antimycotics and preservatives in systems of lower moisture, where the reduced water activity has an inhibiting action per se on microorganisms.

The relative weight percent of water-soluble solids to the moisture content of the animal food, when initially incorporated into the animal food during its manufacture and preparatory to packaging determines the ultimate functionality of the solids in providing the requisite bacteriostatic effect. Usually the level of moisture will range from 15 to 30%. The level of water-soluble solids (principally sugar) may be varied as may the level of moisture incorporated within the desired ranges. However, in varying these levels the relationship of the water-soluble solids in solution to the water should be controlled so as to afford the desired osmotic pressure.

In performing tests to determine the effect of levulinic acid alone, and in combination with potassium sorbate, on nonintermediate moisture and intermediate moisture meat and meat-product animal foods, a system was developed which would maintain most of the characteristics of reduced water meat and meat-product animal foods and at the same time accelerate mold growth to such an extent that useful results could be obtained by inspection for visual growth of generic microorganisms such as Aspergilli and Penicillia, within weeks.

To judge the degree of inhibition, the tests were carried out at 23° and 33°C. The samples were inspected daily and the number of days for visible mold formation determined. In the effectiveness scale, 0 represents no extension of shelf life or extension for less than double the control which contained no antimicrobial or antimycotic; 1 is double to triple the shelf life; 2 is triple to quadruple the shelf life; 3 is quadruple to tenfold the shelf life; 4 is at least tenfold the shelf life; and 5 signifies complete inhibition for approximately 6 months at both storage temperatures.

Example 1: Meat-Product Animal Food —

Ingredients	Parts by Weight
Chopped meat by-products (tripe, udders, cheek trimmings, tongue trimmings, gullets, etc.)	32.0
Defatted soy flakes	31.0
Sucrose	24.0
Flaked soy bean hulls	3.0
Dicalcium phosphate	3.0
Dried nonfat milk solids	2.5
Bleachable fancy tallow	1.0
Mono- and diglycerides	1.0
Sodium chloride	1.0
FD & C red dye	0.006
Garlic	0.2
Vitamin and mineral premix	0.06

An intimate mixture of the ingredients was made by first chopping the meat by-products into small pieces, which were then heated in combination with tallow to 212°F to effect pasteurization and produce a liquefied slurried meat composition. The slurry was then finely ground into a more or less pulpy, pumpable, flowable puree consistency. This hot pureed form of slurry was then proportionately blended with the remaining dry ingredients of the formulation in a steam jacketed cooker where it remained for a period of approximately 1½ minutes at an elevated temperature of 200°F, the product being under continuous agitation throughout this cooking phase. This cooked mixture had a plastic, extrudable, shape-retaining consistency. The moisture content of this composition was 25.0%. The finely comminuted meat by-products and the soya flakes had the aqueous phase evenly distributed throughout, thereby assuring a maximum bacteriological protection to the final product.

The pasteurized mixture was immediately cooled by passage through a refrigerated heat exchanger to an ambient temperature. The samples used in the test are a blend of one-half of a meat-product animal food and one-half of a 45% sucrose solution; however, without potassium sorbate or any other antimycotic. A sucrose solution is used because the liquid phase of the meat-product animal food is a solution, where most of the solute is sucrose. The sample was divided into 3 parts and antimycotics were mixed thoroughly into each for 1 minute at medium speed, and half the paste of each sample was transferred to jars and stored at room temperature (about 23°C) and 33°C. The results of these tests appear in the following table.

Antimycotic	Percentage Tested	Degree of Inhibition at Top Level	Level for Total Inhibition
Potassium sorbate	0.1; 0.3	5	0.3
Levulinic acid	1.0	0	–

	Days of First Visible Mold Growth	
Antimycotic	23°C Storage, days	33°C Storage, days
None	7	4
Levulinic acid 0.1%; potassium sorbate 0.1%	32	170
Levulinic acid 1.0%	8	6
Levulinic acid 0.1%	8	6
Potassium sorbate 0.1%	15	7
Potassium sorbate 0.3%	>70	>70

Inasmuch as the step of diluting the meat-product animal food in half with 45% sucrose solution results in increased A_W, any antimycotic which is active under these more stringent conditions of A_W and temperature could be expected to be considerably more effective in undiluted meat-product animal foods having a low A_W; and such was the case.

Example 2: Same as Example 1, except that the 45% sucrose solution was omitted. In this example the water-soluble sugar solids was maintained between 15 and 35% by weight of the composition, and the weight level of water-soluble solids is at least equal to or greater than that of the moisture of the composition; the moisture weight range being from 15 to 30% of the composition.

Antimycotic	Days of First Visible Mold Growth	
	23°C Storage, days	33°C Storage, days
Levulinic acid 0.1%; potassium sorbate 0.1%	>170	>170
Potassium sorbate 0.3%	>170	>170

The intermediate moisture samples used in Example 2 had a moisture content between 22 and 23.5%. In the above table, it can readily be seen that the relationship between the moisture content of the intermediate moisture meat-product animal food using the various antimycotics, and the speed of visible contamination could not be determined up to 170 days; after which the experiment was terminated.

In Example 1, it may be seen that the use of levulinic acid in amounts as high as 1.0% is not effective in totally inhibiting mold growth, and that 0.3% potassium sorbate must be present to totally inhibit mold growth. However, the combination of one-half levulinic acid and one-half potassium sorbate is more effective than about twice that of either alone in preventing mold growth on diluted or nonintermediate moisture meat-product animal foods.

FISH PRODUCTS

STORAGE IN SULFUR DIOXIDE ATMOSPHERE

One of the most difficult foods to preserve is fresh fish. The seawater bacteria, in conjunction with enzymes and ferments, develop an intensive decomposition activity immediately after killing of the fish. The autolytic processes generally proceed rapidly, producing harmful decomposition products and unpleasant and penetrating fish odor as a result of the formation of organic amines.

Fish are generally harvested far from the location of ultimate consumption, and methods must be used to conserve the fish or to preserve the fish between the time of harvest and use. Pickling, salting, sterilization, freezing, smoking, and antibiotics have been used, but these are either expensive processes or impractical as far as preserving the whole fish is concerned.

The food value of fish may be preserved by deep-freezing the fish immediately upon killing, and various methods and apparatus have been developed for carrying out this process on the vessel on which the fish is harvested. Deep-freezing the fish at sea requires cumbersome and bulky equipment to be carried on shipboard, thereby reducing the capacity of the vessel for cargo. In addition, deep-freezing plants are costly to operate.

E. Levin; U.S. Patent 3,468,674; September 23, 1969 has found that whole or ground fish may be preserved at ambient temperatures when maintained in the presence of an atmosphere of sulfur dioxide gas for periods exceeding a month's time. Sulfur dioxide gas and sulfites have long been known for their preservative properties. Sulfites have been used to preserve sausages and canned meats, but such preservation methods have been discontinued because the concentration of residues left in the product are deemed sufficient to be harmful for human consumption.

He also provides a method of processing wet, fat biological substances which have been preserved or processed in an atmosphere of sulfur dioxide or contain a concentration of sulfurous acid, which results in a product containing only harmless residues in small concentration. The process may be more easily explained by reference to the figures.

Figure 3.1a illustrates a typical conventional fishing trawler 1 which has been equipped for carrying out the processes. The trawler has a hold 2 which contains a plurality of vessels 3, which are adapted to house and contain freshly caught fish. Each of the vessels is provided with an air tight cover 4 which protrudes through the deck of the ship and is accessible from the deck. In addition, each of the vessels 3 has a concave bottom 5 with an aperture at the lowest point thereof connected through a manually actuable valve **6** to a pipeline **7**. The pipeline is connected to a pump **8** located below the vessels **3** and the pipeline and the pump communicate with a port 9 located above the water line.

In accordance with the process, fish are harvested from the sea and thrown directly into one of the vessels **3**. The fish are not preprocessed in any manner whatever, and the fish need not be dead when placed in the vessel. The fish when placed in the vessel, like all fish found in nature, are not sterile, and contain large quantities of seawater bacteria, particularly adjacent to the skin of the fish.

As each vessel is filled, it is sealed and filled with sulfur dioxide gas. A particularly suitable way in which to accomplish this operation is to add bisulfite above the mass of the fish. Any bisulfite which releases sulfur dioxide when contacted with water is suitable, such as sodium bisulfite, metabisulfite, pyrosulfite and other bisulfites. Sodium bisulfite is a preferred sulfite, and may merely be poured in granular form on the mass of fish within the vessel. Sodium bisulfite forms an aqueous solution with the water in and on the harvested fish with the release of sulfur dioxide, as follows.

$$2NaHSO_3 \longrightarrow SO_2 + Na_2SO_3 + H_2O$$

After supplying the granular sulfite, the cover **4** is sealed on the vessel and the valve **6** at the bottom of the vessel is opened. Sulfur dioxide gas also may be directly used from tanks containing the gas under pressure, and it is to be understood that this process is fully equivalent to that of applying granular sulfite.

The pump **8** is utilized to remove any water or other liquid which may be present in the vessel. When sulfur dioxide gas begins to be present at the port, liquid having ceased to be ejected therefrom, the valve located at the bottom of the vessel is closed, and the sulfur dioxide atmosphere within the vessel **3** is secured. The sulfur dioxide penetrates the fish and combines with the water in and about the fish to form sulfurous acid, as follows.

$$SO_2 + H_2O \rightleftharpoons H_2SO_3$$

The reaction is reversible, hence requiring an adequate sulfur dioxide atmosphere to maintain the preserving sulfurous acid. Sulfur dioxide gas in adequate quantity may be formed by sodium bisulfite in the manner described above when applied in a quantity no greater than 4% by weight of the quantity of fish within the vessel. Under normal conditions,

FIGURE 3.1: STORAGE IN SULFUR DIOXIDE ATMOSPHERE

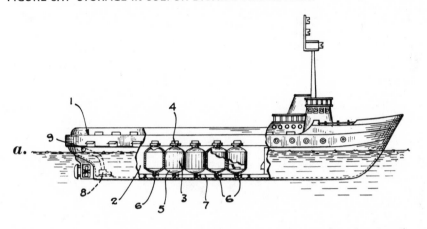

Fishing Vessel for Use in the Process   (continued)

FIGURE 3.1: (continued)

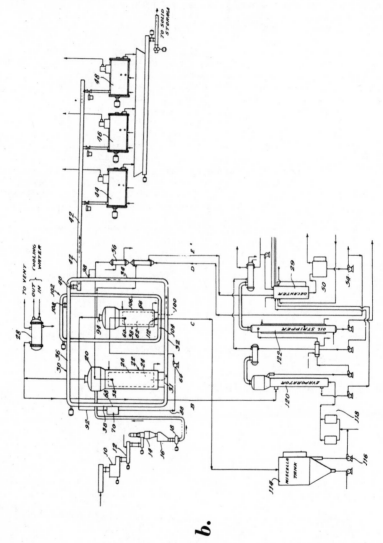

b.

Flow Diagram for Plant Used in the Process

(continued)

FIGURE 3.1: (continued)

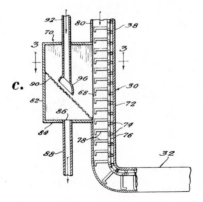

c.

Sectional View of Part of Main Conveyor

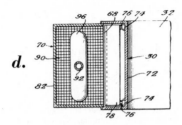

d.

Sectional View Along Line **3–3** of Figure 3.1c

Source: E. Levin; U.S. Patent 3,468,674; September 23, 1969

sodium bisulfite in an amount equal to 2% of the weight of the fish has been found adequate, and in cold weather the quantity of sodium bisulfite may be reduced to 1% by weight of the fish. In warm weather, or hot weather, an additional quantity of sodium bisulfite is usually necessary, generally 3% by weight of the fish present.

When the fish is removed from the vessel, it is not fit for consumption by man or animal and must be processed to remove the sulfurous acid and sulfur dioxide. A particularly suitable manner of processing the fish or meat containing sulfurous acid and sulfur dioxide is to grind the tissue into comminuted particles and subject the particles to azeotropic distillation in an apparatus such as set forth in Figures 3.1b, 3.1c, and 3.1d.

As indicated in Figure 3.1b, biological tissue, such as meat or fish, whether whole or chopped, is transformed into a pumpable fluid by means of a prebreaker **10**, feed screw **12**, disintegrator **14**, and hopper **16**. The small particles of meat or fish are subjected to low pressure steam which is introduced prior to the hopper, specifically into the feed screw. A feed pump **18** is then utilized to pump the pumpable particles into the upper portion or head **20** of a primary cooker **22** or first distillation vessel. The primary cooker is a vertically disposed elongated vessel with the head positioned at the top thereof. A body of substantially water immiscible fat solvent is disposed within the primary cooker below the head thereof. The body of solvent, designated **24**, must have a boiling point under

the conditions of operation of at least 65°C and must form an azeotrope with water preferably boiling below 100°C under operating pressure. The solvent should be selected to
form an azeotrope which will remove substantial portions of water in relation to the amount
of solvent distilled at the operating temperature selected. Among solvents of this class,
ethylene dichloride is a preferred solvent.

Ethylene dichloride has a boiling point at atmospheric pressure of 83°C, and a water-ethyl-
ene dichloride azeotrope boils at 71.5°C. Another example of a particularly suitable solvent
is heptane which boils at 98.4°C at atmospheric pressure. A water-heptane azeotrope boils
at 79.5°C under atmospheric conditions. Other suitable solvents include propylene dichloride, trichloroethylene, perchlcroethylene, and other low boiling chlorinated solvents. Suitable chlorinated solvents may include the bromine, iodine or fluorine derivatives of aliphatic hydrocarbons. In general, a suitable solvent must boil below 120°C under standard
conditions. The hydrocarbon fat solvents including benzene, hexane, toluene, cyclohexane,
heptane, and others are suitable.

The solvent must not be reactive with the tissue constituents or the preservative under
operating conditions and must be capable of being removed by evaporation from the fat
without leaving harmful or toxic residues. The moist particles of fish are continuously
introduced into the boiling solvent within the primary cooker by spraying the particles
into the solvent adjacent to the upper level thereof, or some other suitable means which
will avoid formation of an agglomeration of the particles.

The particles are dehydrated and defatted in the primary cooker, at least to the extent
that they cannot coalesce into lumps or stick to the vessel, but the particles all remain wet
due to the fact that raw tissue particles are continuously introduced, thus maintaining a
water-solvent azeotrope within the vessel. The particles of biological substances within the
first vessel do become heavier than the solvent due to partial drying, however, and tend to
settle toward the bottom of the primary cooker, in spite of the violent boiling of the solvent.

As indicated in Figure 3.1b, the primary cooker is provided with an internal heater **26**
which utilizes low pressure steam as a heat source in order to maintain the body of solvent
under rapid boiling conditions. Vapor from the body of solvent rises through the head of
the primary cooker and is conducted to a condenser **28** which is provided with a flow of
cool water. Both the solvent and the water vapor are condensed to liquid form, and the
solvent and water vapor are separated by a decanter **29**, as is well known in the art. The
hydrogen gas and water vapor are discarded, and the recovered solvent is returned to the
system.

The bottom of the primary cooker is in communication with a runaround main conveyor
30. Granules formed from the fish settle through the boiling body of solvent, pass through
an opening **31** at the bottom of the primary cooker and enter the bottom leg **32** of the
conveyor. The conveyor has four legs **32**, **34**, **36**, and **38** forming a continuous rectangular
path in a vertical plane, and a belt-type conveyor is continuously translated within the four
legs to elevate the granules passing through the opening at the bottom of the primary cooker.
The top leg **36** has an opening **40** which communicates with a horizontal conveyor **42** for
transporting particles to one of a plurality of desolventizers **44**, **46**, or **48**. The desolventizers are used in sequence and operate on the batch system.

A relatively large quantity of solvent is maintained in storage in a work tank **50**, and this
solvent is continuously introduced into a port **52** located near the upper level of the boiling body of solvent in the primary cooker. A pump **54**, and solvent heater **56** are connected in the path between the solvent work tank and the port **52** to provide an adequate
supply of heated solvent to the primary cooker to maintain the level of the body of solvent within the primary cooker.

Operation of the primary cooker results in the body **24** becoming a slurry of solvent,
granules of wet-fat partially dried tissue, and fat which has been extracted from the tissue.

Since raw tissue is continuously being introduced into the primary cooker, and the solvent body is violently boiling, the tissue is not permitted to dry. The relatively wet tissue is present throughout the body of slurry because the azeotropic temperature is maintained while wet tissue is being added continuously. Under these conditions, the miscella is wet, and filtering the miscella would clog the filter. Thus, it is not possible to filter the miscella directly from the primary cooker.

A portion of the slurry which is formed in the primary cooker is continuously introduced into a secondary cooker 58 through a port 60 located in the secondary cooker near the top of a vigorously boiling body 62 of slurry from the primary cooker 22 disposed in a secondary cooker 58. A pump 64 located between the primary cooker and the secondary cooker maintains the body of slurry at a relatively fixed level above the port 60. A heater 66 located within the secondary cooker maintains the slurry under boiling conditions.

The pump 64 is coupled into an opening 68 in the upper portion of the leg 38 of the conveyor. The conveyor travels in a counterclockwise direction and is hence traveling downwardly in the leg 38. The figures illustrate a filter 70 which is utilized to retain as many of the solid particles as possible in the primary cooker and pass only a minimum number of solid particles in the miscella pumped to the secondary cooker.

Figure 3.1c is a sectional view of the filter 70 showing a portion of the leg 38 of the conveyor 30. The conveyor 30 is formed by a continuous tube having a generally rectangular cross section. The inner wall 72 of the tube carries a plurality of rollers 74 which translatably support a plurality of links 76 of a continuous chain. Each of the links carries an L-shaped shoe 78 which catches solid particles passing through the opening 31 from the primary cooker and carries the solid particles through the leg 32 and the leg 34 of the conveyor to deposit them in the outlet 40 in the leg 36 thereof.

As illustrated in Figure 3.1c, the conveyor also has an outer wall 80, and the opening 68 which permits the slurry from the primary cooker to flow to the pump 64 is disposed in this outer wall well below the level of the slurry in the primary cooker. A rectangular fluid-tight box 82 is sealed about the perimeter of the opening, and the box has a bottom 84 with an aperture 86 sealed to a tube 88 which communicates with the inlet of the pump 64.

A screen 90 is sealed within the box 82 on a plane at an angle to the horizontal in order to filter large particles from the flow of miscella to the pump 64, and hence to the secondary cooker. The screen also has the function of limiting the flow of solid particles to the secondary cooker, and hence providing for removal of a large portion of the solid particles impressed upon the system through the conveyor directly from the primary cooker. In practice, approximately 90% of the particles introduced into the primary cooker are removed by means of the primary conveyor, and the screen 90 contributes substantially to this result. If the screen were not present, experience proved that 30 to 40% of the particles would be reworked by pumping the particle laden miscella to the secondary cooker for drying.

The wet miscella causes a glaze to develop and build on the screen, and unless some means is provided, the glaze will clog the screen, even if the perforations of the screen are very large. Even a wire mesh of sufficient size to permit the passage of 30% of the particles from the primary cooker to the secondary cooker will clog unless some means is provided to maintain the screen sufficiently clean to pass the miscella.

A 20-mesh per inch screen will permit flow to the miscella by utilizing dry solvent vapors such as are evaporated from the secondary cooker to keep the screen clean. A tube 92 communicates with an outlet 94 in the head of the secondary cooker 58 and conducts the hot solvent vapors from the secondary cooker to a nozzle 96 confronting the side of the screen opposite the bottom 84 of the box 82. The vapor pressure from the secondary cooker is maintained at about 5 pounds per square inch. The flow of solvent vapors on the screen has two separate functions. The flow of pressurized solvent vapor sweeps the screen clean and open and prevents clogging of any kind to permit the miscella, including the sticky finer particles, to flow through the screen.

The flow of vapor also raises the temperature of any wet particles on the screen to convert the gelatin adhering to the screen by drying into hard solid particles. The solid particles which fail to pass the screen are swept back into the leg **38** of the conveyor, and the conveyor drags the solid particles along the conveyor toward the discharge opening **40** thereof. In this manner, the screen is maintained open for a free flow of miscella in accordance with the demands of the miscella pump.

The slurry in the primary cooker boils at the boiling point of the azeotrope, whereas the slurry in the secondary cooker boils at the boiling point of the solvent. With ethylene dichloride as the solvent, the primary cooker boils at a temperature of approximately 71.5°C at atmospheric pressure, and the secondary cooker boils at a temperature of approximately 83°C at atmospheric pressure. It is thus clear that substantial moisture is present in the primary cooker as a result of the relatively large quantity of wet particles being injected into the slurry of the primary cooker.

Because of the relatively few particles from the primary cooker which enter the secondary cooker and since the particles that do enter the secondary cooker have been partially dried from the raw state, less moisture is introduced into the secondary cooker than is introduced into the primary cooker in the same period of time. As a result, it is feasible and economically practical to supply sufficient heat to the secondary cooker to drive the temperature of the slurry therein to approximately the boiling point of the solvent. Since there is little water present in this secondary cooker, the particles removed therefrom contain very little moisture.

The solid particles from the primary cooker are only partially dried and partially defatted as a result of extraction by the solvent in the slurry. In addition to the fat within the particles, the particles carry with them a quantity of occluded fat. The occluded fat is washed from the particles by a flow of clean solvent introduced into the leg **34** of the conveyor in the upper portion thereof through a port **98**. This flow of clean solvent also extracts fat from the particles, since the quantity of moisture in the particles has been reduced to a level permitting extraction by conventional processes.

A portion of the solvent flowing from the solvent work tank through the solvent pump is used for this purpose. In this manner, the granular solid particles passing through the opening **40** to the horizontal conveyor **42** have a very low fat content. In addition, the countercurrent flow of solvent through the leg **34** and the leg **32** of the conveyor adds to the solvent introduced through the port **52** of the primary cooker to maintain the level of the solvent in the primary cooker and to replace the solvent evaporated by the azeotropic distillation process.

The secondary cooker also is provided with an opening **100** at the bottom thereof, and a second runaround conveyor **102** passes beneath the secondary cooker. The second conveyor has a horizontal leg **104** extending below the secondary cooker, a rising leg **106** which extends to an opening **108** for depositing dried granular meal into the upper leg **36** of the primary conveyor, and hence to the horizontal conveyor. The particles passing through the opening **100** are both dry and defatted, and it is not necessary to introduce a counterflow of fresh solvent in order to wash occluded fat from the particles.

All of the particles in the horizontal conveyor are of low fat content, but those particles from the primary cooker contain substantial moisture. This moisture is removed with the solvent in the desolventizers **44, 46** and **48** to produce a solid product which is granular and contains very little moisture or fat, and hence has great stability. Further, the desolventizers subject the particles to steam for a period of from ½ to 4 hours in order to remove the solvent, and all remaining sulfurous acid is converted to sulfites and sulfates.

A residue of sulfite principally in the forms of calcium sulfite and sodium sulfite is present in the granules after desolventizing, and in addition small quantities of calcium sulfate and sodium sulfate are present in the residue. The granules are then placed in a vessel containing methyl alcohol in an amount equal to approximately 5 times the volume of the granules. The granules are then stirred in the methyl alcohol and the alcohol removed. The process is repeated once again, and the alcohol is removed. Thereafter, the product is dried by the application of heat and vacuum. The use of alcohol is effective to deodorize the product, and at the same time reduces the concentration of the sulfite and sulfate residues. Fish meal produced in this manner has been found to contain approximately 2,000 parts per million of sulfite residues.

The secondary cooker is provided with an outlet port **112** near the bottom thereof, and a miscella is withdrawn through it. It is to be noted that the miscella is highly concentrated in fat, since no fresh solvent is introduced into the secondary cooker, and the secondary cooker concentrates the miscella from the primary cooker. This miscella withdrawn from the secondary cooker is collected in the miscella tank **114**, pumped by a pump **116** through one of two filters **118** to a vacuum evaporator **120**. The solvent is evaporated from the miscella, and the fat is thereafter conducted through an oil stripper **122** to a fat storage tank **124**. The following example further illustrates the process.

Example: One ton of hake is placed in a vessel having a diameter of three feet and a height of approximately nine feet, the vessel containing approximately 63 cubic feet. The fish will fill approximately 50% of the vessel. The fish may be inserted into the vessel without washing or other preliminary treatment. 40 pounds of commercial grade sodium bisulfite is poured on top of the fish when the vessel has been loaded with fish, and the cover of the vessel is then sealed. As an alternative, the atmosphere of the vessel is replaced by an atmosphere of SO_2 gas drawn from a tank containing SO_2 gas under pressure.

All water and other liquid is removed from the interior of the vessel through the port located in the bottom of the vessel, and the port is left open until sulfur dioxide gas begins to flow freely from the port. It takes less than one hour for the sulfur dioxide gas to penetrate the mass of fish within the vessel and begin to flow from the port when sodium bisulfite is used and an even shorter time when applying sulfur dioxide gas from a tank, and at this time the port is closed. The vessel is maintained closed for 4 weeks. Thereafter, the vessel is opened and the fish removed.

The sulfurous acid content of the fish on opening of the vessel is approximately 0.5% by weight. The nutritive value of the fish remains substantially unchanged, that is, the protein biologic quality remains substantially unchanged and the total nitrogen quantity of the fish also remains substantially unchanged throughout the period of storage. There is a slight increase in free fatty acids of the fish during the storage period approximately the same as occurs in deep-freezing over the same period. The fish must be thereafter processed, for example, in a continuous azeotropic dehydrating and defatting process.

The fish is then subjected to pretreatment with steam for approximately 5 seconds to convert part of the collagen to gelatin, and thereafter the whole fish is ground in a prebreaker to particles of less than 0.5" diameter. The 2,000 pounds of hake are introduced into the primary cooker in a continuous stream over a period of approximately 14 minutes. The primary cooker is approximately 65% full of boiling ethylene dichloride, and the pressure within the vessel of the primary cooker is maintained at substantially atmospheric pressure.

The vessel of the primary cooker contains about 800 gallons of ethylene dichloride, and the heating coils of the vessel supplied sufficient heat to maintain boiling at the boiling point of azeotrope, namely approximately 71.5°C. The body of solvent within the primary cooker is kept vigorously boiling. The secondary cooker is identical to the primary cooker in construction and contains a slurry of approximately 800 gallons which is transferred from the primary cooker to the secondary cooker.

The defatted particles are then desolventized in the presence of steam for 4 hours and thereafter deodorized in a methyl alcohol bath with agitation for about 10 minutes. The methyl alcohol bath is repeated, and thereafter the particles are dried by heat and vacuum. The particles are then ground to uniform size. The end product is an odor-free particle of about 1% fat, 1% water, 70% protein, and contains about 2,000 parts per million of sulfite in the form of sodium sulfite and calcium sulfite. The particles are white in color.

AMMONIA PRESERVATIVE

Certain chemical preservatives have been used to preserve fish, such as formaldehyde, sodium chloride, sodium nitrite, sodium hydroxide and various other organic and inorganic preservatives. These preservatives all change the nature of fish drastically. Formaldehyde denatures the protein in the fish; sodium nitrite softens the flesh, as well as having toxic properties, necessitating careful washing of the fish; sodium chloride pickles the fish, changing the taste and dehydrating the flesh; sodium hydroxide pickles the fish, changing the taste and dehydrating the flesh; and sodium hydroxide gelatinizes the tissues, as well as hastening decomposition.

Ammonia has also been proposed for use as a fish preservative. However, this process requires that fish be first eviscerated, immersed in ammonia solution, and enclosed in an airtight storage compartment under a concentrated ammonia atmosphere. Clearly, evisceration is a formidable deterrent to many small fishing vessels, especially when a catch contains many small fish, such as anchovies. For instance, a small fishing vessel is one whose capacity is 100 to 200 tons. At an average weight of 30 grams per anchovy, literally millions of fish are stored; if evisceration is required to preserve each fish, it is much cheaper to take the spoilage loss.

In addition, evisceration tends to prevent decomposition of the fish, even without preservatives. This is due to the removal of the stomach and gut enzymes having rapid digestive activity. The eviscerated fish has none of these enzymes present, and clearly will not decompose as fast as a fish which is not eviscerated.

F.W. Mitchell; U.S. Patent 3,442,661; May 6, 1969; assigned to W.R. Grace & Co. states that when using ammonia as a fish preservative, the weight of ammonia used in relation to the weight of the fish must be closely controlled. When 0.1 to 1% by weight ammonia is used to treat the fish, no evisceration is required to keep the fish in good condition for a reasonable length of time.

The weight of freshly caught fish is estimated, and about 1% by weight of anhydrous ammonia is sprayed over the fish, preferably in an enclosed space, i.e., below deck. The fish are then kept in the hold or fishbox, and are preserved for about 2 weeks or more. At the end of this time, the fish are in excellent condition, firm, yet not tough or gelatinized. The treated fish were not soft and needed no further treatment to restore firmness to the flesh. This shows particularly the criticality of the amount of ammonia used; when greater quantities of ammonia are used with whole (noneviscerated) fish, the fish become soft and remain soft even after being washed.

After storage treatment, the fish are removed and, for many purposes, such as the manufacture of fish meal, may immediately be used without further treatment. However, for fish of the finest quality, with no taste or residue of ammonia discernible, they may be washed with water or seawater. The thus-preserved fish can then be processed by ordinary means into fish meal, fish protein concentrate, or other high protein fish products.

The following table illustrates the advantages of the process. The fish are freshly caught anchoveta, 8 to 10 cm long, about 30 grams each in weight. They were placed in containers and treated with the indicated preservatives. The ammonia used was anhydrous ammonia, although aqueous solutions of ammonia could be used if the weight percent of ammonia used per weight of fish is as indicated. After 48 hours, the liquid was drained off, and enough water was added to cover the fish. After stirring, the liquid containing the soluble and finely divided portions of the fish was drained off and the residue weighed.

Treatment	Residue, wt. %
1% NH_3	86
0.3% NH_3	77
200 ppm formaldehyde	69
400 ppm sodium nitrite	62
Control (untreated fish)	64

By comparison, noneviscerated fish treated with about 1% ammonia and then placed in an air-tight container over 2 N ammonia solution, in a concentrated ammonia atmosphere, had a weight residue of 70%, using the above procedure.

tert-BUTYLHYDROPEROXIDE PRESERVATIVE

J.J. Cavallo and R.A. Reynolds; U.S. Patent 3,622,351; November 23, 1971; assigned to Atlantic Richfield Company reports that tert-butylhydroperoxide (TBHP), which has the following chemical structure:

$$CH_3-\overset{\overset{\displaystyle CH_3}{|}}{\underset{\underset{\displaystyle CH_3}{|}}{C}}-O-OH$$

is an effective bactericidal and/or bacteriostatic agent for the control of bacteria growing on fish. Aqueous solutions of this compound may be sprayed on both the fresh fish and fish fillets. The fish may also be dipped into appropriate aqueous solutions of the compound. The most effective way of utilizing this compound for fresh fish preservation is believed to be by incorporating the compound with ice used for shipboard storage. TBHP in concentrations of from 5 to 50 ppm can economically and effectively be used in preventing the growth of bacteria associated with low temperature (4°C) spoilage of fish. The preferred range is from 20 to 30 ppm dissolved in water.

Example 1: A fillet of halibut was aseptically divided into three approximately equal parts (30-40 g). One part served as a control and was placed unaltered into a petri dish and stored at 4°C. The remaining two parts were dipped in a 20 ppm aqueous solution of TBHP for 1 and 5 minutes, respectively. These samples were also placed in petri dishes and stored at 4°C. Viable cell counts were made on these samples at zero time, 4 days and 12 days. The procedure used for obtaining the viable counts was as follows.

1. A portion of each sample was removed aseptically, weighed (approximately 1.0 gram) and placed in a sterile mortar and pestle.
2. The fish muscle was then ground into very fine particles.
3. 5 ml of sterile saline solution (0.85%) was added and the mixture blended thoroughly.
4. The resultant solution was poured into a centrifuge tube and centrifuged for 15 minutes at 1,500 rpm.
5. The supernatant was transferred to a sterile graduate and diluted with additional sterile saline solution to a volume of 10 ml.
6. Serial 1:10 dilutions were made and plated on Tryptone Glucose Extract Agar.
7. After 2 days incubation at room temperature, the plates were counted. The data obtained from this experiment are presented in the following table.

Viable Cell Titers of Halibut Fillet Treated with TBHP and Stored at 4°C

	---Counts in Viable Cells/Gram Fish---		
Sample	0 Days	4 Days	12 Days
Control—no TBHP treatment	100×10^5	1.2×10^8	5.3×10^8
20 ppm TBHP dipped 1 minute	6.9×10^5	6.9×10^7	4.0×10^7
20 ppm TBHP dipped 5 minutes	1.0×10^5	1.2×10^8	8.3×10^5

These data clearly show that tert-butylhydroperoxide is a good chemical agent for controlling bacterial growth associated with fish spoilage.

Further, organoleptic examination of the fillet samples as to odor, color and texture left little doubt that the tert-butylhydroperoxide had effectively controlled the growth of the active bacterial spoilers. For example, while the control had a very strong and obnoxious rotten fish odor, the TBHP treated samples retained the fresh fish odor originally present. The texture of the control was poor, the flesh being quite soft. The 1 minute dipped sample still had its firm original texture. The 5 minute dipped sample was soft, indicating that the dip time was too long.

Example 2: The following experiment was carried out to demonstrate the effectiveness of tert-butylhydroperoxide against a spectrum of organisms isolated from spoiled fish of many different types. Several concentrations of tert-butylhydroperoxide were incorporated into plates of Tryptone Glucose Extract Agar. These plates were inoculated with 24-hour cultures of 13 bacteria known to be active fish spoilers. The plates were incubated at room temperature (24°C) for 72 hours, after which they were examined for the presence of growth. The results of these studies are presented in the table below.

These data illustrate the bactericidal effect tert-butylhydroperoxide has upon bacteria which are primarily responsible for fish spoilage. The tert-butylhydroperoxide is especially effective against the Pseudomonas species. These organisms are considered by many workers to be the greatest contributors to fish muscle deterioration. In the following table, + designates growth, and – designates no growth.

Effect of TBHP on Bacteria Associated with the Spoilage of Fish
(Agar Incorporated Method)

Bacterium	\-\-\-\-\-\-\-\-\-\-\-\-\-\-Effective Concentration, ppm\-\-\-\-\-\-\-\-\-\-\-\-\-\-								
	0	5	10	15	20	25	50	75	100
Pseudomonas sp. IV-57	+	–	–	–	–	–	–	–	–
Pseudomonas sp. IV-90	+	+	–	–	–	–	–	–	–
Aeromonas sp. 171	+	+	+	+	–	–	–	–	–
Vibrio sp. 80	+	+	+	+	+	–	–	–	–
Achromobacter sp. 16	+	+	+	+	+	+	–	–	–
Achromobacter sp. T-93	+	+	+	+	+	+	–	–	–
Achromobacter sp. C-184	+	+	+	+	+	+	–	–	–
Achromobacter sp. C-205	+	+	+	+	+	+	–	–	–
Pseudomonas sp. I-406	+	+	+	+	–	–	–	–	–
Pseudomonas sp. II-320	+	+	–	–	–	–	–	–	–
Pseudomonas sp. III-322	+	+	–	–	–	–	–	–	–
Acinetobacter moraxella:									
No. 405	+	+	–	–	–	–	–	–	–
No. 4212	+	–	–	–	–	–	–	–	–

PHOSPHATE PEROXIDE PRESERVATIVES

Hydrogen peroxide, due to its oxidizing and bleaching action, has widely been used as a food additive for the purpose of preservation, sterilization, disinfection or bleaching of foodstuff. However, hydrogen peroxide is unstable in itself and is only available in the form of diluted solution. When concentrated, it is accompanied by danger of explosion. This necessarily gives rise to inconvenience in its use, storage or transportation.

For overcoming this inconvenience, an adduct of hydrogen peroxide to alkali metal pyrophosphate, the adduct being available in a crystalline form, has been used, and sodium pyrophosphate or potassium pyrophosphate has also been used as a food additive for the same purpose. These are, in practical use, added to foodstuff or foodstuff material (such as kneading products of meat or fish meat, soybean curd, soybean paste, soysauce, noodles, bread, cereal, starch, gluten, cakes, ice cream, sherbet), and drinking water. It has been found that a dialkali metal hydrogenphosphate peroxide, as expressed by the molecular formula: $M_2HPO_4 \cdot H_2O_2$, where M represents Na or K, is not only more effective and of much longer lasting activity, but is also more conveniently and more safely used than the alkali metal pyrophosphate peroxide. Further details on the preparation and use of these effective preservatives are given by *H. Nakatani and K. Katagiri; U.S. Patent 3,545,982; December 8, 1970; assigned to Takeda Chemical Industries, Ltd., Japan.* The dialkali metal hydrogenphosphate peroxide is produced, for example, by the reaction of 1 mol of the corresponding dialkali metal hydrogenphosphate with 1 or 2 mols of hydrogen peroxide in water or an aqueous solvent, and by concentrating the reaction mixture to dryness by means of evaporation of the solvent under reduced pressure or by spray drying.

For the production of disodium hydrogenphosphate peroxide, 50 grams of crystals of disodium hydrogenphosphate and 35 ml of a 35% aqueous solution of hydrogen peroxide are dissolved in 50 ml of pure water, whereupon reaction takes place. The aqueous reaction mixture is concentrated to dryness under reduced pressure at a temperature not higher than 60°C. The residue is further dried under reduced pressure at 40°C for 10 hours, and is then crushed to powder. Calculated for $Na_2HPO_4 \cdot H_2O_2$ (percent): Na, 26.14; P, 17.61; H_2O_2, 19.32. Found (percent): Na, 26.03; P, 17.31; H_2O_2, 19.50.

For the production of dipotassium hydrogenphosphate peroxide 174 grams of anhydrous dipotassium hydrogenphosphate and 110 ml of a 35% aqueous solution of hydrogen peroxide are dissolved in 500 ml of pure water, whereupon reaction takes place. The aqueous reaction mixture is then spray dried in an air stream at 100°C to give a powdery product. Calculated for $K_2HPO_4 \cdot H_2O_2$ (percent): K, 37.5; P, 14.9; H_2O_2, 16.3. Found (percent): K, 37.0; P, 14.8; H_2O_2, 15.8.

An effective amount of disodium hydrogenphosphate peroxide is about 0.03 to 0.3 weight percent for fish paste, about 0.1 to 0.5 weight percent for meat paste, about 0.2 to 3 weight percent for cheese, about 0.005 to 0.2 weight percent for noodles, about 0.1 to 0.5 weight percent for bread, about 0.05 to 0.2 weight percent for canned peaches, or about 0.01 to 0.1 weight percent for pickles. Any other salt than sodium salt is also used on the same molar basis as above. The peroxide, i.e., dialkali metal hydrogenphosphate peroxide, can directly be added to foodstuff of its material in a solid form or in a composition form with a suitable carrier, such as starch, sugar (e.g., sucrose, lactose or glucose), or proteins (e.g., soya protein, gelatin or casein), as solid carrier and water, ethanol, propylene glycol or in a mixture as liquid carrier. Forms of the foodstuff additive can be in a solid state (e.g., powder, granules) or in a liquid state, but the solid additive composition is usually more convenient in a practical use and for its storage or transportation.

Depending on the kind of foodstuff or the specific purposes involved, the peroxide may be applied to the foodstuff by immersing the foodstuff or its material in the liquid composition or by spraying or sprinkling the compound or its composition onto the surface of the foodstuff or its material. Other foodstuff additives, such as alkali metal phosphate (e.g., sodium dihydrogenphosphate, sodium polyphosphates, potassium hydrogen metaphosphate, sodium hydrogen pyrophosphate, potassium hydrogen tetraphosphate, etc.) and preservation agents, antioxidant agents, coloring agents, etc., can be suitably used in parallel with the dialkali metal hydrogenphosphate peroxide, or as a carrier.

Example 1: The flesh of white croaker was washed with water, dehydrated and minced. To 100 parts by weight of the minced meat were added 3 parts by weight of common table salt, 5 parts by weight of sugar, 5 parts by weight of starch, 0.7 part by weight of monosodium glutamate, 2 parts by weight of mirin (a sweet sake), 0.2 part by weight of sorbic acid, and 3 parts by weight of water. The mixture was crushed to prepare a basic crushed meat.

Sample 1: The basic crushed meat.
Sample 2: The crushed meat, prepared by incorporating sodium pyrophosphate peroxide into the basic crushed meat.
Sample 3: The crushed meat, prepared by incorporating disodium hydrogenphosphate peroxide into the basic crushed meat.

Each sample was cased with a film of rubber hydrochloride and then heated at 85° to 90°C for 45 minutes. Each product thus produced was preserved in a Petri dish under constant conditions of temperature (35°C) and relative humidity (80%). The development of sliminess and of rotten smell of these samples were checked after preservation for 20, 50, 100, 150 and 200 hours. The results are shown below; + indicates development, and - indicates no development.

		0	20	50	100	150	200
Sample 1 (control)	Sliminess	-	+	+	+	+	+
	Smell	-	+	+	+	+	+
Sample 2 (control)	Sliminess	-	-	+	+	+	+
	Smell	-	-	+	+	+	+
Sample 3	Sliminess	-	-	-	+	+	+
	Smell	-	-	-	+	+	+

Example 2: A mixture of pork, beef, mutton and beef tallow, in a proportion of 15, 25, 40 and 20 parts by weight, respectively, is kneaded to obtain a paste. To the kneaded paste is added 0.2 part by weight of sorbic acid. The mixture is again kneaded to prepare a basic meat paste.

Sample 1: The basic meat paste itself; temperature and constant moisture at 30°C, 80%.

Sample 2: The meat paste prepared by incorporating 0.5 part by weight of sodium pyrophosphate peroxide into the basic meat paste.

Sample 3: The meat paste prepared by incorporating 0.5 part by weight of disodium hydrogenphosphate peroxide into the basic meat paste.

Each sample was preserved in a Petri dish under constant conditions of temperature (30°C) and relative humidity (80%). The growth of mold and development of rotten smell or taste of each of the above samples were checked after preservation for 24, 32, 45, 70 and 95 hours. The results are shown below; + indicates mold growth or development, ± indicates faint mold growth or faint development, and - indicates no mold growth or development.

		0	24	32	45	70	95
Sample 1 (control)	Mold	-	±	+	+	+	+
	Rot	-	-	+	+	+	+
Sample 2 (control)	Mold	-	±	+	+	+	+
	Rot	-	±	+	+	+	+
Sample 3	Mold	-	-	-	±	+	+
	Rot	-	-	-	±	+	+

Example 3:

Sample 1: Soybean curd (tofu), commercially available, preserved in tap water at 30°C.

Sample 2: The same soybean curd as in Sample 1, preserved in a 0.2% aqueous solution of potassium pyrophosphate peroxide at 30°C.

Sample 3: The same soybean curd as in Sample 1, preserved in a 2.0% aqueous solution of dipotassium hydrogenphosphate peroxide at 30°C.

The growth of mold and development of bad smell or ill taste caused by putrefaction of each of the above samples were checked after preservation for 24, 32, 48, 72 and 96 hours. The results are shown on the following page; + indicates mold growth or development, and - indicates no mold growth or development.

		\-\-\-\-\-\-\-\-\-\-\-\-\-\-\-\-Time, hrs.\-\-\-\-\-\-\-\-\-\-\-\-\-\-\-\-					
		0	24	32	48	72	96
Sample 1 (control)	Mold	−	+	+	+	+	+
	Rot	−	+	+	+	+	+
Sample 2 (control)	Mold	−	−	−	−	+	+
	Rot	−	−	−	−	+	+
Sample 3	Mold	−	−	−	−	−	+
	Rot	−	−	−	−	−	+

SORBIC ACID–ANTIOXIDANT COMBINATIONS

J.T.R. Nickerson and L.D. Starr; U.S. Patent 2,933,399; April 19, 1960; assigned to Dirigo Sales Corporation state that insofar as oxidative rancidity of processed animal tissue is concerned, antioxidants have been effective, both the fat-soluble type such as butylated hydroxy anisole (BHA), butylated hydroxy toluene (BHT), catechin, quercetin, and 2,6-dimethoxyphenol (DMP), and the water-soluble or synergistic type, such as ascorbic acid, citric acid and phosphoric acid, either singly or in combination. However, such antioxidants have little or no effect in inhibiting the growth of those decomposing and putrefaction-producing bacteria that are mainly responsible for the short storage or shelf life of processed tissue; and it has been found that bactericidal agents while inhibiting the growth of putrefaction-producing bacteria often facilitate oxidative rancidity since they prevent the production of reducing substances formed by bacterial growth.

It was found that when a relatively small amount of sorbic acid, and/or a water-soluble sorbic acid salt is applied to the exposed surface of fresh animal tissue, together or in conjunction with an antioxidant, that the growth of decomposing and putrefaction-producing bacteria is not only inhibited at refrigerator temperatures above freezing, but that sorbic acid and its water-soluble salts, unlike certain bactericidal agents such as chlorine and hydrogen peroxide, do not interfere with the action of antioxidants used in preventing oxidative rancidity, or in any way adversely affect the appearance, color or palatability of the processed tissue. In other words, antioxidants and sorbic acid, its water-soluble salts and esters, when used in conjunction, are synergetic in that neither interferes with the effectiveness of the other and and both cooperate to inhibit deterioration of mechanically processed fresh animal tissue.

The method of incorporation of the sorbic acid compound will depend on the particular type or character of the tissue. For example, with mixtures such as pork sausage which contain seasoning, the dry sorbic acid compound along with the antioxidant may first be mixed with the spice or seasoning which is later incorporated with the pork trimmings. Fresh fish fillets, eviscerated and cut poultry may be sprayed with or immersed in an aqueous solution and/or dispersion of a mixture of the sorbic acid compound and antioxidant for a period of time sufficient to permit the exposed surfaces to absorb the desired amount.

In preparing hamburger and beef sausage, the antioxidant and the dry sorbic acid compound may be sprinkled on or applied to the exposed surface of the beef prior to grinding, but if desired the sorbic acid compound and antioxidant may first be mixed with a suitable carrier so as to assure greater or more uniform distribution.

Where a synergistic or water-soluble antioxidant is to be used in conjunction with one or more fat-soluble types, an aqueous dispersion may be prepared by dissolving the antioxidants in an emulsifier of the sorbitan derivative type, or in propylene glycol, then adding the solution to water containing the sorbic acid compound and dispersing.

The amount of fat soluble antioxidants, i.e., diphenols or compounds of similar electron configuration, may vary from a minimum of the order of 0.005% based on the weight of animal tissue, to the limit of common or practical usage, which is of the order of 0.05%; and the same is true for the water-soluble or synergistic type antioxidants, although as a practical matter it might be desirable to use the latter type in somewhat higher concentrations, e.g., up to 0.2%, but when the fat soluble and water-soluble types are used together, lesser amounts of each are used. The following table shows typical antioxidant formulations.

	A	B	C	D	E	F
*BHA, percent	0.01	0.01	0.01	--	--	--
*BHT, percent	0.01	0.01	0.01	--	--	--
*DMP, percent	--	--	--	0.01	--	--
Catechin, percent	--	--	--	--	0.01	--
Quercetin, percent	--	--	--	--	0.01	--
*Propylene glycol, percent	--	0.1	--	--	--	--
*Sorbitan trioleate, percent	--	--	0.1	--	--	--
Ascorbic acid, percent	--	--	--	--	--	0.1
Citric acid, percent	0.005	0.005	0.005	0.005	0.005	--

*Formulations in which the indicated ingredients were dissolved in fat.

In each of the above formulations phosphoric acid or ascorbic acid may be substituted for the citric acid without materially interfering with the effectiveness of the antioxidant mixture; and other fat soluble antioxidants such as propyl gallate or dihydroguaiuretic acid (NDGA), etc., may be substituted.

Example 1: A fish fillet from one side of a haddock was washed and placed in storage at 36° to 40°F to serve as a control, and the fillet from the other half of the haddock was immersed for 30 seconds in a dispersion comprising 0.25% of sodium sorbate, 0.25% of sorbic acid, and 0.2% of ascorbic acid as the antioxidant. Both control and treated samples were held at 36° to 40°F for a period of 8 days. Bacterial counts and organoleptic observations were made after 3, 5 and 8 days of storage, and the results are indicated in the following.

	Time Held, days	Odor	Standard Plate Count, Bacteria per Gram (average)
Plain fillet	3	Fishy	3,800,000
Fillet treated with sorbic acid and sodium sorbate antioxidant mixture	3	Bland	850,000
Plain fillet	5	Stale	15,000,000
Fillet treated with sorbic acid and sodium sorbate antioxidant mixture	5	No off odor	2,400,000
Plain fillet	8	Very stale	57,000,000
Fillet treated with sorbic acid and sodium sorbate antioxidant mixture	8	Slightly stale	16,000,000

Example 2: A 50-pound block of frozen pork containing about 50% fat and 50% lean trimmings was defrosted at 36° to 38°F and cut up into small pieces. The cut trimmings were then mixed with a seasoning having the following composition: 1.5% salt, 0.375% black pepper, 0.125% rubbed sage, and 0.15% dextrose. The mixture was then passed through a meat grinder having a ⅛" plate.

In the case of the test batch, 0.01% BHA and 0.01% DMP were added to the seasoning along with 0.1% sorbic acid, but with the control batch the sorbic acid was omitted. For test purposes, sample patties of both batches were made and put into sterile Petri dishes for storage at 40° to 43°F. Taste tests, peroxide number tests (Stansby) and standard plate count tests were run prior to storage and periodically thereafter.

No appreciable change in taste was noted after a period of two weeks in samples containing sorbic acid, BHA and BHT, and there was no increase in the peroxide number. The standard plate count tests were as shown on the following page.

| Days in Storage | Standard Plate Count, Bacteria per Gram | |
	Control	Test
1	140,000	80,000
3	200,000	90,000
5	720,000	13,000
9	33,000,000	23,000
12	50,000,000	66,000
14	57,000,000	140,000

Substantially identical results were attained when the test batch also included 0.005% citric acid.

Example 3: A quantity of beef with fat was ground through a plate with ⅜" holes three times to serve as the control ground beef. A second quantity of beef with fat was cut into cubes about 2 cubic inches in size. A mixture of 0.15% by weight of sorbic acid and 0.1% of antioxidant (Formulation F) was then sprinkled over the cubed meat. This material was then ground through a plate with ⅜" holes three times. This served as the ground beef treated with sorbic acid.

Both the control ground beef and the beef treated with sorbic acid were held at 36° to 40°F over a period of 8 days. Bacterial counts and organoleptic observations were made after 3, 5 and 8 days of holding under refrigeration, and the results are indicated in the following.

Sample	Time Held, days	Odor	Standard Plate Count, Bacteria per Gram
Plain ground beef	3	Slightly off	25,000,000
Ground beef with sorbic acid, antioxidant mixture	3	No off odor	23,000
Plain ground beef	5	Sour	260,000,000
Ground beef with sorbic acid, antioxidant mixture	5	No off odor	130,000
Plain ground beef	8	Very sour	9,000,000,000
Ground beef with sorbic acid, antioxidant mixture	8	No off odor	390,000

PROPIONIC–BENZOIC ACID COMBINATIONS

B. Gonthier and J. Mocotte; U.S. Patent 3,600,198; August 17, 1971; assigned to Progil, France found that synergistic effects in biocidal properties are obtained with mixtures of a buffered mixture of propionic acid-metal propionate and of a buffered mixture of benzoic acid-metal benzoate, the total pH between 4 and 5. The following are examples of such compositions:

propionic acid-sodium propionate and benzoic acid-sodium benzoate
propionic acid-magnesium propionate and benzoic acid-sodium benzoate
propionic acid-sodium propionate and benzoic acid-magnesium benzoate
propionic acid-magnesium propionate and benzoic acid-magnesium benzoate

The respective ratios of alkali metal or magnesium salt to the organic acid in the buffered systems depend upon the desired pH. The quantities of each buffered system in the mixture may be chosen within a large range, the volume ratios of propionic acid and propionate to benzoic acid and benzoate being preferably between 0.5:1 and 99:1. The efficient quantities of the biocidal mixtures to obtain protection and to extend fish preservation time depend upon several factors, such as: fish type (from seawater or freshwater); time chosen for treatment (immediately after fishing or N days after it); application mode (incorporation in ice or fish washing water); duration, etc. The use of these compositions may take place in different ways. According to one method of application, the additive is introduced into water which, after freezing, is used as a germicidal ice for fish preservation. In a variant of those techniques, the composition is introduced directly onto ordinary ice, for example, by means of sprinkling. In another method, the required proportions of bactericidal additive are incorporated into the washing water used on whole or eviscerated fish which have been stored in ordinary ice, or in ice already treated as above. Further, compositions can be incorporated into the brine for storing fish pieces or waste intended after drying for flour and powder fabrication for cattle feeding to inhibit putrefaction phenomena and to avoid losses in nutritious substances. It is also possible to treat fish pulverulent fluors directly by spraying, for example, on the food, an adsorbate of the diluted solutions of the buffered mixtures deposited on an inert filler.

The methods used for determining and measuring fish deterioration, in the following tests, are the recommended ones, that is: insulation of bacteria from the fish skin and muscles, then measure the biocidal mixture efficiency according to the capacity of the bacteria to develop or not; examination of fish freshness state by means of macroscopic characteristics (color, smell, stiffness, etc.); organoleptic tests in which there are tastings of reference fishes and treated fishes; and analytical determination of nitrogen-containing bases produced by autolysis and fish bacterial degradation, after the stage of cadaveric stiffness. In the last method, the whole of the steam-volatile bases is determined under the form of total volatile basic nitrogen. In the following examples, the results are expressed under the shortened term TVBN (total volatile basic nitrogen), in mg of nitrogen per 100 g of fish.

Example 1: A germicidal ice was made by freezing water to which had been added 2 grams per liter of a mixture (A) having 95 parts by volume of the buffered propionic acid-sodium propionate of pH 4.5 and 5 parts by volume of the buffered benzoic acid-sodium benzoate of pH 4.5. Simultaneously for purposes of comparison, there were prepared 2 other germicidal ices from a 2 g/l aqueous solution, containing either the buffered mixture (B)—propionic acid-sodium propionate of pH 4.5, or the buffered mixture (C)—benzoic acid-sodium benzoate of pH 4.5.

Whitings eviscerated and stored for 4 to 6 days after being caught were divided into 4 lots which have been stored as follows: the first lot in ordinary ice, free from bactericide; the second lot in ice containing the mixture (A) according to the process; the third and fourth lots in ice containing, respectively, product (B) and product (C).

After 6 days of preservation, the fish lots were separated from their ice and measurements of total volatile basic nitrogen were made as described above. Fishes stored in ordinary ice (blank) presented a TVBN of 36 mg/100 g of fish, and the ones preserved in ices (B) and (C) gave, respectively, the numbers of 23 mg/100 g and 32 mg/100 g. However, fishes kept in the germicidal ice (A) presented a TVBN of only 18. Moreover, fish meat of this last lot was firm and tasting has proven that it had kept all its flavor, while the fishes of the reference group were in a decomposition state and the ones of groups (B) and (C) were consumable only with difficulty.

During a continuation of the tests for a further month, the following TVBN numbers were obtained: blank, 70; ice (B), 40; ice (C), 67; ice (A), 21 (the mixture according to the process).

Example 2: The same types of tests and comparisons as in Example 1 were made with whole trouts, which since catching have been kept in 4 ice lots with bactericide concentrations, which were identical to the ones used in Example 1. After 30 days of storage in a cooler, TVBN measures were made and the following results were obtained.

Blank	45
Group (A) fish	23
Group (B) fish	29
Group (C) fish	32

In other series of longer tests, it was determined that the TVBN number passed only from

23 to 26 after 5 additional days, while the numbers 29 and 32 of the comparative tests reached, respectively, 40 and 65 after the same time.

Example 3: Pieces of whiting muscles caught 8 days earlier were placed for 2 days in aqueous solutions containing 2 g/l of the following products at a temperature of 25°C. Group (D)—buffered system: propionic acid-sodium propionate at pH 4.5. Group (E): propionic acid-sodium propionate plus benzoic acid-sodium benzoate with pH 4.5; the ratio by volume of the two constituents being 70:30. After 48 hours of immersion, the following results were obtained by TVBN measurement (in mg/100 g of fish).

Blank (untreated water)	430
Water treated with (D)	33
Water treated with (E)	27

Example 4: Pieces of whiting muscles, caught 8 days earlier, were immersed for 3 and 7 days, respectively, at a temperature of 25°C in aqueous solutions containing 2 g/l of the following products (pH of 4.5).

Group (C): benzoic acid-sodium benzoate.
Group (F): propionic acid-magnesium propionate.
Group (G): 95 parts by volume of the buffered propionic acid-magnesium propionate and 5 parts of the buffered benzoic acid-sodium benzoate.
Group (H): 50 parts by volume of the buffered propionic acid-magnesium propionate and 50 parts of the buffered benzoic acid-sodium benzoate.

After immersion, TVBN was measured and the following results were obtained.

Days	3	7
Blank (nontreated water)	630	980
Water treated with:		
(C)	168	--
(F)	38	58
(G)	35	51
(H)	32	39

As may be seen from the results of the tests shown in the four examples above, bactericidal mixtures have an efficiency greater than the efficiency of either propionic or benzoic acids buffered by Na or Mg salts. It will be noted also that the buffered solution, propionic acid-Mg propionate, also gives extremely interesting results.

CHELATING AGENTS

T.E. Furia; U.S. Patent 3,563,770; February 16, 1971; assigned to Geigy Chemical Corp. discusses the use of antibiotics and chelating agents for the preservation of fish. Various steps have been taken in an attempt to increase the storage time of fresh fish packed in ice. The most widely employed method until recently consisted in dipping fresh fish for about 20 seconds in an aqueous bath containing approximately 10 parts per million of an antibiotic such as chloro- or oxy-tetracycline. This method usually resulted in increasing the storage life of the fish fillets to about 10 to 20 days when packed in ice.

Because of the potential undesirable consequences possible by the ingestion of uncontrolled quantities of antibiotics by humans, the Food and Drug Administration has banned the application of such antibiotics to fish. This has severely limited the available market area for freshly caught fish of many distant producers because of the greatly contracted permissible storage period and/or increased costs through the necessity of using more rapid transportation facilities. It was found that such chelating agents as ethylenediaminetetra-acetic acid, its sodium salts, sodium citrate, etc., increase the storage time of fish. For instance, about 250 ppm of ethylenediaminetetraacetic acid, when deposited upon fish, have been found to increase the permissible storage time of the fish to approximately six days. Another chelating agent, diethylenetriaminepentaacetic acid, when applied in similar amounts to fresh fish fillets, provides approximately eight to nine days protection.

Combinations of chelating agents and the above-mentioned antibiotics have been used in the processing of fresh fish into various food products obtainable from the fish by a digestive process. The preservative agent, such as oxytetracycline or chlorotetracycline, is added to prevent bacteria growth, to control undesirable loss of nutritive value of the fish products made by breaking the fish down through an enzymatic digestive process. In this instance, the chelating agents are used to maintain the preservative action of the known preservatives, oxytetracycline and chlorotetracycline.

It has been found that the storage life of fresh fish can be increased substantially over the time of storage previously possible by use of chelating agents alone, and without antibiotics. For example, the storage time of iced fresh fish is greatly enhanced by applying to the fish a composition comprising from about 500 parts to about 999 parts of:

(A) A compound of the formula:

$$\text{HOOCCH}_2 \underset{\text{HOOCCH}_2}{\overset{}{\diagdown}} \text{N} \left[\text{—CH}_2\text{CH}_2\text{—N—} \underset{\underset{\text{COOH}}{\overset{|}{\text{CH}_2}}}{} \right]_x \text{CH}_2\text{CH}_2\text{—N} \underset{\text{CH}_2\text{COOH}}{\overset{\text{CH}_2\text{COOH}}{\diagup}}$$

where x is an integer of from 0 to 6, or a water-soluble alkali metal, or ammonium or amine salt; and from about 1 part to about 500 parts by weight of (B), a compound of the formula:

$$R \underset{R_1}{\diagup} \diagdown \text{—O—} \overset{\text{HO}}{\diagup}\diagdown \text{—Cl}$$

where R is a halogen atom, and R_1 is either chlorine or hydrogen. Particularly useful compounds falling within the scope of (A) are ethylenediaminetetraacetic acid and ethylenetriaminepentaacetic acid, and the alkali metal, ammonium, and amine salts. In the compound of (B), R is preferably chlorine and R_1 is preferably chlorine.

Specifically preferred compounds within the scope of (B) are: 2-hydroxy-3',4,4'-trichlorodiphenyl ether, and 2-hydroxy-4,4'-dichlorodiphenyl ether. One of the advantages in using the compounds of (B) in this composition is that they generally volatize at relatively low temperatures and thus under most cooking conditions will be removed from the fish.

In evaluating the comparative effectiveness of various compositions for the preservation of fresh fish in ice, the following procedures were followed. Fresh fillets of haddock caught in Middle Atlantic waters were cut into circular plugs measuring three inches in diameter using sterile techniques. The test plugs were then dipped for 20 seconds in the aqueous test solutions, drained for 30 seconds, and placed in sterile plastic Petri dishes. The Petri dishes were wrapped in polyvinylidene chloride film, packed in crushed ice, and placed in a refrigeration compartment where the temperature was maintained at between 31° and 33°F. Double blind odor evaluation tests were run on each sample once daily by two judges. The odor of each sample was evaluated as acceptable or unacceptable.

The aqueous solution employed in all instances comprises 50% water and 50% food grade propylene glycol. The compounds of (A) are only slightly soluble in water and the use of propylene glycol or a similar agent permits utilization of the compounds in solution levels of over 1% concentration.

In the following table, the effectiveness of two representative agents used previously is indicated. Na_2EDTA is a sodium salt of ethylenediaminetetraacetic acid and Na_2DTPA is a sodium salt of diethylenetriaminepentaacetic acid.

Percent Na_2 EDTA	Percent Na_2 DTPA	Control [1]	Days of protection at 31-33° F.
		x	3
0.1			3
0.3			3
0.5			4
0.7			5
0.9			6
1.0			7
	0.1		3
	0.3		3
	0.5		5
	0.7		6
	0.9		8
	1.0		9

[1] Water or 50:50 propylene glycol (F.G.)/water.

The next table, below, illustrates the lack of protective value of representative compounds of Formula (B) in preserving freshly caught fish from deterioration. The Compound B_1 is 2-hydroxy-3',4,4'-trichlorodiphenyl ether and Compound B_2 is 2-hydroxy-4,4'-dichloro-diphenyl ether. Only at the relatively high concentration of 1.0% do the compounds of Formula (B) exhibit any, even minor degree, of preservative action. Such higher concentrations of Formula (B) compounds are generally not too practical because of higher costs and a tendency to impart a slightly sweet taste.

Percent compound B_1	Percent Compound B_2	Control [1]	Days of protection at 31-33° F.
		x	3
0.001			3
0.01			3
0.10			3
1.0			4
	0.001		3
	0.01		3
	0.10		3
	1.0		4

[1] Water or 50:50 propylene glycol (F.G.)/water.

The graph on the following page (Figure 3.2) dramatically demonstrates the greatly increased storage time imparted to fresh fish fillets by the special treatment. The left-hand side of the graph shows the number of days that fish fillets are preserved when stored in ice at 31° to 33°F. At the bottom of the graph, the various percentages of four representative components of the compositions are indicated. In each instance, the total of solution components is 1.0%. In this test, where the solution comprises 1.0% of either Compound B_1 or B_2, there was no increase in the days preserved over the time for untreated fish.

The sodium salt of EDTA when used alone at 1.0% concentration increases the preservation time to six days and the sodium salt of DPTA increases the preservation time to nine days. As illustrated, the addition of minor amounts of either Compound B_1 or B_2 increases the preservation time up to six additional days in many instances, which is very unexpected when it is realized that neither Compound B_1 or B_2, when present at a full 1% concentration, increases the preservation time over that of the untreated fish. Concentrations of solutions in the range of about 0.5 to about 2.0% by weight of solution are recommended.

A dip time of about 20 seconds with a 1.0% concentration is very satisfactory, but the time suitably may range from about 10 seconds to one minute, depending on the solution concentration and pick-up desired. In general, very satisfactory results are obtained when from 50 to 500 ppm of the composition are applied on the fish.

The fish may be surface sprayed with the composition, or a powder form may be dusted on, or the composition may be included in the ice. Other alternatives include injection into the vascular system of the fish.

FIGURE 3.2: PRESERVATION OF FISH

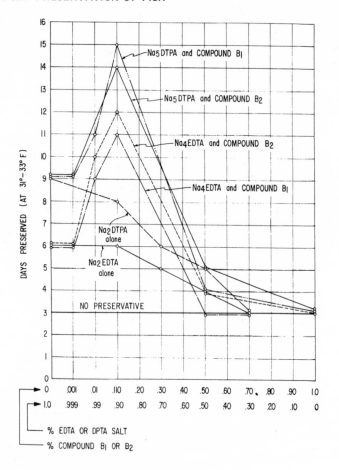

Source: T.E. Furia; U.S. Patent 3,563,770; February 16, 1971

EDTA SALT TABLET

In the processing of canned seafoods such as shellfish, crustaceans, haddock, cod and other members of the Gadidae family, and salmon, it has been previously proposed to treat the seafood prior to canning with an ethylenediaminetetraacetic acid (referred to as EDTA) compound in order to preserve the natural color, flavor and appearance of the canned seafood and to prevent the formation of struvite crystals in the can. Since brine is commonly used in the canning of seafoods, it would be highly desirable to provide to the canning trade a tablet containing predetermined amounts of sodium chloride, EDTA compound, and, if desired, other canning and preserving additives. One or more of such tablets could then be simply added to the seafood in the can or at a suitable stage to provide in situ in the can the desired treating medium containing the sodium chloride and EDTA compound.

The details of the process are described by *C.R. Fellers; U.S. Patent 3,013,884; Dec. 19, 1961;*

assigned to The Blue Channel Corporation. The sodium chloride should be composed of particles of at least about 85%, and preferably at least 95%, of which have a size passing a No. 30 screen which has 0.59 mm openings. The tablet should be substantially free of iron, magnesium and copper, and should preferably contain less than about 1 ppm of copper or iron, and less than about 10 ppm of magnesium. The water-soluble EDTA compound is the calcium disodium salt, but other compounds may be used, such as EDTA and the mono-, di-, tri- or tetra-sodium, potassium, lithium or ammonium salts of EDTA. Such compounds are generally available as fine powders of a more or less sticky nature.

The cohesive agent must provide the particles of the tablet with cohesive forces towards each other exceeding the adhesive forces of the particles towards the wall of the tablet punches or presses, whereby fouling and caking of such punches or presses is eliminated and firm, smooth-surfaced tablets are obtained. Such agents may be selected from among the substances generally referred to as lubricants, mold lubricants and/or binders. Although calcium stearate is preferred, calcium and aluminum salts of higher fatty acids such as stearic, palmitic, oleic, linoleic, behenic and lauric acids may generally be used. Other cohesive agents useful include starch, degraded starches such as dextrin, sugar syrups such as molasses, glucose, sorbose, gelatin, casein, glue, saponine, hemi-cellulose (Tragasol), gum acacia, tragacanth, agar, methyl cellulose, carboxymethyl cellulose, polymerized ethylene glycol (Carbowax), polyvinyl alcohol, polymerized N-vinyl lactams such as polyvinyl-pyrrolidone (PVP), lecithin, hydrated silica, silica gel and silicates.

Improved results in the treatment of seafoods are in some instances obtained by inclusion in the tablets of a water-soluble inorganic aluminum salt such as aluminum sulfate or sodium aluminum sulfate in proportions of 2 to 5% by weight of the sodium chloride. An iron-free aluminum salt should be used which dissolves to a clear solution without cloudiness. It should be composed of particles passing a No. 10 or No. 20 screen, and may be mixed into the composition at any stage of the process prior to tablet pressing. However, unless corrosion resistant equipment is used, it is preferred to add the aluminum salt together with the cohesive agent. In the following examples, all parts are by weight in grams unless otherwise indicated.

Example 1: Into a P–K twin shell blender (with intensifier bar) is charged 6,543 grams of sodium chloride and 226 ml of a 50% solution of the calcium disodium salt of EDTA, and the blender run until a thoroughly mixed, smooth stiff paste is obtained. This paste is dried in a tray type hot air oven at 110°C and the dried mixture ground in a grinding mill into particles about 95% of which pass through a No. 30 stainless steel screen. This ground mixture is then thoroughly mixed in the blender with 7.5 grams of calcium stearate and 141 grams of iron-free aluminum sulfate screened to remove +No. 20 screen particles. The resulting mixture is pressed without fouling or caking of the punches into 75 grain tablets.

Example 2: The procedure of Example 1 is repeated except that the tablets thus produced are reground into particles about 95% of which pass through a No. 30 screen, and the reground mixture again pressed into 75 grain tablets. This additional procedure is found to eliminate the tendency of some of the tablets to cap as a result of an effect akin to case hardening. The resulting tablets are uniformly satisfactory in quality, homogeneity, etc.

Example 3: Into the P–K blender is charged 240 grams of sodium chloride, 15.0 grams of the calcium disodium salt of EDTA, 37.6 grams of iron-free aluminum sulfate screened to remove +No. 20 screen particles, and 10.0 ml of water. The blender is run until a thoroughly mixed smooth stiff paste is obtained. This paste is dried in the hot air oven at 110°C and the dried mixture ground into particles about 95% of which pass through a No. 30 screen. The ground mixture is then thoroughly admixed in the blender with 615 grams of sodium chloride and 0.9 grams of calcium stearate, and the resulting mixture pressed without difficulty into 75 grain tablets of acceptable quality.

In the initial step of forming a homogeneous paste, sufficient water is used, usually at least about 0.05% by weight of the sodium chloride to yield a paste with the remaining components of the mixture, and the water may contain some, or more preferably, all of the EDTA compound. Similarly, the paste may contain all the sodium chloride to be included in the tablets, or only a portion, e.g., in proportions of 1 to 10 parts per part by weight of the EDTA compound, in which case the balance of the salt is added at a stage subsequent to the drying of the paste and prior to the tablet pressing step.

In some instances it is desirable, in the interest of homogeneity and prevention of fouling of the tablet punches, to force the paste prior to drying the same through a No. 10 screen (2 mm openings) or a No. 20 screen (0.84 mm openings). Improved results may in some instances be attained by applying the regrinding and tablet repressing procedure of Example 2 to the tablets produced in accordance with the procedures of Example 3. Any other form of heat drying equipment may be used instead of the hot air oven, supplemented if desired by application of a vacuum. Tablets of any size and shape may be formed, and any suitable type of tablet-making machine may be used. The sodium chloride used in the previous examples has a particle size distribution as follows:

>5% fails to pass through a No. 30 screen with 0.59 mm openings
>25% fails to pass through a No. 40 screen with 0.42 mm openings
>42% fails to pass through a No. 50 screen with 0.297 mm openings
>16% fails to pass through a No. 60 screen with 0.250 mm openings
>8% fails to pass through a No. 70 screen with 0.210 mm openings
>4% passes through a No. 70 screen

The above screen ratings are those of the U.S. sieve series. In the examples, the grinding operations are carried out with any suitable equipment in a manner effective for producing a particle size distribution approximating that of the sodium chloride.

ENZYME–PRESERVATIVE COMBINATIONS

Enzymes and preservatives are used by *C.W. Keyes and W.W. Meinke; U.S. Patent 3,249,442; May 3, 1966* in the process shown in the flow diagram (Figure 3.3).

FIGURE 3.3: FLOWSHEET FOR FISH PRESERVATIVES

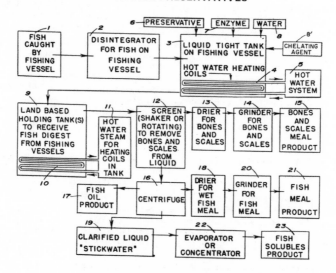

Source: C.W. Keyes and W.W. Meinke; U.S. Patent 3,249,442; May 3, 1966

The fish are caught aboard the fishing vessel as indicated by block **1**. As the fish are caught they are fed to a disintegrator, chopper or grinder, indicated at **2** aboard the fishing vessel. The disintegrated fish are then fed into a liquid-tight tank **3** aboard the fishing vessel which may, however, have an open top. Tank **3** may be fitted with heating or cooling coils **4** in the bottom of the tank. The heat for the hot water system to supply the coils in the tank may be utilized either from the waste heat of the engine of the fishing vessel, or from an independent fuel-fired hot water heater **5**. In many cases, no heat need be added, and alternatively, cooling provided by pumping seawater through the coils at seawater temperature may be utilized. The tank **3** is insulated to conserve heating or cooling and to prevent transfer of heat to or from other parts of the fishing vessel. Some heating may at times be desirable to accelerate the enzymatic process.

In general, the preservation and storage of fish material aboard the fishing boat and the transfer of digested fish to land-based holding tanks may be accomplished at ambient temperatures. However, elevated temperatures are used in the land-based tanks to aid in the separation of fish oil from the other components and to provide for cleaner separation of bones and scales from the liquid digest. Also, elevated temperatures are essential to the drying of the recovered fish solids and for concentrating the stickwater to fish solubles.

Additives introduced into the tank **3** are indicated as the preservative **6**, enzyme **7**, and water **8**, and a chelating agent **8'**. When the fishing vessel reaches port, the contents of the tank on the vessel are pumped to land-based tanks represented at **9** for storage and further processing, it being noted that additional water for pumping purposes need not be added. These tanks **9** are equipped with hot water or low pressure steam coils **10** heated from the heater **11**, which will permit the attainment and holding of a temperature of 70° to 80°C for a period of 30 to 60 minutes. This heating is provided to inactivate enzyme activity and to permit easier removal of bones and scales from oil and liquid digest.

The land-based tanks **9** serve as auxiliary tanks for settling of bones and scales to the bottom of the tank and a top layer of oil. This operation permits decanting of the oil layer, within limitation, from the aqueous liquid layer. Also, decantation of aqueous liquid layer from the bones and scales can be realized. By this process hot rinse water at approximately 70°C can be added to the drained bones and scales and thus remove more of the liquid digest from the bones and scales when they are conveyed to the screen separator **12**. When the process is carried out on a factory ship at sea, a series of tanks to receive the disintegrated fish are utilized both as receiving tanks and settling tanks. The separation is accomplished by a shaker or rotating screen **12**. The bones and scales are dried as indicated at **15**.

The liquid separated from the bones and scales goes to a centrifuge **16** and a threefold separation is realized, i.e., fish oil produce **17**, a wet fish meal **18**, and a clarified liquid **19** known as stickwater are produced. The wet fish meal, delivered from the centrifuge **16**, is dried in the dryer **18**, and ground at **20** to produce a fish meal product **21**. The clarified liquid **19** is concentrated by the removal of water at **22** by an evaporator or concentrator **22** to produce a fish solubles product **23**. This final fish solubles product is acidified to approximately pH 4.5.

The process as outlined yields four products: (a) Bones and scales meal—Protein, 28.2%; moisture, 8.0%; ash, 54.6%; oil, 4.6%. (b) Fish meal—Protein, 60.7%; moisture, 8.0%; ash, 11.8%; oil, 4.8%. (c) Fish solubles—Protein, 32%; total solids, 42 to 50%. (d) Fish oil—Assay values cited are representative of experimental samples prepared from mixed rough or trash fish of the shrimper's trawl.

The two meals (a) and (b) and the fish solubles (c) are designed for use in mixed feed formulations, specifically for poultry feed formulations. Fish oil finds use as a human food and as an industrial raw material, either for direct use or further processing. If the water added to the ground or chopped fish is in the form of distilled water and the fish are reasonably clean, it is not necessary to add a chelating agent, although it may be desirable to do so. On a practical basis, however, the fish may well have some surface dirt and will be wet with seawater. Further, the water used in the process will be ordinary tap water, or more likely, seawater, especially aboard the fishing vessel, both of which introduce metal ions, particularly alkaline earth metal ions and other polyvalent metal ions of chelation potential into the process. The effect of the presence of such ions is to chelate with the preservatives and thus impair the action of the latter.

For the foregoing reason it is essential to the process when water containing metal ions is used, to add a chelating agent which will form coordinates or complexes with the metal ions and so sequester them from the preservatives present as to prevent the formation of chelate complexes between the preservatives and the metal ions.

The function of the preservative is to prevent bacterial growth in the mass of ground fish. An offensive odor, associated with fish which have been subjected to bacterial decomposition, carries through to the finished products. The odor thus is a problem in actual processing of the fish, as well as in feeds formulated with the products derived from fish which have undergone bacterial decomposition.

The function of the enzyme is to produce a liquid digest of the viscera and flesh of the fish. For this action a proteolytic, protein splitting, enzyme is required. As a source of enzyme or enzymes for the process, commercial preparations of animal, vegetable and bacterial origin, fresh fish viscera or offal from fish processing procedures, and whole ground fish (viscera and offal intact) all can supply enzyme activity capable of producing liquid digests of fish viscera and flesh, but quantities of enzyme in excess of that naturally present must be added. The water added to the ground or chopped fish will almost invariably be ordinary tap water or seawater, both of which contain appreciable quantities of metal ions. Under these conditions a chelating agent is to be used.

Polycarboxylic acids (oxalic, succinic), hydroxy acids (citric, tartaric, maleic, ascorbic); polyhydroxy compounds (starch, cellulose), acids (hexametaphosphate, pyrophosphate, phytic acid, adenosine triphosphate); amino acids (glutamic acid, histidine, cysteine, glycine, etc.), peptides, proteins; pteridines, flavinoids, porphyrins, ethylenediaminetetraacetic acid (EDTA), diethylenetriaminepentaacetic acid (DTPA), and (hydroxyethyl)ethylenediaminetetraacetic acid (HEEDTA) are some examples of different types of chelating agents.

The mechanical disintegration or grinding is one of the steps in processing of the fish prior to being added to tanks aboard the ship. Also, whole fish may be added to the tank and then fed to a device or grinding mechanism capable of disintegrating the fish in the tank. Grinding or disintegrating is of a magnitude to produce particle (flesh) sizes of one inch in maximum dimension or smaller. Benefits derived from the preservative-enzyme process are as follows.

(1) Fresh fish are not subjected to bacterial decomposition and therefore odor problems associated with production of feed products from the fish are decreased. Also, products produced have greater nutritional value and formulated feeds containing the fish products do not possess the undesirable odor associated with fish products produced from fish which have undergone bacterial action.

(2) Processing time is gained at sea. That is, fish viscera and fish flesh are being liquified by the enzymes while in the tanks aboard the fishing boat. Thus fishing boats which stay at sea from 1 to 7 days can process fish as they are caught and arrive at land-based operations with a product in their tanks consisting essentially of digested fish liquid covering undigested bones and scales at the bottom of the tank.

(3) Unloading is simple and without loss of fish components. The digested liquid is mechanically pumped to land-based tanks without the aid of pumping water. In general, when whole fish are pumped from boat to land, large volumes of pumping water are used. Fish and water are pumped onto a dewatering screen to separate water from fish. Liquid squeezed from the fish in the hold of the boat is lost by this pumping procedure. This is eliminated by the preservative-enzyme process.

Successful operation of the process involves the following parameters:

1. *Enzyme Concentration*—From 0.1 to 4% based on the weight of fish. Another possible benefit may be significant if chelation is considered. Intact proteins are chelating agents; however, degradation products of protein, peptides and amino acids, are stronger chelating agents than intact proteins. Thus, the addition of enzyme, over and above that supplied by the fish, gives more rapid rise to chelating agents (resulting from enzymic action on the fish protein). Liquid digests prepared in the laboratory with added enzyme and antibiotics (Terramycin) retained their freshness for over 3 months, even in a dilute unacidified state.

Furthermore, the more rapid digestion of fish realized with added enzyme may be beneficial in preservation by an osmotic effect; that is, bacterial cells shrink in a hypertonic solution, swell in a hypotonic solution, and remain essentially unchanged in an isotonic solution. This is true because the bacterial cell attempts to equalize the solute concentration within the cell and the medium surrounding the cell. If excess solute is present outside of the cell, water will pass from the cell to the medium and thereby cause the bacterial cell to shrink.

The opposite is true when a bacterial cell is placed in a hypotonic solution, i.e., distilled water as an extreme. In distilled water the bacterial cell will take on water to try and dilute the solutes within the bacterial cell. Thus, the more rapidly solid protein is put into solution by the enzymes, the more rapid will be the rise in osmotic strength of the solution around the bacterial cell. This would give rise to unfavorable growth conditions of the bacteria by causing the bacterial cell to tend to shrink. Preservation of meats by high levels of table salt and of fruits by high sugar content syrups are two examples of osmotic preservation.

2. *Preservative Concentration*—For soluble Terramycin (20% activity) quantities ranging from 10 to 400 ppm bacterial activity provide adequate preservation for periods up to 12 days.

3. *Temperature Limitations*—Not to exceed 60°C. In some instances with enzymes possessing a high temperature tolerance, the temperature may be as high as 70°C. Above 70°C damage to feed products to be produced from the fish may resut. In general, when a chelating agent is used the process is carried out at ambient temperature.

4. *Water to be Added to Ground Fish*—For economic reasons not to exceed much over 20% of the weight of the fish.

5. *Grinding or Mechanical Disintegration*—To produce fish particle size of 1 inch or less.

6. *pH Limitations*—In general from 6.5 to 7.5, not to exceed 8; for certain enzymes such as pepsin, the pH could be lower (2 to 3).

7. *Chelating Agent*—The chelating agent to be used must have a greater affinity for the cations of magnesium and calcium than for the preservatives (tetracyclines such as Aureomycin and Terramycin). That is, the chelating agent, citric acid for example, must be able to either chelate with the cations and not the antibiotics, or the chelating agent must be sufficiently strong to remove the cations which have chelated the preservative. Citric acid, as tested, was effective in the preservation of the fish. Likewise, EDTA, DTPA, HEEDTA and similar compounds, and the polyphosphates listed above, should accomplish the same result because they have a greater affinity for metal cations, under preservation conditions for the fish process, than do the tetracycline preservatives.

Table 1 shows the result of tests using citric acid as a chelating agent. It is evident from the data that citric acid alone does not provide for preservation. Both the control (Can No. 1) and the fish mixed with citric acid (Can No. 2) gave indications of lack of preservation by the foul odor in less than 18 hours. The addition of as little as 12.5 ppm of Aureomycin (based on weight of fish, approximately 120 pounds) provided for preservation in excess of 8 days when citric acid was added. Earlier runs with Terramycin, another of the tetracycline antibiotics, similar to Aureomycin, failed to provide for preservation when citric acid was not added. Also, as shown by Table 1, 50 ppm Terramycin was effective in preservation in the presence of citric acid; lower levels of Terramycin were not run in this test. The processing procedure in obtaining the data of Table 1 was as follows:

1. Fresh fish taken from the shrimp trawl of a shrimping vessel were used.
2. The fresh fish were ground in a Rietz PreBreaker powered by the fishing vessel engine.
3. 15 gallons of ground fish (approximately 120 lbs) and 3 gallons of seawater were used in each can.
4. Order of addition of ingredients to cans was as follows:
    Can No. 1: Ground fish added to water and mixed.
    Can No. 2: Ground fish added to water containing citric acid and mixed.
    Cans 3 through 6: Aureomycin was added to water containing citric acid. Ground fish was added to citric acid solution and mixed.
    Cans 7 through 10: Aureomycin was added to whole fish and ground. This mixture was then added to water containing the citric acid and mixed.
    Can No. 12: Terramycin used instead of Aureomycin as per procedure for Cans 3 through 6.
5. All containers received 0.5% of the enzyme Rhozyme B-6 (80% activity) after the above addition and components well mixed.

TABLE 1: PRESERVATION IN THE PRESENCE OF CITRIC ACID

Can. No.	Citric Acid, grams	Aureomycin HCl	Terramycin	Observations
		Bacterial Activity		
1	0	0	0	Foul odor in less than 18 hours
2	110	0	0	Foul odor in less than 18 hours
3	110	100	0	
4	110	50	0	Contents of cans 3 through 6 gave
5	110	25	0	no evidence of foul odor in 8 days
6	110	12.5	0	
7	110	100	0	
8	110	50	0	Contents of cans 7 through 10 gave
9	110	25	0	no evidence of foul odor in 8 days
10	110	12.5	0	
12	110	0	50	Contents of can 12 gave no evidence of foul odor in 8 days

The following example describes the process as related to a shrimp boat operation.

Example: The catch, representing a mixture of shrimp and fish, is emptied from the trawl onto the deck of the boat. Shrimp are sorted from fish, deheaded and iced in the hold of the boat. Fish and shrimp heads are fed into a grinder. Ground fish are discharged into a watertight, air-vented tank in the hold of the boat. Water equal to 10% of the weight of the fish containing 10 to 400 ppm bacterial activity soluble Terramycin and 180 to 720 ppm bacterial activity of nf 180 (ppm based on fish weight) along with citric acid at the rate of 1 gram per pound of fish (0.2% by weight based on weight of fish, or 0.1 to 4.0%) is added or was present in the tank before ground fish was added. This procedure is repeated with each trawl catch. The charge of water is determined by anticipated catch of trash fish, 10 tons, 20 tons, etc. In this situation only preservative, chelating agent and enzyme are added as sea operations. Under these conditions the boat can be at sea for a period of 8 days.

The boat returns from sea and ties to the dock of the land-based operations. Here the liquid digest and undigested portions of the fish are pumped into holding tanks on land. Boat tanks are rinsed with small volumes of water and the rinse solution also pumped to land holding tanks. Finally, the storage tanks on the boat are sterilized by steam and/or disinfectants. This eliminates the possibility of development of a strain of bacteria resistant to the preservatives used in the process.

Land based operations involve further digestion at 40° to 50°C, or at higher temperatures such as 70°C, decantation, dewatering of solids by shaker screens, centrifugation, concentration of liquid digest in an evaporator, drying of fish bones and scales for grinding to a meal, drying of undigested fish (sludge from the centrifuge) for grinding to a meal, and oil recovery. Products produced are fish solubles (32% protein), bones and scales meal (approximately 30% protein), fish meal (approximately 60% protein), and fish oil, based on one ton of rough or trash fish caught in shrimper's trawl. Approximately 740 pounds of fish solubles, 137 pounds of bones and scales meal, 58 pounds of fish meal, and 79 pounds of fish oil are derived as products from the preservative-enzyme fish handling process. These yields are from fish assaying 16% protein, 73.5% moisture and 4.5% oil.

5-AMINOHEXAHYDROPYRIMIDINES

The compound 5-aminohexahydropyrimidine is used by *R.G. Sanders; U.S. Patent 2,963,374; December 6, 1960; assigned to Warner-Lambert Pharmaceutical Company* to enhance the storage life of fresh fish. The perishability of fish or other seafood in transit is ordinarily avoided by icing the fish to maintain the temperature below that at which bacterial growth or multiplication is favored. The ice used for this purpose may be prepared by freezing an aqueous solution containing up to about 0.1% by weight of the desired 5-aminohexahydropyrimidine and using the ice so formed for icing purposes. The 5-aminohexahydropyrimidines exhibit a pronounced affinity for protein materials and subsequent icings need only be conducted with ordinary forms of ice. No impairment in the degree of protection obtained is observed where ordinary ice is used to supplement the original icing.

The pyrimidine compounds have the formula:

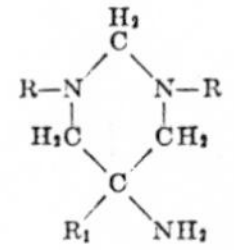

where R is an alkyl, aryl, aralkyl, alkaryl, hydroxyalkyl, aminoalkyl or cycloalkyl radical, and R_1 is hydrogen, or a lower alkyl or hydroxymethyl radical. Among the 5-aminohexahydropyrimidine compounds which have been found effective are:

> 1,3-dimethyl-5-methyl-5-aminohexahydropyrimidine
> 1,3-bis(beta-ethylhexyl)-5-methyl-5-aminohexahydropyrimidine
> 1,3-didodecyl-5-methyl-5-aminohexahydropyrimidine
> 1,3-bis(dicyclohexyl)-5-methyl-5-aminohexahydropyrimidine
> 1,3-dibenzyl-5-methyl-5-aminohexahydropyrimidine

In lieu of employing the 5-aminohexahydropyrimidines above in the form of their free bases, the salts of these amine compounds may also be utilized. Salts are readily formed with acids such as, for example, acetic acid, phosphoric acid, hydrochloric acid, maleic acid, benzoic acid, citric acid, malic acid, oxalic acid, tartaric acid, succinic acid, glutaric acid, gentisic acid, valeric acid, gallic acid, β-resorcylic acid, acetyl salicylic acid and salicylic acid, as well as perchloric acid, barbituric acid, sulfanilic acid, phytic acid and p-nitrobenzoic acid. These amines also form useful salts with long chain aliphatic acids, such as stearic acid, palmitic acid, oleic acid, myristic acid, and lauric acid. The trihydrochloride, for example, is highly water soluble.

In addition to incorporating the 5-aminohexahydropyrimidines in solutions for the treatment of raw foods, it is of great value to use the compounds in the preparation of packaging materials such as the paper and parchment type of wrapping materials employed in packaging fish and similar raw foods. These compounds may also be incorporated in the waxes and similar paper coating compositions which are applied to packaging papers and which are designed to impart greater water resistance to these materials. They may also be incorporated into polyethylene and polyvinylidene chloride materials as well as into the usual cellophane coating compositions.

Example 1: A preservative bath of the following compositions is prepared: 0.1 part by weight of 1,3-bis(beta-ethylhexyl)-5-methyl-5-aminohexahydropyrimidine; 1.4 parts by weight of polyoxyethylene sorbitan monooleate (Tween 80); and water qs to 1,000.0.

Freshly prepared fish fillets are immersed in this bath for 2 minutes and then iced for shipment. This treatment is found to extend the shipping range materially and to reduce the frequency of reicing in transit while maintaining the appearance and acceptability unchanged. This bath is also excellent for the treatment of dressed poultry prior to packaging and for the treatment of meats in carcass form.

Example 2: A 0.005% aqueous solution of 1,3-bis(beta-ethylhexyl)-5-methyl-5-aminohexahydropyrimidine trihydrochloride is frozen into the form of flake ice and used for the icing of fresh shrimp in transit. This treatment effectively controls bacterial action and maintains the product in acceptable market condition, not only in transit, but subsequently with only a minimum of reicing.

EGGS AND EGG PRODUCTS

ANTIFUNGAL ANTIBIOTIC TREATMENT

Cleaning of eggs has been accomplished by any of several means, as, for example, by abrading the shell surface with a fine grit or by washing the eggs with an aqueous, detergent-containing solution. It is evident that washing solutions are frequently carriers for mold spores and are responsible for mold infection in the eggs. Several studies have been made which indicate that hatching embryos are readily contaminated by the organism *Aspergillus fumigatus.* Such infection has been traced to the egg washing operation. Also, there is evidence to show that where the hatching embryos are infected with spores of Aspergillus organisms, the incidence of aspergillosis infection in the hatched chicks is increased.

C.N. Huhtanen; U.S. Patent 3,343,968; September 26, 1967; assigned to American Cyanamid Company uses a cleaning solution for eggs which contains a certain antibiotic for antifungal treatment.

The antibiotic pimaricin has been found to be effective in preventing mold growth in eggs with concentrations as low as 10 ppm while nystatin requires about 25 ppm to give an effective measure of mold control. Amphotericin B, on the other hand, seems to be the most effective agent and has given 68% mold protection with concentrations as low as 0.83 ppm; which concentration is roughly comparable to pimaricin at 25 ppm.

It has also been found that the addition of a solubilizing agent such as dimethylsulfoxide, dimethylformamide, a mixture of methoxyethanol and calcium chloride, formamide, lower alcohols, methylpyrrolidone, glacial acetic acid, or propylene glycol is useful in preparing the solutions and such addition appears to enhance the mold inhibiting activity of the fungicide.

Example 1: The mold cultures used in the following tests were *Aspergillus fumigatus* and *Aspergillus flavus* both of which have been implicated in aspergillosis infections in chicks. The molds were grown on nutrient agar for one week at room temperature and the spores washed off with water. Spore counts were made by plating an orange-serum agar. The eggs used were obtained from local hatcheries before washing and upon delivery were placed in an incubator at 98°F for 4 to 16 hours. After incubation the eggs were dipped in demineralized water at approximately 20°C, the temperature differential between the egg and the dip water insuring diffusion of spores and test materials with the egg. The dip time was ten minutes in all experiments. Dipping was accomplished by placing the eggs in a basket and immersing in 3 gallons of water in a large container. The basket and eggs were agitated several times during the dip period by gently raising and lowering.

Following the immersion period the eggs were immediately placed back in the incubator. The incubator was humidified by means of a tray of water at the bottom and was a forced draft type of egg incubator. The trays containing the eggs were tilted approximately 30° daily. Examination for molds was done by candling at appropriate intervals; the mold usually showed up well as a fuzzy darkened area on the edge of the air space. The green color of *A. fumigatus* could also be detected by candling. The first examination was made six days after treatment (7 days embryo age) at which time infertile eggs and those failing to develop were noted. The dominant characteristic of the latter group was the appearance of a blood ring in the egg indicating initial development followed by death. Subsequent examinations were made at intervals to 20 days of age except for several groups of dead embryos which were followed to 28 days. At 20 days, live chicks were determined by observation of movement in the egg.

Table 1 shows the effect of the spores of *A. fumigatus* and *A. flavus* on fertile and infertile eggs. Nineteen days after dip (20 days embryo age) only 3 out of 70 infertile eggs were infected with *A. fumigatus* while 62 out of 70 of the fertile eggs were found to harbor the organism. *A. flavus* showed a higher rate of infertile egg contamination; however, this was also considerably less than the incidence in the fertile eggs. Prolonged incubation of the infertile eggs increased the number of infected. All calculations of mold infectivity were based on total eggs minus infertiles.

TABLE 1: INCIDENCE OF ASPERGILLOSIS FOLLOWING IMMERSION IN SPORE-INOCULATED WATER

| | | | ---- Days After Treatment ----- | | | | | | |
| Additions to Dip Water | Condition of Eggs | Total Eggs | 15 | 19 | 21 | 22 | 27 | 29 | 33 |
			---- Total Number Molded -----						
None	Infertile	70	0	0	0	0	2	3	3
A. flavus spores, 1.1 x 10⁴/ml	Infertile	70	7	13	16	16	18	25	27
A. fumigatus spores, 1.1 x 10⁴/ml	Infertile	70	1	3	12	13	15	20	23
None	Fertile	70	–	0	–	–	–	–	–
A. flavus spores, 1.1 x 10⁴/ml	Fertile	70	–	22	–	–	–	–	–
A. fumigatus spores, 1.1 x 10⁴/ml	Fertile	70	–	62	–	–	–	–	–

Note: Eggs incubated overnight before dipping. Water used for dipping was demineralized and equilibrated overnight at room temperature.

Example 2: To assess the value of pimaricin in alleviating mold infection in eggs, they were experimentally infected with the mold organism *A. fumigatus* and *A. flavus* by the procedure set forth in Example 1 above. The eggs used in this experiment showed an abnormally high rate of early dead embryos caused possibly by improper temperature during holding at the hatchery. These early dead embryos were included in the calculations of mold growth. Very few molds developed during the first six days after treatment while at 13 days, *A. fumigatus* showed 37% infectivity.

At 19 days, this mold infected 91% of the embryos at the spore concentration used (10,000 spores per ml dip water). Pimaricin at 100 ppm showed 67% protection at 13 days and 22% at 19 days with *A. fumigatus*. The degree of protection against *A. flavus* was higher — 92% at 13 days and 17% at 19 days. It is apparent from this experiment that the detergent normally used for washing eggs does not destroy the spores of *A. fumigatus* or *A. flavus*. The results of these tests appear in Table 2.

TABLE 2: *Aspergillus flavus* AND *A. fumigatus* AS CAUSATIVE AGENTS OF ASPERGILLOSIS IN HATCHING EMBRYOS

								Days Following Treatment				
				6			13			19		
	Total		Early	Molded Protection			Molded Protection			Molded Protection		
Additions to Dip Water	Eggs	Infertile	Dead	No.	Percent*	Percent*	No.	Percent*	Percent*	No.	Percent*	Percent*
No dip, control	70	7	27	0	0	-	0	0	-	0	0	-
None	68	4	29	0	0	-	1	2	-	3	5	-
A. flavus (1.1 × 10^4 spores/ml)	68	3	18	3	5	-	8	12	-	22	35	-
A. flavus, 0.25% detergent	68	3	19	4	6	-	4	6	50	23	35	0
A. flavus, 100 ppm pimaricin	70	1	24	0	0	100	1	1	92	6	9	77
A. fumigatus (1.1 × 10^4 spores/ml)	70	2	24	1	1	-	25	37	-	62	91	-
A. fumigatus, 0.25% detergent	68	4	27	4	6	0	24	37	0	52	81	11
A. fumigatus, 100 ppm pimaricin	69	2	28	1	1	0	8	12	67	48	71	22

*Percent molds and percent based on total eggs minus infertiles.

Note: Eggs incubated overnight before treatment. Water used was demineralized water equilibrated overnight at room temperature.

PHOSPHATE ADDITIVES FOR FRESH EGGS

When freshly shelled eggs are allowed to stand, either at room temperature or under refrigeration for a short period of time, their attractive orange-yellow color changes to a muddy or a pale brown. This color change does not affect the flavor or palatability of the eggs but does detract from the appetite appeal thereof. When eggs which have undergone this color change are used, for instance, in preparing scrambled eggs, the muddy or pale brown color is imparted to the scrambled eggs. This color change also takes place when scrambled eggs made from freshly shelled eggs are allowed to stand for a time, for instance, on steam tables in restaurants.

The problem is described by *R.G.L. Chin and S. Redfern; U.S. Patent 3,383,221; May 14, 1968; assigned to Standard Brands Incorporated* who suggest use of small amounts of soluble phosphorus compounds to prevent this discoloration.

When the egg white portion of freshly shelled eggs is substantially completely separated from the egg yolk portion, little or no color change is observed in the latter on standing but as the proportion of egg whites to egg yolks is increased the color change accordingly increases.

The term egg composition is used below to include both scrambled eggs and liquid eggs which includes freshly shelled eggs, eggs which have turned a muddy brown, whole eggs, and pasteurized eggs.

The liquid egg composition may be produced by incorporating into a mixture of egg white and egg yolk an acid of phosphorus or an acid salt in an amount sufficient to adjust the pH to a value from about 5.5 to about 7. The composition may also be prepared by incorporating a neutral or basic salt of an acid of phosphorus and a sufficient amount of an edible acid to adjust the pH of the composition to the range indicated above, i.e., 5.5 to 7. The phosphorus compound should be distributed uniformly throughout the egg compositions to avoid reduction of the pH in any portion of the composition below about 5.5 since coagulation of the egg protein will occur at this pH.

The fresh egg-like color of the egg composition varies slightly, for instance, yellow with an orange tinge, bright yellow, orange-yellow and various shades thereof. Although these colors, except the orange-yellow color, are not the exact color of fresh egg compositions, they closely resemble the same and are considered superior to the muddy or pale brown color of stored eggs which do not contain the phosphorus compounds. The orange-yellow color is considered to be substantially the same as the characteristic color of fresh egg compositions and is considered to be the best or most desirable color produced.

Generally, only a very small amount of a phosphorus compound is necessary. For example, to impart the desirable color to the liquid egg compositions the preferred proportions of phosphate salts are about 0.3 to 0.7% by weight in the case of anhydrous monosodium phosphate, about 0.3 to about 0.4% by weight in the case of sodium acid pyrophosphate and about 0.3 to about 0.4% by weight in the case of monocalcium phosphate. The most preferred proportion of anhydrous monosodium phosphate is about 0.5%. When orthophosphoric acid is used, the most desirable color is produced when sufficient amount of this acid is incorporated in the eggs to adjust the pH of the same to a range from about 6.4 to about 6.6.

If liquid egg compositions which have undergone the color change to muddy brown or pale brown are treated with these phosphates, their original fresh egg-like color is essentially restored. This reversible effect seems to indicate that the microorganisms naturally present in liquid eggs are not responsible for the muddy or pale brown color change.

The scrambled egg composition may be prepared from the above enumerated liquid egg compositions or from liquid eggs containing a phosphorus compound and having a pH above 7. While such liquid eggs having a pH higher than about 7 will not have the desired color stability and in some instances may be an undesirable muddy or pale brown, scrambled eggs will have the fresh egg-like color of scrambled eggs prepared from freshly shelled eggs. Preferably the pH of the liquid eggs is in the range of from about 5.5 to about 8.25 when the phosphorus compound employed is monosodium phosphate, from about 5.5 to about 7.5 when the phosphorus compound is sodium hexametaphosphate, and from about 5.5 to about 7.7 when the phosphorus compound is sodium tripolyphosphate.

In the following table, it is generally shown that the kind of phosphate compound used in preparing scrambled eggs and the pH of the liquid eggs from which the scrambled eggs are prepared affects both the color and texture of the scrambled eggs.

Phosphate Class	Scrambled Egg Color	Texture Limitation [1]
Ortho-	Acceptable up to pH of about 8.25.[2]	pH above about 8.25.
Pyro-	Acceptable up to pH of about 9.[2]	pH above about 8.5.
Trilinear	Acceptable up to pH of about 9.[2]	pH above about 7.75.
Polylinear	Acceptable up to pH of about 7.25.[2]	pH above about 7.5.
Polycyclic	Acceptable up to pH of about 7.[2]	pH above about 7.25.

[1] Indicates that at pH's higher than shown the scrambled eggs had undesirable textural characteristics.
[2] Indicates a color which is considered better than control egg samples containing no phosphorous compound.

Example 1: This example illustrates the use of six acid phosphate salts for color stabilization of shelled liquid eggs. A quantity of eggs was shelled and placed in stainless steel beaker. The yolks were broken by means of a stainless steel spatula. The eggs were passed through a screen and then blended to ensure substantial homogeneity. The blend was divided into eight portions, and with constant stirring small amounts of acid phosphate salts were added to seven of these portions. Monocalcium phosphate, because of its limited solubility, was added in a water slurry to provide better dispersion in the eggs. The eighth portion served as a control. All eight portions were refrigerated overnight and then allowed to stand at room temperature until they reached such temperature. The pH and color of the portions were as follows:

Additive	Percent Additive In Egg Sample	pH After Refrigeration	Color After Refrigeration
None (control)	---	7.60	Unacceptable.
Sodium Acid Pyrophosphate.	1.0	6.87	Acceptable.
Monopotassium Phosphate.	1.0	6.53	Best.
Monoammonium Phosphate.	1.0	6.52	Do.
Monocalcium Phosphate Monohydrate.	0.3	6.45	Do.
Monosodium Phosphate Monohydrate.	0.5	6.76	Do.
Anhydrous Monosodium Phosphate.	0.43	6.65	Do.
Sodium Hexametaphosphate (Calgon)	0.75	6.93	Acceptable.

Example 2: This example illustrates the use of monosodium phosphate monohydrate in various proportions in whole liquid eggs. The egg samples were prepared according to the procedure described in Example 1 and frozen and stored for six weeks. The samples were then allowed to stand at room temperature until they reached such temperature. The pH and color of the samples were as follows:

Percent Monosodium Phosphate Monohydrate in Samples	pH	Color when Frozen for 6 Weeks
0	7.60	Unacceptable.
0.1	7.26	Do.
0.2	7.07	Acceptable.
0.3	6.92	Best.
0.4	6.85	Do.
0.5	6.76	Do.
0.6	6.70	Do.
0.7	6.65	Do.

Example 3: This example illustrates the use of orthophosphoric acid (1 part 83.7% H_3PO_4 and 2 parts of water) in various proportions for stabilizing the color of whole eggs and scrambled eggs made from such treated whole eggs. The egg samples were prepared in accordance with the procedure described in Example 1. They were refrigerated overnight and then allowed to stand at room temperature until they reached such temperature. The pH and the color of each sample was noted. Scrambled eggs were then made from each sample and their color observed. The results were as follows:

Amount of Ortho-phosphoric Acid in Egg Sample	pH	Color of Refrigerated Eggs	Color of Scrambled Eggs
None (control)	7.45	Unacceptable	Unacceptable.
0.3 ml.	7.10	Acceptable	Acceptable.
0.6 ml.	6.80	....do	Do.
0.8 ml.	6.60	Best	Do.
1.0 ml.	6.55	....do	Best.
1.2 ml.	6.40	....do	Do.
1.7 ml.	6.25	Acceptable	Do.
2.0 ml.	6.05	....do	Acceptable.
3.3 ml.	5.60	....do	Do.

DRIED, DEFATTED EGG PRODUCT CONTAINING ANTIOXIDANT

In the process of *E. Levin; U.S. Patent 3,607,304; September 21, 1971* an egg product is prepared by removing water and fat from whole egg by a low temperature azeotropic distillation. It is required that the distillation temperature not substantially exceed about 100°F (37°C) so that no undesired protein coagulation occurs in the processing. In other words, a dried and defatted raw egg product is desired which can be subsequently hydrated and cooked in the usual way. The dehydration and defatting occurs in a fat organic solvent which is capable of forming the desired azeotrope and, at the same time, defatting the raw egg.

It is an important feature of the process that the substantial content of solvent remaining in the dried and separated product is removed by introducing water in the form of particulate droplets and applying moderate heat at below about 37°C. The solvent in the dried product is then removed or deabsorbed as a mixture of the solvent and water. The fat organic solvent in the dried product can be effectively removed to only trace levels which are retained with certain fat organic solvents to obtain other advantages, as will be later described.

The particulate water droplets need not be of any critical size except that they should be in the small droplet form common to mists or sprays. The purpose is to provide good contact between the water droplets and the solvent in a fine form so that the solvent forms a distillable mixture with the water droplets without the water wetting the body of the dried egg product. It has been found that when introducing water in such a particulate droplet form, the solvent content of the dried and defatted product can be reduced to below 100 ppm.

It is preferred to introduce the raw egg into the heated body of fat organic solvent as an intimate mixture of the egg and the same fat organic solvent. The raw egg is emulsified in a fat organic solvent with the aid of a conventional homogenizer. Different proportions of egg and fat organic solvents may be used to prepare the emulsified mixture but generally an excess volume of solvent is preferred, such as about four volumes of solvent to about one volume of raw egg.

The emulsified egg-fat solvent mixture may then be sprayed below the surface of the heated body of fat solvent heated at moderate temperature levels. The body of fat organic solvent is heated at moderate temperature levels under decreased pressure so that the fat organic solvent boils at the reduced temperature and thereby forms an azeotrope of the water in the raw egg and the fat organic solvent. Pressure levels are applied in pressure vessels to attain the formation of the azeotrope at its selected and desired temperature level. The emulsified egg and solvent mixture may also be introduced in an atomized form by delivering the emulsified mixture under pressure through spray jets underneath the body of boiling fat solvent.

Various fat organic solvents may be used which can form an azeotrope in water such as perchlorethylene, trichlorethylene and, particularly, ethylene dichloride. Effective nontoxic levels of an antioxidant may be added to the boiling fat organic solvent to attain intimate contact of the antioxidant with the individual particles. Such lower levels of antioxidant will more effectively retard oxidation of any trace levels of any fat remaining in the final egg product than if the antioxidant were added after the product were dried and defatted. Such antioxidants may be butylated hydroxytoluene or BHT; butylated hydroxyanisole or BHA; propyl gallate; nordiguaiaretic acid or NDGA; or others.

Example 1: The egg product preparation by azeotropic distillation with ethylene dichloride is as follows. One volume of raw whole egg is mixed with four volumes of ethylene dichloride in a Cherry Burrell homogenizer, and the emulsified mixture is then introduced into a pressure vessel as a fine spray below the level of a boiling body of ethylene dichloride maintained at a temperature of no more than 40°C and at a pressure of about 215 mm of mercury.

An azeotrope of solvent and water is removed by distillation until all the water from the raw egg is withdrawn. The temperature of the egg-solvent mixture is maintained below about 40°C throughout the distillation processing. After substantial dehydration, the ethylene dichloride is removed by draining through a screen opening in the vessel and fresh solvent is added to the product which has been dried of water. The mixture is stirred and the solvent is again removed by draining through the opening in the vessel.

A fine spray of water is introduced into the cold vessel and the temperature is maintained at no more than about 40°C. The content of the ethylene dichloride remaining in the dried residue is removed as a mixture of the water and solvent. The resulting product has trace levels of ethylene dichloride intimately contacted with the individual particles to such a degree that the ethylene dichloride content of the product is from about 60 to about 100 ppm.

Example 2: The process steps of Example 1 are followed except that the emulsified mixture of raw egg and ethylene dichloride is delivered thorugh spray jets under pressure into the body of boiling ethylene dichloride at a point just below the surface. The pressure in the closed pressure heating vessel is about 215 mm of mercury. An azeotrope of solvent and water is removed by distillation until all the water from the raw egg is withdrawn. The temperature of the egg-solvent mixture is maintained below about 40°C throughout the distillation processing. After dehydration, the ethylene dichloride, containing the fat, is removed by draining through a screen opening in the vessel and fresh solvent is added to the water-dried product. The mixture is stirred and the solvent is again removed by draining through the opening in the vessel.

A fine spray of water is introduced into the cold vessel and the temperature is maintained at no more than about 40°C. Substantially all the ethylene dichloride remaining in the dried residue is removed as a mixture of the water and solvent.

Example 3: The egg product prepared according to Example 1 is analyzed to determine its protein and lipoprotein by the methods described in AOAC *Methods of Analysis,* 10th ed.

The Food & Drug bacteriological method is used to determine any Salmonella infection. The cholesterol content is determined by the method described in *Anal. Chem.,* Vol 22, page 1210 (1950). The presence of Salmonella infection is studied by the method described in the *Bacteriological Analytical Manual,* Food and Drug Administration, 1966.

The protein content was found to be about 75% by weight and the lipoprotein content about 9% by weight. The protein content quantity of egg products prepared by a prior method of spray drying was found to have a protein quantity of 45% or substantially lower than the protein quantity of the egg product in this product. The egg product was substantially free of cholesterol as compared to the original cholesterol levels and the cholesterol levels of the prior art spray dried egg products. The egg product is free of any Salmonella infection.

Example 4: The egg product prepared by the method of Example 1 is subjected to Derse Assay for 4 weeks and compared with the recognized standard of casein, (milk protein). The casein had a protein efficiency ratio (PER) of 2.80 whereas the egg product had a PER or protein efficiency ratio of 4.03. The improved egg product by this process permits the product to be provided to the consumer in a packaged form because of its indefinite shelf life. It is recognized that prior egg products prepared by spray drying have limited stability in storage at room temperature.

EPOXIDE-WATER VAPOR TREATMENT FOR SHELL EGGS

A process to sterilize egg shells without overexposure to heat is described by *C.P. Collier and J.E.W. McConnell; U.S. Patent 3,144,342; August 11, 1964.* Fresh eggs, even those which come from the most sanitary poultry farms, are subject to rapid deterioration. This deterioration results in downgrading the eggs even before they have become unfit for human consumption.

Several factors cause deterioration of shell eggs, an important factor being the entry of bacteria or other microorganisms from outside of the shell through the pores of the shell to the interior of the egg. Such microorganisms cause proteolysis and other adverse changes in the egg, particularly the albumen.

It is effective to clean eggs in the sense of removing visible dirt and discoloration provided the eggs are not excessively dirty to begin with, but it does not remove microorganisms effectively. A water wash may have the opposite effect of that intended; that is, it may serve to spread infectious organisms from the infected eggs to uninfected eggs, and it may also serve to carry microorganisms from the exterior to the interior portions of egg shells.

Mineral oil is not itself effective to eradicate microorganisms unless it is applied hot. (The disadvantages of heat will be explained shortly.) In the application of a coat of mineral oil, if the eggs have been previously sterilized by some other means, a sterile coating or film of mineral oil will retard reinfection, but the mineral oil will not itself sterilize egg shells unless it is applied with a sufficient degree of heat.

The use of heat, whether it is applied in the form of a hot mineral oil or otherwise, is difficult because of the difficult balance between a degree of heat and a time of exposure which suffice to destroy heat resistant microorganisms, and the tendency of heat to coagulate the albumen in eggs.

Epoxide-water vapor mixture will kill heat resistant spores of *B. subtilis* at temperatures as low as 130°F and in short time, e.g., 36 minutes at 130°F or 12 seconds at 212°F. *B. subtilis* is representative of the most heat resistant spoilage organisms which contaminate eggs. Exposure to heat and water vapor alone would require exposure periods many times longer to kill this organism.

Shell eggs may be submitted to a precleaning step, for example, sandblasting to remove some or all of the gross impurities, adhering dirt, stains, etc. Also, the shell eggs may be prewashed with an aqueous washing medium. Ordinarily such a wash would spread infection and if followed by a drying period it would carry microorganisms into the interior of the eggs. However, if such a wash is followed quickly by an epoxide-water vapor treatment, the infectious microorganisms on the surface of eggs will be killed.

Following epoxide-water vapor treatment, shell eggs are packaged in sterile, airtight containers or are coated with a sterile coating, as by dipping in or spraying with a sterile mineral oil. It will be understood that the epoxide treatment is effective to kill microorganisms on the shells of eggs but that its beneficial effect will be diminished if, subsequent to the treatment, the eggs are exposed to a nonsterile atmosphere, to contact with nonsterile containers or equipment, or to contact with the hands of operators. Epoxide-water vapor treatment, even if used along without subsequent sterile packaging or sterile coating, will prolong the shelf life of shell eggs but it is preferred to follow the sterilizing step with sterile packaging or coating with a sterile mineral oil. The epoxides are ethylene oxide and propylene oxide which have the following structural formulas:

$$CH_2\!\!-\!\!CH_2 \diagdown_{\!\!O}\diagup \qquad\qquad CH_2\!\!-\!\!CH\cdot CH_3 \diagdown_{\!\!O}\diagup$$

Ethylene oxide Propylene oxide

However, higher homologues may be used, such as isobutylene oxide,

$$(CH_3)_2C\!\!-\!\!CH_2 \diagdown_{\!\!O}\diagup$$

and derivatives such as styrene oxide

$$C_6H_5\!\!-\!\!CH\!\!-\!\!CH_2 \diagdown_{\!\!O}\diagup$$

Also, isomers may be used in which a four-membered oxy ring replaces the three-membered oxy ring. Also such compounds as epichlorohydrin, ethylene imine and ethylene sulfide may be used which have the characteristic heterocyclic ring nucleus

$$C\!\!-\!\!C \diagdown_{\!\!X}\diagup$$

wherein X is oxygen, nitrogen or sulfur. Preferably, an epoxide is used which boils below 212°F, or which at least has a substantial vapor pressure at 212°F.

Shell eggs may be contacted with epoxide-water vapor mixtures by any suitable method, preferably by countercurrent contact by passing shell eggs on a continuous conveyor through a tunnel through which epoxide-water vapor mixture is passed in the opposite direction. The eggs are preheated to or close to the sterilizing temperature prior to contact with the epoxide-water vapor mixture, because epoxide vapor itself does not transfer heat as rapidly as steam. For this reason, mixtures of epoxide and water vapor which are high in water vapor are more efficient heat transfer media. Provided the egg shells are brought up to or near the sterilizing temperature before contact with the epoxide-water vapor mixture, contact times of a few seconds to about 30 minutes and temperatures (of the epoxide-water vapor mixture) of about 130° to 212°F are suitable, the lower temperatures being used with longer exposure and vice versa. Dried egg powder, whole fresh eggs separated from the shell and other egg products may be similarly sterilized.

Shell eggs and other egg products can be sterilized in this manner in short periods of time and at temperatures which do not result in protein coagulation or other forms of degradation caused by heat. Thus shell eggs have been subjected to an 80% propylene oxide, 20% water vapor mixture for 20 minutes at 180°F, during which the outside shells of the eggs reached a temperature of 150°F. The eggs were opened two hours after exposure. No evidence of albumen coagulation was found. Also upon scrambling and cooking the eggs with margarine no off flavor was noticed. A slight medicinal odor lingers for a while after sterilization but disappears within a relatively short time, for example, within twenty-four hours during storage at 40°F.

EGG SALAD

Cooked eggs are prepared according to a special process by *M.W. Miller; U.S. Patent 3,232,769; February 1, 1966; assigned to Western Farmers Association* so that the resulting egg whites and yolks retain the taste, appearance and texture of freshly prepared eggs after storage. An essential step of this process is the use of glucose oxidase-catalase enzyme for removal of glucose from yolks before heating.

After the whites and yolks have been cooked they are mixed, chopped, and cooled and then mixed with various other ingredients which may go into the product. The pH of the mixture is adjusted to a preselected value which has been found to prevent the growth of Clostridium organisms and thus subsequent heating of the product to a high temperature to kill the Clostridium organisms is not required. After the food product has been thoroughly mixed and the pH level adjusted to the proper value, it is placed in cans and sealed. Thereafter the temperature of the mixture in the sealed containers is elevated to a relatively high temperature and maintained at that temperature to eliminate any harmful organisms which might be present.

Figure 4.1 is a flow diagram illustrating schematically the steps to be followed in carrying out the method for producing a product such as egg salad. In general the process involves the steps of: (1) separating the whites and yolks of eggs; (2) cooking the whites and yolks separately at a temperature of 209° to 212°F; (3) mixing and chopping the whites and yolks and adding other ingredients if desired; (4) adjusting the pH of the mixture to a range of 4.45 to 4.55 (inclusive); (5) placing the mixture in containers and sealing the same; and (6) heating the mixture in the container to a temperature sufficient to destroy any harmful agents (approximately 170° to 190°F).

The yolks and whites are first separated by the use of a commercially available separator and then cooked separately. Prior to cooking of the yolks, however, they are desugared with one of the well known techniques. One such technique includes heating the yolks to a temperature of approximately 100°F and holding them at this temperature until there is no trace of glucose. Any one of a number of commercially available dextrose sensitive test papers can be used to test for the presence of sugar in the yolks. Two hundred to three hundred milliliters of a glucose oxidase-catalase enzyme system is added per one thousand pounds of the liquid yolks in the desugaring process. The yolks are gently agitated during the process and at 30 minute intervals over a period of 3½ hours 35% hydrogen peroxide in a water solution is added at the rate of 1,769 milliliters per thousand pounds of yolks.

The desugared yolks and the whites are then cooked separately at an appropriate temperature. The time and temperature is such that the material does not become rubbery. It has been found that if the yolks and the whites are cooked at a temperature of from 209° to 212°F (using steam) for approximately 1.75 minutes the desired texture is obtained.

FIGURE 4.1: METHOD OF PREPARING EGG PRODUCT

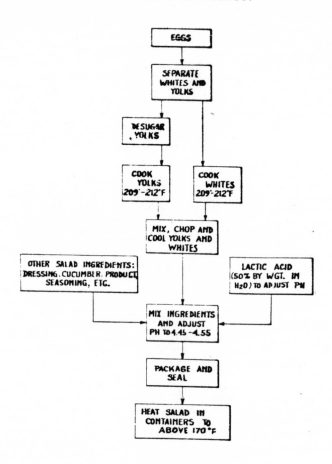

Source: M.W. Miller; U.S. Patent 3,232,769; February 1, 1966

The cooked yolks and whites are then mixed and ground to a suitable size for use in the salad and either simultaneously or later the temperature of the cooked eggs is reduced to approximately 110°F. As a result of this cooling the white and yolks, physical breakdown of the salad dressing used in manufacturing the egg salad is avoided.

The chopped egg is then placed in a large vat and mixed with the various other ingredients which are used in the particular product being produced. In one type of egg salad a cucumber product such as relish or chopped pickles as well as various spices and seasoning agents such as salt, pepper, and sugar are mixed with the chopped yolks and whites. Any one of a number of salad dressings suitable for use in the product is added at this time also. One specific commercial dressing used is composed of starch, sugar, flour, water, vinegar, yolk, salt, seasoning, and oils. The pH level of the mixture is adjusted by the addition of sufficient acid such as a solution of 50% by weight of lactic acid in water to bring the pH to a value which has been found to prevent the growth of Clostridium organisms and thus permit long term storage. The pH level is preferably adjusted to a value in the range of

4.45 to 4.55. It is found that with the pH level adjusted in this range the growth of Clostridium organisms is prevented and a long shelf life achieved.

ALUMINUM FOR COLOR STABILITY

Heat treatments of dried egg white have been employed of varying duration for the reduction of the bacteria count in general and the elimination of pathogenic organisms of the genus salmonella.

The length of the heat treatments resulting in the complete killing of salmonella depends upon the time and temperature of exposure. For example, flake albumen may become free of salmonella in approximately 40 hours at a temperature of 140°F; the time will be approximately 48 hours at a temperature of 130°F; and it will take three or more days to obtain salmonella free albumen at a temperature of 120°F.

Prolonged heat treatments at temperatures over 120°F may result in an undesirable discoloration of some or all of the flake particles. This alteration of color of such dried egg white is exactly the same as that which is observed after a prolonged aging of the dried egg white during storage. Such discoloration in the flake particles is an inherent disadvantage in the saleability of the albumen on the market.

The discoloration of such dried egg white, in which the natural sugar content or glucose has been retained, is attributed to a Maillard type reaction wherein the aldehyde group of the glucose reacts with an amino group from the protein. It is possible that polysaccharides which are present, free or bound, in the egg white may hydrolyze, thereby forming a simpler and more reactive aldose.

In any event, the dried egg white with a natural glucose content would normally discolor, without being subjected to heat treatment, and merely during prolonged storage. This has been attributed to the so-called Maillard reaction arising in aging, and which takes place in the presence of protein and sugar or the natural glucose.

However, with the natural protein and glucose being present and the dried egg white being in a flake form, the albumen darkens and varies in color from orange to red, depending on the aging conditions, or the conditions of the heat treatment and the moisture content of the dried product. When the dried egg white is in a powder form, the albumen becomes an off-white to light tan, again depending on the aging conditions, or the conditions of the heat treatment and the moisture content of the dried product. Thus the same phenomenon with respect to discoloration of the natural dried egg white is observed under either aging or heat treatment conditions.

Critical concentrations and pH values for the use of aluminum salts to inhibit discoloration of egg white albumen is given by *J.J. Epstein; U.S. Patent 3,627,543; December 14, 1971.*

A batch of egg white which may be in a natural liquid condition is subjected to a removal of its glucose which may be effected by conventional fermentation, enzyme action or by other suitable means and through which the glucose content of the liquid egg white may be completely removed so that the albumen is free of glucose as may be determined by usual laboratory methods.

The addition of aluminum salts to the liquid egg white after the step of removing its glucose is effective to inhibit discoloration of the egg white upon the egg white being subjected to subsequent prolonged heat treatment or aging upon storage of the dried albumen product. The pH of the egg white after removal of the glucose is adjusted to less than 5.9, and usually to a range from about 5.3 to about 5.8, before addition of the aluminum salt. A minimum level of 0.035% aluminum ion by weight is necessary in the dried glucose free albumen product to inhibit discoloration with a practical maximum of 0.5% aluminum ion by weight, where the dried egg white product is free of determinable glucose.

The narrow range of from 0.035 to 0.5% may be extended to 0.75% aluminum ion by weight to prevent discoloration when as much as 0.2% glucose by weight remains in the dried glucose free albumen. Thus the so-called Maillard type reaction can be prevented when as much as 0.2% glucose remains in the dried egg white by increasing the amount of the aluminum ion added. However, where the glucose content of the dried glucose free albumen is greater than 0.2% by weight of glucose, the addition of a still greater amount of aluminum ion is ineffective to inhibit discoloration of the dried egg white under prolonged temperature treatment or aging in storage.

The mixing of the aluminum ion in the egg white is accomplished by adding an aqueous solution of the aluminum salt to be used in the liquid egg white. Thus, for example, to 1,000 pounds of liquid egg white there may be added 1.205 pounds of a 25% aqueous solution of aluminum sulfate. The liquid egg white may be rendered free of its glucose in the usual manner before, during or after the addition of the aluminum sulfate so that the liquid egg white is free of determinable glucose. The aluminum sulfate glucose free albumen mixture may then be dried in a conventional manner so as to produce a dried albumen product containing at least 0.035% of aluminum ion by weight of the dried product so as to afford full protection against discoloration of the albumen under prolonged heat treatment or storage conditions.

While, in the example given, an aqueous solution of aluminum sulfate has been specified, the desired result of a dried glucose free albumen product containing at least 0.035% of aluminum ion by weight may be accomplished by the addition to the liquid egg white of any aluminum salt compatible in a food product and which is sufficiently soluble to yield the required quantity of aluminum cations of at least 0.035% by weight to the product.

Thus, for aluminum sulfate, there may be substituted in the liquid egg white other salts of aluminum such as acetate; chloride; phosphate; potassium tartrate; sodium chloride complex; sulfate, both anhydrous and hydrate; ammonium sulfate, and sodium sulfate.

ENZYME SYSTEM FOR MAYONNAISE

Mayonnaise is stabilized by the addition of a small amount of glucose oxidase-catalase preparation in the process of *B.L. Sarett; U.S. Patent 2,940,860; June 14, 1960.* In the manufacture of the mayonnaise, it is preferred that the concentration of the glucose oxidase in the final mayonnaise preparation be about 20 and 200 units per pound. The catalase need be present only in very minute quantities, as little as one unit per pound of mayonnaise being satisfactory, but much larger quantities may, of course, be used without adversely affecting the product.

The commercial glucose oxidase preparations normally contain a small amount of catalase, and it is ordinarily not necessary to add any of this enzyme from another source. Only sufficient catalase is required to catalyze the breakdown of hydrogen peroxide to oxygen and water.

As indicated above, the commercial glucose oxidase-containing preparation is subject to a pretreatment before it is used in the stabilization of mayonnaise. The commercial enzyme preparation is heated in an aqueous dispersion to a temperature between about 50° and 80°C for a period of time and under conditions to remove certain impurities from the glucose oxidase. The conditions of treatment are such that the glucose oxidase and catalase are not destroyed and they thus serve to protect the mayonnaise against oxidative deterioration. On the other hand, certain other biologically active ingredients are removed from the glucose oxidase-catalase preparation by this treatment, and the resulting mayonnaise is thus not subjected to emulsion breakdown or other types of deterioration upon storage. The following ingredients were employed to prepare a 100 pound batch of mayonnaise:

Corn oil	77.5 pounds	Mustard flour	0.5 pounds
Vinegar (10% acetic acid)	4.0 pounds	Sucrose	2.4 pounds
Water	7.0 pounds	Liquid egg yolks	7.3 pounds
Sodium chloride	1.3 pounds		

In formulating the above ingredients, 4 pounds of mayonnaise at 68°F from a previous batch were added to the liquid egg yolk which was at a temperature of 35°F. These ingredients were placed in a conventional mayonnaise mixer or blender and 0.4 pound of vinegar and 0.7 pound of water followed by 2.4 pounds of sucrose, 1.3 pounds of sodium chloride and 0.5 pound of mustard flour were added. The batch was mixed for 2 minutes and 7.7 milliliters of an enzyme preparation. The concentration of glucose oxidase in the enzyme preparation was about 750 units per milliliter.

After the glucose oxidase preparation was added, the composition was mixed for an additional minute and thereafter were gradually added 77.5 pounds of corn oil and the remaining water and vinegar. This was done in such a manner that all of the ingredients in the mixer had been added within about 6.5 minutes. After all the ingredients had been added, the mixture was blended for another 45 seconds in order to produce the desired mayonnaise product. The mixer employed was one that is conventionally used in mayonnaise manufacture and served to emulsify the ingredients and incorporate air in the form of minute bubbles into the product. The product contained about 15% by volume of air. The mayonnaise was then packaged in sealed jars in the usual manner. After over 5 months of storage at room temperature, the product was substantially unchanged and retained its original consistency, flavor and appearance.

The glucose oxidase preparation was a commercial product in the form of a clear amber solution containing catalase and about 750 units of glucose oxidase per milliliter. About 200 milliliters of this solution in a glass bottle were heated in an air oven maintained at 60°C for about 30 minutes during which time a precipitate was formed. This precipitate was separated by filtration and the resulting purified preparation was employed to impart stability to the mayonnaise as herein defined. By this treatment the proteolytic enzyme was removed but the glucose oxidase and catalase was unimpaired.

Storage tests have been made on mayonnaise prepared as indicated in the foregoing example and compared with mayonnaise to which no glucose oxidase preparation was added and with mayonnaise to which an untreated glucose oxidase preparation was added. Mayonnaise containing no added enzyme showed evidences of browning and other oxidative deterioration after 3 to 4 months, and mayonnaise containing added enzyme, but untreated, showed evidence of emulsion separation after this period of time. The mayonnaise preparation containing the treated glucose oxidase was by far the most stable and was virtually unchanged after about 4 months, there being no observable emulsion breakdown, color change or change in flavor.

DAIRY PRODUCTS

MILK PRESERVATIVES

Chloro-Bromo Dimethyl Hydantoin

The growth of bacteria in raw milk is controlled by chloro-bromo dimethyl hydantoin in the process of *C.M. de del Pozo; U.S. Patent 3,499,771; March 10, 1970; assigned to Genhal, SA, Mexico.* Chloro-bromo dimethylglycolyl urea (chloro-bromo dimethyl hydantoin) has the formula:

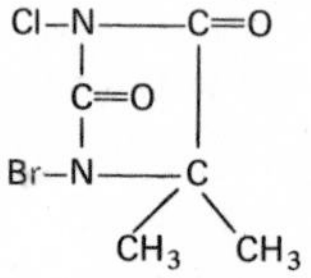

$$Cl-N-C=O$$
$$C=O$$
$$Br-N-C$$
$$CH_3 \quad CH_3$$

The product of this process cannot be considered as an adulterant because it does not affect the physical or chemical properties of the milk, and does not modify the dietetic properties thereof. It cannot be considered as a preserver because it does not alkalize the milk as is the case with all preserves conventionally used such as carbonates, bicarbonates, hydroxides, ammonia and the like.

Finally, the product of this process cannot be considered as a toxic chemical additive because the amount thereof remaining in the milk is negligible and can hardly be chemically detected and also because the halogen derivatives and more particularly those comprising the couple chloro-bromo are quite tolerable to the human organism.

In certain zones in Latin America and Eastern countries there is a lack of suitable facilities to provide for the installation of pasteurizing plants so that some small towns and villages generally consume unpasteurized milk. Some disinfecting chemical products have been proposed for incorporation into the milk, but they have not been acceptable inasmuch as these chemical products constitute an adulterant for the milk and introduce an off-flavor which renders these products unpalatable to most of the consumers.

Therefore, it has been for long a need in the dairy products industry to find a process to aid or to substitute for pasteurization of milk without the detectable incorporation of a disdisinfecting chemical product in the milk. It has also been the need of these industries to look for a germicidal type of chemical product having a killing strength sufficient to be effective in minute doses which will be practically undetectable by the common laboratory methods, which will not constitute an adulterant or a toxic chemical additive.

A germicidal decontaminating pasteurizing-like product comprising chloro-bromo dimethyl-glycolyl urea (chloro-bromo dimethyl hydantoin) is prepared by dissolving a suitable binder, preferably a gum such as gum arabic, gum tragacanth or other binders like carboxymethyl-cellulose or polyvinylpyrrolidone (Plasdone) in a suitable solvent, such as ethanol, adding a large proportion of the chloro-bromo dimethyl-glycolyl urea (chloro-bromo dimethyl hydantoin) which is thoroughly blended to form a paste-like mixture. The latter is granulated, dried and pelletized to tablets or pellets of a suitable size to be used as a contact element towards the milk.

In order to effect the process of decontaminating the milk, a number of these pellets is placed into a foraminous enclosure within a funnel-like receptacle and the milk is passed through. It is highly recommendable that the decontaminating process is effected within 4 hours after the milking operation in order to provide for highest efficiency and killing ability of the chloro-bromo dimethyl-glycolyl urea (chloro-bromo dimethyl hydantoin).

The decontamination process is effected by using an amount of from 10 to 20 grams, preferably 13 grams of the product per about 30,000 liters of milk. As the flow of milk in contact with the tablet or tablets proceeds, the latter are gradually coated with a fatty layer which diminishes the efficiency of the tablet for killing bacteria. Thus, it has been found necessary to periodically stop the flow of milk and rinse the tablet or tablets by a stream of water, preferably shaking the containers for the tablets, in order to remove the fatty layer. It has been found that this operation should be effected after an amount of from 250 to 500 liters of milk have passed in contact with each batch of tablets, depending on the type of milk involved and on other environmental conditions in the processing room.

The amount of germicidal product passing into the milk will be exceedingly small (of the order of 4×10^{-4} g/l). This amount is equivalent to a halogen concentration in the milk of about 6×10^{-6} g/l. The decontamination process is effected by a mere contact of the milk with the tablet and no residual cumulative power remains. The produced milk will have characteristics quite similar to pasteurized whole milk. The decontamination process must be effected on a fairly sweet milk, because if undue lactic fermentation takes place so as to lower the pH value below 5.8, the halogenated product while killing a sufficient amount of bacteria, will not be able to increase the pH of the milk which will then develop more acidity until it finally eventually breaks down.

Example 1: An amount of milk obtained by manual milking without any aseptic precautions was sent to a pasteurizing plant and received there 4 hours after the time of milking. Several samples were taken at the pasteurizing plant and treated in the following manner and with the following results:

(a) One of the samples was fed to the pasteurizing equipment 7 hours after its arrival. This milk was of course completely spoiled by the pasteurizing treatment. (b) Another sample was passed through a cooling screen thereby reducing its temperature from 30°C down to 10°C. The sample was fed to the pasteurizing equipment, its temperature increasing on transportation (1.3 minutes) to 20°C. Ten hours after milking the pasteurization was started. The milk was also completely spoiled on this test. (c) Another sample was cooled by the cooling screen as in (b), previously contacting it with an amount of chloro-bromo dimethyl hydantoin in a test tube (contaminated with dust and by insects) without a stopper and at room temperature (20° to 22°C). After about 14 hours, the pH of the milk was measured and it was found that the initial value of 6.5 was preserved. The alcoholic preservation test was effected on this sample with negative results. At the boiling temperature the milk was not spoiled and it preserved its sweet taste. Upon cooling, a thick consistent scum was formed having normal flavor and odor.

Example 2: A sample of the milk obtained according to the first portion of Example 1 was filtered through a nonsterile clean cloth where a tablet of chloro-bromo dimethyl hydantoin was placed. The thus treated milk was left aside for a period of 16 hours at room temperature. After this period of time, the milk was tested as in Example 1 with very satisfactory results in preservation of chemical and physical properties and pH measurements.

Example 3: 460 liters of milk were stored in a refrigeration chamber after treatment with the chloro-bromo dimethyl hydantoin and precooling in a cooling screen 4 hours after milking. The milk was contained in metallic cans which were kept at a temperature of about 15°C. After a period of 22 hours under refrigeration at the above temperature, a physical and chemical analysis was effected on a sample of milk. Very satisfactory results were obtained.

The milk was then sent to the pasteurizing plant on a nonrefrigerated truck, and was received 1.5 hours later at the pasteurizing plant at a temperature of about 23°C. The pasteurization process was started 24 hours after milking. The duration of the pasteurizing treatment was of about 2 hours. The 460 liters of milk were bottled and then sent on nonrefrigerated trucks to a distributing shop for sale to direct consumers. This trip took about 2 hours at a room temperature of around 27°C.

The bottles were then charged on distributing trucks and sent to the markets and house customers. The time elapsed from the milking time to the consumption time was of an average of about 53 hours. Some of the markets sold some remainders of the milk one day after reception, so that the time for consumption of the milk was increased still more. All the bottles of milk were carefully traced by a group of experts in order to detect the slightest spoilage or deterioration or any change of appearance within the bottles. The reports of the experts were all favorable and there was not a single bottle of milk which could be considered as having an off quality after the abovementioned period of time even when the variations in the temperatures were remarkably frequent and wide. A small portion of the milk was subjected to the same operations except that it was not treated with the chloro-bromo dimethyl hydantoin of the process. All this portion of untreated milk was completely spoiled under the pasteurization treatment.

Example 4: An amount of milk was extracted from the cow's udder without any aseptic conditions and the milk was passed through a funnel-like element containing a number of chloro-bromo dimethyl hydantoin tablets. The thus treated milk was received in a glass flask washed with tap water. This milk was poured into 15 250-ml Erlenmayer flasks and each flask was stoppered with a rubber stopper. The flasks were kept in a refrigerator at 120°C and each of the samples was observed every 24 hours for its condition.

In each observation the flask was opened without any aseptic precaution under room temperature. pH measurements were effected every 24 hours. The time elapsed for the milk to spoil starting from the milking time, was very high, never under 160 hours, as can be clearly seen in Table 1.

TABLE 1

Sample Number	Time to Spoilage	Sample Number	Time to Spoilage
1	168	9	168
2	192	10	184
3	176	11	184
4	193	12	160
5	168	13	168
6	168	14	192
7	160	15	168
8	168		

Another load of milk was treated with the above process with the exception that the step of decontaminating the milk with the chloro-bromo dimethyl hydantoin was omitted. The milk was pasteurized instead. The results obtained are shown in the following Table 2.

TABLE 2

Sample Number	Time to Spoilage	Sample Number	Time to Spoilage
1	72	9	72
2	96	10	88
3	80	11	80
4	96	12	64
5	72	13	72
6	80	14	96
7	64	15	72
8	72		

The bacterial and germicidal power of the product is illustrated by the following example.

Example 5: 20 ml of pasteurized milk were passed into contact with a tablet of chloro-bromo dimethyl hydantoin for about 1 minute, another sample of 20 ml of the same pasteurized milk was not treated with the chloro-bromo dimethyl hydantoin. The treated and nontreated milks were cultivated in Endo medium and incubated in petri dishes. A coli colony count was effected after 48 hours of incubation. The samples were cultivated under the dilutions and gave colony counts as expressed in the following Table 3.

TABLE 3

Dilution	Treated Milk (chloro-bromo dimethyl hydantoin)	Untreated Milk
1 ml, no dilution	222-175 col/ml	Countless
1:10	21-38 col/ml	91-77 col/ml
1:100	3-4 col/ml	10-7 col/ml
1:1,000	2-1 col/ml	1-1 col/ml

From this table, it can be concluded that when heavily diluted, the treated milk could not be differentiated from the nontreated milk. This is probably due to the fact that, as the milk is merely contacted with the chloro-bromo dimethyl hydantoin, the amount contained in the milk under this high dilutions is completely negligible and has no action whatsoever on microorganisms. However, under moderate and low dilutions, the colony growth per milliliter of the treated milk is remarkably less than that obtained with the untreated milk.

Bioflavonoid Preservatives

D.R. Morgan, D.L. Andersen and L. Hankinson; U.S. Patent 3,615,717; October 26, 1971; assigned to the Borden Company found that milk products can be stored for considerably longer periods of time by adding a flavonoid in sufficient amount to inhibit staling.

The flavonoid used may be either naturally occurring or synthetic. Of these the naturally occurring citrus bioflavonoids are preferred, especially hesperidin because in the amounts required to inhibit staling it does not impart any bitter flavor to the milk product. Examples of other suitable flavonoids that can be used are catechol, naringin, phloridzin, rutin, neohesperidin, hesperidin methyl chalcone, quercitin and commercially available flavonoid extracts such as lemon bioflavonoid complex, and lemon-orange flavonate glycoside. Naringin and phloridzin are used in the lower range of permissible concentration since they tend to impart a slightly bitter flavor in the higher concentrations.

The flavonoid is added to the milk product in an amount that corresponds to from 0.002 to 0.04% flavonoid based on the total weight of solids in the finished product, with 0.01 to 0.02% being preferred for liquid milk products. The addition of the flavonoid to the milk products may be made at any time during the processing of the product prior to being packaged for storage since the heating and cooling temperatures used to prepare such products do not affect the flavonoids. It is preferred to add the flavonoid to liquid milk products just prior to sterilization and in the case of powdered milk products just before drying. The bioflavonoids can be added to any milk products which are packaged and conventionally stored for long periods of time; particularly those stored under nonrefrigerated conditons. Examples are evaporated milk, concentrated milk, dried whole milk powder, dried skim milk powder, powdered coffee whiteners, and canned milkshakes.

Example 1: 260 gallons of raw whole milk were preheated in a vat to 165°F for 10 minutes and then further preheated in a tubular heater to 260°F for 35 seconds as it was pumped to a vacuum concentrator and concentrated. The concentrated milk was standardized with water so as to give 26.0% total solids in the finished product. The concentrate was divided into 160-lb batches before further processing. The first batch was run as a control. Catechol was added to the second batch at a rate of 1 g/160 lb to give 0.0053% catechol in the final product (based on weight of total solids). Three succeeding batches and 1.5 g/160 lb were added of phloridzin, hesperidin and naringin, respectively to give 0.0078% in the final product.

These variations were all processed by heating the concentrate to 160°F as it was pumped to a deaeration chamber at a vacuum of 10 to 11 inches. It was further heated to 262°F in a tubular heater with a 7-second holding tube ahead of a steam injector heater. The temperature was increased to 295°F with 2 to 3 second holding time before cooling in a vacuum chamber to 170°F. From the vacuum chamber, it was pumped through a sterile homogenizer at 2,500 psi, cooled to about 65°F in a tubular cooler, and aseptically canned.

The freshly canned concentrates were organoleptically tested and all tasted essentially the same. However, organoleptic testing after 6-month storage at 90° and 75°F, found the samples containing hesperidin to be considerably fresher tasting and to have less of a stale flavor character. The sample containing the catechol was better than the control, albeit not as good as those containing the hesperidin. The samples containing phloridzin and naringin were fresher than the control but slightly bitter at these concentrations.

Example 2: 80 lb of a liquid Imitation coffee whitener (liquid Cremora) were made in the usual manner. The product contained sodium caseinate, coconut fat, minerals, gum, sugars and water. Hesperidin 0.003% (based on finished weight of product) was added to 35 lb of this product and the remainder used as a control. The two products were filled into cans and sterilized HTST in a continuous sterilizer. The freshly canned products all tasted essentially the same. However, after 6-month storage at 75° and 50°F, the samples containing hesperidin were considerably fresher tasting and had less of a stale flavor character as determined by organoleptic testing.

Example 3: The following bioflavonoids were added separately at levels of 0.003% (based on total weight of finished product) to nine lots of a commercial batch of Borden's strawberry milkshake just prior to canning and subsequent sterilization:

Hesperidin (practical)	Hesperidin (complex)
Quercitin (practical)	Hesperidin methyl chalcone
Chalcone	Lemon-orange flavonate glycoside
Hesperidin (reagent grade)	Lemon bioflavonoid complex
Hesperidin (purified)	

The nine lots of strawberry milkshake preparation each containing a different bioflavonoid along with a control were canned and sterilized HTST (220°F for 3 minutes 48 seconds, 262°F for 2 minutes, and then cooled to 80°F) on a continuous sterilizer. These canned milkshakes were stored at 50°F and checked initially and at 3-month intervals for 1 year. Throughout the year's storage and at the end of 1 year the strawberry milkshakes containing the bioflavonoids were tested organoleptically and always judged superior in flavor to the control.

The strawberry milkshakes containing the bioflavonoids were judged to be more fresh and intense in flavor. The color of the milk shakes containing the bioflavonoids had faded to a much less degree than the control at the end of 1 year's storage. The staling and color loss in the control were evident at the end of 3 months.

CHEESE PRESERVATIVES

Sorbic Acid for Cottage Cheese

The process described by *G.A. Perry and R.L. Lawrence; U.S. Patent 2,974,046; March 7, 1961; assigned to Corn Products Company* is predicated upon the fact that sorbic acid, or any of its edible salts such as calcium, potassium and sodium sorbate, which on acidification yield sorbic acid, will, when added to a cultured milk product in small and apparently critical amounts depending upon the pH of the cultured milk product promote growth of useful bacteria, and almost totally inhibit the growth of spoilage bacteria, for periods of time exceeding the maximum periods during which it is likely that such a product would be expected to have a shelf life, or to be kept in the home.

In a series of tests using 31.1 lb samples of creamed cottage cheese taken from regular production in a dairy plant, the control samples were free of sorbic acid and the test samples contained 0.05% sorbic acid, based on the total weight of the creamed cottage cheese which had been uniformly distributed through the cottage cheese by prior dissolution. At the beginning of these tests, samples of cottage cheese from both the control and the test masses were made the subject of bacterial plate counts and microscopic examination, and the number of bacteria present was calculated on a per gram basis.

Since useful bacteria and spoilage bacteria could not, at the beginning of the test, be visually distinguished, the count of 5,000 per gram established at the beginning of the test was assumed arbitrarily to be one-half of each type. The samples were then stored at 50°F for varying periods of time, as indicated in the table which follows. At the end of each period of time a representative interior portion of a control and of a test sample were subjected to bacterial plate counts and microscopic examination of the bacteria contained therein. Each time that a count was made, both on the starting day and on each subsequent day, a gas test was run on an extracted sample, this being the usual gas test employing formate ricinoleate broth.

In the table which follows, the bacterial count is given on a per gram basis for each of the control and test samples for the periods of time set forth in the table, as well as the results of the corresponding gas test.

Control Samples (No Sorbic Acid)

Days	Useful Bacteria	Spoilage Bacteria	Gas Test[a]	pH
0	2,500[b]	2,500[b]	Negative	4.9
6	(c)	500,000	Positive	4.9
10	(c)	2,000,000[e]	Positive	4.8
14	(c)	1,200,000[f]	Positive	4.4
23	(c)	600,000[f]	Positive	4.4

Test Sample (0.05% Sorbic Acid)

Days	Useful Bacteria	Spoilage Bacteria	Gas Test	pH
0	2,500[b]	2,500[b]	Negative	4.9
6	15,000	(d)	Negative	4.9
10	105,000	(d)	Negative	4.9
14	240,000	(d)	Negative	4.8
23	4,000,000	(d)	Negative	4.8

[a]Gas Test — Formate ricinoleate broth.
[b]At day 0 the bacterial count was 5,000 per gram for each of the control and test samples. One-half of these were assumed to be useful bacteria, and the other half spoilage bacteria.
[c]No detectable increase.
[d]No detectable increase in bacterial, yeast or mold counts.
[e]Inedible due to excessive spoilage bacteria coupled with yeast and mold growth.
[f]Decrease in spoilage bacteria after 10 days due to competitive action of yeast and mold growth with attendant lowering of pH.

This demonstrates that sorbic acid has a selective capacity for controlling the growth of useful and spoilage bacteria. It appears to promote the growth of useful bacteria, and to inhibit the growth of spoilage bacteria. In each case, a control mass of creamed cottage cheese, and a mass containing 0.05% of sorbic acid, each of which had been held at 50°F for the number of days indicated, was then subjected to organoleptic examination by a panel of judges. In the case of each test sample, the panel did not report any significant increase in spoilage factors and reported favorably on odor and on flavor retention. On the other hand, each of the control samples, after 10 days of incubation, were found to be objectionable in odor, taste and appearance, and to be inedible.

Selective control of bacterial growth in creamed cottage cheese is effected by adding to the milk product from 0.03 to 0.15% by weight on the total milk product, of sorbic acid or its equivalent in the form of the calcium, potassium or sodium salts of sorbic acid. The lower limit in the stated range is determined by loss of effectiveness. For example, when the sorbic acid is as little as 0.01% there seems to be no discrimination between useful bacteria and spoilage bacteria.

The upper limit of the range is determined by the appearance of an organoleptically notice-able sour taste characteristic of sorbic acid. This sour taste begins to appear in cultured milk products having a pH close to neutrality when sorbic acid is above 0.15%. For exam-ple, when the curd in creamed cottage cheese is made by enzymatic treatment, such as ren-net, the curd and hence the resulting creamed cottage cheese has a pH of about 6.4, and the permissible upper limit of sorbic acid is about 0.15%. However, in the case of cultured milk products having a lower pH, such as a pH of from 4.6 to 5.1, which is characteristic of creamed cottage cheese made with washed cultured curd, the sour taste begins to appear when sorbic acid exceeds about 0.08%. It appears, therefore, that the permissible upper limit of sorbic acid declines when the cultured milk product is slightly acid, and becomes greater as the cultured milk product approaches neutrality.

In adding sorbic acid to a mixed cultured milk product such as creamed cottage cheese, it is preferred that the sorbic acid be mixed first with the cream dressing where it is relatively highly soluble in the fats present and where its inhibiting effect on spoilage bacteria improves the storage characteristics of the cream dressing prior to being mixed with the curd. The thus-treated cream dressing is then mixed with curd in the usual way. Ordinarily, creamed cottage cheese consists of about 4 parts curd and 1 part cream dressing. The amount of sorbic acid added to the cream dressing is calculated upon the total weight desired in the mixed creamed cottage cheese.

The preferred range for sorbic acid is 0.05 to 0.07% based on the total weight of the cul-tured milk product treated. Where calcium, potassium or sodium sorbate is used, an amount is chosen which will release the desired amount of sorbic acid in the final product.

Example 1: To 25 lb of washed cultured cottage cheese curd having a pH of about 4.5, there were added 6 lb of cream dressing containing 0.31 lb of salt to which 0.016 lb of sorbic acid had been added following pasteurization and homogenization while the cream dressing was at a temperature of about 80°F. The cream dressing was cooled to 40°F and the curd and dressing were thoroughly mixed together. In this mixture sorbic acid consti-tuted about 0.05% by weight. Samples of this creamed cottage cheese, preserved at 50°F for 3 weeks, showed no significant spoilage characteristics, had acceptable odor and flavor, and showed a negative gas test.

Example 2: A batch of cultured cottage cheese was made up exactly as in Example 1, ex-cept that the 0.016 lb of sorbic acid was replaced by 0.02 of calcium sorbate. The effect-ive quantity of sorbic acid in the final product remained the same as in Example 1, namely, 0.05% by weight, and the results were the same.

Sorbic Acid Powder

Centralized cutting and wrapping operations and merchandizing methods as practiced in the cheese industry result in a lengthening of the period of time that wrapped consumer-size pieces of cheese remain in trade channels. This situation permits the development of undesirable molds upon the packaged product despite improved sanitation practices in cheese production and wrapping. It is, therefore, necessary to treat the cheese product with harmless food-grade spoilage inhibitors, such as, for example, sorbic acid.

Utilization of sorbic acid for the treatment of process cheese has presented no particular problem because the chemical may be simply added to the blend of cheese going into the processing kettle and thus distributed throughout the cheese mass. This procedure, however, is not suitable for natural Cheddar, Swiss and other cheese in consumer-size pieces, and a problem has long existed with respect to the treatment of such products. Dipping or spraying with solutions is not desirable because the added moisture interferes with wrapping and package sealing. Also, in view of the nature of the fissured surfaces of Cheddar and the eyes in Swiss cheese, there is the possibility of excess liquid antimycotic material collecting in the fissures and eyes of these products.

Treatment with the mold inhibitor in the form of a dry powder further presents a problem because of the difficulty of uniform application and because loosely-applied powder tends to shake off in the conveying system and to be removed in areas during the wrapping and sealing processes. Excessive amounts of the chemical dust or powder on the cheese blocks are not permitted by legal standards, result in a disagreeable taste, and the presence of the dust in the atmosphere is highly irritating to the processing personnel.

Further, substantial losses of the powder acid product are experienced in dusting operations. For effective mold-inhibiting action, it is important to apply to the surfaces of the cheese pieces at least 0.1% of an inhibitor, based on the weight of the cheese, such as sorbic acid, and to apply not more than the legal maximum of 0.3%. If a dry powder is employed in the treating operation, it is extremely difficult to apply to the cheese surface and to retain a coating which is relatively uniform and which lies within this narrow range, and should areas of the cheese have less than 0.05%, effective resistance to mold is not provided, and if certain areas have more than 0.3%, taste and other disadvantages are present.

An apparatus and process was developed by *L.J. Hansen and M.B. Westover; U.S. Patent 3,328,176; June 27, 1967; assigned to Armour and Company* to provide a means and method for treating cheese pieces to obtain effective coating of antimicrobial food spoilage agents. The apparatus is illustrated in Figure 5.1a.

In the illustration a framework **10** is used to support a conveyor housing **11**. The housing provides a chamber having at its bottom a receptable **12** in which is mounted a slide drawer **13**, the drawer being provided for receiving excess powder. In the upper portion of the closed housing is mounted an open mesh conveyor **14** adapted to carry the cheese body to be treated, and the housing provides entrance and discharge ports for receiving the cheese bodies. Connected with the housing are suction ducts **15** and **16** which lead to a fan and to filter bags (not shown). A powder receptacle **17** contains a supply of sorbic acid, etc. in powder form, and the air-powder mixture is delivered to a manifold **18** as compressed air is passed through the powder pump assembly. From the manifold, a series of small tubes **19** lead to various positions about the conveyor **14** so as to discharge the compressed air carrying the powder upon all surfaces of the cheese.

The operating parts are shown more clearly and in greater detail in the schematic view of Figure 5.1b. Compressed air from a source of supply passes through conduit or manifold **18** which is controlled by the solenoid-operated valve **20**. Connecting with the conduit is a powder pump **21** with an air ejector tube extending downwardly into the bottom portion of the powder reservoir of receptacle **17**. The discharge ducts **19** communicate with the manifold conduit through a fitting member **22**. The ducts or tubes terminate in flared discharge nozzles **23** which are designed to apply the powder to the various surfaces of the cheese piece or block **24**. The cheese piece has six sides, and the nozzles are so distributed as to apply the powder evenly with respect to each side, each nozzle being so flared or designed and being so positioned with respect to the cheese block as to cover the area of the face of the cheese toward which the nozzle is directed. Since the conveyor is an open mesh conveyor, the nozzle below the belt is effective in covering the bottom side of the cheese piece simultaneously with the application of the powder to the top and other sides of the cheese piece.

Supported above the conveyor is a microswitch **25** which controls through a relay solenoid **20a** governing valve **20**. An actuating arm **26** from the microswitch is engaged by the cheese block as it moves forwardly, thus actuating the microswitch for the opening of the valve. As the cheese block moves forward during treatment past the arm, the arm swings back to its initial position to bring about the closing of the valve. Connecting with the housing and in order to maintain it under negative pressure are duct means **15** leading to a suction blower **27**. The withdrawn powder is discharged into the chamber **28** and the air is forced through the filter bag **29**. Settled powder may be collected from the lower receptacle **30**.

FIGURE 5.1: CHEESE TREATING METHODS

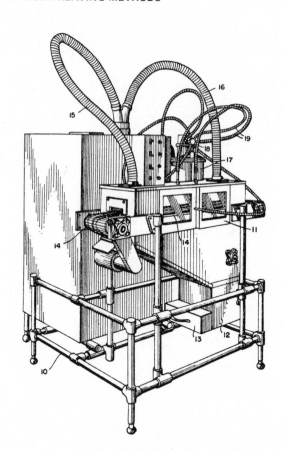

(continued)

FIGURE 5.1: (continued)

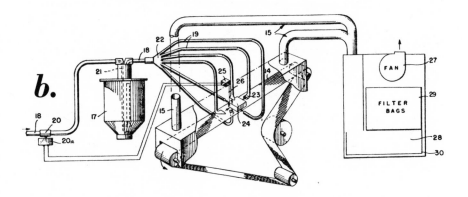

Source: L.J. Hansen and M.B. Westover; U.S. Patent 3,328,176; June 27, 1967

In practice, it is found that the majority of the excess powder is collected in the drawer 13 illustrated in Figure 5.1a, and instead of employing a drawer, connections may be provided for the automatic return of the powder through sieving apparatus to the supply chamber 17 for reuse. Withdrawn air through the ducts 15 and recovered powder from the filter receptacle 29 may also be returned to the supply chamber for reuse.

It is found that an accurate application of the treating powder to all sides of the cheese or plastic body is accomplished so that the body is coated with a uniformly embedded cover of sorbic acid dust at a given chosen level between 0.05 and 0.3% of the weight of the cheese. This is accomplished by impinging the compressed air carrying the powder with force upon the cheese body. The powder particles under the force of the impinging streams stick to the cheese body because a substantial number of particles become embedded in the cheese and serve as lodging means or retainers for the applied powder. Further, the particles thus effectively applied to the cheese body remain thereon during travel on the conveyor and during subsequent wrapping and packaging procedure.

While the pressure may be varied depending upon the treating powder, it is best to use pressures in the range of 10 to 35 psi. Excellent results have been obtained by using about 25 psi at the powder pump. With these pressures, sorbic acid particles are retained upon the cheese body in coatings of little more than 0.1% based on weight of the cheese.

The open mesh conveyor may be supported by any suitable number of rollers, and one or more of the rollers may be driven. The conveyor speed of the apparatus is about 42 feet per minute in order to synchronize with the speed of the wrapping line conveyor, but the speed may be varied, and the powder-applying mechanism can be adjusted to the changed speed of the conveyor. The amount of powder applied can be controlled through adjustment of the powder pump 21 and through the pressure of the air fed through conduit 18.

By providing suction ducts 15 connecting with the housing, excess floating particles of the powdered chemical are removed, and an accurate control of the application of powder to the cheese is maintained by the use of compressed air carrying powder through the ducts 19. If desired, the housing may be maintained under negative pressure so that the air flow is from the outside into the housing and the treating powder does not escape into the operating room. The antimicrobial food spoilage powder or dust may be crystalline sorbic acid, salts of sorbic acid, metal salts of propionic acid, such as sodium and potassium propionates, hydroxyl esters of benzoic acid, and other known food spoilage inhibitors.

Example 1: Cheese pieces at room temperature were passed through the housing on conveyor **14**, as illustrated in Figures 5.1a and 5.1b, the air pressure being maintained at 15 pounds per square inch. Cheese pieces weighing in the aggregate 5,560 lb were treated, using 5.75 lb of sorbic acid. The percent of sorbic acid on each block of cheese was found to be about 0.103% based on the cheese weight. The conveyor belt was 10½ ft long and was operated at 42 ft/min. Good coverage was obtained on all six surfaces of each block of cheese. It was found that about one-third of the sorbic acid particles stuck to the cheese; the remaining two-thirds were recovered from the drawer in the bottom of the housing. The recovered powder was sieved in a 25 mesh screen to remove particles of cheese, and the recovered powder was reused in the operation.

Example 2: Cheese pieces at room temperature were passed through the housing on conveyor **14**, the air pressure being maintained at about 20 psi. Cheese pieces weighing in the aggregate 102.5 lb were treated, and the percent of sorbic acid on each block of cheese was found to be about 0.14% based on the weight of cheese. Good coverage was obtained on all surfaces of each block of cheese and about one-third of the sorbic acid particles stuck to the cheese. The remaining powder was recovered in a manner similar to that described in Example 1.

FATS

Free-Flowing Powders

P.M.T. Hansen and L. Linton-Smith; U.S. Patent 3,271,165; September 6, 1966; assigned to Commonwealth Scientific and Industrial Research Organization, Australia gives the background on the manufacture of dairy fats in powder form and the use of silicates or phosphates to keep the powder free-flowing on storage. Normal butter is a product which by regulation contains at least 80% milk fat, sometimes 82%, and not more than 16% water. Normal butter also contains salt, in an amount which is usually 1 to 2%, and in addition there are milk solids present, amounting to 0.5 to 2% depending on the manufacturing technique. Butter is manufactured either by the conventional churning process, which results in the formation of granules of butter and separation of buttermilk, or by one of the so-called continuous methods.

The plasticity of normal butter is desirable when the butter is to be used for spreading, but is in many ways undesirable when the butter is to be used as a baking ingredient, particularly in prepared dry cake mixes. For use as a baking ingredient the butter needs to be brought into a finely divided state, and it is therefore understandable that the techniques of creaming butter with sugar, or the melting or extrusion of butter for blending with dry ingredients, are tasks which have been accepted as necessary in all domestic and industrial baking methods.

Powders with a substantial fat content can be manufactured for example by drying a homogeneous mixture comprising fat and an aqueous dispersion of nonfatty materials such as proteins, starches or gum. Dry whole milk is an example of such a product; it contains approximately 27% fat, 3% moisture and 70% skim milk solids. Higher ratios of fat to skim milk solids can be obtained by drying milk or cream with a correspondingly higher fat content; however, such powders are usually difficult to manipulate by conventional drying methods, and are not free-flowing.

Conceivably any powder with a substantial fat content could be used as the source of fat in baking and cooking, and such powders would have the advantage over other types of shortening of lending themselves well to blending with other dry ingredients. However, a powder with a low ratio of fat to nonfat milk solids, for example dry whole milk, gives rise to difficulties in adjusting the baking formula to allow for the additional nonfat milk solids which necessarily accompany the fat. Conventional powdered shortenings are frequently observed to give a poor baking performance, however, especially in cakes and similar goods. Cakes made from dried cream or dried modified creams are notably poor since they attain neither the texture nor the volume which normally results. A reason for this is probably that the fat in the powder is in a finely dispersed state, and coated or otherwise associated with the nonfat material, and so is not susceptible to aggregation by agitation and is not in a suitable form for stabilizing incorporated air. Cake batters made from such powders have been observed to possess a high specific gravity, indicating insufficient aeration. Similar disadvantageous features are observed when such emulsifying agents as are usually employed by the food industry are added.

According to this process, a method of making an edible fat-containing composition in powder form suitable for use as a shortening material, comprises the steps of mixing an edible cream in the form of a fat-in-water dispersion containing at least 35% water, with nonfat milk solids and with an edible fat-soluble emulsifying agent in an amount not exceeding 25% of the nonfat solids to form a fat-in-water dispersion system having a fat content comprising at least 80% of the total solids present in the dispersion, and spray-drying the mixture without breaking down the structure thereof to form a powder, the mixing of the emulsifying agent with the other ingredients not involving a homogenizing treatment, whereby a partial deemulsification takes place when water is added to the dried powder. The product may be dried to have a moisture content below 1% by weight. The fat in the dispersion may be milk fat or other edible fat, for example, coconut oil or peanut oil. Thus the dispersion may be made from fats or oils which are solid or liquid at room temperature.

Normally, the dried product is immediately cooled to room temperature or below, and is blended with a finely-divided physiologically acceptable agent to confer free-flowing properties on the powder. Such agents include suitable silicates and phosphates. The cooling is particularly important when the product is based on milk fat because the cooling hardens the fat and induces a desirable crystalline state.

A formulated cream can be made by homogenizing a mixture of fat and nonfat solids, the cream thus produced containing all of the fat but only part of the nonfat solids present in the system prior to drying. It has also been found that the cream must have a water content of at least 35%. In the case of a formulated product, the emulsion would not form, however long the homogenizing treatment, with a water content below 35%. When skim milk solids are used to form a homogenized fat dispersion, the addition of a calcium-sequestering agent is desirable to prevent the mixture from thickening in the homogenizer. When casein is incorporated in the homogenized fat dispersion, the casein used may be an edible caseinate at a neutral or alkaline pH, though it is preferred to disperse an edible grade of dry acid casein in water and/or skim milk with the aid of an alkali. Dispersion of casein may for example readily be effected by warming a mixture of:

    6 parts of 30-mesh casein
    1 part of sodium citrate or an equivalent amount of sodium carbonate or phosphate
    14 parts of skim milk or water at 120°F
    7 parts of 2% sodium hydroxide solution or the equivalent amount of another physio-
        logically acceptable alkali

to a temperature within the range of 140° to 180°F and agitating the mixture until the casein is dispersed. The pH of this mixture is close to that of normal milk (6.7). More liquid can be added to make the dispersion less viscous.

A preferred emulsifying agent for mixture with the fat-in-water dispersion is glycerol monostearate, but mixtures of mono- and diglycerides prepared from hydrogenated fats are quite suitable. The emulsifying agent may be melted, and incorporated with gentle agitation either directly into the fat-in-water dispersion or into a warm dispersion of the milk solids which are to be mixed with the fat-in-water dispersion.

The emulsifying agent is added to the fat-in-water dispersion not only to assist in the mechanical aeration of cake batters in which the dried powder is used, but for its effect on the stability of the dried powder. The method by which the emulsifying agent is incorporated in the concentrate prior to drying is the critical factor with regard to the baking performance of the powder. If all the ingredients were added prior to the homogenizing treatment, the emulsifying agent would sterilize the system to a degree where sufficient free fat would not be released during baking. It is critical that the mixing of the emulsifying agent with the other ingredients does not involve a homogenizing treatment.

Investigations of the microstructure have shown that, in a typical case, the dried powder consists of fat globules (and possibly some free fat) embedded in a mixture of skim milk solids, caseinate end emulsifying agent. It has also been shown that a partial deemulsification takes place when dried powder is mixed with water and this provides the necessary release of the fat to give good baking performance. The emulsifying agent appears to be responsible for this because it has been found that the maximum amount that can be incorporated without danger of the powder breaking down during drying is about 25% of the nonfat solids present.

If desired, antioxidants can be added with the fat or with the emulsifying agent at permissible levels. The milk solids which are mixed with the fat-in-water dispersion are preferably in the form of a dispersion, to facilitate mixing. For example, skim milk itself represents a dispersion of skim milk solids which is highly suitable for admixture with the fat-in-water dispersion.

Powders of substantial fat content tend to create difficulties due to stickiness during spray drying. This may cause some of the powder to be exposed to heat in the spray dryer for prolonged periods of time which is undesirable. More serious, however, would be the accumulation of deposits in the dryer and in the apparatus used for the recovery of air-borne powder particles. Such deposits tend to cake together, thus interfering with the drying process as well as destroying the free-flowing nature of the powder. When this occurs the product becomes unusable for its intended purpose.

By suitable selection of spray drying equipment some of these difficulties may be overcome. If the powder is dried in a cyclone-type dryer, the frictional forces encountered by the powder disrupt the structure of the product and thus destroy its character as a free-flowing powdered shortening material. On the other hand, dryers of the filter bag type avoid the occurrence of these disruptive frictional forces, so that the structure of the powder is preserved and efficient collection can be achieved. It was found that considerable lessening of the difficulties can be achieved by treating the inside walls of a suitable spray dryer with an antisticking agent. An example of such treatment involves applying to the walls a whitewash or spray consisting of a water suspension of finely divided sodium aluminum silicate and allowing the material to dry to give a fine, powdery deposit. The adherence of the silicate to the walls may be improved by the addition of, preferably 20%, calcium hydroxide to the silicate before its suspension in water. Calcium hydroxide by itself is also effective, but is less desirable.

The amount of finely-divided silicate or phosphate which is blended with the spray dried product (after cooling) is preferably ½ to ¾% of the weight of the spray dried product, whereupon the powder so obtained may be screened through a fine sieve. Powder made from milk fat is light and free-flowing and possesses good storage stability when stored under conditions normally observed for milk powders. The powder does not melt when exposed to a temperature at which normal butter would form an oil, and is found to have outstandingly good baking properties.

Example 1: Powder with the following compositions:

	Percent
Milk fat	82.0
Commercial glycerol monostearate	3.5
Skim milk solids	6.7
Sodium caseinate-citrate mixture	6.7
Sodium aluminum silicate	0.5
Moisture	0.6

was produced as follows.

45.0 lb skim milk was pasteurized for 30 minutes at 143°F, whereupon a solution of 0.5 lb of sodium hexametaphosphate in 5.5 lb of water was added to the pasteurized skim milk. 82.0 lb of dehydrated butter fat were melted and mixed with the treated skim milk. The resulting mixture was then homogenized at a pressure of 700 to 800 psi and a temperature of 120°F to form an emulsion. A casein dispersion was produced by warming a mixture of 6.0 lb of dry acid casein, 1.0 lb of sodium citrate, 20.0 lb of skim milk at 120°F and 7.0 lb of 2% sodium hydroxide solution to a temperature within the range of 140° to 180°F and agitating the mixture until the casein was dispersed. The casein dispersion and 3.5 lb of molten commercial glycerol monostearate were then mixed with the emulsion at 150°F.

The mixture so formed was spray dried in a commercial spray drying plant to a moisture content below 1%, and the spray dried powder obtained was, after cooling, blended with 0.5 lb of sodium aluminum silicate. The resulting powder, after being screened through a fine sieve, was light and free-flowing, and highly suitable for use in baking.

Example 2: Powder with the following composition:

	Percent
Milk fat	82.0
Commercial glycerol monostearate	3.5
Sodium caseinate-citrate mixture	13.4
Sodium aluminum silicate	0.5
Moisture	0.6

was produced by a procedure as described in Example 1 but using the following ingredients:

82 lb of dehydrated butter fat	14.7 lb of 2% sodium hydroxide
12.6 lb of dry acid casein	3.5 lb of commercial glycerol monostearate
2.1 lb of sodium citrate	0.5 lb of sodium aluminum silicate
30.0 lb of water	

The butter fat was melted and mixed with 20 lb of the casein dispersion and 35 lb of water and homogenized at a pressure of 700 to 800 psi. The resulting mixture was then blended with the rest of the casein dispersion and with the melted emulsifier before the spray drying.

Example 3: An alternative procedure for obtaining a powder with a composition specified in Example 1 is as follows. 1 lb of sodium citrate is dissolved in 132 lb of pasteurized cream (62% fat) and the mixture homogenized in a double stage homogenizer at 2,000 + 500 psi at 110°F. 6 lb of dry, acid casein is dispersed in 20 lb of skim milk by warming to 140° to 180°F and by gradually adding 7 lb of 2% sodium hydroxide. 3.5 lb of commercial glycerol monostearate is melted with any required antioxidant and blended into the caseinate solution. This mixture is then added to the homogenized cream. The total mix is then prewarmed to 150°F and spray dried under the conditions normally used for dry, whole milk. The total solids before drying is 58 to 59%. The dry powder is cooled and then mixed with 0.5 lb of sodium aluminum silicate before sieving. The product is packed in an atmosphere of nitrogen.

Stabilized Salted Margarine

The process of *D. Melnick; U.S. Patent 2,983,615; May 9, 1961; assigned to Corn Products Company* is concerned with flavor stabilization of salted margarine by the addition of an ethylenediaminetetraacetic acid (EDTA) and citric acid. The combination of the EDTA component and the citric acid component is necessary to produce a synergistic effect as evident from the fact that the result is greater than is expected from the individual effects of the components.

The combination of additives is added to any margarine containing common salt or sodium chloride. According to government regulations, margarine is produced by combining a milk phase with not less than 80% by weight of fat. Usually skim milk in a quantity of

15 to 20% by weight is employed; however, soy milk (a suspension of soy flour in water) may replace in whole or in part the skim milk. Sodium chloride is added to the margarines in an amount of 1 to 4% by weight; this is equal to 5 to 20% in the aqueous phase of the margarine. The greater the quantity of salt in the margarine, the shorter the shelf life from the standpoint of flavor acceptance.

The salt component of the margarine contains trace quantities of salts of metals such as iron and copper. By virtue of the methods of manufacture currently used commercially, these trace metals may be present in an amount of 0.2 to 20 ppm and more frequently from 1 to 3 ppm. Salt manufacturers are selling to the food industry a premium grade salt containing less than 1 ppm of iron. The copper content in the salt is also less than 1 ppm. While it is not known with certainty, it is believed that iron and copper salts which are present as impurities in sodium chloride act as pro-oxidants and thereby have an adverse effect on the flavor of margarine.

It was found that salted margarines have a flavor life which is less than one-half of that obtained with margarine free of salt. In this connection, a simple experiment was performed in which margarine fat was stored under air in a loosely-capped glass jar at 80°F. At the end of a two-week period the first indication of flavor deterioration became evident. The loss of flavor life of the margarine fat was not accelerated by the addition of 20% water. On the other hand, the addition of 2.8% sodium chloride of premium grade to the aqueous phase of the margarine caused flavor defects to appear during the third day of storage and at the end of one week the flavor of the margarine was unacceptable. On the basis of the results obtained in the experimental work, it appears that the iron and copper salts are in some way combined with the additives of this process and thus rendered ineffective in spoilage of margarine flavor.

This explanation is maintained even though peroxide development in foil-wrapped prints of margarine stored at 75°F for a period of up to 10 weeks is small, and there is no significant increase in the concentration of carbonyl compounds in the separated margarine fat. Objective flavor scorings rather than chemical methods are used in determining the value of the process.

The edible citric acid component includes citric acid as well as the monoalkyl esters and edible salts. The alkyl esterifying group contains 3 to 18 carbon atoms. Preferred members of this class of additive are monoisopropyl citrate and monostearyl citrate. Commercially available isopropyl citrate is composed of 65 to 80% of monoisopropyl citrate, about 15 to 30% of diisopropyl citrate, and 5 to 10% of triisopropyl citrate. The stearyl citrate is composed of 10 to 15% of monostearyl citrate, 70 to 80% of distearyl citrate, and 10 to 15% of tristearyl citrate.

In the preparation of the margarine the citric acid may be added to the oil or fat ingredient during the deodorization treatment. The deodorization treatment is conventional and involves subjecting the oil or fat ingredient of the margarine to steam at an elevated temperature of 400° to 500°F to remove undesirable volatile contaminants. The citric acid component is added to the margarine oil or fat following deodorization. The citric acid component may be added, for example, in oil-dispersible form, using a mixture of mono- and di-fatty acid esters of glycerol as the vehicular material for the free acid or for the isopropyl citrate, or it is added directly to the deodorized oil in the form of the stearyl citrate. The citric acid component, so added to provide 0.002 to 0.07% calculated as free citric acid in the margarine, affords protection to the oil under these conditions of treatment even before being combined with the salt in the final preparation of the margarine. Citric acid naturally provided by the milk component of the margarine is as the sodium, potassium and calcium salts and such milk provides 0.02 to 0.08% as free citric acid in addition to that contributed by the oil phase.

The EDTA component includes free EDTA as well as the mono-, di-, tri- and tetra-salts in which the salt forming radical is an edible metallic cation such as, for example, sodium, potassium, or calcium alone or combinations thereof. Specific examples of the EDTA salts are the disodium salt, the tripotassium salt, the disodium monocalcium salt, etc. as such or as their hydrates. In the preparation of the margarine it is preferred to add the EDTA component to the aqueous phase at approximately the same time as the salt addition. If desired, the EDTA component may be added after the margarine has been prepared and is still in a liquid phase. In any case, the EDTA component is added to the margarine under conditions promoting the distribution of the same throughout the aqueous phase of margarine product to provide a concentration of 0.001 to 0.3% in the aqueous phase, equivalent to 0.002 to 0.05% in the overall margarine.

Example: To 80.4 parts by weight of a margarine fat having a melting point of 93.4°F, a setting point of 74.0°F and an iodine value of 85.5 and having dispersed therein 0.01% of isopropyl citrate of which 70% was the monoisopropyl citrate, the customary vitamins, emulsifiers and preservatives were added with 16.8 parts by weight of EDTA-supplemented skim milk and 2.8 parts by weight of salt (sodium chloride) containing 1.6 ppm of iron and 1.1 ppm of copper. The fat consisted of 23 parts of soybean oil selectively hydrogenated to an iodine value of 67.4, 37 parts of soybean oil selectively hydrogenated to an iodine value of 92.8 and 40 parts of cottonseed oil selectively hydrogenated to an iodine value of 89.0.

The fat was deodorized the day before the margarines were made, and it was just after deodorization when the isopropyl citrate in a mono- and diglyceride vehicle (38:62) was dispersed in the warm oil. The milk contributed an additional 0.04% of citric acid to the margarine. The EDTA component added to the margarine was the disodium salt and it was dissolved in the milk to bring the concentration of the additive, expressed as free EDTA, to 0.01% in the margarine, i.e., 0.06% by weight of the skim milk. The emulsion was converted into margarine using the conventional Votator assembly involving a chilling A-unit and a quiescent B-unit. The extruded noodles were packed as ¼ lb prints in a conventional screwdrive packaging (Morpac) machine, the prints being foil wrapped and cartoned.

The packaged margarines were then stored in both the refrigerator (45°F) and at room temperature (75°F) before being flavor-scored by the panel of experts. For control purposes, the same margarine was prepared but this time without the addition of the EDTA component to the milk phase. Whereas there was no significant difference in flavor of the margarine containing the synergistic combination of additives when the margarine was stored at 45°F, there was a striking improvement in the flavor stability of the margarine containing the synergistic combination of the agents when the margarines were held at 75°F.

BREAD SOFTENERS AND MOLD INHIBITORS

ANTISTALING ADDITIVES

Commercial Monoglycerides

G.R. Jackson and J.M. Livingston; U.S. Patent 3,111,409; November 19, 1963; assigned to Top-Scor Products Corporation, Inc. discusses the mechanism of staling in bread. During staling, bread loses its plastic properties, becomes tough and crumbly or hard with a noticeable lack of flavor. Since as bread becomes stale it is harder and feels drier, people are inclined to believe that staling is due to moisture drying out of the loaf. However, studies of relative vapor pressures and water content have been made using fresh and stale bread. Such studies show that differences in water binding by the fresh and stale bread are so small as to be insignificant. It has also been shown that only a small amount of water is lost in the baking process. In fact stale bread can be brought back to a state of temporary freshness by reheating a few minutes in an oven. Thus loss of moisture is not the factor causing staleness.

Even though it is known that the structural changes of staling are not due to a significant decrease in amount of water, it has been shown that if fresh bread is dried quickly to a low moisture content, the structural evidences of staling fail to develop, or do so at a much slower rate. This indicates that during staling some change in structure is developed in bread which involves water in some way. The change in structure during the staling process must, therefore, be due to the formation of some type of cross linkage in the nonaqueous constituent of the bread. Water of fresh bread acts as a plasticizer; that is, it serves as a fluid medium in which motion of solid particles past one another can take place. Apparently, during staling, the water becomes bound to the solid phase and thereby loses its ability to act as a plasticizing medium. This would involve the formation of cross linkages between the nonwater elements of the bread.

As a result of these moisture studies, it was important to determine which component of the flour is responsible for the change in binding of water. Since the three most abundant ingredients in flour are carbohydrates, proteins and water, crystallinity, which occurs during the staling process, must be attributable to either the protein or carbohydrates component of the flour. Protein in wheat flour is primarily gluten which gives the dough its tough plastic and rubbery character. The principal carbohydrate on the other hand is starch. Starch is polymeric carbohydrate. The repeating units, which are glucose, form two separate fractions, a linear component with alpha-1,4 linkages referred to as the amylose component and a branched chain component with alpha-1,6 linkages in addition to the

164

alpha-1,4 linkages, this branched chain component being the amylopectin fraction. Findings made on wheat starch and gluten at moisture contents comparable to those found in bread have been obtained and these show a marked difference between starch and the pure gluten. Starch gels reproduce quantitatively the staling phenomenon in bread, not only in their firming but also in their refreshening behavior on reheating. Gluten gels, on the other hand, show only relatively slight firming with age, and exhibit no refreshening on reheating. Such observations substantiate the general belief that the firming of bread is due mainly to the starch component. The results with starch indicate that the difference in the moisture-binding capacity of stale and fresh bread may be due at least in part, to the retrogradation of the starch in the bread. Observations show, in fact, that a process similar to staling occurs in the retrogradation of starches.

An increase in water binding suggests the possibility that the cross linkages formed actually occur through a water molecule as a part of a bridge or bond between a starch component. The branched nature of amylopectin component of starch would tend to prevent it from forming as great a number of cross linkages per unit weight as would the straight-chained amylose component. It was reasoned therefore, that amylose would form a greater number of cross linkages during retrogradation than would whole starch. The fact that straight-chained amylose acquires during retrogradation a somewhat higher percentage increase in moisture-binding capacity than whole starch substantiates this point of view.

While mechanisms may be involved in bread staling other than the firming of the starch fraction, the crystallinity attributable to starch is the major cause of staling. The observation that a similar process occurs in the retrogradation of starches, and particularly amylose, is considered to indicate that these elements are responsible for the structural evidences of staling in bread.

It was found that molecularly distilled monoglycerides react in some way with the starch component of the flour greatly to retard the rate of bread staling. It may be that the monoglyceride prevents aggregation of the amylopectin or B-fraction of the starch. However, in view of the branched structure of the amylopectin fraction the distilled monoglyceride is believed to react primarily with the amylose or A-fraction of the starch. It has also been found that the molecularly distilled monoglyceride must be provided in a form that will permit its reaction with the amylose component. Commercial monoglycerides have been blended in shortening to be used in bread making and also have been added directly to the dough batch. However, commercial monoglycerides are actually a mixture of mono-, di- and triglycerides.

It has been found that diglycerides and fats have an inhibiting effect on the staling retarding property of the pure monoglyceride. It has been discovered that molecularly distilled monoglycerides must therefore be permitted to contact the starch before it comes in contact with the fats or shortening, that is, the shortening must be used at a noninterfering stage. It has also been found that if molecularly distilled monoglycerides are incorporated per se in the dough, they have no effect on the resulting product. Monoglycerides must, therefore, be provided in a form permitting their reaction in the system. The antistaling ingredient thus, is a composition having the physical properties of cold cream, this semi-liquid consistency permitting it to be readily mixed with flour.

Whereas it has been difficult to make aqueous emulsions of molecularly distilled monoglycerides without either preparing compositions containing only up to 5% monoglyceride in water or using a salt such as calcium propionate in an amount undesirably large for baked goods, it has been found that compositions containing 15 to 60% molecularly distilled monoglycerides can be made. These compositions are made by the application of carefully controlled heating and cooling techniques. The molecularly distilled monoglyceride is gradually heated to about 148°F, being careful to avoid localized overheating. If the temperature exceeds about 155°F, the molecularly distilled monoglyceride will gel. Accordingly, the molecularly distilled monoglyceride-water mixture is carefully heated to 148° to 155°F to bring the monoglyceride into a smooth, creamy dispersion. It has also been found that if the mixture thus heated is merely allowed to cool, the dispersion will probably break. Hence, to form the emulsion the heated dispersion must be gradually cooled with continuous stirring, or more vigorous agitation, until the temperature drops to 90°F. The mixture is stirred or homogenized until it cools to room temperature or lower. As indicated, the final composition is a paste, that is a semi-liquid emulsion. In this form, it immediately enters the water phase to contact the starch fractions. In general, the emulsion composition is added to the water which is added to the flour. However, it can be added to the flour before the water is added. While it is not necessary, the semi-liquid emulsion can, of course, contain a preservative such as propionic, acetic or sorbic acid or an edible salt, for instance, a sodium, potassium or calcium salt in an amount less than 1% based on the emulsion for the purpose of inhibiting mold growth. This preservative can be added during either the heating or the cooling step since it will be inconvenient to incorporate it into the paste.

The molecularly distilled monoglyceride should be given an opportunity to combine with the amylose or perhaps the amylopectin component of the starch prior to the addition of any fatty ingredient. It is generally desirable that the monoglyceride emulsion be added at the sponge stage and that the shortening be added at the dough stage. Preferably, the emulsion is added at the beginning of the sponge and dough stages and in the same proportions of its total as the flour is used to its total in each stage and the shortening held until the end of the dough stage mix. This will expose all of the flour to preferential monoglyceride treatment before the flour is exposed to shortening.

Example 1: To prepare a monoglyceride emulsion, 100 parts of molecularly distilled monoglyceride prepared from hydrogenated lard and 199 parts of water are carefully heated to 150°F in a vessel adapted to provide for uniform application of heat without localized overheating. During the heating stage, the water-monoglyceride mixture is constantly agitated. When the temperature reaches 150°F the heating means are withdrawn and the mixture is permitted to gradually cool. One part of sodium propionate is added during this cooling stage, the mixture being continuously agitated until the temperature reaches 76°F.

Example 2: Following the procedure of Example 1 a semi-fluid emulsion was made using a molecularly distilled monoglyceride obtained from hydrogenated cottonseed oil. Using the sponge dough method and variations in amounts of glyceride and shortening, and in types of glycerides as set forth in the examples which follow, bread was made from the following basic commercial bread formula.

	Percent on Flour Basis	
	Sponge	Dough
Flour	65.0	25.0
Water	56.0	65.0
Yeast	2.5	–
Mineral Yeast Food	5.0	–
Salt	–	2.25
Sugar	–	7.5
Milk	–	3.0
Enr. Tablets	1	–
Lard & Glycerides	As in examples	–

Sponge temperature: 76°F   Floor time: 30 min
Dough temperature: 73°F   Mixing: 5 min at 65 rpm
Fermentation: 4 hrs 30 min   Mixing: 5 min at 65 rpm after clean up

By the use of the above bread formula, three bread doughs A, B, and C, were made in each instance. In type A, the glyceride and all of the shortening were both added to the sponge. In type B, the glyceride and the shortening were both added to the dough. In type C, the glyceride was added to the sponge but all of the shortening was added to the dough. Bread made from each of these three doughs was tested for softness, keeping quality and generally accepted characteristics of good bread after the first and second and third days.

Example 3: By the basic bread formula and 0.5 part of the composition of Example 1, types A, B, and C breads were made. In each case, the sponge ingredients were mixed in the sponge stage and allowed to ferment until a proper degree of maturity was indicated, about 4 hours and 30 minutes at 76°F. The sponge was then put in a mixer and the doughing ingredients added, that is the remaining ingredients called for in the basic formula. This dough was then permitted to ferment for 30 minutes at 78°F. The dough was then divided, rounded, molded, placed in the baking pans and baked at a temperature of 400° to 425°F.

The type C bread in which the monoglyceride was added apart from the shortening was much better than types A and B in both texture and grain. The softness of type C was only slightly better at the end of 12 hours. However, a marked improvement in staleness was evident as determined by odor and texture evaluation at the end of 12 hours. At the end of two days the softness and keeping quality was 10 to 30% better than types A and B. Similar results were obtained using 0.75 part of the composition of Example 1 in A, B and C type breads.

Microcrystalline Stearyl-2-Lactylic Acid Salts

According to *E.F. Bouchard and C.P. Hetzel; U.S. Patent 3,535,120; October 20, 1970; assigned to Chas. Pfizer & Co., Inc.* improvement in texture, antistaling properties and overall shelf-life of bakery products can be obtained by use of microcrystalline forms of salts of stearyl-2-lactylic acid. The specific salts are the sodium, potassium and calcium salts of stearyl-2-lactylic acid, which have been greatly reduced in particle size from the corresponding conventional forms of these salts by either dry-milling or by means of recrystallization from hot fats and oils, etc. The unique form of the stearyl-2-lactylate compound which is most useful here is that which has been subdivided into a microcrystalline structure, preferably by means of dry-milling.

The only other process specifically useful for obtaining the specified salts of stearyl-2-lactylic acid of the desired microcrystalline size and structure involves recrystallization of the corresponding salt form from a molten fat or oil. According to this latter technique, conventional coarsely-sized sodium, potassium or calcium stearyl-2-lactylate is dissolved in an edible oil or in a liquid fat system at an elevated temperature and then recrystallized from the solution by rapid cooling, while agitating to maintain a homogeneous dispersion of the microcrystalline lactylate salt in the lipid system. The resulting mass is then used in the entirety to supply both the required shortening agent as well as lactylate bread softening compound for a yeast-leavened dough, but alternatively the fatty coated, finely-divided lactylate compound may be recovered from the solvent recrystallization system and used as such separately.

In this fat recrystallization process, from about 15 to 25%, of the lactylate compound is used by weight of the lipoid system. The resulting mixture is then heated with good agitation to a temperature in the range of from 115° to 130°C, to solubilize the lactylate compound. Particularly satisfactory lipids for use in this connection include conventional liquid shortening agents such as cottonseed oil, soy bean oil, safflower oil and corn oil, as well as low-melting solid shortenings, e.g., lard and hydrogenated vegetable oil shortenings, etc.

Generally, a fairly low amount, say 20%, of an easily soluble macrocrystalline salt in a vegetable oil which is liquid even at room temperature, when heated with efficient stirring to only about 120°C, is solubilized almost instantaneously. The heating is then stopped and the mass cooled quickly to room temperature, with or without artificial cooling means, as may be indicated, whereupon a thick, relatively clear dispersion soon results.

Especially preferred lactylate salts are calcium stearyl-2-lactylates having a surface area of at least about 1.5 m²/g and consisting of particles with an average size, as measured by means of a Fisher Sub-Sieve Sizer, of from between about 1.0 and about 3.5 microns in spherical diameter. These products are especially well prepared by comminuting the prior art coarsely-sized salt in an impact mill of the Alpine type, or else by the above-described hot fat recrystallization techniques. The finely-divided products, in turn then lend themselves particularly well to incorporation in the yeast-leavened doughs. While up to about 2% by flour weight of this or other products may be incorporated in the doughs with resulting improvement in the shelf-life storage of the ultimate bakery products, best bread softening results with least expense are generally obtained at a maximum lactylate content of about 1.0% by flour weight.

Example 1: Commercial calcium stearyl-2-lactylate, having coarse particle sizes (Fisher average spherical diameter —28μ), was passed through a fluid energy mill. Particle size data on the final product of the single mill pass were then found to be as follows:

Average diameter (Fisher Sub-Sieve Sizer) — 2.0μ
Coulter Counter data for the 50% range — 2.4μ
2% of the particles finer than 0.67μ diameter
5% of the particles finer than 0.85μ diameter
10% of the particles finer than 1.1μ diameter
20% of the particles finer than 1.5μ diameter
100% of the particles finer than 20.0μ diameter

Example 2: Calcium stearyl-2-lactylate as prepared in the above example was tested as a bread softener by the following technique:

4 Hour Sponge	Grams
Flour	1,120
Water	615
Yeast	40
Dough conditioner*	8

*Potassium bromate, 0.3%; ammonium chloride, 9.7%; calcium sulfate, 25%; sodium chloride, 10%; and starch, 55%.

These ingredients were added to the McDuffee bowl of a Hobart mixer in the order listed and mixed for one minute at No. 1 speed, using a 3-spindle fork. The bowl was then scraped down and the sponge again mixed for one minute at the No. 2 speed. The sponge was removed from the bowl at this point, placed in a polyethylene bag and allowed to ferment at room temperature (about 77°F) for four hours. To prepare the ultimate bread dough for the test, sponge so prepared was placed in a ten-quart stainless-steel Hobart mixing bowl and the following ingredients were thereafter added:

Dough Portion	Grams
Flour	480
Water	417
Sugar (granulated)	128
Salt	34
Calcium propionate (preservative)	2
Milk powder	48
Lard	40
Bread softener	8

This mixture was mixed on No. 1 speed for one minute and then on No. 2 speed for seven minutes using a dough hook. The resultant dough was then removed from the bowl, placed in a polyethylene bag and allowed to ferment for 25 minutes at room temperature (about 77°F). It was then scaled into 1 lb portions, with at least four such portions being prepared for each test. These portions were rolled into small balls by hand in order to exclude large air bubbles and gas pockets. Each ball was then run through a sheeter twice, using a ⁵⁄₁₆ inch setting for the first pass and ³⁄₁₆ inch setting for the second pass. The sheeted dough was next molded into a cylinder approximately as long as the pan in which it was to be baked, then placed in a greased pan, and transferred to a proof box and proofed (allowing to rise) for one hour at 120°F and 85% relative humidity. The proofed dough was then baked at 430°F for 25 minutes and the resulting bread subsequently allowed to cool for one hour. All but one of the bread loaves obtained in this manner for each test were then packaged into polyethylene bags and stored either at room temperature or under refrigeration conditions (45°F) for various lengths of time, usually from one day to a week, at the end of which time the staling rate was measured. Each of the unpackaged loaves, on the other hand, was sliced and its initial softness determined as a control.

The determination of staling for each loaf was then made by a standard compression test. Two 1 inch thick slices of bread were cut from each loaf, one slice being taken from the center of the loaf and the other approximately one inch from the end. The compression test was performed with a standard penetrometer using a 1 inch diameter flat stainless-steel disc in place of the usual vaseline cone. A 150 gram weight was used as the load on the end of the compression disc. The load was placed on the slice for a period of ten seconds, after which time the penetration was determined in tenths of millimeters. Three compressions are performed on each slice of bread, two in the bottom corners of the slice and the third at the top center. All these data were recorded and the six values for each loaf were then averaged. In the following table, there are presented the compression data obtained in this manner, not only for the lactylate product but also for the lactylate bread softener as well.

	Bread loaf sp. vol. (cc./g.)	Average compressions (10^{-1} millimeters)			
Softener test sample		Room temp., 2 days	Room temp., 3 days	Room temp., 4 days	Room temp., 5 days
Prior art calcium steryl-2-lactylate	5.68	89	72	64	55
Calcium stearyl-2-lactylate of Example I	5.72	96	82	72	64

Nonionic Surface Active Agents

Bread and similar yeast-raised bakery products can be made significantly softer and the rate of staling considerably decreased by mixing with the ingredients for the bakery product polyoxyalkylene nonionic surface agents. This approach is explained by *S.S. Jackel; U.S. Patent 3,536,497; October 27, 1970; assigned to Wyandotte Chemicals Corporation.* The effective compounds are the water-soluble conjugated polyoxypropylene-polyoxyethylene compounds containing in their structure oxypropylene groups, oxyethylene groups, and a nucleus of propylene glycol. These compounds are prepared by condensing ethylene oxide with a hydrophobic polyoxypropylene glycol base formed by the condensation of propylene oxide with propylene glycol. The average molecular weight of the polyoxypropylene polymer must be at least about 900.

The increasing addition or condensation of ethylene oxide on a given water-insoluble polyoxypropylene glycol base tends to increase its water-solubility and raise the melting point such that the products may be water-soluble, and normally liquid, pasty or solid in physical form. These polyoxyethylene condensates with polyoxypropylene glycol may be designated by the following structure: $HO(C_2H_4O)_a(C_3H_6O)_b(C_2H_4O)_cH$ wherein b is an integer sufficiently high to provide a molecular weight of at least about 900 for the oxypropylene base and wherein a+c is an integer sufficiently high to provide 5 to 90% of the total molecular weight of the compound, e.g., Pluronic series. The following are examples of compounds corresponding to the above formula as disclosed in the brochure.

Name	Mol. wt. polyoxypropylene base	Ethylene oxide content in final product, wt. percent	Calculated mol. wt. of final product
Pluronic P65	1,750	50	3,500
Pluronic F68	1,750	80	8,750
Pluronic P85	2,250	50	4,500
Pluronic P105	3,250	50	6,500
Pluronic F108	3,250	80	16,250

The polyoxyethylene-polyoxypropylene compound may be added to the mix in any suitable manner. For example, it may be thoroughly mixed with the flour prior to the preparation of the dough, mixed with the shortening used prior to its addition to the mix, added with the salt, added directly to the dough, mixed during the mixing operation, or suspended in the water or milk used in preparing the dough. Except for the incorporation of the polyoxyalkylene compound, the normal operations employed in breadmaking or making similar baked goods are followed without any variations from those required when this product is not used.

The resulting baked product will retain its freshness for from one to three days longer than the untreated product. The polyoxyalkylene compounds may be added to the mix for yeast-raised products of any type whether for preparation of white bread, coffee cake, sweet rolls, buns, doughnuts, etc. A small quantity only of the polyoxyalkylene compound is required to make the bread softer and retard its rate of staling. Thus, amounts ranging from 0.20 to 1% on the weight of the flour in the formula have been found to yield particularly good results.

Example 1: This example illustrates the manufacture of a white pan bread by a laboratory version of the well-known Do-Maker system. This system is described in many places in the literature such as the book, *Breadmaking, Its Principles and Practice,* Fourth Edition by Edmund B. Bennion, pp 214 to 216, Oxford University Press, 1967. This process is begun with the production of a yeast brew or broth from the following components:

	Weight percent based on weight of flour	Weight in grams
Water	68.0	3,536.0
Sugar	8.0	416.0
Salt	2.25	117.0
Milk solids	3.00	156.0
Yeast food	0.5	26.0
Calcium acid phosphate	0.1	5.2
Calcium propionate	0.1	5.2
Yeast	2.5	130.0

The above ingredients were fermented for 2½ hours at 86°F. The second step of the process is the Dough or Premix stage wherein the following components were mixed to incorporate them uniformly for 45 seconds at low speed (20 rpm) and 15 seconds at medium speed (100 rpm).

	Weight percent based on weight of flour	Weight in grams	
Flour	100.0	5,200.0	
Broth	(1)		
Lard	2.64	137.28	These ingredients are
Cotton seed flakes	0.14	7.28	melted and blended
Emulsifier	0.22	11.44	before use.
Pluronic P65 polyol	0.25	13.00	
Oxidation solution:			
ppm of KIO_3	12.5		
ppm of $KBrO_3$	50.0		

1 As above.

The above premix dough was then transferred into the loading cylinder of a laboratory developer, wherein the hydraulically operated piston forces the mixture from the loading cylinder into the developer bowl at a constant rate. The developer unit is an oval-shaped bowl in which two counter-rotating impellers are located which subject the dough to a continuous stretching and folding as it passes under pressure to the extrusion point. The development takes place during passage of the mixture through the developer bowl and the developed dough was extruded under pressure as a ribbon which was manually divided. The dough was proofed for about 60 minutes in a proofing cabinet maintained at 110°F with a nearly water-saturated atmosphere, and then baked for 18 minutes at 425°F. The finished loaves had good volume, a very silky texture, and a very white crumb color. Staling was measured by the compressibility of the bread crumb which decreases with staling. A Baker Compressimeter was employed for making compressimeter readings which are the force in grams required to depress a plate 1.25" square, a distance of 2 mm, into a uniform slice 0.5 inches thick. The reading obtained after 73 hours was 13.5.

Example 2: The above process was repeated for the production of bread from four different formulations which were substantially identical with that of Example 1 with the exception that 0.22% of Pluronic polyol was used in each formulation rather than 0.25%. In one case the Pluronic polyol employed was Pluronic F108, in another it was Pluronic P105, in a third it was Pluronic F68, and in the fourth it was Pluronic P85. Bread made from all four formulations had excellent volume, improved texture characterized by a silky feel in the cut section, a very white crumb color, good side-wall strength and good softness retention after 72 hours.

Example 3: This example illustrates the manufacture of 2 batches of white pan bread by the conventional sponge and dough process using the following formula:

	Weight percent based on weight of flour	Weight in grams	
Sponge:			
Flour	62.5	400.0	
Water	46.0	295.0	
Yeast food	0.5	3.2	
Yeast	2.5	16.0	
Total		714.2	
Dough:			
Flour	37.5	240.0	
Water	22.0	140.9	
Sugar (Sucrose)	7.0	45.0	
Lard	3.0	19.0	
N.F.D. milk	2.0	13.0	
Salt	2.0	13.0	
Emulsifier	0.2	1.3	
Mold inhibitor	0.1	0.6	
Pluronic P65 polyol	0.1 and 0.25	0.6 and	1.6
Total dough ingredients		472.5—	473.5
Fermented sponge		707.1—	707.1
		1,179.6—	1,180.6

The sponge was mixed 10 seconds at low speed (20 rpm), 20 seconds at intermediate speed (100 rpm), and 3 minutes at high speed (235 rpm) on a dough-mixing machine. The fermentation time was about 4 hours in a fermentation cabinet at 82°F dry bulb and 79°F wet bulb. The remaining ingredients, that is to say, those listed under Dough above, were then added and the sponge and dough mixed the same as for mixing the sponge except that mixing at the third speed was about 4 minutes.

Next, the dough was placed in a bowl and put in the fermentation cabinet for 24 minutes. Two dough pieces at 430 grams each were scaled off and each sheeted through rollers at ¼ inch and again at 7/32 inch. Each piece was folded in thirds, placed in a baking pan, and returned to the fermentation cabinet for 16 minutes. Each dough piece was then sheeted again at 7/32 inch, folded in half and cross graining sheeted again at 7/32 inch. Each dough piece was curled, the bottom scaled, and then placed in the proof box at 95°F dry bulb and 94°F wet bulb for primary proofing (time: 50 to 60 minutes), and baked for 18 minutes at 425°F. The finished loaves from each batch had excellent volume, improved texture, a cut section had a silky feel, and had good side-wall strength. Each loaf retained softness after 72 hours.

Gum Additives

J.J.R. Andt; U.S. Patent 3,271,164; September 6, 1966; assigned to Stein, Hall & Co., Inc. reports that bread and other baked goods can be improved by the addition of an additive comprising karaya gum and a material selected from the group consisting of algins and carrageenans. The presence of this additive in the bread and other baked goods retards the staling to a substantial degree. At the same time, it provides a number of other advantages both in the final quality of the baked goods and in the processing techniques employed. One of the most important processing advantages is in the improved tolerance of the dough to overmixing. The level of karaya gum present in the baked goods prepared should range from about 0.1% to about 0.9%, based on the weight of the flour in the composition. Amounts in excess of about 0.9% are disadvantageous, in that the resulting baked products loses its strength and tends to crumble. Amounts below about 0.1% do not provide any advantages.

Similar considerations apply to the level of the material selected from the group consisting of carrageenans and algins. No more than about 0.1% of these materials, based on the weight of the flour present, should be used. In a preferred composition, at least about 0.02% should be used. A preferred composition contains 90% karaya gum and 10% carrageenan by weight.

The gum additive may be used in all types of baked goods which are susceptible to staling. Among these are included bread, hamburger and frankfurter rolls, dinner rolls, brown-n-serve rolls, sweet rolls, coffee cakes, donuts and all other types of yeast-raised products. Also included are biscuits, donuts and other chemically leavened baked goods. These bakery products, when prepared with the additive retain their moisture for a longer time and the moisture is more uniformly distributed throughout the article. The products stay softer and fresher tasting for at least 10 to 15% longer. In addition, a live, vital quality is imparted to the crust without darkening the color. The grain has more uniformity, while the texture of the baked goods is improved and the crumb has a truer color and clarity.

Another advantage of considerable importance results from the finding that in baked goods requiring shortening or emulsifier for a desired degree of softness, the amount of shortening or emulsifier can be reduced when the additive is used, with no sacrifice of the softness characteristic. Further, in some instances it has been found that the degree of softness obtained with the combined use of the additive and an emulsifier is greater than would be expected from the use of either one alone. For example, the combined use in bread of calcium stearyl 2-lactylate with an additive consisting of 90% karaya gum and 10% carrageenan, provides a bread having a softness far in excess of that obtained with either ingredient alone.

The additive may be used in all of the three standard methods employed in preparing bread doughs for baking. In the sponge-dough method, the additive is preferably added in the sponge stage although nearly equal results are obtained where the additive is included in the dough. In the so-called continuous process, which utilizes a brew stage and a dough stage, the additive may be included at either point, although it is preferred that it be added to the brew. In the straight dough mixing procedure, the additive is added along with the other dry ingredients.

In the preparation of a dough, an amount of water is generally used which is sufficient to fully hydrate the flour, but is not so large as to cause a wet, sticky dough. This amount is determined empirically from flour samples, and is called the absorption value of the flour. When using the additive, it is necessary that an additional amount of water be added to the dough or batter to insure hydration of the additive. From about five to fifteen or more parts of additional water for one part of additive may be used depending upon the type of dough, the water absorption characteristics of the flour, and the amount of additive employed.

While under normal circumstances the presence of an additional amount of water in the dough would seriously impair its machinability, it has been found that a dough containing the additive and additional water performs satisfactorily during normal processing. In fact, the doughs prepared are easier to handle, in that they are resistant to overmixing. Although the use of the additive requires additional amounts of water in the dough, the final moisture content of the baked product is approximately the same as that of the same formulation without the additive. An instrument called a Farinograph is extensively used in the baking industry to determine physical properties of flours and doughs. The Farinograph is essentially a miniature dough recording mixer by which one may determine a number of important characteristics, which, in turn, help predict the baking performance. The following are the normal characteristics associated with the Farinograph curves and which are used to evaluate the dough runs.

Absorption is defined as the amount of water in a standardized flour-water system required to center a Farinograph curve at the 500 BU (Brabender Units) line. Arrival time is the time for the top of Farinograph curve to reach the 500 BU line, after the mixer is started and water is introduced. This value determines the rate at which the water is absorbed by the flour. Dough development time (or peak time) is the time required for the curve to reach its maximum consistency value. Stability is the difference in time between the arrival of the top of the curve at the 500 BU line, and the time the top of the curve departs from the 500 BU line (departure time). This value indicates the amount of tolerance to mixing the flour will have. Tolerance index (MTI) is defined as the value difference in BU between the BU reached by the top of curve at peak time and the BU reached by the top of curve measured at five minutes after peak time is reached. A related value (TI) is the difference in BU between the 500 line and the center of curve, measured at five minutes after the top of the curve departs from 500 BU line. A high value indicates a weak flour. A low value denotes a strong flour.

The additive used contained 70% karaya gum, 20% guar gum and 10% Seakem 402, a carrageenan. The additive was used at both a 1% and a 0.5% level, based on the weight of the flour. The data obtained from the curves are summarized below. Curves 1 and 4 were controls which were run to determine absorption of the flour. Curves 2 and 5 were run at the same absorption in order to obtain the approximate absorption of the combination. Curves 3 and 6 were run at the calculated absorption. The data obtained from these curves were as follows:

Set 1

Curve No.	1	2	3
Flour	*	*	*
Additive, percent	None	1	1
As is absorption (run), percent	64.7	64.7	67.9
As is absorption (corrected), percent	64.7	67.9	68.6
Arrival time, min	2.75	1.5	2
Stability, min	15.25	14	19
Peak time, min	7.75	3	10.5
MTI, BU	15	10	10
TI, BU	50	65	45

*Spring wheat.

Set 2

Curve No.	4	5	6
Flour	*	*	*
Additive, percent	None	0.5	0.5
As is absorption (run), percent	62.9	62.9	64.4
As is absorption (corrected), percent	62.9	64.4	64.6
Arrival time, min	4	3.5	4.5
Stability, min	15	17	23.75
Peak time, min	7	9.5	9
MTI, BU	15	22	10
TI, BU	50	50	35

*Hi protein spring.

The above data show definite absorption increases occasioned by the use of the additive. Also, arrival time with the additive is generally faster than that of the control, indicating a more rapid hydration and better colloidal dispersion. This partially explains the improved grain uniformity and gas retention obtained using the additive. The stability increase obtained by using the additive amounts to 25 to 58% over the control, indicating an increased tolerance to overmixing. The peak time is also increased over the control by using the additive. This peak time is normally related to optimum mixing and clean-up time. The MTI and TI values show that the flour has been improved in stability and colloidal properties. The TI values are not taken until the doughs show a definite sign of break down, and are therefore a more significant indication of the ability of the dough to withstand overmixing. The following explanation of terms used in the examples will be helpful.

(1) The term clean-up time is the dough stage in the sponge-dough method, and averaged 3.0 minutes using a dough-hook in a Hobart mixer.

(2) Machining refers to the machining of dough when sheeted and formed without flouring of the dough.

(3) Volume was measured by the rapeseed displacement method and is an average of at least two loaves for each test.

(4) Softness, unless otherwise indicated, was determined on the Penetrometer using a Penetrometer cone weighing 102.5 grams. The cone was allowed to depress the baked goods for fifteen seconds and the value recorded. Bread was prepared for the softness determination by slicing it into two 2" thick slices from the center of the loaf. Three values were taken on each surface of each slice, one value in the center of the slice and one value at half the center to the top of the loaf and one value at half center to the bottom of the loaf. In all, twelve values were obtained for each loaf and all values were then averaged. Softness tests were run after forty-eight hours and additionally on either the 4th, 5th or 6th day. These two softness values were then averaged and compared to average values obtained on test samples of a control run in the same series.

(5) Symmetry, oven-spring, grain and other loaf characteristics were judged subjectively by the scoring method set forth below.

(6) The scoring method used throughout was as follows: 4, excellent (superior nature); 3.5, very good (above average results); 3.0, good (average results); 2.0, fair (below average); 1.0, poor (fair below average).

Example: A series of batches of bread were made by the so-called continuous mix equipment. Preliminary tests at various levels of water absorption increase over a control indicated that a 5% absorption increase was optimum for the formula containing the additive. On the basis of these preliminary tests, a 5% increase in water level was used throughout this series. The formula used was as follows. The brew contained 25% flour (50% spring, 50% winter), 62% water, 2.75% compressed yeast, 0.5% yeast food (Arkady), 0.1% emulsifier (T.E.M., tartaric acid ester of glycerol monostearate), 1% sugar and 2% salt. The brews were made up and set at 85°F. They were thereafter fermented for 2¼ hours.

The dough was made of 75% flour, 5% water (when no milk solids used), 8% water (when 4% milk solids used), 3% melted lard, 6% sugar, 0.4% nonfat milk solids, 48 ppm bromate (KBrO$_3$) and 12 ppm iodate (KIO$_3$). The dough was mixed with the brew to optimum consistency, turned out at 102° to 103°F, scaled at 539 grams, proofed to ¾" above pan (approximately 57 minutes) and baked for 17 minutes at 450°F.

The basic formula above was used with either no milk solids or 4% milk solids and an additive as described above. During processing, all of the doughs were dry, smooth and pliable, and all handled and machined well. All of the bread baked was very good in appearance, break and shred, flavor and taste, and excellent in crust color. The addition of the additive of this process improved the volume slightly, gave the grain more uniformity and the texture was slightly silkier. Addition of 0.5% of the additive of this process, together with an increase in water absorption of 5% improves bread softness so long as the level of the emulsifier is kept constant. A decrease in shortening from 10% to 4% does not alter softness.

Amylopectin for Nine Month Shelf Life

In the process of *R. Dehne; U.S. Patent 3,193,389; July 6, 1965;* amylopectin is added to bread ingredients prior to baking with an improvement in retention of freshness. The preparation of the bread is effected in the conventional manner, for example by mixing 75 kg of flour, 50 kg of water and 1 kg of yeast, together with the desired quantity of spices, such as 1 kg of salt, these ingredients being thoroughly kneaded into a dough, and the dough additionally treated with about 5 to 30% by weight of amylopectin based upon the flour quantity used.

The amylopectin must be thoroughly kneaded into the rest of the dough before the baking. It is best to add the required amylopectin quantity to the flour, and only thereafter combine the mixture of these two ingredients with the required amount of water and other dough-forming constituents. The improvement is suitable for the production of all kinds of bread, including rye bread, whole wheat bread, breads composed of mixtures of rye and wheat flour, such as pumpernickel bread, as well as white bread made from white flour.

In general, a bread made from rye flour may be considered as fresh for about four days, while a pure wheat bread, such as white bread, already loses its fresh taste on the second day, and may be considered after an additional two days only of an inferior quality approaching staleness. Significantly, if about 30% amylopectin, based upon the flour content, is added to these bread quantities, prior to baking, the bread will remain fresh for about 3 to 4 times the ordinary period of freshness of the same bread not containing the additive. Specifically, amylopectin-containing rye bread will remain fresh for about 12 to 15 days, while amylopectin-containing wheat bread will remain fresh for about 7 to 9 days. Even where the so-called coarse meal or coarse floured bread is concerned, which is, as known, capable of remaining fresh even without special precautions for about 10 days, by the addition of amylopectin, an extension of the time at which the bread remains fresh by at least three-fold and even more will be attained.

A simple, economical, and efficient process for packaging the freshly baked bread prepared from flour containing amylopectin may be provided so that the packaged bread may be stored advantageously for periods of over 9 months and longer. Thus, amylopectin is added to the flour dough ingredients to be baked, preferably to the bread flour prior to the addition of the water content, and thereafter the ingredients are kneaded, placed in a bread mold, and baked, preferably at about 100°C for a period of about 18 to 30 hours.

These conditions may be varied, as for example, the baking treatment may be performed in ovens of other types, such as gas heated ovens, or using other types of flour or other ingredients. This will be the case if it is desired to prepare rye-wholemeal bread instead of pumpernickel. The freshly baked bread, upon removal from the oven is immediately thereafter aged by allowing the mass to stand for a period of at least about 6 hours immediately after baking to cool and to structurally stabilize or solidify the comparatively hot, soft mass.

The bread is next moistened just on the surface of the mass to soften slightly the outermost crispy surface portion without softening the interior portion of the mass. Then, the mass is hermetically sealed in a flexible, heat-sealable polyolefin foil under a vacuum of at least about 0.8 atmosphere absolute, and the hermetically sealed mass in the foil is sterilized at a temperature between about 70° and 120°C for a period of about 0.5 and 3 hours, and preferably from 0.5 to 2.5 hours, sufficient to effect sterilization of the bread. If desired, in order to ensure storage for periods of 9 to 12 months and longer, without refrigeration, so as to retain the favorable qualities of freshness and taste, the bread, immediately after the sterilization step may be cooled rapidly to at least 10°C in the sterilized hermetically sealed package.

The baked bread mass may be in the form of sliced bread, such as rye bread, or pumpernickel bread, and the foil used may be of such dimensions that the expanded volume of the foil containing the bread is between 25 to 50% greater than the corresponding volume of the bread itself. The aging may be effected for a period between about 6 to 20 hours, depending upon the type of baked bread being treated, and the moistening may be effected for a period between about 1 to 3 hours to take up between about 0.5 and 5% moisture (reckoned on the moisture content of the whole bread) in the bread surface. The hermetic sealing may be effected under a vacuum between about 0.3 and 0.8 atmosphere absolute, which generally corresponds with the preferred range of 400 to 200 mm mercury pressure, in a transparent, flexible foil essentially free from softeners and plasticizers and having a thickness between about 0.03 and 0.1 mm and, if desired, an area of about 1 to 1.2 times the outer surface area of the baked bread mass. The sterilizing may be effected at about 90° to 100°C to achieve a sterilizing temperature of 70° to 87°C, and preferably 77° to 87°C, in the interior of the bread mass, and the subsequent cooling used may be between 10° and 0°C.

The moistening step of the surface of the mass of baked bread may be carried out by wrapping the mass in moist wrappings, such as a moist cloth, the moistening being carried out for a period of from 1 to 3 hours, but preferably about 1 hour. On the other hand, the moistening of the surface of the mass may be carried out by subjecting the mass to contact with steam for about 1 to 2 hours. The steam treatment may be effected conveniently in a closed zone, additionally containing water vapor.

The sterilization may be practiced at a temperature of about 70° to 90°C, using hot air saturated with steam, or, alternatively, a water bath in which the packaged bread is immersed, for this purpose. During the sterilization, the mass may be situated on a support having a low coefficient of heat conduction or heat radiation, each foil package being separated from adjacent foil packages so that uniform heat distribution will be possible during the sterilization. Upon the termination of the sterilization of the foil packaged baked bread mass, the bread may be recovered and stored at room temperature for periods of over 6 months without any significant loss of the flavor, taste, freshness, etc. of the bread.

It is especially advantageous to rapidly cool the foil-package bread immediately after sterilization and while still hot to a temperature below room temperature, i.e. between about 10° to 0°C within about ½ hour, in order to enhance the durability qualities of the mass being stored. This step serves to increase the effect of the sterilization as it has been found that if the rapid cooling step is omitted, a regeneration of micoorganisms may occur during storage of the bread at room temperature to a slight extent.

Example: For the Baking of Pumpernickel Bread — Upon mixing 500 g coarse rye flour of the type 1800, i.e., where 1.8 kg ash (combustion residue) are contained in each 100 kg of coarse flour, 5 to 30% amylopectin based upon the flour content, 300 g water, 1% yeast, and 1.5% salt, 500 g of the dough obtained after kneading are pretreated for 8 hours in hot water, the temperature of the water being more than about 60°C. Then the ingredients are placed in a container and inserted into a steam chamber oven for baking. The baking takes place over a period of 18 hours at 100°C and then the bread is removed from the oven. The baked bread mass has increased in size about 10% as compared with the size of the prebaked mass, and in the 110% baked form has the approximate dimensions 120 x 100 x 70 mm.

For the Aging and Moistening Pretreatment of Fresh-Baked Pumpernickel — Pumpernickel bread having a size of 120 x 100 x 70 mm as taken from the baking oven is immediately cooled from the baking temperature of about 100°C to room temperature for 6 hours and allowed to age, and at the expiration of this time the bread is wrapped with a moist cloth which remains on the bread for 1 hour, so that the bread takes up 0.5 to 1% moisture (reckoned on the moisture content of the whole bread) on its surface. After the aging, the bread is stable and solid to touch with a relatively firm crust containing hard projections, whereas the bread as taken from the oven is soft and pliable. Upon treatment with the moist cloth, the bread is still firm and resilient yet the crust is no longer crispy and the hard projections have been rendered soft and innocuous. The so-treated bread is then sliced and subjected to the following treatment for sealing and sterilizing the same in a polyethylene bag.

For the Packing and Sealing of Fresh-Cut Pumpernickel — The individual slices of pumpernickel which are of equal size are piled so as to form a stack of a size of about 120 x 100 x 70 mm. This corresponds approximately to a weight of 500 grams, i.e. to a weight which is customary for sale. This pile is then placed in a bag consisting of a transparent polyethylene sheet of a thickness of 0.04 mm which is free from softeners and plasticizers and resistant to heat deformation at temperatures up to about 110°C. The polyethylene bag may be made for instance from a seamless tube of a circumference of 350 mm and a length of 230 mm by heat sealing at about 120°C one end with a seam 3 to 5 mm thick spaced 10 mm from the edge.

The other open end of the bag is then clamped on the nozzle of a vacuum pump and evacuated down to 0.5 atmosphere absolute. After this degree of vacuum is reached, this end of the bag is pressed together in a vacuum-tight manner at the portion between the nozzle and the stack of bread, and then hermetically sealed closed at 120°C in a seam of a width of 3 to 5 mm parallel to its opening. Thereupon the bag can be removed from the nozzle of the vacuum pump and the protruding bag end possibly cut off to a length of about 15 mm from the edge of the seam. The pile of bread is then recovered enclosed in a vacuum-tight and moisture-tight manner in a wrapping of an inside length of about 120 mm, an inside width of 100 mm and an inside height of about 70 mm. Due to the vacuum, the wrapping rests closely, under the external pressure, around the bread bag, in which connection all details of the bread can be observed readily through the transparent wrapping.

For the Sterilizing of the Packaged Bread — In this sealed condition, the bag is placed upon an enameled plate with the cut surfaces of the bread parallel to the plate and immersed in a water bath of a temperature of about 90°C for 1½ to 3 hours, depending on the degree of sterilization desired. During this treatment the bag expands somewhat.

For Optional Rapid Cooling of the Sterilized Bread — Immediately after the sterilization, the plate support bearing the packed pumpernickel is set in a refrigerator maintained at 10°C and rapidly cooled, from the sterilization temperature to 10°C within about ½ hour after being placed in such refrigerator. Each bread bag is then removed from the refrigerator and allowed to return to room temperature. The bread may be preserved this way at room temperature for 9 to 12 months, and even longer without impairment of the desired properties of the bread, e.g. freshness, texture, appearance, flavor, etc. In this regard, during the immersing in the water bath, it can already be noted whether the sealed seams are satisfactory. If not air will escape from the wrapping.. After sterilization, the residual air still enclosed in the bag cools down during the rapid cooling step so that the wrapping rests closely against the bread bag, i.e. even after the bread returns to room temperature.

Selected Diglycerides

N.H. Kuhrt and L.J. Swicklik; U.S. Patent 3,068,103; December 11, 1962; assigned to Eastman Kodak Company found they could delay the firming of bread by incorporating into a bread baking mix or dough batch prior to baking a minor proportionate amount of a diglyceride having one acyl radical derived from lactic acid and one acyl radical derived from a saturated fatty acid having 16 to 18 carbon atoms. The higher acyl radical or moiety of the additives can be derived from palmitic acid and stearic acid, and preferably from the fatty acid moieties of hydrogenated fats and oils. The fatty acid moieties of many well-known fatty materials consist essentially of mixtures of palmitic acid and stearic acid, or fatty acids such as palmitoleic acid, oleic acid, linoleic and linolenic acids which can be converted by hydrogenation into palmitic acid and stearic acid. For example, the fatty acid moieties in hydrogenated lard are comprised of about 94% palmitic acid and stearic acid and can be used as the source of the higher acyl moiety in the diglyceride additives. Likewise, the fatty acid moieties of other hydrogenated fats and oils, such as cottonseed oil, peanut oil, soybean oil, palm oil, olive oil, beef tallow and others can be used as the higher acyl portion of the diglyceride additives.

A particularly effective method for preparing the diglyceride additives is to first prepare a monoglyceride by reacting glycerol and a hydrogenated fat or oil, palmitic acid, stearic acid, or the like, in the presence of an esterification catalyst, and thereafter separating from the reaction mixture by distillation, a high purity monoglyceride. The prepared monoglyceride is thereafter reacted with about an equal molar proportion of lactic acid. Usually reaction temperatures from about 110° to 160°C and reaction times from about 1 to 6 hours are employed in reacting the monoglyceride with lactic acid. Maximum amounts of the diglyceride additives can be obtained by correlating such time and temperature conditions. Relatively bitter or sour tasting free lactic acid and such lactic acid esters as monolactin, dilactin, and trilactin usually present in the resulting reaction mixture can be readily removed by washing with water or by subjecting the reaction mixture to vacuum stripping as in a molecular still. Solvent crystallization can be employed to further concentrate the diglyceride additives.

The diglyceride additives can also be prepared by reacting under esterification conditions of usually about 200° to 300°C glycerine, lactic acid and a suitable fatty triglyceride or higher fatty acid. The resulting reaction mixture can be worked up by washing with water or by vacuum stripping, or it can be subjected to thin film high vacuum distillation to separate out a concentrate of the diglyceride additive of the process. Only minor proportionate amounts of the bread additive need be added to the baking mix or dough batch for effective results, with concentrations of 0.1 to 3% being suitable and 0.1 to 1.5% being more generally utilized, the concentrations of additive being based on the weight of the baking mix.

Example 1: A 300 gram portion of molecularly distilled monoglyceride composition having the fatty acid radicals of hydrogenated lard and having a monoglyceride content of about 95% was reacted with a 140 gram portion of 85% lactic acid with stirring at 140°C under nitrogen by five hours. The resulting reaction mixture was stripped of the more volatile materials such as lactic acid, monolactin, dilactin, and trilactin by subjecting the reaction mixture to thin film molecular distillation at a temperature of about 125°C and at a pressure of about 15 microns of mercury. One part by weight of the remaining distilland was then dissolved in a solvent composed of 8 parts by weight of petroleum ether (BP 30° to 60°C) and 1 part by weight of 95% ethanol, chilled to 3°C and left overnight. The resulting mixture was then filtered and the solvent evaporated from the filtrate to yield a concentrate of the diglyceride additive.

The prepared concentrate was further concentrated by dissolving in benzene at a concentration of 1 part by weight of the concentrate for each 9 parts by weight of benzene. The resulting mixture was then filtered and the solvent evaporated from the filtrate to yield a diglyceride composition composed of about 8% by weight of monoglyceride having the acyl radicals of hydrogenated lard and about 92% by weight of diglycerides having one acyl radical derived from the lactic acid and one acyl radical derived from the hydrogenated lard. The diglycerides consisted essentially of a mixture of monolacto monopalmitin and monolacto monostearin.

Example 2: The diglyceride composition prepared as described in Example 1 was tested as a firmness retarder in a typical 2% lard bread formula, the bread formula having the following ingredients:

· · · · · · · · · Sponge · · · · · · · · ·		· · · · · · · · · · · · · ·Dough · · · · · · · · · · · · · ·	
	Parts by Weight		Parts by Weight
Flour	65.0	Flour	35
Water	40.5	Water	27
Yeast	1.75	Sugar	6
Yeast food	0.5	Salt	2.25
		Powdered milk	4
		Lard	2
		Diglyceride composition	0.25

The flour employed in both the sponge and the dough was short-patent, spring-wheat, white, bread flour. The sponge was mixed for 4.5 minutes and allowed to ferment 4.5

hours at 87°F and at a relative humidity of 80%. To the resulting fermented sponge was added the dough ingredients. The sponge and the dough ingredients were mixed for 9.5 minutes, given a 30 minute floor time, made up into loaves, and allowed to rise for 65 minutes at 98°F and at a relative humidity of 95%. The raised dough batch was then baked at 450°F for 20 minutes in loaves weighing about 450 grams. The resulting baked loaves of bread were allowed to cool at room temperature for one hour and thereafter wrapped and sealed in wax paper and stored at 72°F.

For comparative purposes, the same amount of monoglyceride containing the acyl radicals of hydrogenated lard and monoaceto monoglyceride containing the acyl radicals of hydrogenated lard were substituted for the diglyceride composition additive in the above described bread. At intervals the various sample loaves of bread containing the different additives were tested for firmness or softness with a Baker compressimeter. The bread baked with the diglyceride composition retained the softness characteristic of fresh bread for more than 48 hours longer than the bread baked with the aceto monoglyceride or with no additive. The bread baked with the diglyceride composition possessed an appearance of greater whiteness and finer grain than that possessed by the bread baked with the monoglyceride.

Lipase Preparations

In an effort to combat the bread staling problem, most baked goods contain one or more additives which, to a greater or lesser extent, will retard staling. One such additive consists largely of monoglycerides. Although these materials provide improved results, they may give undesirable changes in the cell structure of the bread. This is often associated with the use of monoglycerides made from fats having a high iodine number. The effect is more noticeable as the products containing monoglycerides are permitted to age.

R.H. Johnson and E.A. Welch; U.S. Patent 3,368,903; February 13, 1968; assigned to R.T. Vanderbilt Company, Inc. found that the use of certain lipase preparations in a bread dough mixture will significantly retard the tendency of bread to become stale. At the same time, bread made in this manner exhibits none of the disadvantages usually encountered when monoglycerides are used as bread staling retarding additives. The lipase preparation is added to the bread dough along with flour, water, shortening, yeast and the other conventional ingredients. The dough is mixed until smooth and is thereafter handled in the conventional manner.

Lipases are enzymes which accelerate decomposition of triglyceride fats. The decomposition products are mainly diglycerides, monoglycerides, fatty acids and glycerine, the relative amounts and proportions of each depending upon the characteristics of the lipase preparation and the length of time over which it is permitted to act. The lipases are individualistic in their behavior according to their source, conditions of preparation and use. Some produce an abundance of one end product at the expense of another. Some lipases are more effective in acid medium while others are more effective with fatty acids of a particular type and are relatively or completely ineffective with others. Some may be crystallized as pure materials; others, as extensive trials have shown, are not pure materials but are mixtures of active components which are ineffective when separated. Most commercial lipase preparations contain, in addition to lipase, other types of enzymes in appreciable quantities.

Lipase preparations which are useful must contain lipases of the type which are capable of splitting triglyceride fats containing fatty acids having 12 or more carbon atoms. Also the lipase preparation must be one which is active at the pH of bread, which generally ranges from 4.0 to 6.5. The lipase preparations may contain minimum amounts of extraneous active materials such as enzymes other than the lipases. The lipase preparations are of benefit both in doughs containing added fat as shortening and in those doughs which do not use an added shortening increment. Flour ordinarily contains about 1.5% of lipid materials, most of which can be extracted by solvents such as acetone or ether. It has been noted that once flour has been made into dough by the addition of water and mixing, the lipids apparently become bound by the protein so that only about 0.15% of the lipids can be extracted by ether or acetone. The use of the lipase preparations has a softening effect when used in doughs having no added shortening increment. While the exact mechanism is not known, it is thought that the lipase splits the natural flour lipids to form monoglycerides, and that these monoglycerides are, to some extent, preferentially bound by the flour protein, displacing the lipid component. The increased softness therefore is thought to be the added effect of the monoglyceride on the starch and the possible lipo-protein modifying action resulting from an increase in solvent-extractable lipids.

Example: White pan bread was prepared according to the following recipes:

- - - - - - - - - - Sponge - - - - - - - - - -		- - - - - - - - - - Dough - - - - - - - - - -	
	Pounds		Pounds
Hard wheat flour	70	Hard wheat flour	30
Water	51	Water	15
Yeast (dry)	1¼	Salt	2
Yeast food	¼	Sugar	5
		Dry milk	4
		Shortening	4
		Lipase preparation	⅛

Sponge — The sponge was prepared by dissolving the yeast in a portion of the water at 110°F and the solution was added to the mixer along with flour, yeast food and the balance of water. The materials were mixed just enough to make a homogeneous mass, dumped into a trough and fermented for 3 hours at 78°F.

Dough — The fermented sponge was returned to the mixer, all of the doguh ingredients were added and the batch was mixed until smooth. The dough was allowed to stand about 15 minutes, divided, rounded and allowed to stand again. It was then molded, panned, proofed at 95°F to the top of the pans, and baked at 420°F until uniformly brown, i.e., about 30 minutes, with steam in the oven. The loaves were cooled slowly to room temperature and wrapped in moisture proof paper. Representative samples were taken immediately after baking and analyzed for extractable monoglyceride content. Firmness of the resulting bread was measured objectively after three days using a Baker compressimeter. The results obtained are shown below.

	Lipase	Control (lipase omitted)
Firmness after 3 days as percent of control	73	100
Percent reduction with respect to control	27	0
Extractable monoglyceride content of bread as percent by weight of flour	0.74	0.11

In the foregoing the lipase preparation used was obtained by cultivation of the microorganism *Candida cylindracea* ATCC No. 14830 in accordance with U.S. Patent 3,189,529 to Yamada et al.

Special Dextrans

In the preparation of bakery products, it is desirable to include as much water as possible in the dough. A high percentage of water not only increases the yield of the product, but also increases the moisture content of the product. The product is therefore softer and its keeping quality improved. The amount of water which is absorbed in the dough is largely dependent upon the relative water absorption properties of the particular ingredients used to prepare the dough, most of the water being retained primarily by the farinaceous components of the dough. Although such components as defatted milk solids tend to increase total water absorption of the dough, several other components, such as sugar and fats, tend to decrease this water absorption. Sometimes certain chemicals such as calcium peroxide are added to doughs in small amounts so as to increase the amount of water absorbed by that dough. In the preparation of bakery products which are prepared from doughs containing yeast and gluten, not only is it desirable to achieve a high moisture content in the dough, it is also desierable to effect a softening of the gluten content of that dough. The dough is then more extensible and the yeast action proceeds more effectively. The fermentation time is reduced and the rate of production of bakery products is increased.

R.T. Bohn; U.S. Patent 2,983,613; May 9, 1961; assigned to R.K. Laros Company, Inc. found that by adding small quantities of dextran to doughs containing yeast and gluten, not only are the water absorption properties of the doughs greatly increased, but the gluten in the dough is softened. Moreover, vital gluten, which has been formed into a tough, elastic mass by mixing it with water, is softened by incorporating a small quantity of dextran in the mass. Because of the softening effect exerted by dextran upon the gluten, the handling properties of doughs containing both dextran and gluten are greatly improved, and the dough is rendered more extensible. Consequently, the dough is more readily formed into the desired shape for baking and the rate of production of baked products made from these doughs may be increased. Breads made from these doughs have been found to be softer, whiter, and more moist than ordinary doughs not containing the dextran. Moreover, the volume of bread baked from doughs which contain dextran is greater and the shelf life is appreciably longer than that of bread baked from doughs which do not contain dextran.

It is preferred to use a dextran prepared from the fermentation of cane sugar. This dextran is prepared by growing the microorganism *Leuconostoc mesenteroides* B512 on a suitable medium and in sufficient quantity until it has elicited the enzyme, dextransucrase, following which time a filtrate containing the enzyme is mixed with a 10% sugar solution which undergoes polymerization (under proper conditions) to dextran. The dextran is then precipitated with alcohol. After settling for a number of hours, the supernatant is removed and the precipitated dextran may be reprecipitated with alcohol to completely remove any extraneous materials. After the supernatant has been removed, the precipitated dextran is again dissolved, and the solution dehydrated, resulting in a dextran having a molecular weight of from about 20,000,000 to 40,000,000.

Example: A white pan bread was prepared by baking a dough containing 100 parts by weight of white flour, 60 parts by weight of water, 3 parts by weight of yeast, 5 parts by weight of sugar, and 1 part by weight of salt. A second, white pan bread was then prepared in the same manner, except that in addition to the above ingredients, 3 parts by weight of dextran were added to the dough and an additional 12 parts by weight of water were added to the dough to bring the consistency of the dough back to what it was prior to the addition of the dextran. The addition of the 3 parts by weight of dextran increased the water absorption of the dough by a total of 12%, which represented a 4% by weight increase in the water absorption for each percent of dextran (based on the weight of flour) added to the dough. Both of these doughs were baked under identical conditions. The bread which contained the dextran was about 20% greater in volume than the control, which did not contain dextran. Moreover, the bread which contained the dextran was more moist, had a whiter crumb color, and had a better shelf life than the control.

The firmness of the bread crumb of both breads was measured by a Baker compressimeter, an instrument which is used in the cereal industry for determining the firmness of bread crumb on storage. Using this instrument, it was found that a total stress of 3 grams was required to compress a given thickness of the dextran-containing bread by 2.5 mm, while 7.2 grams of stress was required for a corresponding compression of the control, indicating that the bread crumb of the dextran-containing bread was appreciably softer than that of the control. After three days' storage, these stresses were found to be 5.2 grams for the bread containing dextran and 7.65 grams for the bread not containing dextran, indicating that the former had a longer shelf life than the control.

The incorporation in bread doughs of quantities larger than 1% of dextran, based on the weight of the flour, does not appreciably further increase the volume of the resultant bread beyond 20%. Therefore, when breads were prepared by the sponge and dough method, using a typical white bread formula, and employing from about 1 to 3% dextran, based on the weight of the flour, these breads had approximately a 20% increase in volume over similar breads which contained no dextran.

COMBINED SOFTENERS AND MOLD INHIBITORS

Glyceryl and Isosorbide Propionates

It is current practice in the preparation of yeast-leavened baked products, such as bread, rolls, and sweet doughs, to incorporate a softener or antistalant in the dough and also a fungistat to inhibit microbiological growth in the yeast-leavened products. The softeners most generally used are mixtures of mono- and diglycerides which are plastic materials and are difficult to handle, measure and thoroughly disperse in the dough. Previous attempts to use liquid softeners have been unsuccessful because of adverse effects on dough condition and the resulting baked product.

In addition to the softeners, fungistatic agents are also incorporated into yeast-leavened baked goods. The fungistatic agents most generally used are calcium or sodium propionates. These propionate salts are effective in inhibiting microbiological growth, but also adversely affect the functioning of the yeast as demonstrated by increased proof time and in some instances poorer grain and volume. In view of the long periods of time which may expire between the preparation and consumption of baked goods, such as cakes and cupcakes, it is considered desirable in some instances to incorporate fungistats in this type of baked product. It is important that the fungistatic agent not have a substantial adverse effect on the quality of the final baked product.

A liquid softening agent and fungistat can be added to dough or batter prior to baking according to *W.H. Knightly; U.S. Patent 3,394,009; July 23, 1968 and U.S. Patent 3,369,907; February 20, 1968; both assigned to Atlas Chemical Industries, Inc.* Lower monocarboxylic acid esters of glycerin and isosorbide are generally liquids which are easy to measure and handle. The esters also function as fungistats and do not adversely affect proof time at the recommended levels of use as do the propionate salts which are currently used as fungistats. In addition, the lower monocarboxylic acid esters of glycerin and/or isosorbide do not adversely affect the grain and volume characteristics of the resulting baked products.

The softeners and fungistats are complete and partial esters of glycerin and isosorbide and their mixtures. Though the lower monocarboxylic acid esters may be used in relatively pure form, generally in preparing the esters mixtures are formed, and it is convenient to use the mixture and thereby avoid a purification procedure. The term lower monocarboxylic acid esters of glycerin includes mono-, di- and triesters and mixtures and the term lower monocarboxylic acid esters of isosorbide includes mono- and diesters and mixtures. The term lower monocarboxylic acids includes those monocarboxylic acids having from 2 to 7 carbon atoms. Propionic acid has been found to be particularly effective. In addition to using the acids, their equivalent halides, anhydrides and esters may also be used to carry out the esterification. Conventional esterification procedures may be used to prepare the glycerin and isosorbide esters.

The compositions are usually incorporated into baked products in amounts which are sufficient effectively to inhibit the growth of mold during storage for durations of about a week and also sufficient to preserve the softness of the product during storage. In general, amounts of the lower monocarboxylic acid ester as little as about 0.04% based on the weight of the flour contained in the product are sufficient for effective mold growth inhibition in the product. Since greater amounts of the compositions of the process required to impart the desired softening or antistaling characteristics to the baked product, the critical concentrations of the composition in the baked product are accordingly determined by the amounts required for that purpose. In general, in yeast-leavened products, from from about 0.2% to about 1.0% of the composition based upon the weight of the flour in the product is sufficient to impart both mold growth inhibition and softness or antistaling properties to the baked product; in baked products which do not contain yeast however, the compositions are usually used in amounts within a range of about 0.8% to 20.0% based upon the weight of the flour in the baked product. The following examples demonstrate the preparation of propionate esters of glycerin and isosorbide and their use as softening agents and fungistatic agents.

Example 1: Preparation of Glyceryl Propionate Esters — About 460 grams of glycerin (5 mols) were placed in a 3-necked flask equipped with a mechanical stirrer, condenser, thermometer and dropping funnel. 1,250 grams (9.65 mols) of propionic anhydride was added dropwise at a temperature of 125° to 130°C over a period of 2.25 hours. The reaction mixture was heated for 8 hours at 100° to 110°C after the propionic anhydride had been added. Propionic acid was stripped off under reduced pressure, and the resulting mixture was distilled over a 10 inch vigreux column. The following four liquid cuts were collected:

	Pot Temp. (° C.)	Vap. Temp. (° C.)	Sap. No.	OH No.	Acid No.	Grams Collected	Yield, Percent
Cut #1.....	110-115	98-100	542	298.5	7.45	176	18.2
Cut #2.....	116	107	548	274	1.07	135	14.1
Cut #3.....	114-121	105-110	561	244	0.65	243	25.4
Cut #4.....	117-120	105-110	570	190	0.30	360	38.1
Residue........			561	158	0.66	40	4.2

The approximate composition of each fraction was as follows:

Cut #1 — 5% mono- and 95% dipropionate
Cut #2 — 100% dipropionate
Cut #3 — 2% mono- and tri- and 98% dipropionate
Cut #4 — 70% di- and 30% tripropionate

Example 2: Preparation of Isosorbide Dipropionate — A three liter 3-necked flask equipped with a stirrer, thermometer, dropping funnel and a reflux condenser was charged with 438.3 grams (3 mols) of isosorbide and 2 grams of p-toluene-sulfonic acid. The isosorbide was heated to 80°C and then 858 grams (6 mols plus 10% excess) of propionic anhydride were added dropwise with stirring over a 90 minute period. The temperature during the addition varied between 80° to 90°C. Thereafter, the reaction mixture was heated for 10 hours at 100°C and then for 4 hours at 155° to 160°C, after which propionic acid and excess propionic anhydride were stripped off under vacuum. The resulting liquid product was treated with 10 grams of activated carbon and distilled through a 10 inch vigreux column at 135° to 138°C and 0.5 mm pressure. 540 grams of isosorbide dipropionate was obtained.

Example 3: Preparation of Bread Using Propionate Esters of Glycerin and Isosorbide as Softening Agents — The following bread formulas and procedures were utilized to evaluate several propionate esters as bread softening agents according to the sponge-dough method:

- - - - - - - - - - Sponge - - - - - - - - - -		- - - - - - - - - - - - - - Dough - - - - - - - - - - - - -	
Percent (flour as 100%)		Percent (flour as 100%)	
Flour	65	Flour	35
Water	*Variable	Sugar	8
Yeast	2.5	Salt	2
Yeast food	0.5	Lard	3
		Milk powder	6
		Softening Agent	0.25
		Water	*Variable

*57% of the total water was added to the sponge and 43% was added to the dough. The total amount of water used varied depending upon the flour absorption properties. The amount of water required was determined by a farinograph method.

The sponges were prepared by dissolving the yeast in a portion of water, and this composition was added to a mixer along with flour, yeast food and the balance of the water. These ingredients were mixed for about 3 minutes and thereafter fermented for about 4.5

hours. The fermented sponges were returned to the mixer and all of the dough ingredients were added. The dough-sponge mixtures were mixed to full development (usually about 11 to 13 minutes), fermented for 20 minutes, divided, allowed a 10 minute proof, sheeted, molded, sealed and placed into bread pans. Then the dough was proofed at about 100°F and 85% relative humidity in the usual manner to template height and baked for about 20 minutes at 425°F.

Prior to baking, the bread dough was evaluated for dough conditioning properties and proof time. After baking, the bread was evaluated for volume, grain and softness. Softness studies on ½" thick bread slices were carried out over a period of 6 days using a gelometer to determine softness. The gelometer value represents the number of grams of shot required to depress the gelometer plunger a given distance (4 mm in these evaluations) into the bread slice. The softness data are given for the first day (day following baking), second to fourth day and fifth to seventh day. The data obtained on the second to fourth day are considered to be of greatest significance. Readings obtained on the first day are generally conceded to be a reflection of the extent of baking, while those obtained on the fifth to seventh day are considered of less importance since little bread remains to be consumed this long after baking.

Using the glyceryl propionate esters of Example 1 and the isosorbide dipropionate of Example 2 as softening agents, the baking of three loaves of bread with each softening agent yielded the following results:

Softening Agent	Dough Conditioning	Proof Time (min.)	Volume (cc.)				Gelometer Values			Grain		
			A	B	C	Avg.	1 Day	4 Days	6 Days	1 Day	4 Days	6 Days
Control (no softener)[1]	Very good	75	2,600	2,625	2,650	2,625	215	262	385	Very good	Good	Very good.
Cut #1	do	65	2,700	2,730	2,750	2,727	202	245	373	do	V. good	Do.
Cut #2	do	65	2,775	2,750	2,735	2,753	190	243	364	Good	do	Good.
Cut #3	do	65	2,650	2,675	2,675	2,667	209	245	371	do	Good	Very good.
Cut #4	do	64	2,700	2,675	2,700	2,692	207	249	353	do	do	Do.
								(2 Days)			(2 Days)	
Isosorbide Dipropionate [2]	Good	73	2,650	2,675	2,660	2,663	241	358	375	Good	Good	Good.
Control (no softener)[1]	Very good	90	2,500	2,550	2,525	2,525	261	398	419	Fair	V. Good	Very good.

[1] The controls contained 0.25% calcium propionate bread preservative.
[2] The isosorbide dipropionate and glyceryl propionates were evaluated in separate tests.

These results demonstrate that the propionate esters definitely improved bread volume and bread softness. In addition, the propionate esters did not adversely affect proof time, dough conditioning or the bread grain. The results of further tests demonstrated the fungistatic effect of the glyceryl propionates and isosorbide dipropionates. Loaves of bread prepared as described in Example 3 were exposed to room conditions and thereafter stored at 100°F and 85% relative humidity. No mold had developed even by 10 days.

Low I.V. Monoglycerides and Propionates

In another process, *W.H. Knightly; U.S. Patent 3,485,639; December 23, 1969; assigned to Atlas Chemical Industries, Inc.* uses a blend of a low iodine value monoglyceride and a lower monocarboxylic acid ester of a polyhydric alcohol. In the process a normally hard, low iodine value monoglyceride having an average iodine value in the range of about 6 to 45 is blended with a lower monocarboxylic acid ester of a polyhydric alcohol having from 2 to 6 hydroxyl groups to yield a soft paste or plastic material which is readily dispersed in dough or batter. A monoglyceride prepared from a monocarboxylic acid of about 12 carbon atoms (lauric acid) or its equivalent may have a low iodine number and still be a soft paste or plastic material.

The term normally hard, low iodine value monoglyceride is meant to include monoglycerides prepared from monocarboxylic acids containing at least 12 carbon atoms per molecule, and therefore includes laurates which may be plastic materials. Monoglycerides prepared from lauric acid having iodine values of about 6 to 10 may be sufficiently dispersible as to be useful but in general, the minimum iodine value should be about 15. Generally, as the carbon atom content of the monocarboxylic acid used to prepare the low iodine value monoglyceride increases, the monoglyceride becomes a harder material, and in order to obtain ready dispersibility, the minimum iodine value should be higher. Of course, more of the lower monocarboxylic acid ester may be blended with the normally hard, low iodine value monoglyceride to render it more dispersible, but this reduces the effectiveness of the blend as a softener. It is particularly preferred to use normally hard, low iodine value monoglycerides prepared from monocarboxylic acids containing from about 16 to 20 carbon atoms per molecule for these monoglycerides are particularly effective softeners. When using these monoglycerides, they should have iodine values ranging from about 20 to about 36 in order to insure their dispersibility in dough or batter.

Of particular importance is the fact that the compositions are as effective or more effective than currently used softening agents, and in addition, they function as mold inhibitors without adversely affecting proof time or the volume and grain of the resulting baked product. The normally hard, low iodine value monoglycerides, which term is meant to include mixtures of mono-, di- and/or triglycerides, are those having an average iodine value in the range of about 6 to 45, with an iodine value of about 20 to 36 being particularly preferred. It is possible to use monoglycerides having iodine values outside of this range by blending them with a monoglyceride having an iodine value sufficient to form a monoglyceride mixture having an iodine value within the stated range. If the normally hard, low iodine value monglyceride component has an iodine value less than about 6, it is necessary to add more lower monocarboxylic acid ester to render the composition sufficiently plastic than is required for effective mold inhibition and the effectiveness of the blend as a softener is impaired.

On the other hand if the iodine value of the glyceride is more than about 45, the softening properties of the blended composition will not be as effective as currently used softening agents. It has been found that monoglycerides or monoglyceride mixtures having average iodine values in the range of from about 20 to about 36 are particularly effective softening agents and can be rendered sufficiently plastic by the amount of lower monocarboxylic acid ester required for effective mold inhibition.

The lower monocarboxylic acid esters of a polyhydric alcohol having from 2 to 6 hydroxyl groups are effective mold inhibitors and softening agents. Since these compounds are to be incorporated into baked products which are to be ingested, it is essential that they be edible. The polyhydric alcohol may be completely esterified or partial esters may also be used. Conventional esterification procedures may be used to prepare these esters. Typical of the polyhydric alcohols which may be used are propylene glycol, glycerol, erythritol, sorbitol, and cyclic inner ethers of polyhydric alcohols having at least 2 hydroxyl groups such as isosorbide.

The lower monocarboxylic acids which may be used to esterify the polyhydric alcohol are the monocarboxylic acids having from 2 to 7 carbon atoms. Propionic acid has been found to be particularly effective. In addition to using the acids, their equivalent anhydrides, halides and esters may also be used to prepare the esters of polyhydric alcohols. In general, the lower monocarboxylic acid esters of a polyhydric alcohol having from 2 to 6 hydroxyl groups should be blended with the low iodine value monoglyceride in amounts sufficient to render the mixture a soft paste or plastic material. The blend must be soft enough to be thoroughly dispersed in the dough at room temperature.

The amount of lower monocarboxylic acid ester required will vary depending upon the hardness of the monoglyceride and the particular ester which is used. Usually more ester is required as the iodine value of the monoglyceride decreases and the molecular weight of the lower monocarboxylic acid ester increases. It has been found that blends containing from about 20 to 75 weight percent of lower monocarboxylic acid ester are generally soft enough to be dispersed in the dough. If more than about 75 weight percent of the lower monocarboxylic acid ester is present, the blend will not be as effective as currently used softening agents at comparable levels. The normally hard, low iodine value monoglyceride-lower monocarboxylic acid ester of a polyhydric alcohol composition may be prepared by melting the normally hard, low iodine value monoglyceride and then incorporating the lower monocarboxylic acid ester in the melt. Thereafter, the melt may be mixed thoroughly on a Votator to form a uniform mixture and then cooled to form a plastic, readily dispersible composition. If more than one normally hard, low iodine value monoglyceride or monoglyceride mixture is to be used, one of the monoglycerides may be melted and the other monoglycerides and the lower monocarboxylic acid ester added to the melt in any order, so long as the temperature of the melt is sufficient to melt all of the added components.

The compositions are usually incorporated into baked products in amounts which are sufficient effectively to inhibit the growth of mold during storage for durations of about a week and also sufficient to preserve the softness of the product during storage. In general, amounts of the lower monocarboxylic acid ester as little as about 0.04% based on the weight of the flour contained in the product are sufficient for effective mold growth inhibition in the product.

Example 1: Formation of a Blend of a Normally Hard, Low Iodine Value Monoglyceride and a Glyceryl Propionate — Blends of normally hard, low iodine value monoglycerides blended with glyceryl dipropionate and glyceryl tripropionate were prepared. A mono- and diglyceride mixture having an iodine value of about 2 was heated to a temperature of about 150°F. After a melt was formed, another mono- and diglyceride mixture having an iodine value of about 36 was added to the melt. After both monoglyceride mixtures were completely melted, they were separated into two portions and glyceryl dipropionate was added to one portion and glyceryl tripropionate to the other. Then each portion was votated and cooled to form a plastic composition. The weight ratios of the components of the two compositions was as follows:

Composition A	Parts by Weight
Mono- and diglyceride mixture (I.V. 36)	38
Mono- and diglyceride mixture (I.V. 2)	22
Glyceryl dipropionate	40

Composition B	
Mono- and diglyceride mixture (I.V. 36)	38
Mono- and diglyceride mixture (I.V. 2)	22
Glyceryl tripropionate	40

The blend of the two monoglyceride mixtures used in Compositions A and B had an average iodine value of 23.5.

Example 2: Preparation of Bread Using Softening Agent-Mold Inhibitor Compositions — The bread formula and procedures of the previous patent, U.S. Patent 3,394,009, Example 3, were used to evaluate Compositions A and B of Example 1 as bread softeners and mold inhibitors, with 0.50% of the softener and mold inhibitor substituted for the 0.25% of softening agent used in the previous patent. Compositions A and B of Example 1 were evaluated and were compared to a standard which contained 0.15% calcium propionate as a preservative. The following results were obtained:

Softener and mold inhibitor	Dough conditioning	Proof time (min.)	Volume (cc.)					Gelometer values (avg.)		Grain 6 days	Mold inhibition (days)
			A	B	C	D	Avg.	3 days	6 days		
Standard (no softener).	Very good........	73	2,535	2,460	2,575	2,605	2,544	326	400	Good	>13
Composition A..........do...........		70	2,650	2,535	2,550	2,750	2,621	279	328	Good	>13
Composition B..........do...........		68	2,655	2,575	2,605	2,600	2,609	268	327	Good	>13

These results indicate that both Compositions A and B were effective bread softeners and mold inhibitors.

MOLD INHIBITORS

Dipropionate Additives

Bread, being moist and porous makes an excellent medium for the growth of mold. Almost all of the common molds, *Aspergillus niger, Penicillium glaucus, Penicillus expansum, Rhizopus nigricans, Trichothecium roseum,* etc., under the favorable conditions existing in undried bread, grow rapidly on the surface of the loaf and between the slices if the bread is sliced. The wrapping of the bread and slicing it before wrapping have made the problem of molds more serious since the moisture-proof wrapper retains and holds the moisture of the loaf.

According to *F.R. Benson; U. S. Patent 3,485,638; December 23, 1969; assigned to Atlas Chemical Industries, Inc.* the growth of mold is inhibited by contacting materials, either by coating or by incorporating ethylidene dipropionate or propylidene dipropionate. Ethylidene dipropionate and propylidene dipropionate can be prepared by heating the appropriate aldehyde with propionic anhydride in the presence of an acid catalyst according to the reaction

$$RCHO + \underset{C_2H_5-C}{\overset{C_2H_5-C}{\Big\langle}}\!\!\!\!\!\!\!\underset{O}{\overset{O}{\diagdown}}\!\!\!\!\!\!\!O \longrightarrow C_2H_5-\overset{O}{\overset{\|}{C}}-O-\overset{R}{\underset{H}{C}}-O-\overset{O}{\overset{\|}{C}}-C_2H_5$$

For example, ethylidene dipropionate may be prepared according to the following procedure. Propionic anhydride, 4.4 mols, and 1 gram of sulfuric acid are placed into a three-necked flask equipped with a mechanical stirrer, thermometer, reflux condenser, and dropping funnel. The mixture of anhydride and sulfuric acid is heated to 70°C and then four mols of acetaldehyde are added dropwise through the dropping funnel. The addition requires 2 to 3 hours and then the reaction mixture is heated for an additional 2 to 3 hours at 80° to 90°C. The product is diluted with 300 ml of chloroform and washed several times with a solution of 0.5% sodium hydroxide until the acid number is less than one. The chloroform is removed under vacuum and the final product is either treated with Darco activated carbon or distilled under vacuum. Propylidene dipropionate may be prepared according to the foregoing procedure by using propionaldehyde in place of the acetaldehyde.

Ethylidene and propylidene dipropionates may be applied in any way appropriate to the material treated. The dipropionates may be incorporated directly into the material treated or by spraying the surface with the liquid dipropionate or by dipping the material into the liquid dipropionate. For example, the mold inhibitors may be applied to the surface of cheese, cereals, fruits, vegetables, tobacco, paper, leather, textiles, wood, etc., by spraying or dipping or incorporated into foodstuff such as bread by adding to the ingredients before, during or after mixing.

The amount of ethylidene and propylidene dipropionates used depends to a large extent on the nature of the material treated. In the particular case of inhibition of mold in bread, for example, the use of from 0.12 to 0.18% based on weight of flour is preferred. In the treatment of topical fungal infections, ethylidene and propylidene dipropionates may be applied full strength to the infected area or applied as a dilute solution. Dilute solutions are highly effective at concentrations as low as 0.05% by weight, based on the total weight of the solution.

Example: Bread was prepared by the sponge and dough procedure. The formula is shown in the table on the following page.

Ingredient	Sponge		Dough	
	Percent	Grams	Percent	Grams
Flour	70	700	30	300
Water	41	410	27	270
Yeast	2.5	25		
Yeast food	0.5	5		
Salt			2	20
Sugar			6	60
Non-fat dry milk			4	40
Lard			3	30
Mold inhibitor			Variable	Variable

The procedure is given in the table below.

Mixing time[1]	1 min. at 1st speed, 1 min. at 2nd speed.	1 min. at 1st speed, 4 min. at 2nd speed.
Temperature	78–79° F	80–82° F.
Fermentation time	4 hours	30 minutes;
Intermediate proof		10 minutes.
Scaling weight		18½ oz.
Baking temperature		450° F.
Baking time		20 minutes.

[1] Hobart A-120 with McDuffee bowl and fork.

The loaves were wrapped in waxed paper, sealed, stored overnight, after which they were cut with a sterilized knife and the center slice removed. The center slice was cut vertically in two and both halves were inoculated with mold spores as follows. Two transfer loops of mold spores from actively growing mold were suspended in 75 ml of sterile distilled water in a 250 ml of Erlenmeyer flask. Clumps of spores were broken by vigorous shaking with glass beads. This suspension was further diluted 1:100 with sterile distilled water to provide the working mold suspension. The bread was inoculated with a sterile loop dipped into the working mold suspension to get a standard loop full. This was transferred to a slice, placing the spot about one inch from the one end, with a similar spot at the other end. Inoculated samples were then inserted into cellophane bags which were heat sealed. They were inoculated at 30°C and examined daily for mold growth. The time of appearance of mold was noted and the extent of growth indicated by 1 (slight), 2 (moderate), 3 (moderately heavy) and 4 (heavy). The results are shown in the following table.

Item No.	Mold Inhibitor	Percent of mold[1] inhibitor	3	4	5	6	7	10	11	12
1a	Control	0	1	2	3	4	4	4	4	4
1b	do	0	1	2	4	4	4	4	4	4
2a	Propylidene dipropionate	0.12		1	1	2	3	4	4	4
2b	do	0.12		1	1	1	2	2	3	3
3a	do	0.15			1	1	2	4	4	4
3b	do	0.15			1	1	2	4	4	4
4a	do	0.18						1	1	1
4b	do	0.18								1
5a	Ethylidene dipropionate	0.09			1	2	3	4	4	4
5b	do	0.09			1	1	2	3	4	4
6a	do	0.12				1	1	2	2	3
6b	do	0.12				1	1	1	1	2
7a	do	0.15							1	1
7b	do	0.15								
8a	Calcium propionate	0.12			1	1	2	3	4	4
8b	do	0.12			1	1	2	4	4	4
9a	do	0.15				1	1	2	2	2
9b	do	0.15						1	1	1
10a	do	0.18					1	2	2	2
10b	do	0.18								

[1] Based on flour weight.

Acetate and Coated Acid

R. Ueno. T. Miyazaki, T. Matsuda, S. Inamine and S. Arai; U.S. Patent 3,734,748; May 22, 1973; assigned to Ueno Fine Chemical Industries, Ltd., Japan discuss the problems of spoilage of leavened bread and the use of acetate in combination with coated acid to reduce the occurrence of spoilage. In making bakery products, and especially yeast leavened bread, it has been the practice to use on the order of 0.2 to 0.4%, based on the weight of flour, of calcium propionate. One of the reasons for using this compound resided in preventing the bread product from being subjected to the dreadful consequences of the rope bacterium and another was for preventing the damage that would be caused to the bread by mold. The rope is caused by the growth of microorganisms of the genus Bacillus which possess strong resistance to heat. In the past, countermeasures such as the addition of acetic acid or acid phosphates and the introduction of an acid dough step have been adopted. Thus, most of the methods employed were those which consisted in reducing the pH of the dough to close to that at which the growth of the genus Bacillus takes place. This method, however, had the drawback that the finished product would be adversely affected if the pH was reduced too much.

However, with the appearance of the propionates in recent years the problem of ropy bread was said to be practically solved. There still remained some drawbacks, however. That is, calcium propionate affects the normal leavening action of yeast, and in the case of the recently developed advanced continuous process having a relatively short leavening period, the effects that appear cannot be ignored. Again, if calcium propionate is added in an amount exceeding 0.2%, it gives off an odor which is objectionable.

Immediately subsequent to baking of bread there exists no mold at all. However, in its subsequent handling, i.e., in the removal from the pan, cooling, slicing, wrapping and transporting steps, the opportunities for contaminating bread are very numerous. Thus, when the shelf-life of bread at a retail store and the time that elapses before it is finally consumed by the consumer are considered, it must be kept for a period of at least one week before it spoils. The case as to whether or not mycotoxins such as aflatoxin develop in the molding of bread has not been fully investigated as yet. Although calcium propionate is presently being added to white bread in an amount usually of about 0.2%, based on the weight of flour used, this amount does not possess adequate preservative effects.

The reason calcium propionate is used in a concentration such as indicated is that the addition in excess of this amount results in an aggravation of the effects that are produced by the normal leavening of yeast, as well as that the flavor of the product is adversely affected. Hence, since suitable results cannot be obtained by the use of calcium propionate in its maximum useable amount, the rate of returned bread products has shown no improvement.

The effects on microorganisms of the organic acids having an acid type antiseptic property such as acetic acid, propionic acid and sorbic acid demonstrate a great change depending upon the level of the ambient pH. This is due to the fact that the antimicrobial activity depends upon the non-dissociable molecules of the organic acid. Hence, the antimicrobial activity increases when the pH is on the acid side, the opposite being true when the pH is on the alkaline side. For example, when the acetate is added as sodium acetate alone regardless of its concentration, preservative effects cannot be expected, since the pH of the dough as well as that of the bread increases as the concentration of the sodium acetate is increased. Hence, when an organic acid, is to be used as a preservative for foods, it is most effectively and economically used when the pH is low. When considered from such a viewpoint, the method of adding acetic acid instead of an acetate becomes apparent.

However, this method has shortcomings in that the pH of the dough as well as bread is lowered and also in that an acetic acid odor is imparted. Further, it is necessary in making bread to maintain the pH of the dough within certain ranges so as to ensure the safety of gluten and to prevent the degradation of the texture and volume of the final product. Heretofore when bread had a low pH, it also had a diminished volume due to the interference of the antibacterial activity with the fermentation process.

It was found that it was possible to achieve very pronounced preservative effects as well as an improvement in the quality of bread by using conjointly with the acetate in making bread, a coated crystalline acid compound obtained by coating a crystalline acid compound with a substance which does not melt when cold but melts upon being heated. The coated acid compound, being isolated from the dough system, is not subjected to any change at all during the leavening of the dough, but during the baking stage after proofing of the dough the coating substance surrounding the crystalline acid compound melts and the acid substance leaks to the outside where it acts as a very effective acid in promptly and suitably reducing the pH of the bread. Thus, an effective preservative method has been established by the combination of sodium acetate with a coated acid substance which does not function as an acid in its dough state but works as an effective acid in the final product. Another important feature of this method is the decrease in the retardation of the fermentation, which was encountered without fail in the case where calcium propionate was used. In making bread using this method, an increase in the loaf volume over that of the control experiment demonstrates another important feature. In the case of a propionate, the proofing time must be prolonged for obtaining the usual fermentation and loaf volume, but there is no such necessity whatsoever in the case of this method and bread of a greater loaf volume can still be obtained.

The acid compounds used in the method include the organic acids which are normally crystalline at room temperature such as for example, fumaric acid, citric acid, malic acid, succinic acid, etc. or their acid salts. Further, as acid salts, the crystalline inorganic acid salts, for example, sodium primary phosphate, calcium primary phosphate, sodium metaphosphate and alum are also desirable.

The acetate is used in an amount of 0.2 to 0.5% by weight relative to the wheat flour used in the bread mix. When the amount is less than 0.2% by weight, no preservative effects are accomplished. On the other hand, when the amount exceeds 0.5% by weight, the fermentation of the yeast is hindered and the quality of the bread deteriorates. Further, at greater concentrations the bread making property of the wheat flour gluten is impaired and results in a bread with poor texture. The organic acid to be used with the acetate must be used in an amount of at least 0.03% by weight relative to the wheat flour of the bread mix. When the amount is less than 0.03% by weight, no preservative effects are demonstrated.

On the other hand, the acid salt of an organic acid to be used with the acetate must be in an amount of at least 0.06% by weight relative to the wheat flour. No preservative effects are demonstrated when the amount is less than 0.06% by weight relative to the wheat flour. No preservative effects are demonstrated when the amount is less than 0.06% by weight. In the case of the acid salt of an inorganic acid to be used with the acetate, the amount used must be at least 0.045% by weight based on the wheat flour. No preservative effects are had when the amounts is less than 0.045% by weight.

The amount used of the organic acid, the acid salt of an organic acid, or the acid salt of an inorganic acid becomes less as the strength of an acid of these compounds become stronger. As the coating agent, at least one substance selected from the group consisting of the animal and vegetable oils and fats, monoglycerides, diglycerides, waxes and paraffin, which has a melting point such that it does not melt at room temperature but melts during the heating steps is used.

The coating of the acid compound is carried out, for example, in the following manner. A coating agent predominantly of a hardened oil is heat-melted and held at a temperature of 70° to 80°C. An organic acid or an acid salt comminuted to a particle diameter of 10 microns or less is added to the melted coating agent, after which a surfactant is also added, if necessary. This is followed by thoroughly mixing and dispersing the mixture using a homogenizer. This mixture is then sprayed into a chamber adjusted to a temperature below 40°C, using a rotating disk type spraying apparatus to obtain a powdered product of a particle diameter or less than 500 microns, and preferably 50 to 300 microns. The coating agent must be a substance which can coat acids or acid salts and naturally must be tasteless and odorless as well as harmless from the standpoint of food hygiene. Thus, the hardened oils or mono- or diglycerides of fatty acids, waxes (Japan wax and beeswax) and paraffin are preferred. The coating agent is usually used in an amount of two to fourfold that of the acid or acid salt.

Example and Controls 1 through 5: After kneading a sponge dough for 2 minutes at a low speed, it was fermented at 28°C for 4 hours. The remixing operation was then carried out for 6 minutes, at which time the seasonings as well as the preservatives were added. The recipe of the mix was as shown in Table 1 on the following page.

TABLE 1

	Sponge dough, parts	Remixed dough, parts
Wheat flour	70	30
Yeast	2	
Yeast food	0.1	
Common salt		1.8
Sugar		4
Shortening		4
Water	40	20

After a second fermentation lasting for 30 minutes, the dough was divided into portions each 150 grams in size and a benching time of 15 minutes was allowed. Each of the portioned dough was placed in a mold and proofed for 40 minutes at 37°C in a rotary proofing oven. Next, the baking was carried out for 20 minutes at 230°C in a rotary oven. Six loaves of the product were withdrawn. Of these, three loaves were used and following tests were carried out: a volume measurement after a lapse of one hour, and a sensorial test, measurement of pH and measurement of the crumb hardness by means of a baker's compressimeter after a lapse of 12 hours. Five slices from each of the three remaining loaves, i.e., a total of 15 slices, were obtained and each of these slices of bread was placed in a polyethylene bag and sealed. These bags containing the sliced bread were stored in a constant temperature chamber of 30°C and observations were made of spoilage due to mold. Table 2 shows the effects on the quality of the product when experiments were carried out when the additives indicated were used as preservatives.

TABLE 2

Experiment	Additive	pH		Condition of product		Sensorial test	
		Dough	Product	Volume of bread (ml.)	Crumb hardness (g.), 12 hrs.	Flavor and taste	Hardness
Control:							
1	Preservative not added	5.62	5.40	706	76	Good	Normal.
2	Calcium propionate 0.2%	5.62	5.42	690	75	Acid odor, sour	Do.
3	Acetic acid 0.3%	5.48	5.32	688	84	Strong acid odor	Hard.
4	Sodium acetate 0.4%	5.70	5.40	726	76	Good	Normal.
5	Sodium acetate 0.4% plus 50% lactic acid 0.2%	5.40	5.28	700	78	do	Slightly hard.
Example	Sodium acetate 0.4% plus coated fumaric acid 0.2%.	5.68	5.28	748	73	do	Normal.

When sodium acetate and the coated fumaric acid were used, the pH of the dough did not drop, but a considerable drop took place in the pH of the product bread. While the volume of the bread became small in the experiments in which propionic acid and acetic acid were used, an opposite tendency was noted in the case of the experiment in which sodium acetate was added. As to the hardness of the crumbs, it was seen that the hardness was especially great in the case where acetic acid was added, the pH of which was low from the dough stage.

On the other hand, the results of the sensorial test showed that when calcium propionate was added in an amount of 0.2%, a peculiar propionic acid odor was given off by the bread, which moreover had an unpleasant sour taste. Now, when only acetic acid was added, a peculiar acetic acid odor was evolved, which was attended by a strong acidity. However, in the case where sodium acetate was added, practically no odor was sensed. Moreover, when the coated fumaric acid was added to the sodium acetate, an acid odor could not be detected, and the product bread also had a good flavor. When sodium acetate was conjointly used with other acids, such as lactic acid, the volume was not increased because of the pH from the outset.

Further, as regards the preservative effects, it is apparent from the results shown in Table 3 on the following page, that adequate effects are demonstrated by the conjoint use of sodium acetate and the coated fumaric acid. The product obtained by the use of sodium acetate and the coated fumaric acid also yielded results in the test which compare favorably with the product obtained in Control 1 (wherein a preservative was not added). Thus, it was possible to obtain comparable preservative effects without being troubled at all by the odor that results from the use of propionic acid and without affecting the texture and volume of the final product.

TABLE 3

Experiment	Conditions of experiment	Number of days kept					
		2	3	4	5	6	7
Control:							
1	Preservative not added	[1] 10/15	15/15				
2	Calcium propionate 0.2%				2/15	8/15	15/15
3	Acetic acid 0.3%			4/15	11/15	15/15	
4	Sodium acetate 0.4%	9/15	15/15				
5	Sodium acetate 0.4% plus 50% lactic acid 0.2%				5/15	8/15	15/15
Example	Sodium acetate 0.4% plus coated fumaric acid [2] 0.2%				4/15	9/15	15/15

[1] The numerical values in the table have the following meaning. For exapple, the figure 10/15 denotes that, of the 15 slices tested, 10 were spoiled.
[2] Composed of 33% of fumaric acid and 67% of a hardened oil.

Sorbic Acid and Acid Anhydrides

In many cases antimicrobially active substances are directly mixed in a suitable form with the foodstuffs to be preserved in order to achieve a uniform distribution. In the case of pastry, the simplest mode of incorporation is the addition to the dough. Such an addition of preserving agents may, however, involve difficulties when a microbiological method is used to light the dough, for example with the aid of yeast or leaven. All known preserving agents for bakery goods, such as, for example, acetic acid, propionic acid, dehydroacetic acid, or sorbic acid and the salts impair the yeast fermentation necessary to raise the dough in addition to the growth of undesired mold fungi so that with the use of an agent having a good preserving action against mold fungi the fermentation of yeast is considerably and undesirably inhibited if the agent is incorporated with the dough prior to the baking process.

In practice, the undesired inhibiting action towards bakers' yeast cannot be avoided by simply reducing the amounts of preserving agent added, for in this manner the protecting effect against mold fungi in the finished pastry is suppressed or strongly diminished. Moreover, the known preserving agents do not always ensure a satisfactory protection against mold infestation since they do not have a selective action towards mold fungi.

The preserving agents of *H. Fernholz, E. Lück and H. Neu; U.S. Patent 3,510,317; May 5, 1970; assigned to Farbwerke Hoechst AG, Germany* for breads and pastries are the mixed anhydrides of sorbic acid and unsaturated aliphatic fatty acids. The preserving agents are added to the dough directly in the form of a solution, emulsion or dispersion or in mixture with the baking fat and/or the dough raising agent. The preserving agents can be used without any consideration of the type of dough raising agent used. They are suitable for bakery goods in which the raising of the dough is brought about by fermentation either with leaven or with yeast, as well as for those made with the use of known chemical raising agents, for example baking powder. The amounts of preserving agent required are in the range of from 0.1 to 1.5% by weight, calculated on the weight of the dough.

Example 1: A dough having the following composition:

	Grams
Wheat flour type 405	5,000
Water	3,000
Baking margarine	250
Common salt	80
Sugar	30
Yeast	140

was mixed with 0.5% each of a mixed anhydride of sorbic acid with various organic acids. Each time 100 grams of the dough was filled into a wide beaker having a volume of 600 ml.

The dough was kept at 30°C and the time was determined until the dough had reached a volume of 230 ml. The results obtained are indicated in the following table:

	Raising Time of Dough (minutes)
Control sample	40-45
Sorbic acid/acetic acid anhydride	121-131
Sorbic acid/propionic acid anhydride	118-129
Butyric anhydride	108-120
Sorbic acid/palmitic acid anhydride	38-43
Sorbic acid/undecylenic acid anhydride	40-46
Sorbic acid/stearic acid anhydride	39-47
Sorbic acid/myristic acid anhydride	38-44

The above table reveals that the raising time of the dough is shorter when a mixed anhydride of sorbic acid and a fatty acid with more than 5 carbon atoms is used than with the use of, for example, corresponding anhydrides of sorbic acid with carboxylic acids having shorter chains or butyric anhydride not claimed.

Example 2: A dough as defined in Example 1 was mixed with different amounts of the mixed anhydride of sorbic acid and palmitic acid (I). Each time 100 grams of the dough were filled into a wide beaker having a volume of 600 ml. The dough was kept at 30°C and the time was determined until it had reached a volume of 230 ml. The results are indicated in the following table:

	Raising Time of Dough (minutes)
Control sample	40-45
Addition 0.3% of I	38-43
Addition 0.5% of I	40-43
Addition of 1.0% of I	39-46

The addition of varying amounts of the mixed anhydride of sorbic acid and palmitic acid has no essential influence on the raising time of the dough.

Example 3: Varying amounts of the mixed anhydride of sorbic acid and palmitic acid (I) were added to the dough as defined in Example 1 and a tin loaf was baked in usual manner (time of rest 55 to 65 minutes, baking time 60 minutes, baking temperature 180° to 200°C). As control sample a dough prepared under the same conditions but without addition was baked.

All samples had the same volume of bread. After having been allowed to rest for identical periods of time the bread samples containing the mixed anhydride I had risen to the same volume as the control samples. Differences between the samples containing the mixed anhydride and the control samples could be observed neither with regard to the bread volume, the nature of the crust and the crumb, the pores and the elasticity of the crumb, nor in the organoleptic properties.

The finished bread was sliced and for infestation the slices were exposed to an atmosphere strongly infested with mold fungi. Some of the slices of bread were then kept in an incubator (30°C, relative atmospheric moisture 95 to 100%) while another part was stored at room temperature in bags of polyethylene. The growth of mold fungi in the individual samples is illustrated in the table on the following page. Explanation of symbols in the table:

　　　　　— = no visible formation of mold
　　　　　+ = slight formation of mold, isolated small colonies
　　　　　++ = stronger formation of mold, a greater number of colonies
　　　　　+++ = strong formation of mold
　　　　　++++ = very strong formation of mold

Storage in incubator	Mold formation after 2 to 8 days of storage			
	2	4	6	8
Control sample	+	++++	++++	++++
0.15% of I	−	+	+++	++++
0.3% of I	−	−	++	+++
0.5% of I	−	−	−	−

Storage at room temperature	Mold formation after 2 to 20 days of storage				
	2	4	8	16	20
Control sample	−	+	++	++++	++++
0.15% of I	−	−	−	+	++
0.3% of I	−	−	−	+	+
0.5% of I	−	−	−	−	−

BAKED GOODS AND CEREAL PRODUCTS

SORBIC ACID FOR SLICED BAKED PRODUCTS

In the process of *D. Melnick; U.S. Patent 3,021,219; February 13, 1962; assigned to Corn Products Company* sorbic acid is added to vegetable oil or solid fat to prepare a solution or suspension which is applied to the dough or crust of baked products. In a modified procedure the sorbic acid is applied to the baking pan in a layer of from about 0.1 mm to about 1 mm in depth. At the higher concentrations, viz., 2 to 10%, most of the sorbic acid will be in suspension when the pan grease is at room temperature or when the shortening base is raised to a temperature just above the melting point. However, during the baking operation temperatures in excess of 200°F are attained and at these temperatures practically all the sorbic acid in the pan grease goes into solution.

When sorbic acid is used as a suspension in the pan grease in excess of 10%, the under surfaces of the final baked product will show an undesirable residuum of white sorbic acid crystals. The use of sorbic acid in concentration of less than 0.5% in the pan grease is not adequate to protect these vulnerable lower crust surfaces from mold spoilage. It should be emphasized that of all the crust surfaces of a baked product that which has been in contact with the baking pan is most vulnerable to mold spoilage. This is attributed to the fact that these lower crust surfaces exhibit, in contrast to the upper crust layer, a wide-open porous structure.

When the baked product is removed from the oven, the upper crust may be sprayed or painted with a sorbic acid solution to protect this surface. Any solvent or combination of solvents in which sorbic acid exhibits a solubility of at least about 0.5% and preferably at least about 2% may be used in preparing the sorbic acid solution for treating the upper crust. For this purpose, a solution of sorbic acid in propylene glycol or in a propylene glycol:water mixture has been found to be suitable. In another procedure the cutting blades used in slicing the baked products are allowed to run through a solution of sorbic acid before the food products are cut.

Any solvent in which sorbic acid exhibits a solubility of at least about 2% may be used in preparing the sorbic acid solution for treating the cutting blades. Solvents such as water, vegetable oil (liquid or hydrogenated), propylene glycol or ethanol may be used. Sorbic acid is soluble in these solvents at room temperature to the extent of 0.2%, 0.6%, 5.5%, 12.5% by weight, respectively. Increasing the temperature of the solvents will increase significantly the solubility of sorbic acid. Thus, in the case of the vegetable oils the solubility of sorbic acid is increased to about 8% by raising the temperature from 68°F to 200°F.

Thus, in certain operations where heating of the sorbic acid solution is no deterrent, it is satisfactory to use sorbic acid in suspension and depend upon the subsequent heating operation to bring this sorbic acid into solution. Combinations of solvents are also effective and preference is for those combinations which still exhibit a relatively high solubility for sorbic acid, about 2% by weight or more of sorbic acid in solution.

The composition of the blades used for preparing the sliced baked products should be one which does not react with sorbic acid. Stainless steel, steel hard-surfaced with chrome, or a plastic, such as nylon, may be used for this purpose. Slicing machines for bread come in two basic designs differing essentially in the mode of slicing. The so-called reciprocating slicer makes use of slicing blades mounted in two frames which move up and down at high speed in opposite directions, slicing the bread as it is gently fed through the moving blades.

In this operation the sorbic acid in a solvent, viz., 5% in propylene glycol, is fed on to the blades in drop-wise fashion from a reservoir mounted above the blades. Another type of slicer is the so-called band slicer. These have endless band knives which are run at high speed through the baked product to be sliced. In this operation the sorbic acid in solution, viz., 5% in propylene glycol, is held in a well through which the endless band knives pass thereby picking up the fungistatic solution continuously and depositing it into the exposed inner portions of the baked products.

The above process is effective in protecting baked products such as sliced pound cake, sliced fruit cake, it is particularly effective and useful in protecting sliced yeast-raised bread and rolls, particularly when these products are exposed to high humidity. The process is the most economical way for protecting the latter type of baked products against mold spoilage when sorbic acid is used as the fungistatic agent. When sorbic acid is introduced directly into the dough of a yeast-raised product in concentrations of 0.05% or greater, expressed on the flour basis, it interferes with yeast fermentation.

Lower concentrations of sorbic acid fail to give significant fungistatic protection to sliced yeast-raised baked products. With the higher concentrations of 0.05% or greater, dough maturation is impaired, proofing time is extended, and baked products low in volume and poor in grain and texture are obtained. By this process, all vulnerable areas of the yeast-raised baked products are protected with sorbic acid, the sorbic acid being introduced so late in the baking operation that no interference with yeast fermentation occurs.

Example 1: Conventional marble pound cake was prepared from a combination of white and chocolate cake batters. The cakes contained 29% moisture and 15% fat. The baking pans had been greased by brushing the bottom and walls with a shortening of 112°F melting point, heated to about 120°F and containing 2% sorbic acid in solution. The top crusts of the pound cakes were washed with an 0.5% solution of sorbic acid in water at about 120°F as the cakes were removed from the oven. The cooled cakes were then sliced in a reciprocating slicing machine, the blades of which had fed on to them in a controlled drop-wise fashion the same sorbic acid solution in water described above.

The individual slices of pound cake were wrapped in moistureproof, heat-sealable cellophane. It required 10 days at room temperature before the cake slices exhibited the first evidence of mold spoilage. The control cake slices, which were similarly prepared but without sorbic acid treatment, showed the first signs of becoming moldy after 6 days of storage. In both the treated and control cakes, mold growth was first detected on the crumb surface. The nonsliced control pound cake exhibited the first evidence of mold spoilage after 8 days of storage.

Example 2: Conventional yeast-raised bread was prepared using the sponge dough method. The bread contained 35% moisture and 3% fat. The baking pans had been greased by brushing the bottom and walls with a shortening of 112°F melting point, heated to about 120°F and containing a total of 6% sorbic acid, 2% in solution and the remaining 4% in suspension. The top crusts of the bread were washed (actually painted) with a 5% solution of sorbic acid in propylene glycol as the breads came from the oven. The cooled baked breads were then sliced using endless band knives which passed through a well containing 5% sorbic acid in propylene glycol solution. The sliced breads were wrapped in moisture-proof, heat-sealable cellophane. It required fully 10 days at room temperature before the breads exhibited the first evidence of mold spoilage, noted at the break on the top crust and in the crumb of some slices.

The control breads duplicated the above, but were baked in pans greased with the melted fat alone, given a simple water wash as they left the oven, and sliced without benefit of the sorbic acid treatment of the band knives. These control breads also wrapped in cellophane and stored at room temperature, exhibited the first evidence of mold spoilage after 4 days of storage; mold growth was noted at this time only in the crumb of some slices. The control breads which were similarly stored, but not sliced, showed the first evidence of mold spoilage after 7 days; mold growth occurred at the break on the top crust and on the wall crust surfaces.

Example 3: The bread of Example 2, but without the sorbic acid wash of the upper crust surfaces, was prepared. These sliced breads required 8 days at room temperature before the breads showed the first signs of mold growth. This occurred at the break on the top crust.

Example 4: The bread of Example 2, but without the greasing of the pans with the shortening containing the sorbic acid and without the sorbic acid wash of the upper crust surface, was prepared. Silicone-coated pans, requiring no greasing, were employed to hold the dough during proofing and baking. Thus, the only sorbic acid treatment in this case was passage of the band knives through a well containing the sorbic acid solution, 5% in propylene glycol. These sliced breads required 7 days for the first signs of mold growth to become apparent.

NISIN AND SUBTILIN

Refrigerated doughs for making such products as biscuits, pastry, rolls and related products are offered for sale in cans having metal ends and sidewalls of fiber and foil composition. These cans are designed to withstand internal pressure developed by the leavening agents plus a safety factor which permits unavoidable processing variations in leavening strength and provides for additional pressure resulting from gas production by yeast and bacteria contaminants during storage. However, failure or rupture of the can body will ultimately occur after lengthy storage as the pressure resulting from gas production exceeds the strength of the can wall. Failure may be shown as an actual explosion of the can or it may be more gradual with a progressive breaking of the layers of the can body. In either case, total failure may be preceded by extrusion of a syrupy exudate from the interior onto the surface of the can. This exudate is also a result of bacterial action. Any of these phenomena make the can of dough unsalable.

In nonrefrigerated packaged doughs, batters, puff pastes and the like, the problem of microbial spoilage and deterioration of the product is even more aggravated. In the manufacture of bakery products, even under the most stringent sanitary conditions, the result is a final unbaked product contaminated with microflora which are capable of multiplying even at refrigerated temperatures (40° to 50°F). In time these grow to numbers sufficient to bring about deterioration of the dough. Side reactions brought about by the microbial growth release gases, lead to increased internal can pressure with subsequent exudation and can ruptures. Also, the microorganisms because of their enzyme activity cause deterioration of the quality of the product.

In order to overcome these difficulties a small amount of an antibiotic peptide is used by *S.A. Matz, E.G. Linke and D.E. Mook; U.S. Patent 3,295,989; January 3, 1967; assigned to The Borden Company*. The antibiotic peptide used is selected from the group consisting of nisin and subtilin. The levels of peptide added varies dependent upon the microbial content of the product. Thus, doughs prepared under the most stringent conditions will require only about 30 units of nisin per gram of final bakery product to achieve the necessary storage life. This appears to be a minimum amount of nisin with the most commercially suitable amount which takes into account the average contamination of bakery products being about 100 units of nisin per gram of final unbaked product. Equivalent amounts of the subtilin are used. Nisin and subtilin have no effect on the unbaked bakery products and do not alter any of the characteristics, such as texture, taste, etc., after they have been baked.

Dough, batter or puff paste is made in the conventional manner by mixing the ingredients, developing the product if needed, and shaping the product as desired. The antibiotic peptide is added to the dry ingredients just prior to the addition of the water and thoroughly mixed to distribute it substantially uniformly through the product. The dough is then placed in a container, allowed to proof if necessary, and then transported and sold in the normal channels of commerce. The purchaser need only open the container and bake the product.

Example 1: Two cereal doughs were made of the following formula:

	Parts
Flour	100
Sucrose	6.25
Salt	1.6
Shortening	6.25
Nonfat milk solids	6.25
Sodium bicarbonate	2.0
Sodium acid pyrophosphate	3.5
Roll-in shortening	30.0
Water	55.5

Both were made by mixing the dry ingredients in a standard mixing bowl used for dough development. The shortening was blended with the dry ingredients and then the water was added. The dough was mixed until thoroughly developed and the roll-in shortening incorporated into the dough in such a manner as to give a dough with a laminar structure composed of alternate layers of dough and shortening. The dough was sheeted out to about ½ inch thickness, dusted with rice powder and cut into hexagonal pieces. The pieces were placed in cans made to handle preleavened doughs, allowed to proof, and stored.

Batch 1 was stored at 70° to 75°F with no treatment whatever. Batch 2 was identical to Batch 1 except that the antibiotic nisin was added to the dry ingredients just prior to the addition of water at a level of 100 units per gram of final product. A total of 40 cans per batch was stored at 70° to 75°F. All 40 cans of the untreated batch exploded between the third and fifth week of storage. All 40 cans of dough treated with nisin remained perfectly intact over a full 8-week storage period. There was no exudate on the outside of the cans, and there were no points of rupture on the cans themselves. The cans used were those conventionally used for unbaked refrigerated products having metal ends and sidewalls of fiber and foil laminated construction.

Example 2: Two cereal doughs were made of the following formula:

	Parts
Flour	100
Cerelose	3.75
Salt	3.5
Shortening	12.5
Nonfat milk solids	4.0

(continued)

	Parts
Sodium bicarbonate	2.0
Sodium acid pyrophosphate	1.0
Fumaric acid	3.5
Water	61.0

Both were made using the same procedure as described in Example 1, except that no roll-in shortening was used and no laminar structure formed. Batches 1 and 2 were identical, except that Batch 2 contained 100 units of nisin per gram of final product. Both batches were stored at 55°F. All five cans of Batch 1 exploded by the fourth week of storage. The 5 cans of Batch 2 were intact after 8 weeks' storage, with only a few places on some of the cans showing incipient rupture.

COLOR STABILIZATION

Citric Acid-Ascorbic Acid

Semolina, fabrina, durum wheat flour or unbleached flour milled from hard spring wheat or hard winter wheat are used for the preparation of macaroni and noodle products. In preparing such products the ingredients which may be one or more of the above mentioned wheat materials and enriching substances, or other common additives, such as vitamins, salt and disodium phosphate, are dry mixed and then made in a dough or paste with water or milk, extruded into the desired shape and dried.

This processing, as well as storage of the flour or other wheat material after milling, is attended by a loss of the golden yellow color of the unbleached milled products of wheat, particularly durum. This desirable bright yellow color is identified with the absence of brown pigments and with the presence of a relatively high carotenoid pigment content, mainly Xanthophylls. The resulting macaroni and noodle products prepared from these materials consequently do not possess their characteristic and desirable lemon yellow color, which is indicative of a better quality in the finished products.

During the processing of the wheat milled products, some brown pigment is formed because of a complex chemical reaction as yet not completely understood. Furthermore, some carotenoid pigment is destroyed because of lipid oxidation reactions. The brown color arising from brown pigments formed in the finished products is undesirable, as is the loss of carotenoid pigments through oxidation. Both these effects are intensified on the aging of wheat and/or its milled products and are more pronounced in wheat and its products than are high in moisture. Oxidative loss of carotenoid pigment is greater when a milled product is derived from hard spring or hard winter wheat because such wheats have a higher pro-oxidant stress due to increased amounts of the enzyme lipoxidase.

Unbleached flour or other milled wheat material used for macaroni and noodle products may be stabilized against color loss due to the loss of carotenoid pigments by the incorporation of a small quantity of L-ascorbic acid. As yet, however, there is no known method of stabilizing wheat products against color loss due to brown pigment formation.

It has been found according to L.K. Dahle; U.S. Patent 3,503,753; March 31, 1970; assigned to Peavey Company that unbleached flour, semolina or other milled wheat products used for macaroni and noodle products may be stabilized against the color change caused by brown pigment formation by the incorporation of small, but effective, quantities of citric acid, the amount required varying with the amount of browning potential in the particular milled product stabilized. It has been further found that small, but effective, quantities of citric acid enhances the effect of ascorbic acid in the stabilization against carotenoid pigment loss.

Wheat crops vary from season to season, from region to region and even with a particular region; therefore, the particular amount of citric acid necessary to accomplish the desired result must be determined with respect to the particular wheat products sought to be stabilized. Normally, the incorporation of no more than about 0.1% of citric acid will accomplish the desired result. In some instances as little as 0.03% or less will be effective to stabilize a wheat product against browning. In cases where both citric acid and ascorbic acid are added to stabilize the wheat product against both brown pigment formation and carotenoid pigment loss, the same factors above must be considered. Normally, quantities of about 0.03 to 0.1% of citric acid are necessary.

Other substances may be used which supply citric acid in the quantities above indicated and which do not exert any deleterious effect on the flour or the macaroni and noodle product prepared from it. Furthermore, it must be pointed out that citric acid does not achieve the results produced by the addition of ascorbic acid only to wheat products. Furthermore, it is to be understood that the addition of ascorbic acid alone to wheat products does not achieve the results produced by the addition of citric acid to milled wheat products. Rather, the presence of each acid enhances the effect of the other in stabilizing the milled wheat product against color loss; the presence of ascorbic acid inhibiting carotenoid pigment loss and the presence of citric acid inhibiting the brown pigment formation.

Any desired method for incorporating citric acid and ascorbic acid in the wheat products may be employed. For example, citric acid may be mixed with a small portion of the milled wheat product to be stabilized followed by incorporating this small portion into the bulk of the milled product with mixing and agitating; the proportion of citric acid to wheat product being selected so that the mixing of the smaller portion of wheat product with a predetermined larger amount of wheat product will yield a final product containing the ultimately desired percentage of citric acid. However, any other desired method such as spraying a solution on the milled stock may be used to add the citric acid.

Wheat products treated during periods of storage retain the characteristic golden yellow color. Upon subsequent preparation of dough from the mixture by the addition of water in a mixing chamber followed by extrusion through dies to form shaped units of dough and then drying, the retained color of the wheat provides the desired yellow shade in the final product.

The process is applicable where various enriching ingredients are utilized such as vitamins and minerals, e.g., B vitamins (riboflavin, thiamine, niacin, niacinamide), vitamin D, iron and calcium.

Example 1: Citric acid in the form of a fine powder is first diluted by mixing with semolina [prepared by grinding and bolting cleaned durum wheat to such fineness that it passes a 20 mesh sieve (U.S. standard screen) but not more than 3% passes a 100 mesh sieve, the wheat being bran coat and germ freed to an ash content on a moisture freed basis of not more than 0.92% and having a moisture content of not more than 15%]. The proportions of citric acid and semolina are such that when one ounce of the resulting diluted mixture is added to 100 lb of semolina, a citric acid content of 0.001% is obtained. The dilute mixture is added to the bulk of semolina with the aid of a powder feed, and the product is subjected immediately to complete mixing for uniform distribution.

During periods of storage, the golden yellow color of the durum wheat is retained. The treated product was then processed into spaghetti in the conventional methods currently employed. The spaghetti was then tested for color, cooking quality and taste. The data obtained from tests on spaghetti processed from flours treated as above containing varying amounts of citric acid are tabulated below:

Citric acid, p.p.m.	Color		Cooking quality	Taste
	Brightness index	Visual appearance		
0	10.0	Dull yellow	Satisfactory	Good.
500	10.5	Bright vivid yellow.	do	Do.
1,000	10.0	do	do	Do.

Measurement of the carotenoid pigment content indicated that the presence of citric acid did not affect the oxidative loss of carotenoid pigments and, thus, the increasingly vivid and brighter color is attributed to inhibition to the browning reaction. Citric acid did not alter the cooking quality or the taste of the spaghetti at the levels employed.

Example 2: This example shows the effect of citric acid on the brown pigment forming reaction which occurs when ground wheaten materials are processed into macaroni and noodle products. An indication of the extent of browning is given by the absorbance at 275 mμ of a 95% ethanol extract of ground spaghetti as measured by a Beckmann D. U. spectrophotometer. This test is more fully explained in an article by R.J. Stenberg, et al, entitled "Studies on Accelerated Browning in Starch Pastes Containing Various Bread Ingredients" in *Cereal Chemistry,* volume 37, pages 623 through 637 (1960). Basically, the test involves grinding spaghetti made from ground wheaten products and extracting with a 95% ethanol solution and determining its absorbance at a wave length of 275 mμ. The higher the absorbance at this particular wave length, the greater the extent of brown pigment formation.

In this example, citric acid, ascorbic acid and a combination of the two were incorporated in durum flour which was then processed into spaghetti in the same manner as Example 1. The spaghetti was ground and subjected to the above described measurement for browning. The data is tabulated below:

Sample	Citric acid p.p.m.	Ascorbic acid, p.p.m.	Absorbance 275mμ
1	0	0	0.629
	0	100	0.625
	300	0	0.602
	500	0	0.600
2	300	100	0.613
	0	0	0.650
	400	200	0.560

From this data, it is apparent that brown pigment formation is at a maximum where there is no citric acid present. It is also apparent that the presence of ascorbic acid in the absence of citric acid likewise has no effect on the brown pigment formation. However, where citric acid is present and where citric acid in combination with ascorbic acid is present, there is a reduction in brown pigment formation.

Example 3: Ascorbic acid alone and a combination of ascorbic acid plus citric acid were incorporated in an aged flour of high oxidative stress and high susceptibility to browning. The samples were processed into spaghetti in the same manner as in Example 1 and tested for pigment content. The color difference due to browning was readily apparent. The data is tabulated below:

Ascorbic acid p.p.m.	Citric acid p.p.m.	Carotenoid Pigment retained, percent	Color
0	0	55.0	Dark brown.
200	0	64.0	Do.
200	400	73.0	Light brown.

The desirable yellow color, as reflected by the amount of carotenoid pigment retained in the spaghetti after processing, was best in the sample containing ascorbic acid plus citric acid, thus indicating an enhancing effect of citric acid on ascorbic acid. The lighter brown color observed in the sample containing ascorbic acid plus citric acid is attributable to inhibition of the browning reaction effected by the citric acid.

Example 4: Flour was milled from durum wheat of high sprout damage. This flour is susceptible to browning and to oxidative loss of carotenoid pigment during aging and processing into spaghetti. Two samples of this flour were processed into spaghetti in the method of Example 1; one a control containing no additives and the other containing 100 ppm of ascorbic acid plus 1,000 ppm of citric acid. The resultant spaghetti was compared for color and carotenoid pigment content with the following results:

Additive	Color	Carotenoid Pigment Content, ppm
None	Dark, dull yellow	5.00
Ascorbic acid, 100 ppm, citric acid, 1,000 ppm	Bright yellow	5.33

The sample containing ascorbic acid and citric acid produced spaghetti of a brighter yellow color which is attributable to the inhibition of the browning effected by the citric acid. Furthermore, the sample produced spaghetti of a higher carotenoid pigment content which is attributable to the antioxidant effect of ascorbic acid and citric acid.

Ascorbic Acid for Macaroni

It is customary to use wheaten materials such as semolina, farina, durum wheat flour, or unbleached flour milled from hard spring wheat or hard winter wheat for the preparation of alimentary pastes, such as macaroni and noodle products. In preparing such products, wheat materials and enriching substances, or other common additives, such as vitamins, salt, and disodium phosphate, are dry mixed and then made in a dough or paste with water (or milk), extruded into desired shape and dried. This processing, as well as storage of the flour after milling, is attended by a loss of desirable characteristic golden yellow color of the unbleached wheat and wheat flour. This desirable color is identified with the relatively high carotenoid pigment content of these wheat materials. The resulting macaroni and noodle, or other alimentary paste food products consequently do not possess their characteristic and desirable lemon yellow color, which is indicative of a better quality in the paste products.

Avoidance of this color change can be provided by incorporation of l-ascorbic acid according to *F.D. Schmalz and N.W. Risdal; U.S. Patent 3,043,699; July 10, 1962; assigned to F.H. Peavey & Company*. The ascorbic acid addition at the same time improves the cooking qualities of the alimentary paste by increasing stability to deformation of the shaped pieces when cooked, and by improving its resistance to stickiness. It was determined that as little as 0.001% ascorbic acid based on the weight of the flour prevented the loss of color and conversely resulted in retention of the desirable golden yellow color during storage and alimentary paste product preparation.

In view of variations occurring in durum wheat and the other wheat raw materials used, due to growing conditions and other varietal influences, the most effective quantities of ascorbic acid may vary from even less than 0.001 to somewhat greater than 0.005% (from about 10 to about 50 parts per million of flour). Quantities as high as 0.01% are not required for the purpose intended and may adversely affect the cooking quality of the alimentary paste product. An effective amount below 0.01% is suitable and 0.005% or less is preferred.

Although l-ascorbic acid as such is preferred, other substances may be used which supply this compound $C_6H_8O_6$ and which do not exert any deleterious effect on the flour or the alimentary paste food product prepared from it. However, salts of l-ascorbic acid have been found not suitable for the purpose.

Example: l-ascorbic acid in the form of an ultra fine powder is first diluted by mixing with semolina [prepared by grinding and bolting cleaned durum wheat to such fineness that it passes a 20 mesh sieve (U.S. standard screen) but not more than 3% passes a 100 mesh sieve, the wheat being bran coat and germ freed to an ash content on a moisture-free basis of not more than 0.92% and having a moisture content of not more than 15%]. The proportions of ascorbic acid and semolina are such that when one ounce of the resulting diluted mixture is added to one hundred pounds of semolina, an ascorbic acid content of 0.001% is obtained. The dilute mixture is added to the bulk of the semolina with the aid of a powder feeder, and the product is subjected immediately to complete mixing and agitation. During periods of storage, the golden yellow color of the durum wheat is retained. Upon subsequent preparation of a paste from the mixture by addition of water in a mixing vat followed by extrusion through dies to form shaped units of dough and drying, the retained color of the durum wheat provides the desired yellow shade in the final product.

The process is applicable to any wheaten product used in the manufacture of alimentary pastes, such as the macaroni and noodle products. Also, the process is fully applicable where various enriching ingredients are utilized, such as vitamins and minerals, e.g., B vitamins (thiamine, riboflavin, niacin or niacinamide), vitamin D, iron and calcium. It is noted that ascorbic acid and alkali metal ascorbates have been utilized for their physiological vitamin C value and to increase baking strength in flour for bread making. In this process, however, the ascorbic acid functions differently to stabilize against loss of the desirable golden yellow color in unbleached flour or other wheat products used for making alimentary paste products.

POLYVINYL PYRROLIDONE ADDITIVE

It has been observed that when dry prepared mixes are aged certain changes occur in the mixes which affect the ultimate quality and appearance of the goods baked therefrom. It has further been observed that some dry prepared cake mixes tend to degrade in batter making and baking procedures as evidenced by the loss of moisture and flavor or the resultant baked product. Additionally, it has been observed that the mixes dry out and lose flavor during periods of storage. In general, the lack of sufficient keeping qualities of dry prepared mixes results in baked products that are unattractive in appearance, dry and, in some instances, tasteless.

The process of *H.H. Tiedemann; U.S. Patent 3,153,595; October 20, 1964; assigned to General Aniline & Film Corporation* comprises dry prepared cake mixes containing sugar, farinaceous materials and shortening, wherein the mixes contain uniformly distributed polyvinyl pyrrolidone. The polyvinyl pyrrolidones are prepared by the bulk or solution polymerization of N-vinyl pyrrolidone using any of the well-known conventional catalysts such as peroxides, ultraviolet light and heat to form polymers of varying molecular weights. The polymers so produced are usually available as a white or nearly white powder which is readily soluble in water and many common organic solvents.

The amount of polyvinyl pyrrolidone may vary for purposes of broadest usage within the range 0.5 to 1.5% by dry weight of the total solids of the mix. Higher amounts may tend to impart a gummy or slimy impression in the mouth during mastication of the cakes prepared from such mixes. The polyvinyl pyrrolidone is in the form supplied by the manufacturer, namely, as a white powder or aqueous solution. It is not necessary that these particles be dissolved in water; in fact, it is more satisfactory for the distribution of the polyvinyl pyrrolidone to add the same in the shortening since the shortening is preferably reduced to a liquid form by heating prior to batter making. Except for the addition of polyvinyl pyrrolidone the mixes are formulated in the same ways and with the same ingredients that are commonly used. The procedure used in compounding the mix may be that which is generally observed. Essentially, the sequence of steps involves:

(1) The major ingredients, i.e., flour, sugar and polyvinyl pyrrolidone containing shortening, are blended into a homogeneous premix;

(2) The premix from step 1 is passed through an impact grinder to eliminate lumps or agglomerates;

(3) The delumped premix is subjected to the shearing and crushing treatment;

(continued)

(4) The minor ingredients are uniformly incorporated; and
(5) The total mixture is subjected to impact grinding to eliminate lumps from final product.

Example: A typical yellow cake of the basic formulation containing 1.0 weight percent of polyvinyl pyrrolidone based on the dry weight of the total solids is prepared as follows:

Ingredients	Parts by Weight
Sugar, industrial fine	43.5
Cake flour	40.5
Emulsified shortening	11.0
Nonfat milk solids	1.5
Sodium bicarbonate	0.9
Monocalcium phosphate	0.4
Sodium acid pyrophosphate	0.8
Salt	0.7
Dextrose	0.3
Flavoring	0.2

After baking, the cake will be characterized by greater moistness and of satisfactory eating quality.

PHOSPHATES FOR BREAKFAST CEREALS

In the investigations to develop anticariogenic agents for sugar coated ready-to-eat breakfast cereals, *J.C. Muhler; U.S. Patent 3,467,529; September 16, 1969; assigned to Indiana University Foundation* found that the incorporation of at least a minor, but anticariogenically effective, amount of monosodium dihydrogen phosphate, NaH_2PO_4 improves the oxidative stability and other properties of the product.

When monosodium dihydrogen phosphate, NaH_2PO_4, is incorporated in a comestible, especially sugar-containing comestibles and sugar, at a level of at least about 0.2% by weight, the cariogenic potential is substantially reduced and the overall dental health of the individual is enhanced. Indeed, the regular ingestion of a tablet or capsule comprising at least about 50 mg monosodium dihydrogen phosphate, NaH_2PO_4, provides an effective method of reducing the incidence of dental caries, particularly in individuals maintained on a dietary regimen having a high cariogenic potential.

The anticariogenic agent may be added directly to the comestible. Where a comestible is sugar-coated at least a part of the NaH_2PO_4 is provided in close proximity to the sugar coating. Thus, at least a part of the agent may be provided in the form of a coating in close proximity to the sugar coating. Substantially any method may be used to provide a food product in which sugar and NaH_2PO_4 are released together in the mouth to give a significant, system anticariogenic effect.

The orthophosphates which are operable to reduce the oxidative rancidity of breakfast cereal products are members of the class consisting of monosodium dihydrogen phosphate, NaH_2PO_4, disodium monohydrogen phosphate, Na_2HPO_4, and mixtures. In the event mixtures or blends of the phosphates are used, it is preferred that they be blended in proportions from about 76 to 74% of the monosodium salt with 24 to 53% of the disodium salt. The total phosphate content may vary from 0.5 to about 1.5%, the percentage being by weight and based on the weight of the total composition.

In order to determine the stability of a breakfast cereal product against oxidative rancidity, the following test procedure was carried out. Samples to be tested were sealed in a glass jar with a metal screw cap. Three such samples were placed in an oven maintained at a constant temperature of 113°F. Each sample was tested daily for rancid odors. The time taken for each sample to become rancid was noted. When all three of the samples had become rancid, the average time in days was calculated. Results are reported as the number of days necessary for samples to become rancid at the temperature of the test. The following examples illustrate the effectiveness of inorganic orthophosphates in increasing the stability of breakfast cereal products against oxidative rancidity.

Example 1: A cereal dough was prepared by mixing 30 pounds of 80% oat flour, 20% corn flour blend with 15.9 pounds of a flavoring syrup having the following composition:

	Percent
Water	75.92
Liquid sugar (69% solids)	6.63
Salt brine (saturated)	17.38
Color	0.07

After thorough mixing, the material was extruded, pelletized into alphabet letter form, and dried for 4 minutes at 260°F. The pellets were then tempered for about 20 minutes and puffed in a cereal puffing gun at 170 psi after preheating for 56 seconds at 350°F. Liquid brown sugar of 67.5° Brix was heated to 232°F and poured over a 5-pound quantity of the puffs as prepared as described above. They were allowed to tumble in a reel heated to 200°F until they became free-running. The product was then dried for 4 minutes at 280°F.

The above example was repeated except that 137.1 grams NaH_2PO_4 and 170.9 grams Na_2HPO_4 were added to the flavoring syrup used to prepare the cereal dough. In addition, 21.6 grams of NaH_2PO_4 and 7.3 grams of Na_2HPO_4 were added to the brown sugar syrup used to coat the puffs. The resulting product contained 1.0 weight percent of total phosphate, dry basis, from a blend of 47% monosodium salt and 53% disodium salt. When subjected to the oxidative stability test as described above, the cereal puffs which did not contain phosphate became rancid after 29 days, whereas the phosphate containing samples became rancid only after 51 days.

Example 2: A cereal dough was prepared by mixing 30 pounds of a blend of 80% oat flour, 15% corn flour, and 5% wheat flour with 15.9 pounds of the flavoring syrup described in connection with Example 1 above. After thorough mixing, the flavored dough was extruded, pelletized into the desired form, and dried for 4 minutes at 260°F. The pellets were then tempered for about 20 minutes and puffed in a puffing gun at 170 psi after preheating for 56 seconds at 350°F. Five pounds of a liquid brown sugar of 67.5° Brix was heated to 232°F and poured over 5 pounds of the puffs and prepared as described above. The puffs were then allowed to tumble in a reel heated to 200°F until they became free-running. The product was then dried for 4 minutes at 280°F.

A second cereal sample was prepared in accordance with the same procedure, except that 137.1 grams of NaH_2PO_4 and 170.9 grams of Na_2HPO_4 were added in the flavoring syrup used to prepare the cereal dough, and 21.6 grams of NaH_2PO_4 and 7.8 grams of Na_2HPO_4 were added in the brown sugar coating syrup. The resulting product contained 1.0 weight percent of total phosphate, dry basis, from a blend of 47% NaH_2PO_4 and 53% Na_2HPO_4. When subjected to the oxidative stability test as described above, the cereal containing the phosphate added was stable for a total of 31 days. In contrast, the control sample gave evidence of rancidity after 27 days.

Example 3: 1,700 pounds of corn grits was placed in a rotating pressure cooker with 632 pounds of a flavoring syrup having the following formulation:

	Percent	
Water	33.53	
Liquid sugar (69% solids)	31.79	(continued)

	Percent
Malt syrup	2.15
Iron sulfate	0.02
Salt brine (saturated)	32.51

The mixture was cooked for 2½ hours at 18 psig using live steam. After cooking, the grits were dried at 180°F to a moisture content of 18%. They were then tempered for 5 hours, flaked and toasted at 500°F for 90 seconds. The toasted flakes were then cooled to room temperature. The procedure described above was repeated using in the flavoring syrup 12.87 pounds of NaH_2PO_4 and 4.07 pounds of Na_2HPO_4. The resulting product contained 1.0 weight percent of total phosphate, dry basis, from a blend of 76% NaH_2PO_4 and 24% Na_2HPO_4. Samples of the corn flakes with and without added phosphate were subjected to the oxidative stability test described above. The control sample became rancid at the end of 25 days. The sample containing added phosphate was stable for 30 days.

PRESERVATIVES FOR TORTILLAS

Aluminum Hydroxide Gel

Many expedients which apply for preventing staling and preventing microbiological spoilage to the usual wheat, whole wheat or rye bread, commonly eaten, are not effective when applied to tortillas, because of the marked differences in their composition from the breads mentioned above. Tortillas, when normally prepared without additives of any kind, have a maximum shelf life of 12 to 15 hours. After such time they are spoiled by microorganisms and become hard or stale.

It is known that tortillas when kept under conditions in which no moisture is lost, nevertheless become hard and inflexible with the passage of time and break or crumble easily when flexed or bent. This effect increases with time. Freshly made tortillas are very flexible but lose their flexibility with the passage of time. Hardening is appreciable after 24 hours, marked after 48 hours and almost complete after 72 hours if the product is kept at room temperature. It should be noted that the hardening or staling effect increases with decreasing temperature until the freezing point of water in the product is reached. At temperatures below room temperature but above the freezing point of water in the tortillas, therefore, hardening proceeds at a faster rate than at room temperature and vice-versa.

In determining the flexibility index a tortilla is bent around a bar of known radius, and this is tried with successively smaller bars until a bar is found which is the smallest around which the tortilla just breaks when it is flexed. A more flexible tortilla will just break when it is bent around a smaller bar than a less flexible tortilla. A higher flexibility index corresponds therefore to bars of lower radii and indicates higher flexibility.

TABLE 1: VARIATION OF THE FLEXIBILITY INDEX OF TORTILLAS WITH TIME AT ROOM TEMPERATURE

Elapsed Time	Flexibility Index
0 hr	8.5
24 hr	7.7
48 hr	6.5
72 hr	6.0
96 hr	5.5

Table 1 gives typical values. Hardening or loss of flexibility of tortillas is believed to be due to a physico-chemical change in the starch constituent of tortillas which is known as retrogradation.

An additive is reported by *M.J. Rubio; U.S. Patent 3,709,696; January 9, 1973; assigned to Roberto Gonzalez Barrera, Mexico* to retard the loss of flexibility of tortillas in storage and at the same time increase the resistance to microbiological spoilage. The additive is aluminum hydroxide gel.

Typical moisture contents of dough and tortillas and typical shelf lives are given in the table below:

TABLE 2: LIMED CORN DOUGH AND TORTILLAS WITHOUT ADDITIVES OF ANY KIND

Product	Moisture Content	Shelf Life at 25°C
Dough	55 – 60%	6 hours
Tortillas	42 – 48%	12 hours

The shelf life is the time required to detect unmistakable signs of bacterial spoilage in the product. These signs include production of off-flavors and odors, production of rope (a polysaccharide resulting from the growth of certain bacteria such as *B. mesentericus*) and appearance of moldy spots. Obviously, the shelf life of a product depends upon the temperature at which it is stored, as higher temperatures, normally produce lower shelf life and vice versa. Also, the type of spoilage which first occurs depends on the temperature. In the limed corn dough and tortillas the first sign of spoilage at higher temperatures (above 30°C) are usually off-flavors due to the growth of bacteria, while below 30°C the first signs are usually moldy spots. Table 3 gives typical shelf life of limed corn dough and tortillas at different temperatures.

TABLE 3: TYPICAL SHELF LIVES OF LIMED CORN DOUGH AND TORTILLAS AT DIFFERENT TEMPERATURES

Product	Storage Temperature	Shelf Life
Dough	37°C	3 hours
	25°C	6 hours
	15°C	12 hours
Tortillas	37°C	6 hours
	25°C	12 hours
	15°C	18 hours

It has been known that the yields of tortillas depend upon the ability of the dough or of the tortillas to retain water. The additive increases the water-binding ability of the dough and of the tortilla.

When tortilla dough is prepared from limed corn or limed corn flour, it has a certain consistency. When the additive is mixed with the dough, unless more water is added, the dough becomes stiffer and in order to obtain the consistency of tortilla dough without the additive, it is necessary to add more water to the dough. Thus, the yield of dough obtained per unit weight of limed corn or limed corn flour is increased. In the case of tortilla dough and tortillas the yield is the number of kilos of dough per kilo of corn, or the number of kilos of tortillas per kilo of corn, or the number of kilos of dough per kilo of limed corn flour or the number of kilos of tortillas per kilo of limed corn flour.

Similarly, when tortillas are prepared from ordinary limed corn dough, a certain amount of water is lost when they are cooked. If, however, tortillas are prepared from limed corn dough which contains the additive, an amount of water which is less than that from ordinary dough will be lost from the tortillas when they are cooked. The resulting yield of tortillas obtained per unit weight of limed corn or limed corn flour will be increased.

Thus, the additive increases the yield of dough made from limed corn or limed corn flour and increases the yield of tortillas made from limed corn or limed corn flour. The additive may be introduced into the dough in any of several different ways. It may be added to the dough as an aqueous suspension or dispersion which is thoroughly mixed in the dough to form a uniform distribution of the additive. Since this involves the addition of water to the dough, allowance must be made for the water incorporated with the additive, and compensation may be required in respect to the quantity of other water added. If the dough is made from limed corn flour, the additive may be dispersed or suspended in the water which is subsequently to be mixed with the flour in order to make the dough.

Table 4 shows typical values for the flexibility of tortillas after different times at room temperature with different additions of aluminum hydroxide gel. It will be noted that the flexibility is markedly increased by increasing additions of the additive.

TABLE 4: FLEXIBILITY INDEX OBTAINED WITH USE OF WET, FRESHLY PREPARED ALUMINUM HYDROXIDE GEL

Dose of Gel Based on Weight of Tortillas	- - - - - - - - - - - Flexibility Index - - - - - - - - - - - -				
	0 hr	24 hr	48 hr	72 hr	96 hr
0.0% (Control)	8.5	7.7	6.5	6.3	6.0
0.5%	8.8	8.3	8.0	7.8	7.5
1.0%	9.3	8.8	8.6	8.3	8.0
2.0%	9.8	9.3	9.0	8.7	8.5

Aluminum hydroxide gel has been found to be one of the most effective agents for retaining the flexibility of tortillas during storage. Aluminum hydroxide gel also greatly promotes the yield of tortillas as shown in Table 5. Both the dough and the tortilla yields were increased markedly as the doses of additive were increased where the tortillas were made from limed corn flour.

TABLE 5: YIELDS OBTAINED USING WET ALUMINUM HYDROXIDE GEL, FRESHLY PREPARED

Dose of Gel Based on Weight of Tortillas	Yield of Dough*	Yield of Tortillas*
0.0% (Control)	2.258	1.753
0.5%	2.314	1.803
1.0%	2.377	1.842
2.0%	2.472	1.924

*Kilos per kilo of flour.

Drying of hydrophilic inorganic gels such as aluminum hydroxide gel reduces their water-binding capacity considerably so that they lose most of their effectiveness. For this reason the gels must be freshly prepared at the time they are inserted in the tortillas and must be retained wet. Since the aluminum hydroxide gel contains an appreciable amount of water, the presence of this water must be considered and compensated for in preparation of the dough.

Example 1: Aluminum hydroxide gel is prepared by reacting 100 grams of sodium aluminate in a 3.33% solution with 50 grams of hydrochloric acid in a 20% solution and adjusting the final pH to a value of approximately 4. The gel, still wet, is separated from the supernatant liquid by filtration or decanting and kept wet until used in the tortillas. It is important that the final pH be on the acid side so that it will not affect the color of the tortillas. The pH range should be between 4 and 6. The pH should not be so acid that it will render the tortillas inedible.

Example 2: As an alternate procedure for making aluminum hydroxide gel, 100 grams of aluminum chloride in a solution having a concentration of 3.33% by weight and 90 grams of sodium hydroxide in a solution of 20% by weight are reacted together throwing down aluminum hydroxide as a precipitate.

The final pH is adjusted to a value of approximately 4, and the aluminum hydroxide gel is separated either by decantation or filtration and maintained wet continuously preparatory for use in the tortilla dough.

Example 3: Nixtamalized corn flour is mixed with 140% on the dry weight of water including 1% of the dry weight of aluminum hydroxide gel freshly prepared and added with the water.

The ingredients are mixed in with the water in a dough mixer until the mix is homogeneous and the dough is sold as dough in containers that prevent loss of moisture or cooked into tortillas at a temperature of about 290° to 410°F at sea level.

Example 4: Tortilla dough is mixed in a dough mixer by introducing nixtamalized corn flour in 100% of the weight of the dry ingredient of water and after the dough is completely homogenized, 1% of aluminum hydroxide gel in 40% of water, both of the weight of the flour, is added in the dough mixer and incorporated in the dough.

In Examples 3 and 4 the tortillas when stored under conditions in which loss of moisture is prevented, retain their flexibility unusually during storage as set forth in Table 2 and the yield is improved as set forth in Table 3.

As shown in the following table, ammonium hydroxide gel promotes the effectiveness of additives to control microbiological spoilage very effectively. The example is given of propionic acid as such an additive but similar results are obtained in the other additives referred to.

Table 6 shows the effect of combinations of propionic acid and aluminum hydroxide gel of pH 4 in increasing the shelf life of tortillas at 37°C.

TABLE 6:

Additive Combination and Dose	Shelf Life, hours
0.2% propionic acid + 2% aluminum hydroxide	84
0.4% propionic acid + 2% aluminum hydroxide	192
Control 1, 0.2% propionic acid alone	72
Control 2, 0.4% propionic acid alone	144
Control 3, no additive	6

Epichlorohydrin

In another process by *M.J. Rubio; U.S. Patent 3,690,893; September 12, 1972; assigned to Roberto Gonzalez Barrera, Mexico* epichlorohydrin is added to tortilla dough to retard staling.

The additive may be introduced into the dough in any of several different ways. It may be added to the dough as an aqueous solution or suspension which is thoroughly mixed in the dough to form a uniform distribution of the additive. Since this involves the addition of water to the dough, allowance must be made for the water incorporated with the additive, and compensation may be required in respect to the quantity of other water added.

If the dough is made from limed corn flour, the additive may be dissolved or suspended in the water which is subsequently to be mixed with the flour in order to make the dough. It will be understood that where the additive acts to increase the yield, the total amount of water used to make the dough of a certain consistency is greater than if the additive was not employed. In some cases it is preferred to mix the additive with the dry limed corn flour prior to mixing the flour with the water.

Epichlorohydrin cross links or cross bonds with starch and is effective to increase the flexibility of tortillas after storage for a period of time without loss of moisture if incorporated in the dough in the range between 0.25% and 1% of the weight of the tortillas. Epichlorohydrin must be added under alkaline conditions at a pH of 8 to 10 in the dough. The table shows the increase in flexibility at room temperature of tortillas due to the addition of epichlorohydrin.

Dose of Epichlorohydrin Based on Weight of Tortillas, percent	Flexibility Index			
	0 hr	24 hr	48 hr	72 hr
0.0 (Control)	8.5	7.7	6.5	6.0
0.5	8.7	8.0	7.0	6.7
1.0	8.8	8.3	7.5	7.0

The epichlorohydrin may be incorporated directly into the tortilla flour which is used in making up the dough. In such case, the alkaline material, usually sodium hydroxide, is added dissolved in the water which is mixed with the flour in order to make up the dough. Again, the epichlorohydrin may be dissolved, together with the alkaline material, in the water which is subsequently mixed with the flour in order to make up the dough. Finally, the epichlorohydrin may also be dissolved, together with the alkaline material, in a portion of water which is subsequently mixed with dough previously made either directly from limed corn or from limed corn flour. In such case due compensation must be made for the water added to the dough in this manner.

Example 1: Nixtamalized corn flour is mixed in a dough mixer with about 130% of water on the weight of the dry materials, the water having incorporated within it 1% of epichlorohydrin on the weight of the tortillas, and sufficient sodium hydroxide to adjust the pH of the dough to a value of 9.5. The result is shown in the table.

Example 2: The procedure of Example 1 is followed except that the dough is made and then a 2% water solution of epichlorohydrin and sufficient sodium hydroxide to adjust the pH of the dough to 9.5 is incorporated into the dough mixer, due allowance being made for the water.

Polycarboxylic Acids

The process of preparing tortillas containing an aliphatic polycarboxylic acid as a preservative is described by *M.J. Rubio; U.S. Patent 3,694,224; September 26, 1972; assigned to Roberto Gonzalez Barrera, Mexico.* The additives may all be incorporated in the following ways:

(1) They can be added to the dough as an aqueous solution, dispersion or suspension thoroughly mixed with the dough. Allowance is made for any water incorporated with them in the water to be used in the dough.

(2) If the dough is made from limed corn flour, the additive may be dissolved, suspended or dispersed in the water which is subsequently to be mixed

with the flour to make the dough for the tortillas.
(3) In some cases it is preferable to mix the additive with the dry limed corn flour prior to mixing the flour with the water.

The additive is an acid which will lower the pH to 5.5 to 5 and which may be any edible organic or inorganic acid such as hydrochloric acid, sulfuric acid, citric acid and monocalcium phosphate in doses of from 0.01 to 4%. This must be used with an agent which has the property of inhibiting microbiological spoilage, the other agent functioning as an enhancer. A typical description of the agent which can be used with the acid of this process is a lower fatty acid having from 1 to 4 carbon atoms, its anhydride, the sodium, potassium or calcium salt, or the sodium, potassium or calcium diacetates in doses of 0.01 to 0.8%. An example of the effect is given in Table 1.

TABLE 1:

Additive Combination and Dose	Shelf Life, hours
0.15% propionic acid + 0.10% citric acid	60
0.15% propionic acid + 0.15% citric acid	72
0.20% propionic acid + 0.05% citric acid	72
0.20% propionic acid + 0.10% citric acid	96
Control 1, 0.15% propionic acid alone	36
Control 2, 0.20% propionic acid alone	48
Control 3, no additive	12

Edible organic aliphatic polycarboxylic acids and their anhydrides having 3 to 6 carbon atoms in their carbon chain can to advantage be incorporated in tortilla dough in concentrations of 0.25 to 2%, of the weight of the tortillas, and they increase the retention of flexibility when tortillas are stored without much loss of moisture. The results are shown in Table 2.

TABLE 2:

Dose of Additive Based on Weight of Tortillas	Flexibility Index			
	0 hr	24 hr	48 hr	72 hr
Adipic acid, percent:				
0.5	8.6	7.6	6.7	6.3
1.0	8.8	8.3	8.0	7.8
2.0	9.3	8.8	8.6	8.2
Control	8.5	6.5	5.7	5.7

Typical examples of such acids and anhydrides are citric, succinic, adipic and glutaric acids and anhydrides.

Example 1: Nixtamalized corn flour is mixed with water in a dough mixer, the water containing adipic acid dispersed therein to the extent of 2% of the tortillas. The product after cooking the tortillas produces the results shown in Table 2.

Example 2: Dry powdered adipic acid is mixed with nixtamalized corn flour to the extent of 2% of the tortillas, and incorporated with 120% of water as above set forth in a dough mixer. The results are the same as those set forth in Table 2.

SHELF STABLE PANCAKES AND WAFFLES

Microbiologically stable fully cooked pancakes and waffles, packaged without hot packing were developed by *M. Kaplow and R.E. Klose; U.S. Patent 3,753,734; August 21, 1973;*

assigned to General Foods Corporation. The batter used to produce the shelf stable pancakes and waffles is formulated on the principles of Aw, that is, the ability of the soluble solids of the batter to limit the amount of free water available to bacteria; the bacteria's inability to survive this condition; and the subsequent shelf stability or product stability obtained by virtue of this condition. An Aw value is a direct measure of water vapor which is a function of unbound water. It is determined by dividing the mols of water plus mols of soluble solids into the mols of water. The lower the Aw value, the more stable the pancake and waffle products against microbiological decomposition, e.g., 0.80 in a product indicates more stability than 0.90 in a similar type product.

To compute the complete Aw, the Aw lowering of the calculated ingredients are added together and subtracted from 1, 1 being equivalent to 100% water vapor or maximum water vapor which would be produced if none of the free water were bound by soluble solids. Thus a calculated Aw of 0.96 indicates an Aw lowering of 0.04.

The relative weight percent of water-soluble solids to the moisture content of the pancake and waffle products, when initially incorporated into the products during their manufacture and preparatory to packaging determines the ultimate functionality of the solids in providing the requisite bacteriostatic effect. Usually the level of moisture will range from 14% to 40%. The level of water-soluble solids may be varied as may the level of moisture initially incorporated within the desired ranges.

However, in varying these levels the relationship of the water-soluble solids in solution to the water should be controlled so as to afford the desired osmotic pressure. A good rule to observe in this connection is to be sure that the water-soluble solids available for solution are at least equal to the weight of moisture present, although in some cases it is possible that a lower level of water-soluble solids might afford some protection against microbiological decomposition provided an equivalent degree of osmotic pressure is available to protect the product. It will be found, however, that the water-soluble solids will constitute a major percent by weight of the pancake and waffle products.

As a precautionary measure against the growth of yeast and molds certain antimycotic agents are incorporated in the batter at sufficient levels to prevent the growth of such organisms. Sorbate salts such as potassium sorbate as well as sorbic acid can be used either separately or in combination. Propylene glycol which may be used alone or with other humectants like sorbitol to impart a degree of product softness or tenderness has also been found to serve as an antimycotic. The amount of antimycotic agent added is selected so as to produce the desired results and will constitute a minor proportion of the product, from about 0.1 to about 2.5% of the total weight, depending on the particular antimycotic and the particular product composition, although even lower levels in the order of 50 ppm can be used in the case of some antimycotics such as pimaricin.

Potassium sorbate in a water solution can be sprayed onto the surface of the pancake or waffle product, or the product can be dipped in this solution; other antimycotics lend themselves to such surface application as esters of the parabens (para-hydroxy benzoate) such as propyl and methyl parabens. Cellophane and other enwrapments for the food can be spray-coated with a sorbic acid solution but impregnation or dusting with sorbic acid or potassium sorbate is preferred. Antimycotics which can generally be used are benzoic acid, sodium benzoates, propionic acid, sodium and calcium propionate, sorbic acid, potassium and calcium sorbate, propylene glycol, diethyl pyrocarbonate, menadione sodium bisulfite.

The shelf stable pancake or waffle product has an appearance, texture, color and aroma not unlike conventional pancakes and waffles. The product can be packaged using a nonhermetic packaging material such as cellophane. When removed from its pouch, the moist and soft product may be warmed just prior to consumption using a toaster or grill. The product has sufficient cohesive strength so that when it is removed from the toaster or grill, it will not tear, nor adhere to the toaster or grill.

Edible polyhydric alcohols constitute the principal source of water-soluble solids of the Aw emulsion and may range from about 20 to 35% of the batter depending upon the particular polyhydric alcohol or polyhydric alcohol mixture, to provide the desired bacteriostatic protection. As the moisture content of the product increases in the intermediate moisture range, the level of a given edible polyhydric alcohol will correspondingly increase in order to maintain a sufficient bacteriostatic effect. The quantity of edible polyhydric alcohols chosen will vary depending upon the presence and level of auxiliary soluble solids which produce a similar increase in osmotic pressure to the batter. Thus a variety of low molecular weight polyhydric alcohols having two or more hydroxyl groups, including glycerol, sorbitol, propylene glycol, mannitol, mixtures thereof and the like may be used. The polyhydric alcohols further assist in depleting the moisture of the pancake and waffle products by substituting for a portion of the moisture present in the interior of the product and causing moisture transfer to the exterior thereof.

In general the shelf stable pancake and/or waffle product is formulated by blending milk and an edible polyhydric alcohol into pancake or waffle mix, adding eggs, shortening and an antimycotic to the mix to form a shelf stable batter, and cooking the batter on a griddle. If desired, syrup may be added along with the edible polyhydric alcohol and milk.

Example:

Ingredients	Parts by Weight	Percent
Pancake Mix*	140.0	29.4
Milk	120.0	25.3
Glycerol	140.0	29.4
Egg	60.0	12.6
Cottonseed oil	5.0	1.1
Propylene glycol	9.0	1.9
Potassium sorbate	1.5	0.3

*Pancake Mix

Ingredients	Percent
Wheat flour (Bleached)	43.0
Wheat flour (Unbleached)	20.0
Corn flour	16.0
Sucrose	5.6
Rice flour	5.0
Dextrose	4.0
Salt	2.4
Sodium bicarbonate	2.0
Sodium aluminum phosphate	2.0

Milk and glycerol were added to pancake mix in a Mixmaster bowl and blended for two minutes at medium speed. Next, egg, cottonseed oil, propylene glycol and potassium sorbate were added and the mixture was mixed at high speed for two minutes to form a homogeneous batter. The batter having a moisture level of 14 to 40% was then cooked on a griddle at about 375°F to form a dark brown product having the texture and appearance of conventional pancake products. The Aw of the batter was about 0.83, and the Aw of the cooked pancake product was about 0.74. Microbiological evaluation of the batter and the cooked pancake product of Example 1 show a standard plate count, mold count and yeast count of less than 10 when tested after four weeks storage in a nonhermetic package of 100°F. The test for salmonella under these conditions disclosed none present.

NONHYGROSCOPIC BAKING POWDER

Sodium aluminum phosphate (SAP) is a well recognized and approved food additive used, e.g., as a leavening acid in baked products, as a melt controlling additive in cheese, and as a meat binding agent. SAP has its major use in baking products, being utilized in biscuit mixes, pancake mixes, waffle mixes, cake mixes, doughnut mixes, muffin mixes, canned biscuits, self-rising flours and frozen rolls. Because of its compatibility with the highly efficient, lactylated mono- and diglyceride or propylene glycol monostearate type shortenings (presently being used in cakes), it has become a standard leavening acid for cake baking. Sodium aluminum phosphate is, however, an inherently hygroscopic material which will absorb a large quantity of atmospheric moisture, usually about 28 to 29% by weight.

Originally produced, SAP is a dry, white crystalline product. If permitted to stand exposed in a hot, humid atmosphere it rapidly absorbs moisture, first forming water droplets or caking at the surface, then becoming what may be termed a viscous semifluid. Commercially, this phenomenon is minimized somewhat by the use of sealed, airtight containers. Nevertheless, the precautions required are time consuming and expensive, and in practical applications, the problem remains a significant disadvantage.

J.E. Blanch and F. McCollough, Jr.; U.S. Patent 3,205,073; September 7, 1965; assigned to Stauffer Chemical Company found that if the original SAP molecule is modified by the introduction of potassium, a product results which has an extremely low level of hygroscopicity and which will remain substantially uncaked after long periods of exposure to humid, hot atmospheric conditions. Potassium atoms replace hydrogen atoms in the crystalline lattice of SAP. Moreover, the potassium compound used must be one capable of ionization. Addition of the potassium may be accomplished either during manufacture of the SAP, even before the SAP-forming reaction, or as a final step after preparation. The preferred potassium modified sodium aluminum phosphate of the process has all of the desirable reactivity, taste, and other baking characteristics of the prior art SAP plus the advantage of greatly decreased hygroscopicity.

The term sodium aluminum phosphate describes a crystalline compound of the empirical formula $NaAl_3H_{14}(PO_4)_8 \cdot 4H_2O$. Other terms by which this compound is known include sodium aluminum phosphate, tetrahydrate; regular sodium aluminum phosphate; and sodium aluminum acid phosphate. Often the compound is represented symbolically as NALP, SALP, $SAP \cdot H_2O$, $SAP \cdot 4H_2O$, or SAP. The term nonhygroscopic, as it is applied to the compositions of the process is intended to define a potassium modified form of the normally hygroscopic SAP which will not increase in weight by more than about 10% of its original weight during continued exposure at $35°C$ and 75% relative humidity for 140 hours. By comparison, the unmodified material increases in weight by about 28 to 29% under the same conditions of temperature and humidity.

The manufacture of SAP usually consisted of reacting a reactive aluminum compound such as the metal itself or its oxide or hydroxide, etc., with a stoichiometric or excess amount of phosphoric acid, adding sodium hydroxide or carbonate in an atomic ratio of one sodium to three aluminum, and concentrating the resulting viscous liquor until crystallization occurred. As water was removed during the concentration step, the liquor gradually became more viscous until finely divided particles of $NaAl_{14}H_{14}(PO_4)_8 \cdot 4H_2O$ were formed.

The potassium compound may be added at any stage during manufacture or after crystallization. Thus a potassium compound may be added to the phosphoric acid together with the sodium and aluminum ingredients or, in the alternative, after addition of the sodium and aluminum ingredients. Another method is to add potassium ion from a solution by spraying the solution on an agitated or thin static bed of SAP crystals. Spraying is preferably accomplished during crystallization while the SAP crystals are still moist, but may also be carried out after the crystals are fully dried.

When it is desirable to introduce large amounts of potassium ion into the original SAP molecules, addition of the potassium is usually accomplished before crystallization. The spraying technique is more applicable to the inclusion of small amounts of potassium at the surface of preformed SAP particles. Another technique which has been found satisfactory is to physically mix a dry potassium compound with preformed SAP particles. Generally, the potassium compound may be dissolved, if readily soluble, and added from solution, or it may be added in the dry form during manufacture or by physical mixture to dry SAP (whether soluble or insoluble). Potassium has been found to be readily reactive with SAP (and the viscous liquid phase present before crystallization) without any necessity for controlling the temperature. However, intimate contact between the potassium ion and the reactive SAP molecule is necessary as is the presence of an ionizing medium, such as water. In-process addition of a potassium compound during the manufacture of SAP will generally not require any extra mixing (over that normally provided) to achieve suitable dispersion of the potassium ion. On the other hand, dry physical mixtures require vigorous mixing.

The amount of potassium required to produce nonhygroscopic compounds and compositions is largely influenced by the method of addition. In this regard it is believed that the potassium may situate at the surface of SAP particles, or distribute throughout the particles, depending upon the particular method of preparation. For example, a substantially nonhygroscopic compound may be formed by the inclusion of as little as 0.1% by weight chemically bonded potassium, providing the potassium is added during the late stages of crystallization or thereafter.

This method of addition obviously concentrates a large percentage of the bonded potassium at the outer surface of the particles. By this method a baking acid may be prepared in which at least about 0.1% and not more than about 2.0% by weight potassium, based on the weight of the final product, is chemically bonded to SAP. A preferred baking acid contains between 0.3% and 1.0% by weight potassium. Larger quantities of potassium are generally required to achieve equivalent nonhygroscopicity where the potassium is added to the liquid phase before crystallization. As much as 10.0% by weight potassium may be used to effect a satisfactory hygroscopicity by this latter method.

Desirable acid-reacting compositions, suitable for use as baking acids, are prepared by intimately mixing unreacted SAP with a potassium compound, preferably an edible potassium salt. A preferred dry mix baking acid comprises from 0.5% to 4.0% by weight of potassium compound. Where desirable, a large excess of potassium compound, more than that which will effectively react with SAP, and up to about 20.0% by weight of the dry mix composition, may be added.

Furthermore, it is often desirable to use excess potassium by the other methods of addition described herein. From a consideration of the chemical formula of SAP (containing fourteen acidic hydrogen atoms) it can be readily seen that the calculated theoretical limit of chemically bound potassium cannot exceed about 37% of the total weight of the potassium modified sodium aluminum phosphate, based on a one for one exchange of K^+ for H^+. When manufactured for use as a leavening agent, the potassium modified sodium aluminum phosphate will normally not contain more than about 10% by weight of potassium as chemically bound potassium atoms.

The preferred nonhygroscopic baking acids of the process, those comprising at least about 0.3% and not more than about 1.0% potassium, based on the total weight of the product, will normally have neutralizing values within the range of 96 to 103. Regular SAP has neutralizing values of from about 99 to 103. The neutralizing value is a standard quantitative measurement of the acidic strength of a baking acid and is measured as the parts by weight of bicarbonate of soda which will be neutralized by exactly 100 parts by weight of the baking acid.

Example: One gram (0.0135 mol) of KCl was dissolved in 50 to 75 ml of a 50% water-alcohol solution (volume percentages) in a 100 ml volumetric flask. After the KCl dissolved completely, additional solution was added to bring the volume up to 100 ml. The resulting solution was transferred to a 250 ml beaker. Thirty grams of SAP was then added to the beaker and the slurry stirred for one minute. The slurry was filtered on a Buchner funnel and sucked as dry as possible with the aid of vacuum. The SAP filter cake was not washed. The filter cake was then transferred to a 250 ml suction flask and placed under a vacuum of 10 to 15 mm Hg overnight for drying. The same exact procedure, using the same quantities of salt (on a mol basis) was used to treat SAP with KBr, KI, and KF. Five grams of each of the SAP samples prepared by the above described procedure, plus one control sample of untreated SAP, were weighed into previously weighed aluminum dishes having 2 inch diameters and ½ inch depth. The sample dishes were tapped lightly to insure an even distribution of material and then placed in a humidor maintained at 75% relative humidity and 35°C. Humidity was maintained by a saturated NaCl solution. Temperature control was ±1°C. The dishes were removed periodically, weighed, and the percentage weight gain calculated.

Visual inspections were also made. After about ten hours in the humidor, the untreated SAP began to feel sticky to the touch, at which time it had absorbed about 7% of its original weight of water. After 90 hours in the humidor, the untreated sample had absorbed sufficient water to become virtually a viscous liquid. All of the potassium halide-treated samples remained essentially dry to the touch, picking up 1% or less of water in the first 10 hours, and not more than about 5% water during the entire period of humidification (140 hours).

Figure 4.1 sets forth graphically the results of the example. Percentage weight gain, or the percentage increase in weights of the test samples based on their original weights, is plotted against time of retention under the humidifying conditions of 75% relative humidity and 35°C. Graphically, the differences in hygroscopicity between untreated SAP and the potassium modified compounds are quickly seen. Whereas untreated material absorbs moistures at a very rapid initial rate, reaching equilibrium in the neighborhood of 28% weight gain, the treated material slowly increases in weight, and reaches equilibrium at a much lower point.

FIGURE 4.1: NONHYGROSCOPIC MODIFIED SODIUM ALUMINUM ACID PHOSPHATE

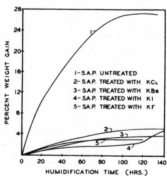

Source: J.E. Blanch and F. McCollough, Jr.; U.S. Patent 3,205,073; September 7, 1965

FRESH FRUITS AND VEGETABLES

SODIUM BISULFITE-PHOSPHORIC ACID

Newly peeled fruits, e.g., apples, bananas, and pears, will quickly darken at their cut surfaces upon standing in air. The darkening effect, initially observed as a brown or black surface discoloration, is thought to be largely enzymatic, involving an interaction between oxygen and polyphenoloxidases. Nonenzymatic darkening may also accompany the more pronounced enzymatic reactions. Further, upon prolonged storage of the fruits, their sugars may ferment to liberate carbon dioxide and form various organic acids (the latter indicated by a fall in pH). These changes in the stored fruit are undesirable since they result in a shorter shelf life, spoilage, and consumer rejection. Enzymatic spoilage has been known to be delayed somewhat by certain chemical treatments and by blanching.

A synergistic mixture was identified by *W.J. Sutton and R.M. Lauck; U.S. Patent 3,305,366; February 21, 1967; assigned to Stauffer Chemical Company* to inhibit enzymatic darkening and fermentation of fresh fruit. The synergistic mixture consists of phosphoric acid and at least one sulfur compound selected from the group consisting of sodium sulfite and sodium bisulfite. The phosphoric acid which may be used will have an analysis between 68.8 and 86.3% by weight of P_2O_5 (based on the undiluted acid).

The synergistic mixture is added to the raw fruit (after peeling, trimming, slicing, etc.) either as a dilute cover solution or as a dry mixture. This latter procedure is perhaps preferable when preparing a puree, such as banana puree. In applications where a sliced or diced fruit is to be canned and stored for long periods, a cover solution is used. Where solutions of the active compounds are used, no advantage is seen in either heating or cooling the solution prior to application. After addition of the active compounds, the fruit may be packaged, canned, refrigerated, or frozen.

The over-all weight ratio of phosphoric acid (basis, undiluted acid) to sulfur compound is between 4:1 and 1:2. With the preferred pyrophosphoric acid additive, where the sulfur compound is sodium bisulfite, a weight ratio of the former to the latter of approximately 2:1 furnishes optimum stabilization.

The total amount of stabilizing compounds added to the fruit must be controlled. Higher amounts produce undesirable properties in the final product, e.g., poor flavor, etc., while lower amounts will yield ineffective stabilization. The maximum amount of total additives (not including any water used in preparing cover solutions) will be approximately 1.0% by weight of the fruit. The minimum amount of total additives for a high degree of stability is approximately 0.01% by weight of the final fruit. Within this range, apples and, to a lesser extent, bananas require a lower total additive content for maximum effectiveness.

Example 1: Ripe bananas were peeled, mashed, stirred well with a metal spoon; and a small amount of the resulting puree was rapidly placed in screw cap jars. Using a 50 ml pipette, cover solutions comprising various candidate stabilizing compounds and mixtures were added (above the puree) to the jars. The bananas so treated were stored at room temperature (25°C) and were periodically examined to observe color changes and fermentation characteristics.

In these treatments, one standard tablespoon of banana puree was mixed well with 50 ml of deionized water (or solution). The testing procedures showed that 0.20% citric acid or 0.20% pyrophosphoric acid used alone has little effect; the products brown after one day. Testing for storage life, i.e., until fermentation was apparent in the product is stated in days at 25°C. The results of these tests are presented in the following table.

Cover Solution Used	Storage Life	Observations
Deionized water	1	Browning on surface (1 cm thick)
0.10% Ascorbic acid	4	No brown color by 4 days, but gas production
0.10% Ascorbic acid and 0.20% pyrophosphoric acid	4	No brown color by 4 days, but gas production
0.10% Sodium bisulfite	11	No brown color by 11 days, but gas production
0.10% Sodium bisulfite and 0.20% citric acid	13	No brown color by 13 days, but gas production
0.10% Sodium bisulfite, 0.20% citric acid and 0.20% pyro-phosphoric acid	20	No brown color by 20 days, but gas production
0.10% Sodium bisulfite and 0.20% pyrophosphoric acid	33+	Original condition after 33 days

Example 2: Banana purees and cover solutions were prepared as in Example 1. These purees with various cover solutions were placed in storage and examined visually from time to time. Also, each puree was tested using catechol solution after standing at least one day at about 25°C (room temperature). The puree was stirred and filtered and then 0.10 ml of filtered extract was added to 10 ml of a 1.0% catechol solution in deionized water in a glass vial.

Oxidation with a resultant discoloration of the catechol was measured at 400 mμ in a 19 x 105 ml cuvette at 0.5 minute intervals for a period of 5 minutes at about 25°C. For this measurement a Spectronic-20 instrument was used. The reaction rate in OD (optical density) units per minute was taken over a 1 to 3 minute reaction time. The results of these tests are presented in the table on the following page.

From the table it may be observed that pyrophosphoric acid, when used alone, is relatively ineffective for stabilizing the color of banana puree. Even at a 1% level of pyrophosphoric acid, the puree will brown after only 3 days. On the other hand, pyrophosphoric acid is highly effective in decreasing the polyphenoloxidase activity of the product.

Sodium bisulfite may be seen to be somewhat effective for inhibiting browning although considerably less effective than sodium bisulfite-pyrophosphoric acid mixtures. Significantly, when used alone, 0.2% pyrophosphoric acid or 0.1% sodium bisulfite will stabilize banana puree for 1 day and 14 days, respectively, while a mixture of 0.2% pyrophosphoric acid and 0.1% sodium bisulfite will permit storage of the puree for at least 21 days without adverse effect.

Cover Solution Used*	Storage Life (Days at 25°C)	Oxidase Enzyme Activity, mμ**	Observations
Deionized water	1	0.320	Brown after 1 day
0.20% pyrophosphoric acid	1	0.160	Brown after 1 day
0.50% pyrophosphoric acid	2	0.030	Brown after 2 days
1.00% pyrophosphoric acid	3	0.000	Brown after 3 days
0.002% sodium bisulfite	1	0.320	Brown after 1 day
0.005% sodium bisulfite	1	0.245	Brown after 1 day
0.010% sodium bisulfite	2	0.220	Brown after 2 days
0.020% sodium bisulfite	3	0.190	Brown after 3 days
0.050% sodium bisulfite	3	0.010	Brown after 3 days
0.100% sodium bisulfite	14	0.001	At 14 days, gas production and blue mold forms; no brown color
0.002% sodium bisulfite and 0.20% pyrophosphoric acid	1	0.190	Brown after 1 day
0.005% sodium bisulfite and 0.20% pyrophosphoric acid	2	0.065	Brown after 2 days
0.010% sodium bisulfite and 0.20% pyrophosphoric acid	2	0.040	Brown after 2 days
0.020% sodium bisulfite and 0.20% pyrophosphoric acid	7	0.004	Brown after 7 days
0.050% sodium bisulfite and 0.20% pyrophosphoric acid	21	0.010	Fermentation at 21 days; no brown color or fermentation up to 21 days
0.100% sodium bisulfite and 0.20% pyrophosphoric acid	21	0.001	Fermentation at 21 days; no brown color or fermentation up to 21 days

*Using one tablespoon of banana puree and 50 ml of cover solution.

**0.1 ml filtered banana extract and 10 ml 1.0% catechol solution. Enzyme activity measurements made on banana puree filtrate on day browning occurred or at later times when fermentation occurred.

Example 3: A group of Jonathan apples were peeled, sliced and placed in small (approximately 50 ml) glass vials. To each glass vial was then added a cover solution consisting of distilled water in which was dissolved a minor amount of sodium bisulfite, or sodium bisulfite and pyrophosphoric acid. As a control, distilled water containing no additives was used as a cover solution in one vial. After being sealed, the vials were placed in boiling water for 15 minutes and then cooled to room temperature.

Upon reaching room temperature, each vial was examined visually to ascertain the color of the apples. The color was judged on a scale of from 0 for pure white to 10 for dark brown. Without breaking the seals the vials were then placed in an oven maintained at 50°C to accelerate degradation. At 5 days after canning, the vials were removed from the oven and the color of the apples was evaluated in accordance with the color scale. The results are set forth in the following table.

	Canned Apple Color, 50°C Storage	
	0 Days	5 Days
No additive (control)	8	10
0.026% sodium bisulfite	2	6
0.026% sodium bisulfite and 0.052% pyrophosphoric acid	0	3
0.012% sodium bisulfite	4	8
0.012% sodium bisulfite and 0.024% pyrophosphoric acid	3	6
0.006% sodium bisulfite	6	10
0.006% sodium bisulfite and 0.012% pyrophosphoric acid	0	6

The color scale for the preceding table is as follows: 0 = white; 2 = trace brown; 4 = very slightly brown; 6 = slightly brown; 8 = moderately brown; 10 = dark brown. Intermediate intensities are assigned odd numbers.

STANNOUS CHLORIDE

W.K. Higby and D.E. Pritchett; U.S. Patent 3,219,458; November 23, 1965; assigned to Sunkist Growers, Inc. discuss the problem of nonenzymic anaerobic browning of citrus products on storage. Citrus juice and citrus juice products are known to develop a brown or dark color in a relatively short period of time when exposed to air, due in part at least to oxidation changes. In the absence of air, citrus juices also turn brown, but more slowly, apparently from interreaction of juice constituents, to form dark colored products. The latter form of browning is the most serious type in preserved citrus juices packaged in closed containers since several means are known by which atmospheric oxygen can be reduced to a very low level at the time of packaging.

Previous efforts to eliminate or retard the nonenzymic anaerobic browning include ascorbic acid addition, addition of sulfur dioxide and packaging in containers with exposed tin surfaces. Ascorbic acid addition is relatively expensive and not always effective. For example, in concentrated lemon juice ascorbic acid promotes rather than retards browning. Sulfur dioxide cannot be used in cans because of its corrosive action and further is limited in application by its characteristic odor. In reconstituted bottled lemon juice, sulfur dioxide very effectively prevents browning but eventually gives rise to a unique and unpleasant off-taste described as "skunky". Acid foods react with tin surfaces to produce a reducing effect which prevents browning but at the same time gives rise to an unpleasant metallic off-taste and in some cases hydrogen gas is produced.

It was found that the addition of small amounts of stannous ions to citrus juice products such as single strength lemon, orange and grapefruit juices, concentrated lemon, orange and grapefruit juices and concentrates for orangeade and lemonade retards or inhibits browning of such products. It has been found that by the addition of greater amounts of stannous ions, the reconstituted beverages made from such concentrates are also preserved against darkening.

A number of tin salts exist which can serve as sources of stannous ions for this purpose such as, for example, stannous chloride, stannous sulfate and stannous tartrate, which are water-soluble, and stannous oxalate and stannous oxide, which are soluble in hydrochloric acid. Stannous chloride is preferred since it is readily available as a reagent grade chemical of known purity and therefore is suitable as such for use in foods.

The stannous salt is simply added to the fresh citrus juice or concentrate. Stannous chloride forms an insoluble oxychloride upon standing, and upon dilution with much water, it forms an insoluble basic salt. Relatively concentrated, freshly prepared aqueous or citric acid solutions of stannous chloride can be used, but considerable care must be taken to avoid or minimize the formation of the insoluble oxychloride or basic salt, neither of which appears to be effective to prevent browning. Best results in minimizing the formation of the insoluble compounds are obtained by dissolving the stannous chloride in concentrated hydrochloric acid and subsequently diluting the solution with water.

To keep to a minimum the amount of strong acid added to the juice, only enough hydrochloric acid is used to dissolve the stannous chloride to form a clear solution. Thus, clear solutions can be obtained with as little as 2.5 ml of hydrochloric acid for 5 grams of salt.

Stannous ion additions as low as 25 parts per million retard browning moderately, increased amounts bringing about a marked retardation until a concentration of 200 to 300 ppm is reached. Beyond this level, relatively small additional gain is realized. Taste tests with lemonade indicate that stannous ions can be detected at about 250 ppm in the fresh product. After storage and reconstitution of concentrate for lemonade, 90 ppm of stannous ions can be detected as a metallic off-taste, but taste differences between the treated and control samples are obscured by the generally improved flavor of the treated samples. Improved retention of the original flavor is an added benefit to be derived from addition of stannous ions.

Example 1:   Concentrate for Orangeade — 0.579 gram of $SnCl_2 \cdot 2H_2O$ was dissolved in 0.29 ml concentrated HCl. This solution was diluted with about 20 ml water and then added to 3,808 grams of concentrate giving 80 ppm Sn^{++}. One volume of this concentrate is intended for dilution with water to 5.333 volumes of beverage. The concentrate was then pasteurized into enamel cans at 180° to 190°F. This product normally consists of orange juice, sugar, citric acid, sodium citrate and orange oil.

Concentrate for Lemonade — 0.599 gram $SnCl_2 \cdot 2H_2O$ was dissolved in 0.3 ml concentrated HCl. This solution was washed with a few ml of water into 4,639 grams of concentrate intended for dilution of 1 part to 5.333 parts by volume, giving 68 ppm Sn^{++} in the concentrate. The concentrate was pasteurized into enameled cans at 170° to 180°F.

Concentrated Grapefruit Juice — 0.572 gram $SnCl_2 \cdot 2H_2O$ was dissolved in 0.29 ml concentrated HCl and then washed with a few ml of water into 3,773 grams of 5.25 fold concentrated grapefruit juice to give 80 ppm Sn^{++}. The concentrate was then pasteurized at approximately 160°F into enameled cans.

Concentrated Lemon Juice — 0.541 gram $SnCl_2 \cdot 2H_2O$ was dissolved in 0.28 ml concentrated HCl and then washed with a few ml of water into 3,557 grams of 5.7 fold concentrated lemon juice. The concentrate was then pasteurized at 170° to 180°F into enameled cans.

Single Strength Lemon Juice — 9.5 grams $SnCl_2 \cdot 2H_2O$ were dissolved in 4.5 ml concentrated HCl and then diluted to 100 ml with distilled water to provide a solution containing 0.050 gram Sn^{++} per ml. 3.12 ml of this solution was added to 1,000 ml (1,034 grams) vacuum deaerated lemon juice (pasteurized, natural strength, 6.10% acid as citric, 8.95° Brix, specific gravity 1.03382). 0.1% sodium benzoate was also added to protect against microbial spoilage, and the juice was stored in bottles.

Single Strength Orange Juice — One ml of a $SnCl_2 \cdot 2H_2O$ solution made up by dissolving 2 grams in 1 ml concentrated HCl and diluting to 10 ml with water was added to 1,045 grams of vacuum deaerated orange juice to give 100 ppm Sn^{++}. The orange juice was preserved against microbial spoilage with 0.1% sodium benzoate and was stored in bottles.

The effect of the stannous ion on shelf life, from a color standpoint, of these products was determined by examining them for color, initially and at periodic intervals during storage. Color of single strength lemon juice, concentrated lemon juice concentrate for lemonade and concentrated grapefruit juice was measured on a filtered mixture of equal parts isopropyl alcohol and the juice product or a dilution of the juice product using the Klett-Summerson photo electric colorimeter with the blue, No. 42 filter and a water-isopropanol blank single strength lemon juice or concentrated lemon juice diluted to single strength was judged to be unacceptable at a reading of 125.

Concentrate for lemonade diluted 75.04 grams to 100 ml was unacceptable at a reading of 98 and concentrated grapefruit juice diluted 23 to 89 grams to 100 ml was unacceptable at a reading of 175. Reflectance of concentrate for orangeade was measured using an Agtron reflectance meter on which a reading of 25 corresponded to unacceptable loss of orange color. Color of single strength orange juice was estimated visually and samples more brown than orange were considered unacceptable. The results are set forth in the table shown on the following page.

It will be noted from the table that in all instances except for the concentrate for orangeade, a significant improvement in shelf life was obtained. This product and some of the other orange juice products behave differently from the yellow colored lemon and grapefruit products in that the effect of the stannous ions is less a matter of prolongation of shelf life than it is one of maintenance of brighter, more acceptable color during the effective shelf life. An additional benefit derived from color preservation with stannous ions is improved retention of ascorbic acid. This appears to be true in all citrus juice products.

Product	Storage Temperature, °F	Untreated Storage Life, Days	Treated Storage Life		
			Sn^{++}, ppm	Days	Percent Improvement
Concentrate for orangeade	100	77	80	78	1
Concentrate for lemonade	82	150	68	181	21
Concentrated grapefruit juice	82	69	80	150	117
Concentrated lemon juice	82	62	80	183	195
Single strength lemon juice	82	97	150	299	208
Single strength orange juice	100	30	100	>90	>200

Example 2: In order to demonstrate that sufficient stannous ions can be added to concentrated lemon juice to preserve the juice both before and after reconstitution, 6.00 grams $SnCl_2 \cdot 2H_2O$ were dissolved in 3.00 ml concentrated HCl and added to 2,500 ml (3,154 grams) of 8.42 concentrated lemon juice, giving 1,263 ppm of Sn^{++} in the concentrate, calculated on a weight-volume basis. The concentrate was stored in polyethylene bags sealed in cans. At periodic intervals, samples of the concentrate were removed, reconstituted, the reconstituted juice was bottled and it in turn was stored. The reconstituted juice contained 150 ppm Sn^{++} on a weight-volume basis. The stored reconstituted juice was examined periodically for color as described in Example 1, with the following results.

	Shelf Life of Bottled Reconstituted Juice at 82°F	
Days Concentrate Stored at 35°F	Without Sn^{++}, Days	With Sn^{++}, Days*
0	82	>179
35	35	----------
39	----	>136
95	14	----------
104	----	>135
184	31	----------
195	----	>86

*Because of the slow rate of darkening, it was not possible to make any estimate of final shelf life.

It is to be noted that the color stability obtained in this manner is superior to that obtained by adding the same amount of stannous ions to natural strength lemon juice. The mechanism by which the stannous ions prevent browning in citrus juice products is not entirely clear, but it is believed that it involves more than a simple reaction with oxygen to prevent oxidative reactions with the natural juice constituents, since other means of oxygen removal do not produce results comparable with those obtained by this process. However, a direct reaction between the stannous ions and oxygen is possible and the presence of sufficient oxygen in the sealed containers will reduce or nullify the effect of the stannous ions. Accordingly, for best results, the product should be packaged in substantially oxygen free containers.

AMINO ACID PRESERVATIVES

Amino acids are used to preserve the freshness of fruit juice, vegetable juice and processed fruits and vegetables in the process of *Y. Harada, Y. Kanzaki, H. Furukawa, K. Yamazaki and H. Matsuo; U.S. Patent 3,615,703; October 26, 1971; assigned to Kyowa Hakko Kogyo*

KK, Japan. The amino acids which may be used for this purpose are lysine, ornithine, histidine and arginine. They may be used either in the free form or in the salt form as formed with an acid, such as hydrochloride, sulfate, phosphate, inosinate or succinate, or in the form of salt formed with an alkali metal such as sodium. Amino acids are stable to heat and inert to various other compositions commonly included in foodstuffs so that it is possible to add the amino acids to foodstuff at any stage of its preparation.

When the amino acids are used for the preparation of sparkling beverages such as beer or carbonated drinks, the gas maintenance in carbonated drinks and the foam retention of sparkling wines can also be greatly improved. When the amino acids are used for the preparation of sparkling wines, for example, it is preferred to use 10 to 200 grams of lysine in the salt form per 10 kg of the dried malt, but it is also possible to use lysine (in the salt form) in an amount of from 5 to 100 grams per 50 liters of beer obtained. For instance, the amino acids can be added to the green malt, or if desired, it is also possible to add it after the malt drying step, although better results may be obtained by using the amino acid in the earlier stage.

Example 1: Samples of canned peaches were prepared in the following manner. After removing the core, a peach was soaked in water containing 3% NaCl (by weight of water). After peeling, the peach (weight 275 grams) was packed in a can (capacity 454 cc) into which a syrup having the following composition was poured in an amount of 160 grams. Air was exhausted from the can at 95°C for 5 minutes. After sealing in vacuo, the can was heated to 90° to 95°C for 20 minutes for sterilization and was then cooled. Similar procedures were performed by using 4 types of the syrups which have the following compositions in percent by weight.

Syrup	Sugar	Sodium Isoascorbinate	L-Lysine Hydrochloride
1	43	-----	-----
2	43	0.1	-----
3	43	0.1	0.1
4	43	-----	0.1

Each sample was stored for 6 months at room temperature and was then opened to determine freshness regarding taste, flavor and appearance. For this purpose, a functional test was carried out using Kramer's method. Results are shown in the following table, in which a test panel consisting of 15 testers was used.

Sample	1	2	3	4
Total Ranking	60	45	17**	26*

*Significant difference at 5%
**Significant difference at 1%

As shown in the table, the flavor was significantly preserved by using L-lysine HCl, in particular, by using L-lysine HCl together with sodium isoascorbinate owing to their synergistic effects. The combined use of L-lysine HCl with sodium isoascorbinate was superior at a level of significance of 1% to the sole use of sodium isoascorbinate and to the control sample, and the sole use of L-lysine HCl was superior to the samples 1 and 2. With regard to the outlook and maintenance of shape, samples 3 and 4 were superior to those of samples 1 and 2. It was also determined that by the addition of L-lysine HCl, the flavor was largely improved and can smell was entirely removed, while samples 1 and 2 had more or less a can smell. With respect to taste, the samples containing L-lysine HCl had a rich and complicated taste, while the control sample had a monotonous taste. The sample containing sodium isoascorbinate had a strongly acidic taste, and the syrup with L-lysine HCl added had a higher clarity than those without the addition of L-lysine. The control syrup had a considerable turbidity.

Example 2: Samples of canned mandarin orange juice were prepared in the following manner. Raw oranges were washed well with water and squeezed to obtain orange juice which was centrifuged and packed in an amount of 250 grams in a can (capacity 273 cc).

After sealing in vacuo, the ingredients were sterilized at 90°C for 10 minutes. Ten kilograms of the product obtained had a concentration of juice of 70% by weight and its composition was as follows: mandarin orange, 7 kg; sugar, 0.84 kg and water, 2.16 kg.

L-lysine HCl (0.1% by weight) was added to the juice. The juice was canned and stored for 6 months. After this, the sample was subjected to a functional test by means of two-point discrimination method using 15 testers and was compared with a control sample prepared in a similar manner to that described above without addition of L-lysine HCl. The obtained significance differences are shown in the following table.

From this table, it is apparent that the addition of L-lysine HCl resulted in good preservation of the mandarin orange juice with regard to freshness and color changes. In addition, the control sample had a can smell and inferior freshness, which could be improved by the use of L-lysine HCl.

Sample	Total Ranking
With addition of lysine	12
Without addition of lysine	3

Example 3: In preparing a canned tomato juice, tomatoes of market grade were washed well with water, removed of carpophore, etc. and crushed by means of a conventional crusher. After heating to 80°C, they were squeezed and table salt was added. The juice was canned in a can (capacity 208 cc) which was sealed in vacuo and sterilized at 95°C for 25 minutes to prepare a sample. The composition of the sample per 10 kg was tomato juice, 9.95 kg and salt 0.05 kg.

L-lysine monophosphate, L-ornithine HCl, L-arginine HCl and L-histidine HCl (0.05% by weight each) were added, respectively, to individual samples. The cans were stored for 6 months and a functional test was carried out in a manner similar to that described in Example 2. The treatment helped preserve tomato juice.

EUGENOL PRESERVATIVES

A simple process is used to preserve fresh fruits and vegetables by *G.C. Layton; U.S. Patent 3,518,096; June 30, 1970.* A solution containing essence of cloves or eugenol is used to cover the exterior of the food and permit the moisture of the solution to evaporate. In some instances it is desirable to include ascorbic acid in the solution.

By treating fresh fruits, vegetables and flowers by this process, the retention of weight, texture and resistance to rot, mold and discoloration can be appreciably increased so that the food and plants retain their fresh characteristics for extended periods without refrigeration or special handling. This, of course, is valuable in the shipment and distribution of fresh fruits, vegetables and flowers by greatly reducing losses by spoilage and the process is particularly useful with respect to desiccative fruits such as strawberries, blackberries, blueberries, grapes and the like.

In carrying out the process, 1½ ounces of clove claws are boiled 1 hour in 1 gallon of water. It appears that a limited amount of the oil of cloves from the claws will be dissolved in the water and therefore the solution will be saturated with clove oil after the boiling process described. After the clove claws have been boiled in the water, 2 ounces of ascorbic acid is dissolved in the solution. Alternatively, in place of boiling the clove claws in the water, eugenol, which has the formula may be added to water in the proportions of about 20 to 25 drops per gallon of water, and as mentioned with respect to the preparation of the solution by the use of clove claws, 2 ounces of ascorbic acid is added to each gallon of the water.

As soon as practical following picking of the fruit, vegetables or flowers to be preserved, the surfaces are wetted with the prepared solution. This wetting may be accomplished in any convenient manner, as by immersion, spray or fogging. The food or plant articles so treated may then be packaged or not, as is desired. The residue retained on the treated articles inhibits rot, mold, drying and loss of weight of the articles so that rapid handling and culling of the articles by the merchandizer selling these products is obviated and the articles retain a fresh condition and appearance for many days without special handling. The residue may be easily removed from the articles by rinsing in water, and the taste or flavor of the food articles is not affected.

THIOISOMALTOL DERIVATIVES

Thioisomaltol is the common name for the chemical compound, 2-acetyl-3-hydroxy-thio-phene.

$$\text{structure of 2-acetyl-3-hydroxy-thiophene, showing a thiophene ring with S, an OH group and a } C\text{--}CH_3 \text{ group with } =O$$

It has previously been stated that isomaltol will preserve foodstuffs against microorganisms. It has now been found that thioisomaltol, which is a more stable compound than isomaltol, can be used for similar purposes and that thioisomaltol is about 3 times more effective as an antimicrobial agent than is isomaltol.

The use of thioisomaltol to preserve foodstuffs is described by *C.-H. Lee and J.R. Feldman; U.S. Patent 3,695,899; October 3, 1972; assigned to General Foods Corporation.* Thioiso-maltol is normally added to the foodstuff in amounts ranging from 0.005 to 0.5% by weight of the food. The low level of thioisomaltol which is able to impart microbial stability to the foodstuff is a particularly advantageous feature since the permissible amount of food additives is usually set at a relatively low level by the controlling governmental agencies. Low levels of antimicrobial agents are also desirable since the presence of off-tastes and flavors will not be imparted to the foodstuff.

Thioisomaltol is meant to include thioisomaltol itself as well as the derivatives of thioiso-maltol that readily hydrolyze to yield thioisomaltol, such as the nontoxic salts and esters or mixtures of any of these. Among the salts of thioisomaltol which may be used are the alkali metal salts such as sodium and potassium, the alkaline earth salts such as mag-nesium and calcium, as well as the zinc, iron, manganese, and ammonium salts. Also included are the salts of organic amines such as the salts of sec-butylamine and triethanol-amine. The esters of thioisomaltol which may be used include both the inorganic esters such as the sulfate and phosphate esters as well as the organic esters such as the formate ester.

Foods containing small amounts of thioisomaltol can be stored at room temperatures for much longer periods than untreated foods without becoming spoiled. Freezing or storing at low temperatures will protect the foods for even longer periods.

Applications by spraying or dipping the food in dilute solutions of thioisomaltol or its salts have been found to be convenient, effective and economical. The addition of these compounds to processed foods by simply mixing a solution of thioisomaltol into the food has also proven most convenient. Other methods or combinations of methods which are known in the food processing industry are also applicable. For example, in mixing large commercial quantities of cake mix, the thioisomaltol may be blended with the shortening before mixing the shortening with the other cake ingredients. In this manner, the thioisomaltol is used to preserve the shortening in storage, the cake mix on the grocer's shelves and the final finished cake as well.

The application of this process to the preservation of fresh fruits, including berries and vegetables, involves dipping the fruits, berries and vegetables in an aqueous solution of thioisomaltol or one of its salts. The foods can then be stored at room temperature, frozen or put in cold storage areas. Another convenient method of applying these preservatives is by spraying a solution of thioisomaltol on the fruits, berries or vegetables.

For convenience and economy, a 1% solution can be effectively applied to large batches of fruit or vegetables in open containers with a hand sprayer. Where the application must be performed in the shortest possible time, a 10% solution of the sodium or potassium salts of thioisomaltol is most conveniently used.

Where large quantities of foodstuffs are stored such as in railroad cars or commercial storage bins, the vapor phase method of preserving food with thioisomaltol is very convenient. The simplest method of vaporizing the thioisomaltol is to drop aqueous solutions on a heated electrode within the sealed or semiclosed storage area. For use in refrigerated railroad cars, the thioisomaltol vapor can be metered into the atmosphere by a suitable device. In this manner, a 0.1% concentration based on the weight of food in storage may be readily maintained in the atmosphere of the storage area. The use of such devices also allows the thioisomaltol vapor to be used in conjunction with a high carbon dioxide concentration in the air. The use of carbon dioxide as a preserving atmosphere is well known in the food preserving and shipping trades.

Another method which was found effective in applying thioisomaltol to foods comprises applying solutions of thioisomaltol or one of its salts to wrapping papers, paper cartons or carton liners used for storing food. For this type of application, the packing papers or paper cartons are weighed and dipped in a 10% solution of a salt of thioisomaltol. The papers are then inserted into the cartons so as to give an effective concentration of thioisomaltol based on the weight of the foodstuff. The papers need not be in direct contact with the surface of the food to be effective. Similarly, the food may be placed in paper boxes treated with either thioisomaltol or one of its salts. A modification of the method for applying thioisomaltol to paper food wrappers and cartons is by milling thioisomaltol or one of its nontoxic salts into a plastic film such as polyvinylidene chloride, polyethylene or polyvinyl chloride.

Example: The antimicrobial activity of thioisomaltol was tested on the microflora of corn flour. Corn flour was adjusted to a moisture content of 25% and incubated at room temperature for 48 hours to increase the microbial count. This flour was used to prepare a slurry consisting of 1 gram of corn flour per 100 ml of isotonic saline. 2 ml of this slurry were plated together with 1.0 ml of a solution of 1% thioisomaltol into 8 ml of nutrient (agar). This gives a final concentration of thioisomaltol of 0.09% by weight.

After 4 days these plates were free of microbial growth. Control plates without the addition of thioisomaltol were completely overgrown with bacteria and molds. This shows that thioisomaltol substantially inhibits the growth of both bacteria and molds. Comparable antimicrobial activity is observed with the use of the sodium salt of thioisomaltol. Comparison tests were made and it was found that a final concentration of 0.3% by weight of isomaltol was necessary in order to effectively inhibit microbial growth.

COBALT SALT WASH

A.J. Kraght and H.C. Marks; U.S. Patent 3,347,683; October 17, 1967; assigned to Wallace & Tiernan Inc. state that an aqueous wash containing a cobalt compound is effective as a post-harvest treatment of raw fruits to inhibit decay during storage. Raw fruits have been treated by dipping in a tank of an aqueous wash containing a decay preventing agent or by spraying or foaming with an aqueous wash containing the decay preventing agent. Washes usually contain orthophenylphenol or a water-soluble derivative as the decay preventing agent or, in the case of the treatment of citrus fruits, soda ash or borax have been used. Also it is the general practice to include in the wash a surfactant or detergent to improve the washing or treating operation.

An improved method to treat raw fruits is provided by adding a cobalt compound to water to form a dispersion. Ordinarily, even with the so-called insoluble cobalt compounds, at least a very small concentration of the cobalt compound (and the cobalt portion thereof) is present in molecular dispersion, that is, dissolved in the aqueous wash. For example, cobaltous carbonate which is substantially water-insoluble and which has a room temperature solubility in water yielding a dissolved cobalt concentration of the order of one part per million by weight is highly effective to inhibit decay. In any event, with the insoluble cobalt compounds, an amount of the solid undissolved cobalt compound should remain dispersed in the wash or at least with the portion of the wash which is being used to treat the fruit.

It has been found that when solid particles of such less soluble cobalt compounds are maintained in dispersion in the wash used for treatment, decay is prevented or substantially inhibited whether by continual replenishment of the small amounts of cobalt actually dissolved in the wash or by contact of the solid particles themselves with the fruit surfaces. With the more soluble cobalt compounds, there is no difficulty in maintaining an effective concentration of cobalt in molecular dispersion (solution) in the wash.

Cobalt-containing compounds which are useful in this process include cobaltous acetate, cobaltous ammonium phosphate, cobaltous ammonium sulfate, cobaltous carbonate, basic cobaltous carbonate, cobaltous chloride, cobaltous citrate, cobaltous formate, cobaltous hydroxide, cobaltous nitrate, cobaltous oxide, cobaltous phosphate, cobaltous potassium sulfate, cobaltous sulfate and cobaltous tartrate. In general, the cobalt compounds selected from the group consisting of cobalt oxides, cobalt hydroxides and the cobalt salts of acids whose anions are nontoxic are preferred.

The process is generally applicable to all fruits including citrus fruits such as lemons, oranges, grapefruit, tangerines and limes, also stone fruits such as peaches, nectarines, cherries, apricots and plums, also pome fruits such as apples, pears and quinces, also grapes and other berry fruits such as tomatoes. Further the process is applicable to the treatment of melons such as cantaloupes, watermelons, honeydew melons, Persian melons and Cranshaw melons.

Various tests described in the examples were carried out to demonstrate the effectiveness of the treatment to inhibit decay. In all the tests, unless otherwise specified the fruit was inoculated with a water suspension of *Penicillium digitatum* or *Penicillium italicum* spores. After inoculation the fruit was permitted to dry and then selected at random into lots for treatment. All treatments involve an aqueous wash containing a cobalt compound followed by a fresh water rinse. After treatment the fruit was stored in a room maintained at a temperature of about 70°F at about 90 to 95% relative humidity for periods as long as 3 to 4 weeks. Under such conditions the inoculated, untreated fruit developed most of the decay within 4 to 7 days.

Example 1: Tests were carried out to show the effectiveness for decay control of the cobalt-containing aqueous treating washes. In these tests inoculated lemons were dipped for 2 minutes in various cobalt-containing washes containing 0.02% of a surfactant or detergent such as an alkyl aryl sulfonate dissolved in it. The temperature of the wash solution during the dipping operation was in the range of 117° to 119°F. The results of these tests are set forth on the following page. In the test results presented in the table, No. 1 amounts of the above identified cobalt compounds were employed to give cobalt concentrations in the test washes at two levels of cobalt content, 0.05% and 0.1% by weight. The cobaltous carbonate employed had an average particle size in the range of 2 to 3 microns.

	Aqueous Wash Containing	- - - - - - - - - - - Percent Decay - - - - - - - - - -			
		5 Days	10 Days	18 Days	24 Days
1	Cobalt free	96	- - -	- - -	- - -
2	3% Soda ash	48	68	72	72
3	0.2% Cobaltous chloride	16	28	56	60
4	0.4% Cobaltous chloride	12	12	32	32
5	0.2% Cobaltous acetate	20	44	52	52
6	0.42% Cobaltous acetate	4	4	12	20
7	0.24% Cobaltous sulfate	12	20	32	36
8	0.48% Cobaltous sulfate	8	12	24	40
9	0.1% Cobaltous carbonate	4	20	24	24
10	0.2% Cobaltous carbonate	12	24	24	28
11	0.16% Cobaltous formate	- - -	16	32	44
12	0.32% Cobaltous formate	4	4	16	16
13	0.335% Cobaltous ammonium sulfate	4	12	36	44
14	0.67% Cobaltous ammonium sulfate	4	16	24	32

Example 2: Tests were carried out to show that the process may be used whereby the fruit, instead of being dipped into a treating solution is sprayed with the wash or passed through a curtain of foam of the wash as the fruit passes on brushes or roller conveyor beneath or through the spray or foam. Usually in a spraying operation the fruit is in contact with the wash for a shorter period of time than in the dipping method. The results of these tests on lemons treated with 0.5% $CoSO_4$ are set forth below.

Treatment	Percent Decay After 17 Days
Cobalt free wash	74
2-3 second spray	58
5 second spray	48
15 second spray	42
30 second spray	26

Example 3: Further tests were carried out to show the effectiveness of cobalt washes at various temperatures. The results of these tests showed that the special cobalt containing washes are effective at a relatively high temperature, about 110°F, and at room temperature, about 70°F. This is not true of conventional treating agents, such as borax and soda ash which are ineffective at room temperature.

In fact, the cobalt-containing washes are effective at a temperature as low as about 32°F, the freezing point of water. Accordingly, it may be convenient to refrigerate the cobalt-containing wash and to use the cooled wash, as in a Hydrocooler apparatus, to effect preliminary or at least partial cooling or refrigeration of the treated fruit before storage or shipment. Thus, the washing operation may be carried out at a temperature in the range of 35° to 120°F.

When the cobalt wash is employed in the Hydrocooler apparatus the contact time of the fruit undergoing treatment is rather long, up to about 20 to 30 minutes, more or less. However, the effectiveness of the wash treatment, as compared with treatment at a temperature from about room temperature up to about 110° to 115°F, is relatively less at lower temperatures. In the Hydrocooler treatment it may be desirable to use a relatively insoluble cobalt compound, such as cobaltous carbonate, since the fruit being treated is in contact with the cobalt wash for a relatively long period of time.

In general, the higher the cobalt content of the wash, particularly the dissolved cobalt content, the shorter the contact time effective to prevent decay. Aqueous washes having a cobalt content as low as about 0.005% by weight and as high as about 2% by weight are effective to prevent fruit decay.

COLOR STABILIZATION — PIGEON PEAS

Pigeon peas *(Cajanus indicus)* are also known as Congo peas or green gandures. This vegetable is well known in India, Hawaii, Puerto Rico and the British West Indies. It has long been used in these areas as a foodstuff by the population. A general characteristic of this type of pea that has prevented its general acceptability is that this product discolors on canning from its natural green color to a greyish brown.

The use of sulfur dioxide has long been established as a means of preventing, retarding, or even reversing the browning reaction of fruit and fruit products. It has, however, been confined mainly to dehydrated vegetables or fruits, to fruit juices and concentrates and to bleaching or preserving onions and cherries.

For green vegetables and for ordinary green peas, *(Pisum sativum)* in particular, it has long been known that the chemical changes which result in the destruction of the chlorophyll are accompanied by a notable increase in the acidity of the product. A solution to the color change was therefore found in controlling the pH of the product by addition, for example, of suitable alkaline reagents.

The process of *J.W. Barlow; U.S. Patent 3,583,873; June 8, 1971; assigned to Catelli Food Products Ltd., Canada* gives the details on addition of sulfur dioxide and adjustment of pH in the application of stabilizing the color of pigeon peas. To be effective, the reagent used to give off SO_2 must be used in an amount sufficient to obtain at least 100 parts per million of SO_2 in the finished product. Little advantage is obtained by using more than 1,000 ppm. A preferred range is one that results in the peas, after processing, containing 500 to 700 ppm.

It has been found that the effect of sulfur dioxide is enhanced by adding an edible organic acid to lower and control the pH. Suitable organic acids include citric acid, malic acid, tartaric acid, fumaric acid and adipic acid. Although a compound capable of yielding sulfur dioxide can be used alone in treating pigeon peas, it is preferred to take advantage of the combined effect referred to above by using both reagents. A most preferred combination is sodium metabisulfite and citric acid.

Ascorbic acid is widely used in the food processing industry as an additive. It can be used in addition to other edible organic acids and sulfur dioxide to effect additional flavor improvement. In one example, the peas, after being made ready for canning, i.e., vined, cleaned and washed in accordance with standard procedure, are blanched in an aqueous medium containing the aforesaid acid reagent or reagents at the usual boiling or near boiling temperature for a suitable length of time, usually a few minutes. The blanched peas may then be cooled if desired. The peas are then canned in brine. Sugar may be added to the brine for flavor as desired.

In another example, the peas, after being made ready for canning, are subjected to soaking in an aqueous medium containing the acid reagent or reagents at room temperature or at a temperature slightly above, e.g., 70°F for an hour or more. Following this soaking step, the peas are drained and canned in brine which may contain sugar as desired.

Example 1: One thousand grams of frozen pigeon peas were blanched for 5 minutes at boiling temperature in 800 grams of a solution of 2.4 grams sodium metabisulfite ($Na_2S_2O_5$), 2.4 grams citric acid ($C_6H_8O_7 \cdot H_2O$) and 795.2 grams of water. At the completion of the blanching period, the peas were drained, cooled and canned in a solution of 1,000 grams water, 20 grams salt (NaCl) and 40 grams sugar (sucrose). The cans were placed in a static retort at 240°F for 35 minutes for sterilization. On completion of the sterilization cycle, they were rapidly cooled.

Example 2: Frozen pigeon peas were washed and placed directly in cans to ½ inch from the top of the can. The peas were then covered with a solution of the following composition: water, 93.90%; sugar, 3.76%; salt, 1.88%; sodium metabisulfite, 0.33%; citric acid, 0.13%. The cans were sealed, placed in a static retort and sterilized for 35 minutes at 240°F. At the end of the sterilization period, they were rapidly cooled. In this example, the addition of the metabisulfite, citric acid, sugar and salt may be made in tablet form, prior to sealing. The addition of one or several tablets in the above manner greatly facilitates continuous operation.

SO₂ RELEASE IN PACKAGE

Fresh fruits and vegetables are preserved in a special package which contains a bactericidal gas generator which releases sulfur dioxide. The process is described by *C. Illouze; U.S. Patent 3,502,485; March 24, 1970.*

The term plant products includes all types of fruits including citrus fruits such as grapefruit, oranges, mandarins and lemons; stone fruits such as apricots, plums, peaches and cherries; seed fruits such as grapes, tomatoes and pears. Other fruits are avocados, mangoes and bananas. Fresh vegetables are beans, shelled or unshelled peas, carrots, peeled or unpeeled potatoes, and edible stalks such as asparagus, celery and rhubarb.

In the storage and in the transportation abroad of plant products, many difficulties have to be overcome. The action of mildew and fungi such as penicillium, phytophthora and *Bacillus putrificus* may do considerable damage to plant products during their storage in cold rooms even at a maximum degree of refrigeration for each category of fruits, vegetables or flowers. A deteriorating action may also occur during the transportation of these plant products at a relatively warm temperature.

The main constituents of plant products consist mostly of diastases, glucosides, lipids, vitamins, soluble sugar, acids and salts. The plant products also contain a relative amount of water and other substances. When they are stored in cold rooms, there is a decrease of their water content, which is detrimental to the constituents referred to above and consequently to the quality of the plant products.

The transformations and deteriorations occurring in the plant products are mainly due to the transpiration, respiration and fermentation of these plant products. It is to be noted that they always stay alive. A high water and pigment content is an indication of freshness and gives a nice appearance to the plant products. The respiratory phenomena are in the nature of a combustion. The respiration, transpiration and fermentation are more active when the ambient temperature is slightly elevated.

The respiration is related to the normal life of the organs of the plant products. If respiration is prevented by limiting the availability of oxygen, the life of the plant products is made difficult. Respiration is dependent on the variations of the respiratory quotient of each fruit, vegetable or flower.

Fermentation per se results in the production of products such as ethyl alcohol and acetaldehyde to produce particular taste and smell in the plant products. The cells die and are then quickly destroyed by the action of bacteria, mildew and other microorganisms. During the storage of these plant products in cold rooms, the respiratory phenomena are limited to a strict minimum. However, cold temperature does not prevent the microorganisms from developing and after some time, they invade the plant products and render them unfit for consumption.

In the process the plant products are enclosed in bags or wrappers made of plastic materials, preferably polyethylene. The plant products so wrapped are disposed in crates, wooden cases or cartons for an easier storage of the plant products in cold rooms. When these plant products are in cold rooms, means are provided to circulate within these wrappers or plastic bags a chemical gas, preferably a sulfurous gas around these plant products.

In these bags or wrappers are provided openings which could vary in size according to the plant products. The remaining portions of these bags or wrappers are made impervious. The opening is used to assure a proper respiration which will be reduced to a bare minimum. The plant products must have a supply of oxygen and they discharge carbon dioxide. The fumigation within these bags or wrappers enclosing these plant products may be induced by any solid or liquid chemical substance which is enclosed separately either in a small bag made of plastic material or in any other wrapping allowing the evaporation of a chemical gas.

This chemical substance may be a metabisulfite, preferably of an alkali or alkali-earth metal. The most suitable compound is $K_2S_2O_5$ which will slowly and continuously release sulfur dioxide. This substance may be associated with a compound such as alum to regulate the release of sulfur dioxide. However, any other compound which will provide a permanent fumigation may be used. The fumigating material constantly and permanently liberates a vapor which surrounds the plant products to prevent the fermentation of mildew during the storage and transportation of the plant products. After some time has elapsed there is no difficulty in introducing another sachet of the chemical substance through the above mentioned opening so that the sulfur dioxide can be renewed.

The quantity of $K_2S_2O_5$ which is used within the range of 0.001 to 5% of the weight of the plant products. It is obvious that the plant products must be put in cold rooms in a controlled atmosphere. A high degree of humidity will be necessary to compensate for the loss of water in all these fruits, vegetables or flowers.

Another advantage of the process is in the fact that the degree of humidity is as high as possible. This is accomplished by wrapping the stalks of the plant products with a layer of an absorbing material such as cellulose, cotton, cotton waste impregnated with distilled water. This applies particularly to cut flowers, bunches of grapes having vine shoots, the trunks of banana clusters or bunches of bananas or other plant products having stalks which may vary according to length. By this method there will be an additional supply of liquid food for all the above plant products. The combined humidity inside and outside the wrappings are essential to the metabolism of such plant products in order to store them under the best possible conditions for an indefinite period so that they are always fresh.

The action of microorganisms on fruits and vegetables would appear to comprise in the initial stage a fixation at the surface of the fruit and a penetration in depth to the interior of the fruit. It has been observed that by the action of the fumigating agent, i.e., conversion of all the chemical substances which are either volatizable or capable of producing gases by the decomposition, oxidation, sublimation, reduction or any other physical or chemical action converting the solid or liquid chemical product to gas or vapors, gives a depth penetration property which is therefore much more effective on the microorganisms which attack the fruit.

It has also been found that the retarding agent enables the evolution of these gases or vapors to be restrained so that they have contact for a longer period and hence give a much more effective action against these microorganisms. It has also been found that with the fumigating agent, such as the metabisulfite, it is possible to maintain a delayed evolution of sulfur dioxide. A curative action can be obtained by the addition of certain essential elements already present in each of the fruits or vegetables and which are compatible with the plant products and the forms in which they are to be used, and this effect is due to the supplementary action of these essential elements. Such essential elements may include lipids used alone or in combination with alkali or alkaline earth metal. Microorganisms attack fruits only if there is a deficiency of some of the elements of the fruits.

It was found that the so-called "red" fruits or vegetables such as raspberries, strawberries, cherries, red currants, tomatoes and red beets have a very high sorbates content. The following acids may be used as auxiliary agents in combination with the fumigating composition: sorbic acid, citric acid, benzoic acid, pectic acid, quinic acid, tannic acid, malic acid, tartaric acid, stearic acid and para-aminobenzoic acid.

Example 1: Approximately 5 kilograms of grape clusters provided with vine shoot portions, each of the clusters being covered by a cotton pad impregnated with water, are inserted into a bag of plastic material. Separately, in a corner of a crate there is disposed a small bag made of a perforated plastic material containing 0.005% sodium or potassium metabisulfite and 0.001% of alum which is being used as a retarding agent. This impervious wrapping comprises an opening on the upper part thereof for a proper respiration of the grapes with the outside. The wrapper is well covered with the plastic bag crate and is sent for storage.

Example 2: A banana cluster is inserted into a plastic bag. The stem of this cluster is surrounded by a layer of cotton impregnated with distilled water. Separately, there is provided a small bag of perforated plastic material containing a metabisulfite and 0.001% of alum. This plastic bag is completely impervious except for an opening provided at the top portion thereof to maintain a proper respiration of the bananas with the outside. The imperviousness of this plastic material may be made according to any known sealing method.

Example 3: Twenty kilograms of pears are put into a crate. At the four corners of this crate there is disposed a small bag comprising 0.008 part of potassium metabisulfite and 0.002 part of alum. The crate is covered with a bag of plastic material which has an opening at the top portion thereof to permit a proper respiration of these fruits with the outside.

Example 4: One kilogram of shelled peas are placed in a carton. Separately, beside this package, a small bag of perforated plastic material is provided for fumigation with sulfur dioxide by means of 0.002 part of metabisulfite and 0.001 part of alum. The carton and the small bag are wrapped in a plastic bag having a small hole to permit respiration with the outside.

GASEOUS EXCHANGE METHOD

In the process of *A. Kramer; U.S. Patent 3,597,235; August 3, 1971; assigned to Food Technology Corporation* fruit and vegetable produce is first subjected to a subatmospheric pressure to extract substantially all of the gaseous oxygen content. The produce is then treated by exposure to an atmosphere of a gas having enzymocidal activity. Thereafter or simultaneously the produce is also exposed to an atmosphere of a gas having bactericidal characteristics. It has been found that on storage the treated produce retains good color and texture and is biostatic. Storage of the so treated produce may be at normal and ambient temperatures and may even be in an oxygen atmosphere provided that recontamination of the produce with enzyme or bacteria does not occur.

It is to be emphasized that the operativeness of the process depends upon the particular order of treatment and that inadequate preservation of the produce results if a bactericidal gas is first used followed by an enzymocidal gas treatment. However, the enzymocidal and bactericidal gas treatments may take place simultaneously with good results. Furthermore, if only an enzymocidal, or a bactericidal, gas is used, preservation of the produce will not be satisfactory.

The process is applicable to any harvested fruits and vegetables. It is particularly important to the Rose family fruits, e.g., pears, peaches, cherries and apples, the root vegetables, e.g., carrots, turnips, beets and sweet potatoes, and the tuber vegetables, e.g., potatoes.

The produce must first be prepared in such a manner that evacuation of the produce may be accomplished without unduly long time periods. The produce is comminuted to a suitable size by dicing, slicing, cubing, cutting or grinding. The more surface area of the produce exposed and the thinner the section of produce, the more quickly the evacuation of the natural gases in the produce will take place. Conveniently, $\frac{1}{16}$ to $\frac{1}{2}$ inch slices are used, especially ¼ to ⅜ inch. Other geometric sections are suitable, e.g., French style strips, rings or wedges. While the time necessary to extract the natural gases of the produce will depend on the particular produce, its comminuted dimensions and the degree of vacuum applied, in general, the time necessary to attain the extraction will vary between 2 and 12 minutes. Since many fruits and vegetables brown rapidly after being cut and exposed to air, it is preferred that the comminuting be done in a nonoxygen inert atmosphere, e.g., under nitrogen or any other like nonoxygen gas or vapor.

After the produce is comminuted to a suitable size, it is placed in a suitable evacuating apparatus which may establish and maintain a low vacuum. The ultimate content of oxygen in the remaining space occupied in the evacuated produce should be no more than 0.05% by volume. A preferred range of oxygen content which is readily obtainable in a reasonable time period and produces good results is from 0.01 to 0.05% by volume. Further, in order to extract the oxygen content to the necessary level, the vacuum must be low enough to create a pressure difference capable of pulling out a sufficient amount of oxygen. It has been found that to insure the removal of oxygen, a vacuum of at least 29.0 inches of Hg is necessary.

After the produce is evacuated, it is stored under a nonoxygen-containing atmosphere, e.g., nitrogen, until such time as it is treated with the gases. The pressure of the gases may be at atmospheric pressure or above. The greater the pressure, the more rapidly will be the absorption of the gases. As an alternate, the produce may be exposed to the gases immediately after evacuation while the produce is still under vacuum. In this case, after evacuation, the gases are merely fed into the evacuation apparatus. Suitably, the amount of gases fed into the apparatus is sufficient to effect a pressure equal to the atmospheric pressure or greater, i.e., 1 to 20 atmospheres.

The time necessary to expose the produce will depend on the particular produce, the comminuted size, the pressure of the gases in relation to the initial pressure on the produce, the type and amount of enzymes and bacteria in and on the produce and the particular gases being used. Time periods of gas exposure of from about 1 to 30 minutes will produce good results with most produce at typical conditions, e.g., ¼ to ⅜ inch discs, with easily obtainable pressure differentials between the pressure of the produce and gases and with typical amounts of enzymes and bacteria normally encountered.

Since the purpose of the gas treatment is to inactivate the enzymes and substantially destroy the bacteria, suitable gas exposure times may be determined directly for any produce, gases and conditions by measuring the resulting enzyme and bacteria activity of a processed produce. However, since these determinations are subject to some error, the best method is to simply test a plurality of samples under various conditions and time periods of the process and observe the length of time for which preservation is obtained.

Carbon monoxide has been found to be a very effective enzymocidal gas and ethylene oxide has been found to be a very effective bactericidal gas. Sulfur dioxide has been found to be both enzymocidal and bactericidal. Hydrogen cyanide and ozone are also enzymocidal and bactericidal, respectively, but require special and careful handling. Nitrogen is also an enzymocidal gas, but less effective for long time preservation than the foregoing gases.

The comminuted produce may be treated with the gases in substantially pure form or, if desired, diluted with an inert gas such as nitrogen. Two or more gases and mixtures may be used for each of the two treatments, provided that the enzyme inactivation precedes the bacteria inactivation or the two are at least simultaneous. The amount of dilution and ratio of members of a mixture are not critical but dilution will require substantial increases in exposure time, i.e., the exposure time is inversely proportional to the concentration of the active gases in a diluted mixture.

After the enzymocidal and bactericidal treatments, the residual gases contained in the produce may be removed, if desired, at least to a low level. In some cases, the substantial removal of these residual gases is most desirable or necessary from the taste and toxicity point of view, e.g., when sulfur dioxide or hydrogen cyanide gases are used. This may be accomplished whereby the produce may again be evacuated and then resaturated with sterile air, or the produce may be simply flushed with sterile air. Carbon monoxide and ethylene oxide are not substantially absorbed by the tissues of produce and consequently leave practically no trace after being exposed to an atmosphere or a flush of sterile air. Sulfur dioxide, on the other hand, is absorbed rapidly by the tissues and must be aerated by vacuum or flushing in order to provide an acceptable taste. In general, the sulfur dioxide content of the produce should be reduced to about 50 ppm or less.

Packaging of the produce may be as desired, e.g., plastic bags, cans, paper wrapping, etc. However, whatever packaging is selected, it should be accomplished in sterile air and with sterile packaging materials and containers to prevent contamination by bacteria and other noxious microorganisms.

It may be assumed that when a food product is sealed in a container there will be included microorganisms which, unless they are destroyed, will thrive under the environmental conditions afforded and cause spoilage of the food. Furthermore, when most fruit and vegetable tissues are injured in any way or cut, a darkening of the tissues, called the browning reaction, occurs. Some browning reactions are enzymatic, and take place when the tissues still contain active enzymes.

Most evidence suggests that oxidation of phenols or polyphenols by enzymes is the principal reaction in enzymatic browning. The nomenclature for the enzyme which causes oxidation of phenols or of polyphenols is not standardized and is alternately called phenol oxidase, polyphenoloxidase, phenolase and polyphenolase. To prevent browning, therefore, it is necessary to inactivate the enzymes present in the tissues.

Food also may act both as a vehicle carrying pathogenic microorganisms and as a culture medium for the growth of pathogens; consequently, it is essential that food products be treated in such a way that these disease causing microorganisms are killed or their growth is inhibited. Some of the more common microorganisms are *Escherichia coli, Clostridium botulinum, Streptococcus faecalis, Salmonella typhimurium, Staphylococcus aureus,* and molds, e.g., *Aspergillus niger.*

A number of compounds may act as substrate for the polyphenoloxidase in browning reactions, including catechol, hydroquinone, anthocyanins and flavonoids. The course of the reaction is not fully known. Oxygen acts as hydrogen acceptor, carbon dioxide is often evolved, a quinone forms and the final pigment is a polymer. Although there are several enzyme systems present in plant tissues, destruction of the polyphenoloxidase indicates that the remaining enzymes have also been inactivated.

Example: Pure cultures of microorganisms that are ubiquitous and are implicated in spoilage of food products as well as being the main causative agents of food poisonings and infections were selected as test organisms. These include *Escherichia coli* ATCC 11229, an index of sanitation and spoilage organism; *Aspergillus niger,* a mold that had previously been isolated from samples treated with gases; the sporeformer *Clostridium botulinum/I;* a coagulase positive *Staphylococcus aureus* ATCC 6538; *Streptococcus faecalis* ATCC 6057; and *Salmonella typhimurium.*

The test bacteria were grown in Brain Heart Infusion broth (Difco) for 1 hour at 37°C The bacterial cells were harvested by centrifugation, washed twice with distilled water and suspended in buffered dilution water. The inoculum was adjusted to a population of approximately 1,000,000 cells per gram of product. The standardized spore suspension of the *Clostridium botulinum* was diluted to an equivalent of 100,000 spores per gram of product.

Apples and potatoes were selected because they represent important crops, are consumed in large quantities and have active enzyme systems. A number of apples and potatoes purchased from a local supermarket were thoroughly washed and then placed in the preparation chamber. All cutting tools were sterilized prior to use. The chamber was made of heavy plastic film on an aluminum frame. Several outlets and parts allowed the connection of the chamber to a vacuum pump and a nitrogen gas source as well as permitting the introduction of the raw materials, presterilized bottles, and stoppers, instruments used in the cutting of the products, and the removal of the filled jars and of the waste material. The chamber was so constructed as to allow the operator's hand to work inside the chamber.

The chamber was evacuated of air and pure nitrogen was replaced. A slight positive pressure of nitrogen was maintained inside the chamber throughout the preparation period. The samples were prepared in an atmosphere approaching 100% nitrogen in order that oxidative enzymatic browning would not be initiated prior to the final treatment with the appropriate gases. The produce was sliced into sections of about 1 gram weight and about 10 sections were placed in each of 50 ml capacity, rubber stoppered glass containers and were then ready for further treatment. The bottles were inoculated inside the chamber with the test microorganisms as described above.

A modified Case anaerobic jar was used as the chamber for the evacuation and gas exposure of the samples. The jar had a capacity of 16 liters and was equipped with a vacuum pressure gauge. A vacuum of about 29 to 30 inches of Hg in the jar was drawn and maintained for about 10 minutes until substantially total evacuation of air from the environment and from within the produce was accomplished. The desired gas or gases was then allowed to flow into the chamber until the gauge returned to the zero position. The gas remained in contact with the produce for various lengths of time. The chamber was then evacuated for about 10 minutes and the product was subsequently washed with sterile air for an additional ten minutes. A bacteriological glass filter was used in filtering the air. The chamber was then opened and the stoppers were securely adjusted in place to avoid recontamination.

In all cases, a gas analysis of the headspace of the bottled samples by the use of the Fisher-Hamilton-Partitioner operating with helium as the carrier gas failed to register the presence of the tested gas and therefore showed that the residual gas was removed by this evacuation washing treatment.

The ten 1-gram samples were transferred aseptically from each treatment bottle and tested after 7 days storage at room temperature for microorganisms. The nontreated control samples packaged in air became dark and liquefied at the end of 7 days. Similarly, the samples treated with nitrous oxide displayed darkening, liquefaction and putrefaction. Samples treated in nitrogen remained light while microorganisms continued to grow resulting in liquefaction of the product accompanied by putrefaction. Potato disks packaged in sulfur dioxide remained white and whole demonstrating the effectiveness of sulfur dioxide against enzymatic and microbial agents. Treatment with ethylene oxide and carbon dioxide resulted in complete inactivation of the microorganism but had no effect on browning enzymes. Carbon monoxide had no effect on microorganisms but afforded substantial freedom from color deterioration. The carbon monoxide-ethylene oxide treatment proved satisfactory with only slight discoloration occurring. The product remained whole and the flavor was clean.

The results obtained with apple disks indicate the effect of the various gases on color were comparable to the effect observed with potato disks with sulfur dioxide, nitrogen and carbon monoxide preventing color deterioration. Sulfur dioxide and the combinations of carbon monoxide and ethylene oxide showed good results in preserving the apples by inactivating the enzymes and the bacteria.

TETRAIODOETHYLENE TREATMENT

A.G. Hess; U.S. Patent 3,558,329; January 26, 1971; assigned to Hess Research and

Development Corporation gives a process for preserving melons to obtain an extended storage life 4 to 5 times the usual time. Melons are treated with a water suspension of tetraiodoethylene, with or without a preliminary treatment step using a bacteriostat, and then wrapped and sealed while wet with a suitable container which is waterproof and presents a barrier to water either entering or leaving the container. The term melons refers to various types of *Cucumis melo* and includes canteloupes, crenshaw melons, honeydew melons and other closely related varieties.

In general, the melons are treated by spraying or submerging them in a water suspension of tetraiodoethylene for a time sufficient to thoroughly wet the outside surface of the melon and then by wrapping and sealing the melons either individually or in bulk in a suitable material which is waterproof. A bacteriostat of various types may be used in water solution or suspension, either by submerging or spraying, as a pretreatment step to the treatment with tetraiodoethylene and the bagging.

Example 1: Upon ripening, a number of melons were selected and treated by being immersed in an aqueous suspension of tetraiodoethylene for about 5 minutes. The tetraiodoethylene suspension had an approximate concentration of about 1 gram per liter of water. This mixture was in the form of a slurry.

The melons were removed from the tetraiodoethylene suspension and while still wet, they were placed in individual polyethylene bags. The bags were sealed and the bagged melons were kept at a temperature of 10°C (45°F) throughout the remainder of their life after treatment. Observations of these treated melons indicated that their useful life was extended substantially. These treated melons were still edible up to 50 days after treatment. On the other hand, control melons which were untreated and also kept at 10°C had a life of only 10 to 12 days.

Example 2: A quantity of ripe melons was immersed in a water suspension of tetraiodoethylene for about 5 minutes. After this time, the melons were removed from the tetraiodoethylene suspension and while still wet were individually wrapped in a relatively close fitting polyethylene bag. The tetraiodoethylene suspension had an approximate concentration of about 1 gram per liter and was in the form of a slurry. After being wrapped, the treated melons were sealed in the bags and kept at a temperature of 2°C (35°F). Observations thereafter indicated that these treated melons had a useful life, i.e., were edible for a period of over 56 days after treatment.

In comparison, untreated control melons kept at the same temperature, 2°C, had a useful life of only 15 days. A pretreatment step of beginning the above process by subjecting the melons to the action of a bacteriostat had the effect of increasing the useful life of the treated melons as shown by the following example.

Example 3: The bacteriostat chosen for this example was a water solution of sodium hypochlorite. Four tablespoons of 5% sodium hypochlorite solution were added per gallon of water used as the immersing solution for this pretreatment step. A quantity of ripe melons was first immersed in this sodium hypochlorite solution for about 5 minutes. Immediately thereafter, the melons were removed and while still wet, immersed in a water suspension of tetraiodoethylene for about 5 minutes. Immediately thereafter and while still wet, the melons were individually wrapped in relatively close fitting polyethylene bags. After this treatment, the bagged melons were sealed in their bags and kept at a temperature of 2°C. These melons had a useful life and were edible for a period of over 76 days.

SURFACTANT-IODINE COMPLEX

Harvested fruits and vegetables are sprayed or dipped into an aqueous solution containing an iodine-releasing complex in the process of *R.P. Gabriel; U.S. Patent 3,620,773; November 16, 1971.* The biocidal effect of the aqueous solution is provided by the action of the iodine from the surfactant-iodine complex which is capable of destroying fungical spores and bacteria of those plant pathogens usually involved in economic losses of stored agricultural products. The surfactant-iodine complex is used instead of pure iodine, which is highly toxic and can cause severe skin irritations and blisters in warm-blooded animals, particularly humans. This process, therefore, uses the iodine in a complex with a compatible surfactant which has the capability of releasing iodine in a non-toxic form over an extended period of time. When used in this form, the irritating and sensitizing effects of the iodine is significantly reduced although the biocidal and germicidal properties are substantially unimpaired.

One form of the surfactant-iodine complex is a polyvinylpyrrolidone iodine complex formed by mixing and heating elemental iodine or an iodine containing compound with polyvinylpyrrolidone. The iodine may be in the form of pure iodine, Lugol's solution, colloidal iodine suspensions, salts of iodine, iodides, iodates and iodites or acids of iodines such as HI and HIO. Regardless of the form, iodine mixed with the polymer in amounts between 1.5 and 25% by weight based on the weight of the complex will in turn enable certain dilutions to provide adequate biocidal action on the harvested product. The iodine may be incorporated into the surfactant either by a dry mixing or grinding process or by mixing the iodine and surfactant in an aqueous solution under conditions of mild heating.

Polyvinylpyrrolidone-iodine complexes are preferred because of the physiological acceptance of polyvinylpyrrolidone. The vapor pressure of the iodine when complexed with polyvinyl-pyrrolidone is reduced to a degree that the bactericidal and germicidal properties of the complex can be retained over a particularly long period of time. Moreover, the foaming characteristics of polyvinylpyrrolidone make it particularly advantageous for use in spray solutions or dip solutions.

While polyvinylpyrrolidone is the preferred form of surfactant-iodine complex, other complexes can similarly be prepared and used. For example, water-soluble interpolymers of polyvinylpyrrolidone with other vinyl phthalamide, vinyl pyridine, acrylamide or vinyl caprolactam may be used. Homopolymers, chemically related to polyvinylpyrrolidone, may be used such as polyvinyl caprolactam, polymerized gamma-valerolactam and poly-vinyl-ϵ-valerolactam. Other complexes which are suitable due to their safe nontoxicity for most warmblooded animals may be formed with nonylphenoxy-polyoxyethylene-ethanol or alpha-(p-nonylphenol)-omega-hydroxypoly(oxyethylene).

The surface active agent used with the surfactant-iodine complex functions to reduce the surface tension of the solution so that maximum wetting and draining of the treated surface will be accomplished. For this purpose, any compatible surfactant can be used, but particularly good results are obtainable with anionic and nonionic surface active agents such as the alkyl aryl sulfates, the alkyl aryl sulfonates, the higher fatty alcohol sulfates, the higher fatty alcohol sulfonates, the polyglycol ethers of alkyl phenols, the higher fatty alcohols and the polyglycol ether esters of higher fatty acids. Especially good results are obtainable with the polyoxyethylated nonylphenols condensed with 9 to 12 mols of ethylene oxide, since it is especially safe and nontoxic to warmblooded animals.

The acid and the aqueous solution acts primarily as a buffer to maintain the solution in at least a slightly acidic condition. The specific pH is dependent upon the particular harvested products being treated and on the degree of biocidal action required. It has been found, however, that for most harvested products, a pH of approximately 3 to 4.5 provides good results. Suitable acids for this purpose include the compatible organic and mineral acids such as phosphoric acid, hydrochloric acid, glycolic acid, citric acid, acetic acid, tartaric acid, fumaric acid and salicylic acid. Phosphoric acid and glycolic acid are particularly desirable when used in amounts sufficient to provide a pH of about 3.3 since they provide synergistic results in combination with the surfactant-iodine complex to inhibit the development of *Fusarium roseum* spores. Glycolic acid will also provide synergistic effects in controlling the development of *Verticillium albo-atrum* spores and *Corynebacterium sepedonicum*.

Example: The following specific formulations are suitable for use in the treatment of harvested, nonchlorophyllous produce.

Formula 1:

Synthetic surfactant-iodine complex (as VRO-20) (Nonylphenoxypolyoxyethyleneethanol-iodine complex provides 1.75% titratable iodine)	9%
Glycolic acid (100%)	5%
Surfactant (as Igepal 880) (polyoxyethylated nonylphenol)	7%
Water	79%

Formula 2:

Synthetic surfactant-iodine complex (as VRO-20) (Nonylphenoxypolyoxyethyleneethanol-iodine complex provides 1.75% titratable iodine)	9%
Phosphoric acid (100%)	5%
Surfactant (as Igepal 880) (polyoxyethylated nonylphenol)	7%
Water	79%

Formula 3:

Polyvinylpyrrolidone-iodine complex (as PVP Iodine 10) (provides 1% iodine)	10%
Phosphoric acid (100%)	2%
Surfactant (as Igepal 630) (polyoxyethylated nonylphenol)	1.5%
Water	86.5%

Formula 4:

Polyvinylpyrrolidone-iodine complex (as PVP Iodine 10) (provides 1% titratable iodine)	10%
Glycolic acid (100%)	2%
Surfactant (as Igepal 630) (polyoxyethylated nonylphenol)	1.5%
Water	86.5%

Formula 5:

Polyvinylpyrrolidone-iodine complex (as PVP Iodine 10) (provides 2% titratable iodine)	20%
Phosphoric acid (100%)	5%
Surfactant (as Igepal 630) (polyoxyethylated nonylphenol)	3%
Water	72%

Bacteriological tests using these formulas show a high degree of efficacy in the treatment of various plantborne, waterborne, and airborne pathogens. In conducting these tests, the following procedure was followed. Two milliliters of Czapek-Dox broth were added to test tubes containing 2 ml of Formula 1, yielding a 75 ppm iodine solution. One loopful of the test organism from the stock cultures were added to each tube and agitated. The controls were prepared in a similar manner with the exception that 2 ml of sterile distilled water were added to the tubes in place of Formula 1. After 2 minutes a loopful of bacteria plus liquid were withdrawn from each tube aseptically and inoculated into a flask containing 100 ml of the Czapek-Dox broth. The inoculated flasks were placed on a shaker for 24 hours for *Erwinia atrospectica* and *Pseudomonas fluorescens*. *Corynebacterium sepedonicum* was maintained in the shaker for 3 to 5 days because of the slow growth rate. Bacterial growth was determined by measuring turbulence of the broth as optical density on a Spectronic 20 at 555 mμ and the treatments were compared with controls.

One ml of 4x concentration of Conidia was added to 1 ml of Formula 4. At the end of 2 minutes, 2 ml of 0.1 N sodium thiosulfate were added to each of the 3 tubes and the tubes were agitated. The sample was withdrawn from each tube by a pipette and a drop placed on either end of the glass slide. The slide was placed on a Petri dish and incubated at room temperature for 24 hours. The drops containing the spore suspensions were then examined microscopically. Spores were considered to be germinated if the germ tube was longer than the width of the spore. Controls were treated in a similar manner except that 1 ml of sterile distilled water was substituted for Formula 4. Germination was determined by counting 100 spores per drop reported as percent germination. The following table summarizes the results of these tests.

Formulation Diluted to Make Available 75 ppm Iodine	Verticillium albo-atrum (Verticillium Wilt)	Fusarium roseum (Tuber Rot)	Corynebacterium sepedonicum (Ring Rot)	Erwinia atrospectica	Pseudomonas fluorescens
Formula 1 growth after 2 min exposure	0	0	0	0	0
Control ter	84*	0.180**	0.21**	0.105**	0.02**
Formula 4 growth after 2 min exposure	0	0	0	***	***
Control	39*	4.54*	94*	......	......

*Average percent germination. **Optical density. ***No data.

Five day old Czapek-Dox broth cultures of *Corynebacterium sepedonicum* and 10 to 14 day old cultures of *Fusarium roseum* grown on potato dextrose agar were used in the experiments. The control formulation of agar used in these tests contained infusion from 200 grams (20%) autoclaved potatoes, 15 grams of dextrose, 15 grams of agar per liter. A 1 to 1,000 ml dilution of *Corynebacterium sepedonicum* was prepared and 1 ml was added to 999 ml of sterile distilled water under aseptic conditions. A sterile loopful of this solution was streaked across each plate containing the test material in the agar. For inoculation of *Fusarium roseum*, sterile distilled water was added to a pure culture of *Fusarium roseum*. With a sterile loop, the spores were loosened from the mycellium by agitation. A loopful of the spore streaked across the plates. Plates of all the tests were examined for evidence of growth 2 to 4 days after inoculation. The results of this test are summarized below.

Formulation diluted to make available 150 p.p.m. iodine	Fusarium roseum (tuber rot)	Organism Verticillium albo-atrum (Verticillium wilt)	Corynebacterium sepedonicum (ring rot)
Formula II growth in 100% potatoe agar.	0	Not tested	0
Control	(1)		(1)
Formula III growth in 100% potatoe agar.	0	do	0
Control	(1)		(1)

1 Heavy growth.

In using these solutions for treating harvested produce, it has been found that 1 gallon of concentrate will effectively inhibit the growth of deleterious pathogenic microorganisms in 20,000 pounds of potatoes. The preferred dilution of the concentrate is about 65 gallons of water for each gallon of concentrate.

ENZYME INACTIVATION IN ONIONS

Fresh onions are treated by a process of *G. Kruse; U.S. Patent 3,138,464; June 23, 1964* with a special salt-acid solution to inactivate the enzymes which cause the development of aftertaste and discoloration. It has been found that the influencing or diminishing of long lasting and undesirable aftertaste and the unpleasant mouth odor which results from the consumption of fresh, raw onions is due to allyl mustard oil, i.e., one of the volatile oils of the onions. It appeared to be an obvious step to diminish this unpleasant aftertaste by diminishing the content of volatile oil.

It is possible, however, to reduce the unpleasant aftertaste without diminishing the content of volatile oil since the highest possible content is desirable if the onion taste is to be retained completely. From the theoretical and experimental work which has been done, it would appear that specific, characteristic bisulfides such as allyl propyl bisulfide and thiocyanogen compounds such as allyl thiocyanate or thiocyanic acid are responsible for the unpleasant aftertaste of the onion. It is extremely probable that a part of these substances is present in glycoside bond and, in the same way as sinigrin in mustard, requires to be released by enzymatic processes. In this connection, ferments of the type of the glycosides play the decisive part. On the basis of these presuppositions, it has been found that by influencing the enzymatic reaction during and after the processing of the onions, the long lasting, unpleasant aftertaste may be diminished without detrimentally influencing the immediate taste of the fresh, raw onions.

Freshly peeled onions are comminuted in such a manner that less than 5% of the cells in their microscopic structure is destroyed and the comminution is carried out until a piece size of approximately 5 mm lateral length is achieved and considerable or complete inactivation of the enzymes in the onions is achieved. This is done by adding to the onion preparation during comminution approximately 7% sodium chloride, 1.1% pectin, 0.17% ascorbic acid and 0.5% citric acid.

MUSHROOM SULFITE TREATMENT

Mushrooms, even under refrigeration, are subject to discoloration, flavor loss and premature spoilage particularly where they are sliced or cut up for packaging purposes. To prevent loss of flavor and discoloration and to prevent premature spoilage, a special method of preservation is used by *C.C. Molsberry; U.S. Patent 3,342,610; September 19, 1967.*

The mushrooms are treated initially in a sealer composition consisting essentially of a mixture of three inorganic salts: sodium sulfate (anhydrous), 27.5% by weight; disodium phosphate (anhydrous), 25.8% by weight and sodium metabisulfite, 45.7% by weight. These compounds are readily soluble or dispersible in selected liquid media. In addition to the sealer, there is a bleaching mixture of the following composition: sodium sulfate (anhydrous), 27.3% by weight; sodium chloride, 29.0% by weight and sodium metabisulfite, 41.7% by weight. The following steps are given for the process.

Step 1: The mushrooms are dipped in an aqueous solution containing 1½ ounces of the sealer composition mixed with 6 gallons of water and left in the solution for 1 to 1½ minutes.

Step 2: The mushrooms are dipped in an aqueous solution containing 3 ounces of the bleaching mixture mixed with 6 gallons of water and left in that solution for 3 to 4 minutes.

Step 3: The mushrooms are sliced or cut into pieces.

Step 4: The mushroom slices or pieces are then given a further treatment according to Step 1 above.

Step 5: The mushroom slices or pieces are then given a further dipping and soaking according to Step 2 above.

The five steps follow each other in rapid sequence and afterwards the mushrooms are subjected to flash freezing or refrigeration for 3 to 4 minutes and are then placed under ordinary refrigeration. During processing the mushrooms increase in weight by about 15 to 20%. It was found on examination that the treated raw mushrooms had an SO_2 residual of only 74 parts per million, well below the required SO_2 residual count recorded as being suitable for distribution and public consumption. When the treated mushrooms are fried or otherwise cooked, they are found to be of excellent flavor and color and contain the insignificantly low SO_2 residual count of 8 parts per million.

FUNGICIDES TO PREVENT DECAY

Diphenylamine

Apple scald is a physiological disfunction which occurs in many varieties of apples at some time after harvesting the fruit, and in some seasons it may be particularly severe on fruit which has been placed in refrigerated storage. It may also occur on fruit which has been held at room or ambient temperatures. Apple storage scald is characterized by a superficial browning of the apple surface affecting a few layers of cells directly below the cuticle. The unsightly appearance of the scalded apples lowers their market value, and the damage to the skin increases susceptibility to invasion by fungi. While scald affects most varieties of apples customarily held in refrigerated storage, some varieties are noticeably more susceptible than others. Rhode Island Greening, Northwestern Greening, Cortland, Stayman, Rome and Grimes are among the varieties highly susceptible to scald. Other varieties may be fairly resistant, yet in some seasons they are also subject to the disorder.

The magnitude of this problem becomes more apparent upon consideration of the facts that the occurrence of this disfunction is entirely unpredictable both as to its latent presence and as to the severity of the outbreak. It is not possible to determine in advance by any known tests or procedures whether or not scald will occur, and it is only after removing the fruit from storage and holding it or shipping it through the normal channels of commerce that the scald malady shows up. Thus the value of the entire contents of a cold storage fruit warehouse may be in jeopardy with no known remedy and with no possible way of predicting whether or not the fruit is storable at the time of placing it in storage and whether or not the fruit will be salable shortly after removing it from storage.

Since storage is basic to the apple industry and is necessary for the orderly and profitable use of each season's crop, it can be seen that scald presents a problem of great magnitude and importance and that any methods whereby its very costly losses can be controlled is of tremendous nutritional and economic significance.

A special aqueous suspension of diphenylamine is used by *M. Kleiman; U.S. Patent 3,034,904; May 15, 1962* to treat apples to reduce scald and invasion by fungi. The diphenylamine is in the form of a suspension or emulsion in water with water as the continuous phase, the diphenylamine in finely divided state forming the disperse phase, in substantially the complete absence of any added organic solvent.

The finely divided diphenylamine should be of such particle size that essentially the maximum dimension of the particles is below 150 microns. It is best to use diphenylamine particle sizes which essentially have an approximate 75 microns maximum dimension or smaller in order to provide a large area of exposed total surface for coating or leaving a protective residue on the surfaces of the apples to be treated and to facilitate maintenance of a suspension or emulsion which is free from separation and settling for reasonably prolonged periods sufficient for proper use and application. The finely divided diphenylamine is applied to the apples to be protected in a prewetted state since it has been found that under such conditions it is extremely active in producing a protective residual film.

The aqueous suspensions or emulsions can be prepared in a number of ways. For example, the diphenylamine and water mixture can be put through a colloid mill, also ultrasonic vibrations can be made use of or any other intensive mechanical agitation and/or grinding serving to produce and disperse the finely divided diphenylamine in the water to result in a sufficiently stable dispersion of desired particle size can be used. If desired, suspending or emulsifying agents can be added to render the dispersions more stable, but such additions are not necessary as even without such additions dispersions of useful stability can be attained.

Suitable suspending or emulsifying agents, for example, are the synthetic commercially available surface active agents of the sulfate or sulfonate type such as, for example, sodium lauryl sulfate, sodium dodecylbenzene sulfonate, sodium alkyl naphthalene sulfonate, long chain alkyl sulfonates, nonionic surfactants such as the polyethoxyethanol esters and ethers of fatty acids and alcohols.

Such aqueous suspensions or emulsions may be made up in a concentrated form containing diphenylamine up to 85 to 90%. Alternatively, the diphenylamine may be finely powdered and reduced to particle sizes and this fine powder may be mixed with other powdered inert ingredients which may improve the free flowing and shelf storage characteristics of the material. Such inert ingredients may, for example, consist of various finely divided clays, talcs, kaolins, synthetic calcium silicates and calcium alumino-silicates, bentonites, fuller's earths, pumice, silica, chalk, magnesium carbonate, and other commonly available inert carriers. Surfactants may be added to render the compositions more rapidly dispersible in water. Such dry compositions may likewise be prepared in the range of concentrations up to about 85 to 90% diphenylamine and they are diluted down with and dispersed in water for use.

The diphenylamine suspensions or emulsions are prepared simply by diluting down the more concentrated preparations with water so as to result in a suspension or emulsion containing from about 0.02 to 1.0% of diphenylamine. In most instances, the preferred range of useful concentrations will be from about 0.1 to 0.3%.

These dilute diphenylamine suspensions or emulsions can be applied to the surfaces of apples in a number of ways, such as, for example, by spraying the suspensions or emulsions on the apples while the latter are still on the tree prior to harvest, or by spraying the suspensions or emulsions on the apples after they have been harvested, or during grading, or after packing, or the apples may simply be dipped in the suspensions or emulsions and then allowed to drain and dry or placed wet in the storage rooms.

The particular manner in which these suspensions or emulsions are applied will in some degree determine the most effective concentration of finely divided active diphenylamine suspension or emulsion to be used. Thus, for example, in using tree sprays prior to harvest, where there is bound to be less efficient use of the material being applied as compared with sprays directed on the harvested fruit itself, or as compared to the use of a fruit dip as the means of fruit treatment, higher concentrations are generally required than when more efficient means of application are chosen. In general, for tree sprays, concentrations of 0.3% or higher may be desirable, whereas for dip treatments concentrations of 0.1 to 0.2% are preferred. Even higher concentrations may be required for effective prevention or reduction of scald incidence in the case of tree sprays in some instances, depending upon such factors as the output of the spray machinery, the pattern of coverage, and the speed at which the spray equipment is drawn through the orchard.

Example: Diphenylamine crystals (2 lb) of over 99 mol percent purity were placed in a ball mill with 38 lb of water, and the mixture was tumbled for 48 hours. The resulting fine suspension was screened to separate particles larger than 250 microns. Only a minute quantity was retained on the screen. The suspension of diphenylamine settled slowly upon standing and redispersed readily upon gentle agitation.

A 100 gallon batch of dilute diphenylamine suspension was prepared by diluting 3.3 lb of this 5% concentrate with water to result in a final diphenylamine concentration of 0.02% by weight. One hundred freshly harvested Rhode Island Greening apples were sprayed with this suspension until thoroughly wet to the point of runoff, and placed in storage at 0°C and 90% relative humidity. The apples were removed after 125 days, held at room temperature (20° to 22°C) for 7 days and then were examined for scald incidence.

For comparison, 100 freshly harvested untreated apples of the same variety (control #1), and 100 freshly harvested apples of the same variety sprayed with a 0.02% diphenylamine solution prepared by diluting a solution of 75 grams diphenylamine in 1,500 ml of 95% ethyl alcohol with 100 gallons of water (control #2) were placed under identical storage conditions and examined for scald incidence in the identical manner as the test apples. The following results were observed.

	Percent Scald	Percent Scald Control
Test apples	44	55.6
Control #1	99	0
Control #2	78	21.2

Resin and Diphenyls

A.F. Kalmar; U.S. Patent 3,189,467; June 15, 1965; assigned to FMC Corporation describes the benefits and procedure of using a resin in combination with diphenyl as a fungicide for prevention of fruit decay. The resin appears to regulate the release of the diphenyl from the solution so that the latter slowly diffuses from the resin film at such a rate as to produce a fungistatic "skin" of diphenyl vapor around the fruit, the presence of which skin may last for a considerable time, and thus increase the normal market life of the fruit. The resin should be present in the solution in a concentration of 3 parts resin per 1 part diphenyl.

The factors which will govern the particular concentration of fungicide are the solubility of the fungicide in the particular resin solution employed, the phytotoxicity of the fungicide, the residue of fungicide on the fruit permitted by government regulation, the effect of the fungicide on the appearance, taste and aroma of the fruit, and the economic factors involved. Approximately 10 ppm of fungicide, based on the whole fruit weight, should usually be applied for a minimum degree of mold control.

The organic solvent used for the solution of the fungicide and resin is a petroleum distillate or an alcohol type solvent. If the coating is to be applied as a spray, a solvent having an initial boiling point of 100°F and a final boiling point of 350°F is preferred. Representative resins which are suitable are the hydrocarbon terpenes, polyindenes, coumarone indenes, phenolic resins, polymerized hydrocarbon resins, chlorinated paraffins, terpene phenolics, alkyds and polyamides. Rosin and rosin compounds and resins are also suitable.

Example 1: A solution of resin and fungicide was prepared by dissolving 10% of the resin known commercially as Piccovar FM resin and 2% diphenyl in 88% of the petroleum solvent known commercially as SOCAL 100-L. The solution was applied to lemons by a spraying process such as disclosed in U.S. Patents 2,212,621 and 2,342,063, spray nozzles known commercially as #4.00 Monarch nozzles operating at 40 lb pressure, the conveyor speed being 126 valleys per minute, and approximately 70 ppm or 1.4 grams of coating per carton being deposited.

Example 2: A resin-fungicide solution was prepared following the method of Example 1 except that 2% of ortho-phenylphenol was used instead of diphenyl. The manner of application was the same as in Example 1.

Example 3: A resin-fungicide solution was prepared following the method of Example 1 except that 2% of diphenyl and 2% of ortho-phenylphenol were dissolved in 86% of the solvent SOCAL 100-L. The solution was applied as in Example 1.

Example 4: For comparison purposes, a solution of 10% Piccovar FM resin alone was dissolved in 90% SOCAL 100-L and applied to lemons in the manner set forth in Example 1. The Piccovar FM resin, a coumarone-indene resin, has the following characteristics.

Color (coal tar scale)	maximum 2
Ball and ring softening point, °C	158 to 165
Resin mineral spirits solubility, °C	98
Specific gravity at 25/15.6	1.10
Refractive index (n^{20}/D)	1.62
Approximate molecular weight	800

Iodine numbers of the Piccovar FM resin by a variety of methods indicate a very low order of reactivity and the resin shows little unsaturation. SOCAL 100-L is a petroleum solvent having the following characteristics from which the solvent can be made.

Gravity, °API	49.7
Aniline point, °F	65.5
Kauri-butanol value	57

Distillation, ASTM D-86, percent recovered, °F:

Initial boiling point	190
5	200
10	208
20	214
30	221
40	228
50	234
60	240
70	250
80	260
90	268
95	272
Dry point	282
End point	302

The following table sets forth the results of tests illustrating the effectiveness of solutions prepared and applied in controlling mold growth at the surface of lemons.

Type of Coating	Percent Mold After 13 Days
Example 5	11.4
Example 1	0.18
Example 2	1.17

The following table sets forth test results comparing the effectiveness of solutions prepared and applied in controlling mold growth in lemons which have been injured and mold inoculated.

Type of Coating	Percent Mold After 1 Week
Example 4	98.0
Example 2	18.4
Example 3	6.2

As is apparent from the tables, solutions containing a resin and fungicide substantially limit mold growth such as would occur when solutions containing a resin alone are used.

Dibromotetrachloroethane

In another process by *A.F. Kalmar; U.S. Patent 3,189,468; June 15, 1965; assigned to FMC Corporation* vaporizable fungicidal agent DBTCE is used with a waxy release material to preserve fruits and vegetables. The fungicidal agent is incorporated along with a waxy material in a volatile organic solvent and the solution is applied to the surface of perishable products such as fruits and vegetables.

As a result of its application directly to the surface of the fruit or vegetable, the fungicide acts in several ways. First, there is direct contact with the mold spores on the surface. Secondly, the fungicide vaporizes and diffuses into the cuts and punctures in the surface and into other areas where there may not have been direct contact. There is also evidence that the fungicide diffuses through the waxy cutin of citrus fruit and into the peel tissue thereby providing further protection against mold growth. Since the fungicide is incorporated in a solution and no emulsifier is present, the fungicide intimately contacts the mold spores.

The waxy material present in the solution along with the fungicide provides conditions of fungicidal action not possible when the fungicide is used alone. The waxy material appears to regulate the amount of DBTCE which comes into intimate contact with the fruit surface at any one time. The waxy material appears to controllably release the DBTCE so that the latter slowly diffuses from the waxy film at a rate so as to produce a fungistatic "skin" of DBTCE vapor around the fruit, the presence of which skin may last for a considerable time and thus increase the normal market life of the fruit. DBTCE has a high vapor pressure and will not persist on the surface of the fruit if it is deposited by itself from a solvent solution. Accordingly, the use of a waxy material in a solvent solution provides a medium by which the DBTCE can be effectively applied to fruits and vegetables.

The volatile organic solvent used for the solution of fungicide and waxy film former is preferably a petroleum distillate. Other volatile solvents may be used such as ethyl and isopropyl alcohol. In treating citrus fruit by the process, petroleum solvents having boiling points respectively in the ranges of 210° to 280°F, 150° to 230°F, and 180° to 380°F have been employed. Generally, at least about 90% of the solvent should have a boiling point below 300°F.

The waxy material should be completely dissolved in the solvent. The amount of waxy material which can be held in solution varies with the temperature of the solution and it is preferable to use about 10% by weight, or more, of waxy material in solution. Representative waxy film formers which are suitable are the coumarone-indene resins, polyindene resins, hydrocarbon terpene resins, phenolic resins, alkyd resins, polyamide resins, rosin and modified rosins, paraffin wax, spermacetic wax and shellac.

Example 1: A solution of resin and fungicide was prepared by dissolving 10% by weight of the resin known commercially as Nevindene R-1 and 4% by weight DBTCE in 86% by weight of the petroleum solvent VM & P Naphtha. This solution was applied to lemons by a spraying process using spray nozzles known commercially as #4.00 Monarch nozzles operating at 40 pounds pressure. The conveyor speed was 126 valleys per minute and approximately 140 parts per million were applied on each lemon, that is, for each lemon of approximately 100 grams weight, approximately 0.014 gram of solid coating of resin and DBTCE was applied.

Example 2: A solution of resin and fungicide was prepared by dissolving 12% by weight of the modified rosin, Limed Poly-pale, and 3% by weight DBTCE in 85% by weight of the petroleum solvent VM & P Naphtha. This solution was applied to oranges under the conditions set forth in Example 1. Limed Poly-pale is formed by reacting a poly-pale resin with lime.

As is known, rosin is composed largely of unsaturated isomeric resin acids such as pimaric and abietic acids. The process used for preparing Poly-pale resin causes a portion of these unsaturated resin acids to react with each other through their double bonds to form polymers. The concentration of the polymers, as in Poly-pale resin, is approximately 40%, with the remaining portion being the normal constituents of rosins. To make the Limed Poly-pale resin, the Poly-pale resin is reacted with hydrated lime under controlled temperature conditions.

The following table sets forth the results of a test illustrating the effectiveness of a process and coating as described in Example 1. Fresh lemons were artificially injured and inoculated with spores of *Penicillium digitatum*. The fruit was then treated as described in Example 1. After storage at 57°F for 10 days and at 72°F for 7 days, the fruit was examined for mold with the following results.

Treatment	Percent Decay After 17 Days
10% by weight Nevindene resin in VM & P naphtha	96.5
10% by weight Nevindene resin plus 4% by weight DBTCE in VM & P naphtha	16.0
10% by weight Nevindene resin plus 6% by weight DBTCE in VM & P naphtha	11.5

It is apparent from this test that a waxy material coating containing the fungicide DBTCE has a marked effect on control of decay as compared with a coating containing the waxy material without DBTCE.

Results of tests using the solutions prepared above indicate that the process and coating enable good control of *Penicillium digitatum, Penicillium italicum,* Botrytis, Rhizopus, Diplodia, Phomopsis, Trichoderma, Geotrichum, and Alternaria. The presence of the DBTCE in the coating film enveloping the fruit and vegetables does not significantly affect the appearance, odor or flavor characteristics of the treated produce.

It has been determined that by the process the waxy material is so deposited on the surface of the fruit that not only is the desired control of decay obtained, but also the evaporation of moisture from the fruit is reduced by about 30 to 60% as compared with the moisture that would be lost if the fruit was marketed without the subject coating. Furthermore, the waxy material coating also improves the appearance by imparting to it a desirable gloss.

Thiazolyl Benzimidazole and Phenylphenol

All fruit are subject to attack by organisms such as molds or fungi which cause decay of the picked fruit. In addition to decay of fresh fruit, there is a problem of the sporulation of molds and fungi which are growing on the fruit. Molds and fungi will attack and cause decay of the specific fruit which they are in contact with; and if they sporulate, the molds and fungi spoil adjacent fruit and render them unsalable.

In the past, generally one compound was applied to the fruit in order to prevent decay of the fruit due to molds and fungi and another compound was applied to the surface of the fruit in order to prevent or inhibit sporulation of the molds and fungi. It is self evident that utilizing two separate operations to prevent decay and sporulation is uneconomical and, in addition, the most prevalent compound (biphenyl) utilized in preventing sporulation of fruit has many objectionable disadvantages, such as the very strong odor found on fruit treated with biphenyl.

J.R. Bice and P.J. Lewis; U.S. Patent 3,674,510; July 4, 1972; assigned to Brogdex Company use a mixture of 2-(4-thiazolyl) benzimidazole (TBZ) and an alkali metal salt of ortho-phenyl-phenol tetrahydrate (AOPP) for the simultaneous prevention of decay due to mold and fungi and to inhibit sporulation.

Both TBZ and AOPP have been utilized before in preventing decay of fruit. The Food and Drug Administration has approved the use of these two compounds in preventing decay of fruit providing that the TBZ residue left on the fruit does not exceed 2 parts per million by weight and the AOPP residue (expressed as ortho-phenylphenol) does not exceed 10 parts per million.

Since these two compounds do not adversely affect the appearance or color of the fruit and do not have objectionable odors, it would be desirable if these compounds could replace biphenyl and the other objectionable compounds utilized in preventing sporulation of molds and fungi growing on the fruit. However, when TBZ and AOPP are utilized in dosages permissible by the Food and Drug Administration neither compound will inhibit sporulation of molds and fungi to any significant extent.

This process gives the critical amounts and procedures to make compositions containing TBZ and AOPP effective by a synergistic result. The TBZ and AOPP can be applied to the fruit separately or in the same composition. That is, the AOPP may be applied first, for example, in solution in a foam for washing the fruit and the TBZ applied later. The TBZ can also be applied in a so-called water eliminator rinse, which removes the foam from the fruit. In the alternative, the TBZ can be first applied in the form of an aqueous dispersion and thereafter a composition containing both TBZ and AOPP can be applied, providing that the total amount of TBZ remaining on the fruit does not exceed about 2 parts per million by weight, based on the weight of the fruit.

It should be noted that in the following examples all parts are by weight unless expressly stated otherwise. In addition, the residue of AOPP is expressed in parts by weight based on the amount of ortho-phenylphenol (OPP) rather than on the hydrated alkali metal form. The reason for this is because this is the manner in which the Food and Drug Administration sets a limit on the amount of residue permissible on the fruit and because the weight obviously depends upon the particular metal utilized and the amount of water combined in the basic compound.

Therefore, in speaking of the residue of AOPP it will be understood that throughout the specification the amount of AOPP is expressed as the amount of OPP left on the fruit. However, when speaking of the amount of AOPP in a particular composition, the amount is by weight based on the AOPP unless expressly indicated otherwise.

Salts of ortho-phenylphenol which have proven particularly effective are the alkali metal salts such as potassium and sodium. The alkali metal salts are all water-soluble and therefore it is convenient to utilize these alkali metal salts in aqueous compositions so that the AOPP is in solution.

TBZ is not water-soluble to any significant extent and therefore when utilized in an aqueous composition, it is preferred to have a dispersion. However, since TBZ is soluble in a number of organic solvents, the use of such solvents are not excluded and it is contemplated that the solvent can be miscible in water thereby having a solution of TBZ and AOPP. But it should be emphasized that AOPP and TBZ can be applied directly to the fruit as a solution, suspension, dispersion, foam or any other manner. If the TBZ and AOPP are applied directly to the fruit in the amounts specified, such treatment will inhibit organisms which cause decay and concurrently prevent formation of spores on the surface of any decayed fruit thereby preventing spoilage of adjacent sound fruit by the spores.

The TBZ and AOPP, either separately or together, can be formulated with waxes, resins, or coloring material used to enhance appearance or to improve gloss or to retard shrinkage of the fruit in question. Various methods of application may be used, such as washing, dipping, spraying or rubbing.

It is generally preferred to have a residue of TBZ on the fruit as close as possible to the legal permissible limit. It was found that if the TBZ is present from 1 to 2 parts per million, satisfactory results are obtained, although the higher the concentration, the better, e.g., from 1.4 to 2 parts per million. The particularly preferred range would be to have a TBZ residue on the fruit between 1.3 and 2 parts per million, (e.g., over 1.5 parts per million).

It is similarly desirable to have the residue of AOPP on the fruit be at its permissible legal limit, i.e., 10 parts per million. Here again, very good results have been obtained by utilizing an amount of AOPP which results in a residue of from about 5 to 7, (e.g., 6) parts per million on the fruit, but in certain instances better sporulation control has been obtained when a residue of from 8 to 10 parts per million remains on the fruit.

In the following example, TBZ and sodium ortho-phenylphenol (SOPP) were added to a wax formulation having the following formula.

Constituent	Amount, Parts by Weight
Water	76
Protein	1.3
PW*	17
Shellac	3

*PW is a polyethyleneoxide wax composed of 40 parts by
weight of Epolene 45 (low molecular polyethylene oxide
resin), 7 parts by weight of oleic acid and 5 parts by weight
or morpholine.

From this wax was made three compositions: one containing the wax and 3.5 parts by weight of TBZ, a second containing about 2 parts by weight of SOPP and a third containing 2 parts by weight of SOPP and 0.35 part by weight of TBZ.

Two hundred forty navel oranges were inoculated with *Penicillium digitatum* and the fruit divided into four separate groups containing 60 oranges each. To Group 1 was applied the wax formulation, to Group 2 was applied the wax formulation containing TBZ, to Group 3 was applied the wax formulation containing SOPP, and to Group 4 was applied the wax formulation containing TBZ and SOPP. After the wax formulation had been applied to the fruit, the fruit were stored for three weeks at between 45° and 50°F and for one week at 65° to 75°F. At the end of four weeks, the fruits were inspected for sporulation control and decay control and analyzed for the residue of SOPP and TBZ on the fruit. The results of these tests are given in the following table.

| | Percent | | Residue of— | |
Formulation	Decay	Sporulation control	TBZ (p.p.m.)	AOPP (p.p.m.)
1	49.5	0	0	0
2	7.1	25	1.5-2	0
3	25	0	0	5-7
4	8.6	90	1.5-2	5-7

It should be noted that in order to control decay and sporulation it is necessary that the correct residue of TBZ and AOPP be left on the fruit and therefore the percentages of AOPP and TBZ in the compositions are not too important (except that the weight ratio of TBZ and AOPP must be correct). However, it should be noted that it is preferred that TBZ be present in an amount of from 0.25 to 0.5% (preferably from 0.3 to 0.4%) and that the AOPP be present in an amount of from 1.0 or 1.5% to 2.5 or 3%. The amount of resin can range from 10 to 20% and the amount of water from 60 to 90%.

In another example oranges were scratch inoculated with both *Penicillium italicum* and *Penicillium digitatum*. The fruit were divided into three substantially equal groups and labeled Groups 1, 2 and 3. To the Group 1 oranges was applied the wax formulation given above, the Group 2 oranges were sprayed with water containing 0.125 part of TBZ and thereafter there was applied the wax formulation containing 2 parts of AOPP, and to the Group 3 oranges was applied the wax formulation containing 0.35 part of TBZ and 2 parts of AOPP. All three groups of oranges were stored at 56°F and 95 to 100% humidity for 11 days. The fruit were then inspected and the following results obtained.

| | Percent | | Residue of— | |
Formulation	Decay	Sporulation control	TBZ (p.p.m.)	AOPP (p.p.m.)
1	31.25	5	0	0
2	3	20	0.4	6-7
3	3	97	1.8-2	6-7

There is no difference between Groups 2 and 3 in decay control but Group 3 gave 77% better sporulation control on decaying fruit. It should be noted that the foregoing tests demonstrate that utilizing TBZ and AOPP in the correct amounts inhibit sporulation to a much greater extent than utilizing biphenyl (which inhibits sporulation about 40%).

These compositions and methods are equally effective on other fruit such as lemons, as demonstrated by the following example. Lemons were inoculated and held overnight at 58°F. The lemons were then divided into two groups labeled Group 1 and Group 2. The Group 1 lemons were sprayed with an aqueous composition containing 2% dissolved SOPP. The Group 2 oranges were also sprayed with the 2% SOPP composition and immediately thereafter sprayed with the wax composition described above containing 0.35% TBZ and 2% SOPP. The fruit were stored at 58°F and 90% relative humidity and examined. The results are shown in the table on the following page.

Formulation	Percent		Residue of—	
	Decay	Sporulation control	TBZ (p.p.m.)	SOPP (p.p.m.)
1	7.15	0	0	5
2	0.2	75	1.7	6-7

5-Acetyl-8-Hydroxy-Quinoline Salts

A series of microbicidal active substances for preserving fruit is known from the literature, e.g., diphenyl, alkali metal salts of o-hydroxydiphenyl and salicylanilide, also thiourea, ammonia, sulfur dioxide, carbon dioxide, etc. Such substances, however, only partially meet the desired requirements as either the breadth and duration of action is too slight or they are too toxic to warm-blooded animals or they have an unpleasant smell.

E. Hodel and K. Gatzi; U.S. Patent 3,759,719; September 18, 1973; assigned to Ciba-Geigy Corporation uses microbicidal amounts of the sulfate and dihydrogen phosphate salts of 5-acetyl-8-hydroxy-quinoline corresponding to the formulas:

Melting Point 237° to 238°C

Melting Point 230° to 235°C
With Decomposition

They are active against a broad spectrum of microorganisms which attack fruits, roots or other edible parts of plants after harvesting, among which microorganisms there are especially bacteria, yeasts and fungi. The salts are particularly valuable as active substances in preserving agents which are used, e.g., for citrus fruits such as grapefruit, oranges, lemons, also bananas, pineapple, avocados, mango, guanabana, nut kernels, apples, pears, peaches, apricots, plums, grapes, strawberries, guava, sweet potatoes, potatoes, yams, carrots, turnips, sugar beets, onions, lettuce, artichokes, cabbage, asparagus, edible mushrooms, etc.

In addition, these salts also have the other properties required of active substances for food preservatives, i.e., in the concentration necessary to preserve foodstuffs they are nontoxic to man and they have no adverse effect on the taste and smell of the foodstuff. These salts are practically neutral compounds and do not alter the pH of the substrate; furthermore, they do not, or to a much smaller extent, irritate the skin and mucous membranes of humans. Also they are nontoxic to man under the above described conditions and are, therefore, particularly suitable for practical uses as food preservatives.

The salts can be produced by conventional processes, e.g., by reaction of 5-acetyl-8-hydroxy-quinoline with the equivalent amount of acid. They are compounds which crystallize well. They are very stable in air and, in contrast to 5-acetyl-8-hydroxy-quinoline itself, also in neutral or weakly acid media. These two salts are distinguished by good fungicidal properties. They are soluble to a certain extent in water and in organic solvents.

As preserving agents these salts are applied in an amount of from about 0.01 to 0.5 milligram per square centimeter of the surface of the fruit, plant or plant part to be treated. The fungitoxic action of the salts and other substances was determined by the so-called "spore germination test" on the following types of fungi: *Penicillium italicum, Penicillium digitatum, Rhizopus nigricans, Aspergillus niger,* and *Botrytis cinerea.*

1 cc of a 1, 0.5, 0.1 and 0.01% solution of each active substance in a suitable solvent such as acetone is placed on two glass slides (26 x 76 mm) under the same conditions. The solvent is evaporated off and a uniform coating of active substance is obtained on the glass slides. The slides are inoculated with spores of the above fungi and then kept in dishes at room temperature in an atmosphere which is almost saturated with steam. The germinated spores are counted twice, once after 48 hours and the second time after 72 hours. The results are summarized in the following tables.

Compound	Concen-tration, percent	Action against—					
		Penicillium italicum		*Penicillium digitatum*		*Aspergillus niger*	
		48 hr.	72 hr.	48 hr.	72 hr.	48 hr.	72 hr.
5-acetyl-8-hydroxy-quinoline dihydrogen phosphate	0.1	++	++	++	++	++	++
	0.01	−	−	++	++	−	−
Bis-(5-acetyl-8-hydroxy-quinoline) sulfate	0.1	++	++	++	++	++	++
	0.01	−	−	+	+	−	−
Diphenyl	0.1	−	−	+	−	−	−
	0.01	−	−	−	−	−	−
Sodium-o-chlorophenyl phenolate	0.1	−	−	+	−	−	−
	0.01	−	−	+	−	−	−
Tetramethyl thiuram di-sulfide	0.1	−	−	+	+	−	−
	0.01	−	−	+	−	−	−
Control		−	−	−	−	−	−

Compound	Concen-tration, percent	*Botrytis cinerea*	
		48 hr.	72 hr.
Bis-(5-acetyl-8-hydroxy-quinoline) sulfate	0.1	++	++
	0.01	−	−
Diphenyl	0.1	−	−
	0.01	−	−
Tetramethyl thiuram di-sulfide	0.1	−	−
	0.01	−	−
Control		−	−

LEGEND: ++ = Complete inhibition of germination caused by residue of 1 ccm. of solution of active substance. + = A 60 to 90% inhibition of germination caused by residue of 1 ccm. of solution of active substance, − = No inhibition of germination.

Fruit and other edible parts of plants are preserved by providing them with a protective coating of the salts by either dipping, spraying, washing or painting them with liquid or waxy preparations containing a microbicidally effective amount of the active substances. In many cases, because of their advantageous solubility in water, the microbicidal salts can be used in the form of aqueous solutions. On the other hand, it is possible to mix the active substances together with suitable noninjurious dispersing agents such as vegetable oils, fats or waxes and to use these dispersible preparations as aqueous or anhydrous dispersions. Solvents suitable for use in such dispersions are nontoxic, low boiling, organic solvents such as low molecular hydrocarbons such as pentane, hexane, ketones, alcohols and ethers such as methoxy ethanol or ethoxy ethanol.

A test to preserve oranges was made with the bis-(5-acetyl-8-hydroxy-quinoline) sulfate and two substances known from the prior art. The test was run with healthy, completely ripened oranges and for each run 20 fruits were used. The orange peel was injured by means of a small board fitted with needles with which 30 pinpricks about 2 mm deep were applied to each orange. The fruits were then dipped during one minute at room temperature into an aqueous suspension of spores of *Penicillium digitatum* containing about 100,000 spores per milliliter and left to dry.

Then they were dipped for one minute into another aqueous solution or suspensions, also at room temperature, containing 1% by weight of the substance to be tested. The fruits were then left to dry and packed into two paper bags each containing 10 oranges. The bags were stored at room temperature. The evaluations of the fruits were made the fifth and eighth day after treatment and the oranges were inspected for attack by *Penicillium digitatum.* A control run of 20 oranges was injured and exposed to the suspension of spores of *Penicillium digitatum* the same way but not treated with active substance. The results are shown in the following table.

Substance to be tested	Evaluation of oranges—			
	5 days		8 days	
	After infection by *Penicillium digitatum*			
	Infected	Healthy	Infected	Healthy
Bis-(5-acetyl-8-hy-droxy-quinoline) sulfate.................	1	19	10	10
5-acetyl-8-hydroxy-quinoline (base).......	9	11	14	6
Bis-(8-hydroxy-quino-line) sulfate (Chinosol).............	2	18	20	0
None (control)...........	9	11	20	0

The salts may be applied directly as an aqueous solution or preferably in the form of a ready made application form. Such a paste, a pulverulant concentrate, an emulsion concentrate and a wax are described below.

Paste: Twenty parts of bis-(5-acetyl-8-hydroxy-quinoline) sulfate, 20 parts of a 1% aqueous hydroxyethyl cellulose solution, 3 parts of ditertiary acetylene glycol and 57 parts of distilled water are milled and homogenized in a mixer into a thinly flowing 20% paste. On direct addition to water, the paste forms a milky dispersion. Freshly harvested oranges are dipped into a 1% dispersion and allowed to drip dry. About 0.01 to 0.1 mg of active substance remained on the fruit per sq cm of surface. It was observed that the fruit showed no damage whatever due to microorganisms after several days.

Pulverulent Concentrate: Ninety five parts of bis-(5-acetyl-8-hydroxy-quinoline) sulfate and 5 parts of a condensation product of nonylphenol and ethylene oxide (molar ratio 1:9 to 1:10) or another suitable, preferably nonionic, wetting agent are homogenized and finely milled in a suitable mill. This pulverulant concentrate is not dusty and quickly dissolves in water. A suspension in 1% concentration evenly wets fruit treated therewith.

Emulsion Concentrate:

Composition	Parts
5-Acetyl-8-hydroxy-quinoline dihydrogen phosphate	10
Ethoxyethanol	15
Dimethyl formamide	15
Emulsifying agent consisting of a mixture of calcium dodecylbenzene sulfonate and the condensation product of nonylphenol and ethylene oxide (molar ratio about 1:8 to 1:10)	10
Petroleum (boiling range 230° to 270°C)	50

This mixture is a 10% emulsion concentrate which can be diluted with water to form emulsions of any concentration desired. A 1% aqueous emulsion is used for the treatment of mold, (e.g., *Penicillium italicum*) on grapefruit. For this purpose, the fruit is dipped in the emulsion and left there for a few seconds. It is then allowed to drip dry. About 0.01 to 0.1 mg of active substance remains on the fruit per sq cm of surface. The emulsion can also be sprayed on the grapefruit or the latter can be washed therewith or can be applied thereto with a brush. The attack by fungi is greatly reduced by this treatment and any new infection is prevented.

Paraffin Blocks: Eight hundred and ninety parts of melted soft paraffin which has a melting point of about 41° to about 43°C are mixed with one hundred parts of paraffin oil at about 60°C, and ten parts of very finely ground (of the average particle size of

20 to 50 microns) bis-(5-acetyl-8-hydroxy-quinoline) sulfate are worked in until a homogeneous substance is obtained which is poured into forms and then allowed to harden. The resulting blocks are then used in a molten state in suitable apparatus with slight heating to coat citrus fruit with a thin film of bis-(5-acetyl-8-hydroxy-quinoline) sulfate in paraffin and thus protect them against spoilage. This apparatus may wax the surface of the citrus fruit, e.g., by a rubbing action.

The consistency of the blocks can be varied as desired by increasing the amount of paraffin oil and also optionally adding surfactants such as sorbitan fatty acid esters, e.g., sorbitan-sesquioleate, sorbitan-monooleate, sorbitan-trioleate, polyoxyethylene-sorbitan fatty acid esters to obtain an increasingly softer block thus adjusting the consistency to the requirements of the particular coating apparatus used.

Paraffin Emulsion: 440 parts of soft paraffin (melting point 41° to 43°C), 220 parts of paraffin oil, and 50 parts of Arlacel 83 (sorbitan-sesquioleate) are melted together while stirring at 50° to 60°C. A hot solution (80°C) of 300 parts of distilled water and 10 parts of bis-(5-acetyl-8-hydroxy-quinoline) sulfate is stirred into the melt which is still warm (50°C) with a suitable stirring apparatus, (e.g., a Homorex-mixer) until a homogeneous substance is obtained. The resulting water-in-oil emulsion is stirred while allowing to cool to about 30°C. A creamy substance of soft consistency is obtained which can be applied in suitable apparatus, e.g., by brushing or lightly rubbing the peel of citrus fruit.

The concentration of active substance in the composition can also be varied, (e.g., 0.5 or 1.5%). In lieu of Arlacel 83, other paraffin emulsifying agents, e.g., mixtures of Span 80 and 85 (sorbitan-monooleate and sorbitan-trioleate) can be used as well as other inert substances suitable for preparing similar water-in-oil emulsions of end products having optimal consistency whereby the consistency can be adapted to the requirements of the particular apparatus used for wax coating the citrus fruit.

ANTIOXIDANT FOR DEHYDRATED VEGETABLES

A special natural antioxidant is used to retard oxidative changes in dehydrated vegetables in place of usual treatment with sulfur dioxide. The process is described by *A. Patron and K. Schreckling; U.S. Patent 3,497,362; February 24, 1970; assigned to Maggi-Unternehmungen AG, Switzerland.* Vegetables such as carrots, potatoes, spinach, leeks, beans and celery are sometimes preserved by dehydration, the vegetables usually being reduced to slices, cubes, strips or flakes. Upon a more or less extended period of storage, the vegetables tend to alter in color and flavor and, in extreme cases, become unfit for human consumption.

Such alterations in flavor and appearance are most frequently caused by oxidation of different substances present in the vegetables and hence, it has become standard practice in the food industry to treat dried vegetables with various antioxidants the most common of which is sulfur dioxide. Other antioxidants which are also used are phenolic substances such as BHA (butyl hydroxy anisole), BHT (butyl hydroxy toluene), NDGA (nordihydroguiaretic acid) and DPPD (propyl gallate).

Although in a number of cases the use of antioxidants gives satisfactory results, their use in many ways is restricted. For example, vegetables which are in small pieces such as cubes, slices or strips have a large surface area which has to be treated and consequently large quantities of antioxidant have to be used to ensure thorough penetration. When sulfur dioxide is used, a high dosage leads to a concentration in the final product which is above the taste threshold and the vegetable acquires an unpleasant flavor. Furthermore, the use of artificial antioxidants is strictly governed by food legislations and in some countries is totally prohibited.

The active antioxidant is contained in an extract of an aromatic plant of the Labiatae family and 0.1 to 1.0% by weight of a molecularly dehydrated phosphate containing at least one P—O—P linkage in its molecule and having the empirical formula $(M_2O)_m \cdot (X_2O)_n \cdot P_2O_5$ (in which M represents an alkali metal atom, X represents a hydrogen or alkali metal atom, $1 \leqslant m \leqslant 2$, $0 \leqslant n \leqslant 1$ and $m + n \leqslant 2$). Molecularly dehydrated phosphates are sometimes referred to as polyphosphates and the P—O—P linkage is one of their characteristic features. These substances may exist in polymeric forms.

Examples of suitable phosphates are tetrasodium pyrophosphate, disodium dihydrogen pyrophosphate, sodium hexametaphosphate and sodium trimetaphosphate. Commercial grades of phosphates are usually mixtures of different substances. The extract of an aromatic plant of the Labiatae family may, for example, be prepared from rosemary, sage, origanum, thyme or marjoram. This extract may be obtained by conventional extraction of the selected plant, which is preferably rosemary or sage, with an organic solvent such as methanol, ethanol or petroleum ether. The plant is extracted in dry, ground condition. It is also possible to prepare the extract from the residues which are obtained in the preparation of essential oils by distillation of the plants. When the extraction has been completed, the solvent may be eliminated and the extract is obtained in powdered form which is substantially odorless.

The aromatic plant extract is itself insoluble in water but it is solubilized to a certain extent by the phosphate. The extract may first be dissolved in a solvent such as methanol, ethanol or acetic acid and this solution may be added in appropriate quantities to an aqueous solution of the phosphate to provide a bath in which the vegetables are then dipped. In this manner, the extract precipitates in the aqueous solution as a very fine suspension. Ethanol is the preferred solvent for dispersing the extract.

The aqueous antioxidant composition contains at least 0.010% of aromatic plant extract. Since the extract may possess a faint odor, the quantity employed should be such that its presence remains imperceptible. In general, extract concentrations in the composition of 0.015 to 0.025% give satisfactory results, that is, the treated vegetables are still acceptable after 9 months storage under normal atmospheric conditions.

The duration of treatment with the composition will depend on the vegetable and its state of subdivision and can be determined by a simple test. For vegetables in small pieces such as potatoes in the form of strips about 30 to 35 mm long and 3 x 6 mm in cross section, a treatment time of 2 to 5 minutes is satisfactory. For larger species, longer times, up to 10 minutes, should be sufficient. Although it is preferable to effect the treatment of the vegetable at a temperature of 40° to 60°C, in general, temperatures between 0° and 80°C are satisfactory.

It is best to effect the contacting of the vegetable with the aqueous antioxidant composition before the vegetable is subjected to any blanching treatment. It is usual practice to treat vegetables with an antioxidant immediately before final drying. It would appear that in the treatment the antioxidant acts immediately on the vegetable enzyme and protects the tissues from the action of occluded oxygen. An explanation of this phenomenon may be found in the observation that the plant extract is insoluble in water and doesn't volatilize in steam.

The treated vegetables may be precooked as desired and dried by conventional methods, such as air drying or freeze drying. Alternatively, the vegetables, especially potatoes, may be partially (40 to 60% moisture) or more or less completely dried (10 to 15% moisture) after precooking and then fried in a suitable fat. The fried vegetable preferably has a moisture content of 4 to 6%. The treatment may be used in the dehydration of a large variety of vegetables, especially carrots, potatoes, spinach, green beans and celery.

Example 1: Preparation of Aromatic Plant Extract — Sixty grams of dried ground rosemary, from which the essential oil has been removed by distillation, are extracted for 8 hours in a Soxhlet apparatus with 250 ml of 95% ethanol. After separation of the solid matter, about 200 cc of alcoholic extract, which is greenish brown in color, are obtained. The extract contains 7 to 8 grams of solids which represents a yield of 10 to 12%. Extracts of other aromatic plants of the Labiatae may be prepared in a similar manner.

Example 2: Carrots are peeled, washed and machine cut into 8 mm cubes. A bath is prepared containing, per 10 liters of water, 40 grams of crystalline neutral sodium pyrophosphate and 50 ml of alcoholic rosemary extract containing 40 grams of dry matter per liter. The bath is warmed to 50°C. The cubed carrots are dipped in the bath for 3 minutes and drained. Ten liters of bath are used for 1,500 grams of carrot cubes. The cubes are then blanched for 5 minutes in steam at 98°C and cooled by a water spray during 30 seconds. The carrots are then freeze dried to a final moisture content of about 1%. A control sample is also prepared under conditions which are identical except that the antioxidant treatment is omitted. After 4 months of storage at 37°C in air, the control sample is completely discolored and has an unpleasant sharp flavor which renders it unfit for consumption whereas the product which has been treated with the antioxidant still has the normal orange color and typical carrot odor and flavor.

Example 3: Celery is treated as described in Example 2 except that 10 liters of bath are used for treating 2,500 grams of celery cubes. An untreated control sample is also prepared. After 6 months of storage in air at 37°C, the celery cubes which were treated with antioxidant have a characteristic celery flavor and odor and are perfectly edible whereas the control sample has a rancid odor when dry; and upon reconstitution, the flavor is unrecognizable and even unpleasant.

Example 4: Spinach is tailed, washed in cold water, drained and cut into pieces 2 to 5 centimeters square. These pieces are dipped in a bath containing, per liter of water, 10 cc of alcoholic rosemary extract prepared as described in Example 1 and 4 grams of crystalline tetrasodium pyrophosphate. The bath temperature is 50° to 55°C and the residence time of the pieces is 4 minutes. After draining, the pieces are blanched and dried in air to a final moisture content of 3 to 4%. A control sample is prepared in the same manner, but the antioxidant treatment is omitted. After 3 weeks of storage in air at 20°C, the control sample has an odor of hay whereas the treated sample has retained a pleasant aroma and flavor.

Example 5: Potatoes are washed, peeled, sorted and cut up into strips about 30 to 35 millimeters long and 3 x 6 mm in cross section. These strips are washed in cold water to remove surface starch and are dipped for about 2 to 4 minutes in a bath containing, per liter of water, 3 to 4 cc of rosemary extract prepared as described in Example 1 except that the powder is dissolved in acetic acid and 4 grams of tetrasodium pyrophosphate. The bath temperature is 55°C and the residence time of the strips is about 3 minutes. Thereafter the strips are blanched or precooked in steam at atmospheric pressure and are then dried to a final moisture content of 3 to 5%. After 9 months of storage in air at room temperature, the appearance, odor and flavor of the product are unchanged.

Example 6: Washed, peeled and sorted potatoes are cut into thin chips or slices. The pieces are washed in water and dipped for 2.5 to 3.5 minutes in a bath containing, per liter of water, 2.7 grams of tetrasodium pyrophosphate ($Na_4P_2O_7$) and 4.5 cc of alcoholic rosemary extract prepared as described in Example 1. The bath temperature is 55°C, however, it may vary from 45° to 60°C. The treated potato pieces are blanched or precooked in steam for 4 minutes, sprinkled with water and predried in hot air (60° to 85°C) to a moisture content of 50%. The predried product is left to equilibrate in a closed chamber for 4 hours and then is fried for 2 minutes in groundnut oil at 180° to 200°C. The final moisture content of the fried potatoes is 4 to 6%.

ADDITIVES FOR APPLE SLICES

For sale to the bakery trade, institutions, etc., it is conventional for food processors to put up apples in a form wherein they may conveniently be used in preparing pies, tarts, and similar products. The conventional procedure involves the following steps. Apples are washed, peeled, cored and cut into slices. The slices are then sulfited, that is, they are dipped into an aqueous solution containing sodium bisulfite, or other soluble sulfite, or sulfurous acid. The sulfited slices are packed, for example, in plastic bags, then cooled to refrigeration temperature (about 35° to 45°F) and held at such temperature during shipping and storage.

The products are not intended to be preserved indefinitely; they are meant to be used within about 7 to 30 days after preparation. It may be noted that the sulfiting treatment is an essential element to preserve the color of the fruit; without such treatment the slices would rapidly turn brown and be unfit for use. Although sulfited fresh apple slices are used in industry on a substantial scale, they suffer from the disadvantage that the sulfiting treatment gives the fruit an undesirable taste and odor.

A special preservative solution and process are given by *J.D. Ponting; U.S. Patent 3,754,938; August 28, 1973; assigned to the U.S. Secretary of Agriculture.* Basically, the process involves the following steps: (1) using conventional techniques, apples are washed, peeled, cored and sliced (2) the apple slices are dipped into a special preservative solution as described and (3) the treated slices are drained, packaged in conventional manner, cooled to refrigeration temperature and held at such temperature until ready for use.

The preservative solution contains the following ingredients dissolved in water: (a) ascorbic acid in a concentration of 0.5 to 1%, preferably 1%, (b) calcium chloride in a concentration of 0.05 to 0.1% (as calcium), preferably 0.1% and (c) for best results, a solution containing enough sodium bicarbonate to maintain it at a pH of 7 to 9 during use. It has been observed that where sodium bicarbonate is not used, the solution becomes quite acid (due to leaching of malic acid from the slices) with the result that the preservative effect of the solution is lessened.

It is acknowledged that treatments are known wherein either ascorbic acid alone or calcium chloride alone are applied to apple slices. However, these agents have only slight affect by themselves so that such treatments cannot maintain quality for the required distribution time. It was found that where the agents are used in combination, a synergistic effect is attained in that the products retain their quality for extended periods of time, for example, as long as two months or more.

Example 1: Newtown Pippin apples were peeled, cored and sliced. The apple slices were divided into several lots, each lot being treated as follows. The apple slices were dipped for 3 minutes in an aqueous solution containing certain proportions of ascorbic acid and/or calcium chloride. In each case the solutions were adjusted to pH 7.0 by addition of sodium bicarbonate as necessary. After the dipping treatment, the apple slices were drained, packed into plastic bags and stored for 13 weeks in a refrigerator at 34°F. At the end of this time the slices were removed and examined.

Using a reflectance meter, measurements were made of the reflectance of the fresh (untreated) slices and the stored products. From these measurements there was calculated the percent loss in reflectance of the stored products versus that of the fresh slices. The resulting data is an indication of the color of the products in that a lower figure denotes a product lighter in color, that is, closer to the natural color of the fresh fruit. The results are tabulated below.

| | Contents of Dipping Solution | | Condition of Product After Storage | |
| | Ascorbic Acid, Percent | Calcium Chloride, Percent Ca | Loss in Reflectance, Percent | Color (Visual) |
Lot				
1 (Control)	0	0	43.5	Dark brown
2 (Control)	1.0	0	40.6	Medium brown
3 (Control)	0	0.10	15.7	Slightly brown
4 (Test)	1.0	0.1	2.1	Light, natural color

Example 2: The procedure of Example 1 was repeated, except that in this case no attempt was made to adjust the pH of the solutions. The results are tabulated below.

Lot	Contents of Dipping Solution		Condition of Product After Storage	
	Ascorbic Acid, Percent	Calcium Chloride, Percent Ca	Loss in Reflectance, Percent	Color (Visual)
1 (Control)	0	0	39.7	Dark brown
2 (Control)	1.0	0	8.3	Slightly brown
3 (Control)	0	0.1	12.1	Slightly brown
4 (Test)	1.0	0.1	5.4	Light colored

VITAMINS AND NATURAL COLORS

VITAMINS

Tocopherol-Synergist Mixture

Antioxidants are employed to delay the decomposition of oxidation-sensitive materials such as vitamin A. The most frequently employed antioxidants are tocopherols such as α- and γ-tocopherol, and related compounds such as α-tocopheramine and N-methyl-γ-tocopheramine.

H. Klaui and W. Schlegel; U.S. Patent 3,637,772; January 25, 1972; assigned to Hoffmann-La Roche Inc. describe antioxidant compositions containing an antioxidant synergist mixture. Further, it has been discovered that it is possible to increase the activity of antioxidants so that it exceeds the activity of known compositions containing a single synergist by utilizing two synergists in the antioxidant composition. Specifically the antioxidant composition contains (a) an antioxidant, (b) colamine and/or a higher fatty acid salt thereof and (c) a higher fatty acid ester of ascorbic acid.

It has been found that a particularly strong synergistic action is achieved with tocopherols, which in view of the physiological acceptability of these materials is of extraordinary importance. Colamine can be used as the base or as a salt with a higher fatty acid having 10 to 20 carbon atoms. The palmitate and the stearate are the preferred salts. The ascorbic acid esters used are esters of higher fatty acids having from 10 to 20 carbon atoms, preferably palmitic or stearic acid. For every part by weight of antioxidant, the amount of colamine or colamine salt can suitably be from about 1 to 10 parts by weight, and the amount of ascorbic acid ester from about 1 to 10 parts by weight. The amount of antioxidation composition utilized depends on the requirements of the substrate to be stabilized and amounts between 0.01 and 0.1% by weight are generally sufficient with oils and fats, and amounts between 0.5 and 30% by weight are sufficient in the case of vitamin A preparations.

Many foodstuffs contain such oxidation-sensitive materials only as ingredients, often only in small concentrations. Such foods are hydrophilic in their nature and are denoted as water-based foods. Examples of such foods are, for example, vegetables, especially in comminuted form, vegetable meals and conversion products thereof, and meat commodities. The good dispersability in aqueous environment of the antioxidant mixture is of advantage in such cases. The dispersability is particularly marked when colamine and the ascorbic acid ester are used in about stoichiometric amounts.

Example 1: 2 g of ascorbyl palmitate, 3 g of sodium ascorbate, 0.5 g of α-tocopherol, 0.6 g of colamine and 1 g of dextrin are dispersed in 20 ml of water at 50°C. The emulsion obtained is blended into 10 kg of sausage meat for hard sausages. The sausages manufactured therefrom have stability equivalent to those whose filling contains double the amount of ascorbyl palmitate and no colamine or α-tocopherol.

Example 2: 1.5 g of vitamin A palmitate, 1.5 g of sunflower oil, 50 mg of α-tocopherol, 250 mg of ascorbyl palmitate, 250 mg of colamine palmitate, 200 mg of polyoxyethylene (2) oleyl ether (HLB value 4.9), and 0.3 mg of copper oleate are heated to 60°C and homogeneously mixed. This mixture is stored in air at 45°C and the vitamin A retention is measured and compared to compositions without colamine and compositions without ascorbyl palmitate. After a storage time of 600 hours, the vitamin A retention in the composition prepared according to this example is 96%. In the composition without colamine, the vitamin A retention is 0% and in compositions without ascorbyl palmitate the vitamin A retention is about 10%.

Example 3: 1.5 g of vitamin A palmitate, 1.5 g of sunflower oil, 50 mg of γ-tocopherol, 250 mg of ascorbyl stearate and 250 mg of colamine palmitate are heated to 60°C and homogeneously mixed. This mixture is stored in air at 45°C and the vitamin A content determined after 600 hours and compared to compositions without colamine palmitate and compositions without ascorbyl stearate. The vitamin A retention in the composition prepared according to this example is 95%, while in the composition without colamine palmitate, vitamin A retention is 0% and in the composition without ascorbyl stearate the vitamin A retention is about 10%.

Example 4: 50 mg of β-carotene, 10 mg of α-tocopherol, 10 mg of ascorbyl palmitate and 5 mg of colamine palmitate are dissolved in 100 g of sunflower oil with heating to 80°C. The content of β-carotene is measured after a storage of 750 hours at 45°C in air and compared to compositions without colamine palmitate and compositions without ascorbyl palmitate. The β-carotene retention in the composition prepared in this example is 66%. The β-carotene retention in the composition without colamine palmitate is 33% and in the composition without ascorbyl palmitate the β-carotene retention is 50%.

Vitamin A-Fat Antioxidant

It is known that phenolic compounds in general have antioxidant properties. However, there are a number of phenolic compounds which actually promote oxidative degradation. Moreover, most phenolic compounds have only slight antioxidant qualities and are quite inferior to certain specific phenolic compounds which have enjoyed commercial utility in recent years. The well-known phenolic antioxidants available commercially include butylated hydroxyanisole, butylated hydroxytoluene, N-butylated p-aminophenol, 2,2'-methylenebis(4-methyl-6-tert-butylphenol), etc. Among the phenolic antioxidants is 4,4'-dihydroxydiphenyl ether which has only slight antioxidant properties and is generally inferior to the commercially available antioxidants. Thus, it would be expected that related ethers would possess similarly unattractive properties.

A. Bell, M.B. Knowles and C.E. Tholstrup; U.S. Patent 2,967,774; January 10, 1961; assigned to Eastman Kodak Company found a special group of tetra-oxy phenolic compounds that have improved antioxidant potency. These compounds have the formula:

where each of R and R' represents a member selected from the group consisting of a hydrogen atom and an alkyl-organic radical containing from 1 to 12 carbon atoms, the alkyl-organic radical encompassing members selected from the group consisting of alkyl, cyclo-alkyl, alkenyl and aralkyl radicals, and X represents an alkylene radical containing from 1 to 10 carbon atoms.

These phenolic compounds can be used for stabilizing any of the materials which are normally subject to oxidation. These antioxidants contribute their highest order of potency in the stabilization of hydrocarbons such as paraffin wax, petroleum oils including gasoline and other motor fuels, polyethylene, polypropylene and other normally solid polymers of alpha-monoolefins containing from 2 to 7 carbon atoms, petroleum derivatives of various types such as lubricating oils, transformer oils, etc.

In addition, the antioxidants can be used in stabilizing polymeric compositions such as polyesters including the linear polyesters and alkyd resins, synthetic rubber compositions, natural rubber compositions, etc. These antioxidants can be used for stabilizing fatty triglycerides such as lard, vegetable oils, animal fats, fish oils, etc. The antioxidants are particularly valuable in the stabilization of fatty oils or other compositions containing vitamin A. Specific materials which can be stabilized with these antioxidants include margarine, cottonseed oil, corn oil, peanut oil, free fatty acids such as oleic acid and similar readily oxidizable acids.

In most cases the stabilizers can be used in concentrations of from 0.01 to 2% by weight based on the total weight of the composition being stabilized. In addition, the antioxidants can be employed in conjunction with other antioxidants known to be useful for the stabilization of compositions normally subject to oxidative deterioration. When required, metal deactivators and other additives can be included. Thus, the antioxidant or stabilizer compounds provided can be used in conjunction with other compounds such as butylated hydroxytoluene, butylated hydroxyanisole, propyl gallate, citric acid, N,N'-di-sec-butyl-p-phenyl-enediamine, disalicylalpropylenediimine, 5-acenaphthenol, 2,4,5-trihydroxybutyrophenone, octadecyl gallate, 2-tert-butyl-4-dodecoxyphenol, tocopherol, phosphoric acid, lecithin, etc.

Example 1: Preparation of 4,4'-Tetramethylenedioxybis(2-tert-Butylmethylbutylphenol) — A mixture of 44 g (0.2 mol) of 2-(1,1,3,3-tetramethylbutyl)hydroquinone (tert-octylhydroquinone), 19 g (0.1 mol) of 1,2-dibromoethane, 0.2 g of zinc dust, and 75 ml of ethanol was gently refluxed under a nitrogen atmosphere. A solution of 13 g (0.2 mol) of potassium hydroxide in 95 ml of 75% ethanol was added dropwise. After refluxing for one-half hour the mixture was cooled and then filtered to remove zinc and potassium bromide. Dilution with water caused a sticky, crystalline material to separate. Two recrystallizations from acetic acid-water and one from methanol-water gave a white crystalline material which melted at 125 to 128°C. The yield was 3.5 g (4%).

Example 2: Preparation of 4,4'-Tetramethylenedioxybis(2-tert-Butylphenol) — This compound was prepared in a fashion similar to that of Example 1 starting with 33 g (0.2 mol) of tert-butylhydroquinone, 21.6 g (0.1 mol) of 1,4-dibromobutane, and 13 g (0.2 mol) of potassium hydroxide. The white crystalline product, crystallized from benzene-hexane, melted at 143° to 148°C. A yield of 6.5 g (17%) was obtained.

Example 3: Preparation of 4,4'-Decamethylenedioxybis(2-tert-Butylphenol) — This compound was prepared by the procedure used in Example 1 starting with 23 g (0.14 mol) of tert-butylhydroquinone, 9.2 g (0.14 mol) of potassium hydroxide, and 21 g (0.07 mol) of 1,10-dibromodecane. The product was crystallized from acetic acid-water, then twice from naphtha. A yield of 2 g (6%), which melted at 132° to 137°C., was obtained.

The antioxidants are effective for vitamin stabilization. The compounds were evaluated for stabilizers for vitamin A in pollock liver oil and the data obtained is presented below.

Additive	Weight Percent	Days Until 50% Loss of Activity
Control	0	2
4,4'-Ethylenedioxybis(2-tert-butylphenol)	0.02	13
4,4'-Decamethylenedioxybis(2-tert-butylphenol)	0.02	14

The procedure used in determining this data involves adding the designated percentage of the antioxidant to a petroleum ether solution of pollock liver oil so as to have present in the solution the appropriate amount of antioxidant based on the weight of the pollock liver oil. Aliquots of the stabilized petroleum ether solution were pipetted into a series of 50 ml beakers containing a single layer of 4 mm glass beads. The petroleum ether was then allowed to evaporate at room temperature in the dark. The beakers were stored at 100°F. The vitamin A content of the pollock liver oil in the beakers was determined at suitable intervals by the Carr-Price method until a 50% loss of original vitamin A activity resulted. It can be seen from the data that the antioxidants of this process are quite effective in the stabilization of the vitamin A content of fish liver oils.

Stabilized Vitamin Supplements

The product described by *E. Halin; U.S. Patent 3,338,717; August 29, 1967* is classified as a stabilized nutritional food. It is formulated with soybean meal, vegetable oil, arrowroot starch, saccharide, hydrogenated oil, vitamin C, vitamin B compounds, iodized salt, calcium phosphate and iron salt with a saccharide. The steps of the process are shown in Figure 9.1 and described below. It is claimed that the vitamins have been stabilized in this process to enhance shelf life.

FIGURE 9.1: PREPARATION OF STABILIZED VITAMIN SUPPLEMENTS

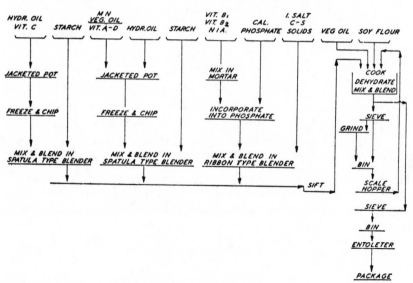

Source: E. Halin; U.S. Patent 3,338,717; August 29, 1967

Step 1 involves introducing a predetermined amount of vitamin C into a stated amount of edible hydrogenated oil heated sufficiently to be substantially in the liquid state. Generally, temperatures of 40° to 60°C, preferably about 45°C, are adequate for this purpose, depending largely on the particular oil employed. Suitable hydrogenated oils include hydrogenated cottonseed oil, hydrogenated soybean oil, hydrogenated safflower oil, hydrogenated olive oil and hydrogenated peanut oil, among others. Also suitable in this respect is coconut oil, which is a naturally occurring saturated oil and is, in that sense, already the equivalent of the oils that occur naturally in the unsaturated condition and have to be hydrogenated for these purposes.

The mixture of vitamin C and the edible hydrogenated oil is stirred, preferably slowly, to keep the vitamin C in substantially uniform suspension and then cooled until it solidifies. It is preferred to keep this mixture in a closed container at a temperature of 5° to 10°C until it is completely solidified. After complete solidification, the mixture is cut, chipped or shredded, as desired, in order to get it into the form of reasonably small particles. In this particular form, it is introduced into a blender, along with a weighed portion of finely particled arrowroot starch. In place of the arrowroot starch, there may be used potato starch or mixtures of these, if desired. The specificity of this component is based on the presence of a wide range of desired nutritional factors, along with a high degree of ready assimilability. The mixture of the Step 1 components is then ready for blending with any or all of the components of the other four steps of the process.

In Step 2, a weighed portion of 2-methyl-1-naphthoquinone (vitamin K) is added to and dissolved in a defined amount of an edible vegetable oil. There is added to the oil solution a predetermined amount of vitamin A and vitamin D dissolved in a weighed portion of an edible hydrogenated oil, which has been heated to 40° to 60°C, preferably about 45°C, in order to convert it into the liquid state. This edible hydrogenated oil is the same kind as used in Step 1. The mixture is stirred in order to promote solution. The system is then chilled to 5° to 10°C and the solution stored, preferably in a closed container, until it completely solidifies. After solidification, this material is cut, chipped or shredded, as desired, into reasonably small particles and introduced into a blender. A weighed portion of arrowroot starch is also added to the blender and thoroughly mixed with the previously described material. While the type of blender in Step 1 is not particularly critical, the commercially available spatula type blenders produce satisfactory blends. The components prepared according to Step 2 are then ready to be blended with components of Step 1 and the components of the subsequent steps.

In Step 3 there are intimately ground together in a mortar or similar device, predetermined amounts of at least vitamin B_1 (thiamin hydrochloride), vitamin B_2 (riboflavin) and vitamin B_5 (nicotinamide or niacin). In most instances, particularly when the final food product is intended for adults, there are also incorporated defined amounts of vitamin B_6 (pyridoxine), vitamin B_4 or H (biotin) and vitamin B_{12} (pantothenic acid or cobalamine). To these vitamins is added a defined amount of finely comminuted calcium phosphate. If it is desired to incorporate iron into this formulation, there is also added a weighed portion of finely divided ferrous sulfate or ferrous gluconate. In many instances, particularly when the requirements of adults are contemplated, there are also incorporated other mineral sources, such as manganese sulfate, potassium chloride, zinc sulfate, magnesium sulfate, among others.

The above described vitamins and salt or salts are added to a blender and intimately mixed with weighed amounts of iodized salt and a saccharide, preferably a monosaccharide, such as glucose. The glucose may be conveniently obtained in the form of dried corn syrup solids. All of these components are blended to a uniform consistency and distribution in a blender, such as the ribbon type. The components of Step 3 are then ready for mixing and blending with the components of the other steps.

In Step 4 there is introduced into a preheated pressurizable vessel at 10 pounds steam pressure, such as an autoclave or other pressure regulable cooking vessel, a predetermined amount of soybean meal or flour. The soybean meal is heated with agitation at 210° to 220°F with live steam introduced into the inner jacket of the autoclave, and with steam at lower pressure, about 4 to 5.5 pounds introduced inside the autoclave for 12 to 15 minutes. In order to remove the condensed steam, a vacuum is applied, 25 to 27 inches of vacuum being satisfactory for this purpose, for about three minutes, although greater vacuum may be employed, if needed. The soybean flour is then cooled in the autoclave and passed through a screen, preferably 100 mesh. The portions held on the screen are processed through a mill or grinder to comminute finely the larger particles of the processed soybean meal and thereby provide the soybean meal in a substantially uniform particle size. The soybean flour is then ready to be blended with the other components of the other steps and can conveniently be readied for the final blending by storing in a bin.

Step 5 consists of introducing a weighed amount of an edible vegetable oil into a mixing or blending vessel. This edible vegetable oil is the same kind as defined under Step 2. The components of the various five steps may be introduced in any order into the final mixer or blender. It is preferred to introduce first the edible vegetable oil of Step 5, followed by a blend of the components of Steps 1, 2 and 3, and finally by the processed soybean meal of Step 4, or the latter two may be reversed. In any case, the components of Steps 1 through 5 are intimately mixed and blended. The total blend is then screened to provide a uniform particle size in the product, processed, if desired, through an entoleter or similar device, and then finally packaged.

The following components should be employed within particular percentages by weight:

Processed soybean meal	30 to 62
Edible vegetable oil	12 to 40
Arrowroot starch or equivalent	9 to 25
Saccharide	6 to 10
Edible hydrogenated oil	1 to 5
Calcium phosphate	3.16 to 4.5
Iodized salt	0.75 to 1.5
Vitamins, as noted below	

The above components are to be calculated to total 100%, then sufficient vitamin and mineral supplements are added to supply the known amounts required for contemplated users. In this respect, for instance, for children, generally about 25,000 USP units of vitamin A per day is adequate and about 1,380 USP units of vitamin D daily is generally sufficient. Similarly with the other vitamins and minerals. Adults and especially those with known nutritional inadequacies or deficiencies may require up to 25,000 to 30,000 or more USP units of vitamin A daily and up to 275 to 300 USP units of vitamin D per day. The other vitamins and minerals are also usually required in greater amounts, as is known. Furthermore, it is important that in a food material containing predetermined amounts of nutritive values for particular users, those values remain constant for prolonged periods of storage without appreciable diminution or degradation.

As a complete daily food material and, therefore, valuable in diet control, the above composition showing percentages by weight can be determined on the basis of 20 to 40 grams of soybean meal with vitamins and minerals added, as desired. Of course, it is known already that the required components of the food material previously defined contain appreciable vitamin and mineral values. However, in most instances, additional vitamin and mineral values are desired and are, therefore, added to and become part of the food material. The word "processed" used with respect to the soybean meal refers to soybean meal that has been subjected to direct steam pressures reaching the stated temperatures of 200° to 230°F for 10 to 18 minutes, then vacuum is applied to remove the condensate water ranging from just below normal atmospheric pressure down to about 25 inches.

A specific formulation for children from infancy to 10 or 12 years of age is, in parts by weight, as follows:

Processed soybean meal	30
Edible vegetable oil	15
Arrowroot starch or equivalent	16
Saccharide	18
Calcium phosphate	3.2
Iodized salt	1
Vitamins	1

For older persons beyond about 40 years of age, the same specific formulation will be used except for variation in the vitamin content, and particularly an increase of the ascorbic acid which is important in the development of connective tissues. For persons in the age range from about 12 to 40 years of age, the same specific formulation can be employed but with the vitamin content modified. The relationship among vitamins for a daily administration of useful therapeutic formulation is as follows:

Vitamin A	25,000 USP units
Vitamin B	1,000 USP units
Thiamine hydrochloride	10 mg
Riboflavin	5 mg
Nicotinamide	100 mg
Pyridoxine hydrochloride	5 mg
Cobalamine (B$_{12}$)	20 mcg
Panthothenate	20 mg
Ascorbic acid	200 mg

The quantity of ascorbic acid is preferably increased for aged persons, ascorbic acid being important in the development of connective tissues.

Dihydroquinoline Antioxidants

E.G. Jaworski and G.J. Marco; U.S. Patent 3,347,677; October 17, 1967; assigned to Monsanto Company report that the antioxidant properties of well-known antioxidants are significantly enhanced by mixture with certain nitrogen-containing compounds. The improved antioxidant compositions contain a 2,2,4-trimethyl-1,2-dihydroquinoline antioxidant of the formula

where n is an integer from 0 to 2 inclusive and X is selected from the group consisting of chlorine, bromine and RO—, where R is selected from the group consisting of hydrocarbyl and substituted hydrocarbyl where the hydrocarbyl is selected from the class consisting of alkyl having not more than 12 carbon atoms, alkenyl of not more than 12 carbon atoms, alkynyl of not more than 12 carbon atoms, cycloalkyl having at least 4 and not more than 7 carbon atoms, benzyl and phenyl, and where the substituents of not more than 4 carbon atoms on the hydrocarbyl are selected from the class consisting of alkyl, alkynyl and alkoxy, and a nitrogen-containing compound present in an amount sufficient to provide an enhanced antioxidant property.

The following are representative of the nitrogen-containing compounds useful in this process: tetraethyldithiooxamide; N,N'-dimethyl thiooxamide; N,N'-dicyclohexyl dithiooxamide; 3-methoxydithiophthalimide; thiophthalimide; 1,4-piperazine-bis(thioacetamide); dithiobiuret; thiourea; thioacetamide; etc.

The amount of nitrogen-containing compound necessary to provide enhanced antioxidant properties in the antioxidant compositions varies to some extent, depending upon the specific dihydroquinoline antioxidant employed. The antioxidant compositions usually contain from 0.33 to 10 parts by weight of nitrogen-containing compound per 1 part by weight of dihydroquinoline antioxidant. The improved antioxidant compositions can be incorporated into any substance which is subject to oxidative degradation in the presence of air, ozone, oxygen or various other oxidizing agents under the influence of heat, natural or artificial light, or various metals. The improved antioxidant compositions are added to the substance to be stabilized in any conventional manner. The components of the improved antioxidant compositions can also be added separately to the substances to be stabilized followed by conventional mixing to form the stabilized substance. Substances subject to oxidative degradation in which the improved antioxidant compositions are useful include, for example, food products, particularly those containing unsaturated oils, vitamins and vitamin-containing components. The antioxidant compositions are particularly useful for the stabilization of animal feeds containing components subject to oxidative degradation such as unsaturated oils, dehydrated alfalfa and other forage crops, vitamin concentrates, meal from oil processing industries, proteins and antibiotics.

It is well known that the carotene in dehydrated alfalfa and other forage crops is subject to oxidative degradation and loss of vitamins, particularly vitamin A, under normal processing and storage conditions. Furthermore, animal feeds which include dehydrated forage crops and other vitamin-containing components can undergo further decomposition, such that animals to which they are fed are subject to malnutrition and more serious disabilities attributable to deficiencies in essential vitamins. It has also been found that many animals have the ability to store vitamins, and that the retention of these vitamins for future use is extended if antioxidants are included in the animal diets. The animal industry regularly includes antioxidants as a conventional component in animal feeds.

The improved antioxidant compositions, when used in place of conventional amounts of dihydroquinoline antioxidants, give an equivalent level of antioxidant protection with less antioxidant or a more effective level of antioxidant protection with the same amount of antioxidant. Thus, the compositions are usually employed in an amount equal to or less than the amount of antioxidant which is conventionally employed.

The antioxidant properties of nitrogen-containing compounds are determined by a standardized procedure using carotene emulsions and microanalysis techniques. The 6-ethoxy-2,2,4-trimethyl-1,2-dihydroquinoline is used as a standard for control and its antioxidant activity under identical conditions is assigned a rating of 100. The antioxidant activity of each nitrogen-containing compound tested, both alone and in a one to three part mixture with the 6-ethoxy-2,2,4-trimethyl-1,2-dihydroquinoline standard, is expressed as a percentage of the antioxidant rating of the standard. Results and further details are given below.

| | - - - - - - Antioxidant Activity - - - - - - | |
| | | 1 Part Compound plus |
Compound	Alone	3 Parts Standard
Standard 6-ethoxy-2,2,4-trimethyl-1,2-dihydroquinoline	100	–
Dithiooxamide	-3.0	184.0
N,N'-dimethyl thiooxamide	1.5	138.6
N,N'-dimethyl dithiooxamide	3.0	163.6
Thiooxanilinonitrile	0.9	126.3
Alpha-anilino thioacetamide	-4.4	126.0
Dithiobiuret	-3.3	204.9
Thiophthalimide	0.8	135.8

The antioxidant compositions can be incorporated into any or all components of animal feed compositions in any convenient manner. Conventional animal feed compositions generally comprise at least about 2% plant ingredients such as hay, straw, silage, yellow corn, pasturage, ground corn cobs, cottonseed hulls, cotton mill wastes, beet pulp, cornmeal, soybean oil meal, wheat bran, wheat middlings, dehydrated alfalfa, ground oats, millet, linseed oil meal, coconut oil meal, distillers dried grains, peanut oil meal, cottonseed meal and the like plant products. Most animal feed compositions also contain up to about 2% mineral ingredients such as bone meal, limestone, salt (NaCl) and the various trace minerals including salts of zinc, copper, manganese, cobalt, iodine, and iron.

Other materials which can be incorporated into animal feed compositions in varying amounts include animal ingredients such as fish meal, meat and bone scraps, dried milk, animal fats, dried whey solubles; vitaminaceous ingredients such as vitamins A, B_{12}, D and K, as well as the B vitamins such as riboflavin, niacin, pantothenic acid or salts, choline, pyridoxine, thiamine, nicotinic acid or salts thereof, biotin, folic acid; amino acids such as methionine, phenylalanine, arginine, glycine, histidine, isoleucine, leucine, lysine, threonine, tryptophan, valine; and medicaments such as antibiotics, steriods, arsenicals and anthelmintics.

Dihydroquinoline and Sulfur Amino Acids

In another process *E.G. Jaworski; U.S. Patent 3,279,921; October 18, 1966; assigned to Monsanto Company* describes the use of sulfur-containing amino acids to activate the antioxidant properties of 2,2,4-trimethyl-1,2-dihydroquinolines.

Example: The active compounds included within the scope of the generic structure, and near homologues of little or no activity, were studied by a standardized procedure to determine their ability to prevent or minimize the oxidation of carotene. The stabilizing ability of the standard antioxidant 6-ethoxy-2,2,4-trimethyl-1,2-dihydroquinoline was also determined to provide a suitable control. The data set forth below is the percentage of the antioxidant property of the standard antioxidant. The compounds were studied both alone and in a 1 to 3 proportion with the standard antioxidant. The following observations were made.

	Compound	Alone	1 to 3 of Standard
—	6-Ethoxy-2,2,4-trimethyl-1,2-dihydroquinoline	100	-
(1)	Cysteine	-4.1	185.1
(2)	2-Aminomercaptobutyric acid	-7.6	129.3
(3)	2-Mercaptoethylamine	-3.3	141.3
(4)	2-Hydroxyethylamine	1.1	111.5
(5)	Serine	1.1	106
(6)	4-Hydroxy-2-aminobutyric acid	0	98.5
(7)	Cystine	12.2	110.8
(8)	2-Amino-4-thiobutyrolactone	-1.1	107.6

Of the above, numbers (1), (2) and (3) are valuable activators for antioxidants. Compounds (4), (5), (6), (7) and (8) have little or no activity and are not included within the scope of the generic definition of useful activators. It is apparent that the replacement of the mercapto group with a hydroxyl radical reduces or destroys the antioxidant activator's property of the adjuvant.

Mixtures of the antioxidants and the activators as above described may be prepared in advance and used as such in preventing or controlling undersired oxidation reactions. When added to the substances which are subject to deleterious oxidation, they are useful in stabilizing the substances during preparation, storage or use. Generally, they are gradually consumed while they are effecting the stabilization. The compounds may be added to the substances being stabilized as separate components and mixing the composition to effect an intimate dispersion. Many compositions contain the substances to be stabilized as only minor components and the small amount of antioxidant used is difficult or impossible to attain the necessary degree of dispersion for effective control of the oxidation reactions.

It has been found that the antioxidant compositions are prepared by use of the principal antioxidant in quantities such that the maximum is from 80 to 95% and the minimum from 5 to 20%, the balance of these compositions being the activating compound. It has been found that very large amounts of nonantioxidant activators may produce a diluting effect without a complete compensatory activation of the antioxidant. Some of the activators are in themselves antioxidants and these may provide additional antioxidant capacity as well as providing the more important activation effect.

STABILIZED CAROTENOID PREPARATIONS

Special Antioxidant Mixture

The successful use of carotene-containing compositions as food coloring additives has been achieved only when the carotene-containing compositions were not subjected to high temperatures. For example, carotene has been used to impart a yellow color to such foodstuffs as margarine, butter, shortenings, egg yolk products, processed cheeses, orange drink bases and concentrates, cream and dry prepared cake mixes. In the coloring of these products, the carotene is not subjected to high temperatures. On the other hand, in the case of high temperature applications, appreciable decomposition of the carotene was found to occur at temperatures from 325° to 375°F. As a result excess carotene was, of necessity, used with resultant economic disadvantages.

At temperatures of 400°F and above, carotene has proved useless as a food coloring additive because nearly complete carotene decomposition occurs, and excess amounts of carotene will not cure this defect. Note that even where prepared cake mixes are concerned, the carotene is present primarily to impart a pleasing yellow color to the cake batter which the consumer prepares. In preparing yellow cakes, egg yolks are added to the batter. These egg yolks contribute to the color of the yellow cake. When the cake batter is placed in the oven, temperatures as high as 400°F are never used and the temperature of the cake while in the oven never approaches the oven temperature, except at the surface of the cake which is generally brown. Even under these favorable conditions, at least a 25% loss of carotene can be accurately anticipated.

In the preparation of yellow popped corn, by the wet process, carotene has not been used because the majority of commercial corn poppers operate at temperatures of from 400° to 475°F and higher. Furthermore, satisfactory yellow coloring additives were used until the use of these specific additives was prohibited by the Food and Drug Administration.

A feasible substitute for Yellow 3 and Yellow 4 with regard to coloring ability is carotene. However, as indicated previously, carotene decomposes causing loss of its yellow color. In the case of low temperature decomposition the cause is believed to be due to free radical attack on the long chain unsaturated carotene molecule, and that this free radical attack causes the carotene molecule to split and form decomposition products such as peroxides which in turn may form aldehydes. These decomposition products impart no color and hence all coloring value of the original carotene is lost.

The mechanism of carotene decomposition at higher temperatures, viz 325°F and above, is not known. It is possible that at lower temperatures, carotene decomposes preferentially and at a temperature of 325°F and above, there is no preferential decomposition. Preferential decomposition means that there are several decomposition reactions or routes possible, and at the lower temperatures, the carotene may decompose along the lines of certain preferred reactions. However, at the higher temperatures, viz 325°F and above, this preference for certain decomposition reactions or routes may be destroyed and the carotene may decompose in a completely unfamiliar or unknown manner.

A color stable carotene that can be used at 325°F and above is reported by *B. Borenstein; U.S. Patent 3,039,877; June 19, 1962; assigned to Nopco Chemical Company*. The process for preparing yellow popped corn is also given. The carotene stabilizers which can be used are butylated hydroxy anisole, n-propyl gallate, nordihydroguaiaretic acid and butylated hydroxy toluene. These stabilizers can be used in any combination with one another so that any amount of one stabilizer can be used with any amount of one or more of the other stabilizers, as long as the quantity of the carotene stabilizer composition is between 0.005 and 0.03% based on the total weight of oil incorporated therewith or to be incorporated therewith.

If desired, a metal chelator such as citric acid or another suitable material may be incorporated into the carotene-containing stabilizer composition. It is known that heavy metals such as iron act as prooxidants towards carotene. These heavy metals may be present in the equipment used to prepare popped corn, as rust or in some other manner.

In this process carotene in any of its various forms can be used. For example, naturally occurring mixtures containing primarily alpha- and beta-carotene derived from palm oil, carrots, or any other natural source can be used, or synthetic alpha- or beta-carotene or mixtures can be used. In addition, the entire unsaponifiable material from palm oil can also be used as a carotene source.

The stabilized carotene composition or parts upon addition to the popping oil can be in the form of a concentrate in any edible liquid such as propylene glycol and vegetable oils, or in the form of a suspension in known edible materials or a paste of known edible materials. In addition, crystalline carotene can be used either as a solution, suspension, paste or as a dry solid. It is not advisable to use crystalline carotene as a dry solid because it is believed that exposure to atmospheric conditions will degrade the carotene, but when carotene is utilized as a paste, solution or suspension, it is protected from the atmosphere and will not degrade due to atmospheric conditions.

The stabilized carotene composition can be dissolved in the popping oil or the individual components of the composition can be added directly to the popping oil. If the stabilized carotene composition is in the form of a concentrated solution, then whatever solvent or solvent mixture is used to dissolve the components must be sufficiently polar to dissolve them and must be sufficiently soluble in the popping oil or oils used to assure that the stabilizer or stabilizers and a metal chelator, if used, will dissolve in the popping oil or oils used. When propyl gallate or citric acid is used, a solvent for these materials must be present, such as propylene glycol. If desired, palm oil may be used which contains carotene, and merely added to the stabilizers described in order to prepare the stabilized compositions.

All of the solutions referred to above may contain some undissolved carotene particles; however, the small amount of carotene which is undissolved is not critical. Carotene can be present in the popping oil in an amount of from 0.007 to 0.02% based upon the weight of the popping oil. Amounts of carotene which are below 0.007% are ineffective as colorants. If an amount of carotene in excess of 0.02% is used, the popped corn may have undesirable odors.

The carotene stabilizer should contain, whether one or a mixture of stabilizers is used, from 0.005 to 0.03% of stabilizer based on the weight of the popping oil to be used. If the carotene stabilizer or stabilizers are present in amounts less than 0.005%, then such stabilizer or stabilizers will be ineffective in preparing a commercially acceptable yellow popped corn, although amounts as low as 0.001% are somewhat effective in stabilizing carotene. If the stabilizer or stabilizers are present in amounts exceeding 0.03% based on the weight of the popping oil, these higher levels may produce undesirable odors in the popped corn.

Popped corn can be prepared by either the wet or dry process. Only the wet process can be used in this method. In the wet process of preparing yellow popped corn, the oil, popcorn, carotene and stabilizer or stabilizers are placed in any one of the known apparatuses for preparing popped corn by this method. The entire mixture is then heated to a temperature between 325° and 550°F. The kernels of popcorn then explode to give the finished yellow popped corn. Salt and flavorings such as butter flavoring, caramel, and the like, can be added to the popped corn, if desired.

The fluffy white irregular mass which is the kernel of popped corn represents the expanded endosperm of the original grain. The apparent volume of the kernel increases thirty or more times when it undergoes popping. The popping of corn is related to a sudden expansion of moisture in the intercellular spaces of the endosperm. A possible explanation for the popping of corn may be that the very tough, thick and continuous bran coat of the popcorn kernel retains the steam until pressures in the kernel of popcorn are built up, at which time it ruptures, suddenly releasing the endosperm with its contents of superheated water vapor.

Example 1: The carotene used in this example was derived from palm oil and dissolved in coconut oil. The following general procedure was used in determining the stability of each of the carotene-containing compositions both before heating and after heating at temperatures of from 325°F and above. The carotene and stabilizer were dissolved in 40 cc of coconut oil and the solution was divided into two 20 cc portions. One 20 cc portion was placed in a test tube. The other 20 cc portion was set aside and not heated in order to compare the absorbance of the unheated solutions with the absorbance of the corresponding heated solution. The sample was immersed in an oil bath maintained at a temperature of 455°F and was agitated continuously for 180 seconds. The tube was then removed from the bath and air cooled. One cc samples were then taken from the solutions, both heated and unheated, and added to 15 cc portions of chloroform.

The stability of the stabilized carotene solution was determined by measuring the absorbance of the solution, before and after heating, at the wavelength of maximum absorbance of carotene. The wavelength of maximum absorbance of carotene is 460 millimicrons, and all absorbances of the sample solutions were measured by placing the samples in 12 millimeter diameter optically matched tubes using a Coleman Junior Spectrophotometer to read the absorbance. The absorbance after heating divided by the absorbance before heating multiplied by 100 is called Percent Color Retained. It is well known that absorbance is directly related to concentration. Thus, a reading of half the absorbance after heating as compared to before heating would mean that one-half of the carotene had been destroyed.

In the following table, the amounts of the various materials used are expressed in percents based upon the weight of the oil carrier used. In this example, 0.00292% of citric acid, based upon the weight of the oil used, was added to each carotene-containing stabilizer composition. The amount of carotene used was 0.0140% by weight, based upon the weight of popping oil used. Where a control is used, no citric acid was added. The following table presents the results obtained.

Heat Stability of 0.0140% of Carotene Derived from Palm Oil Dissolved in Coconut Oil

Percent by Weight of Stabilizer Used	Absorbance Before Heating	Absorbance After Heating	Percent Color Retained
None	1.38	0.595	43
0.0140% butylated hydroxy anisole and 0.0035% n-propyl gallate	1.38	0.82	59.5
0.020% nordihydroguaiaretic acid	1.38	0.86	62.3
0.01225% n-propyl gallate	1.30	0.69	53.1
0.0175% butylated hydroxy anisole	1.38	0.75	56.5
0.0200% butylated hydroxy toluene	1.38	0.70	51
0.010% butylated hydroxy anisole and 0.010% butylated hydroxy toluene	1.38	0.745	54
0.0070% butylated hydroxy toluene and 0.0070% butylated hydroxy anisole and 0.0035% n-propyl gallate	1.38	0.73	53
None	1.25	0.505	40.4
0.001% nordihydroguaiaretic acid	1.25	0.535	42.8
0.004% nordihydroguaiaretic acid	1.25	0.600	48.0

Heat Stability of 0.0140% by Weight of Synthetic Beta-Carotene Dissolved in Coconut Oil

Percent by Weight of Stabilizer Used	Absorbance Before Heating	Absorbance After Heating	Percent Color Retained
None	1.30	0.49	37.7
0.00875% butylated hydroxy anisole and 0.0035% n-propyl gallate	1.30	0.675	52.0
0.014% butylated hydroxy anisole and 0.0035% n-propyl gallate	1.30	0.70	53.8

As the above data illustrate, a carotene composition is more stable at higher temperatures when compared with a carotene composition which does not contain the stabilizers. The enhanced stability of the carotene enables one to utilize the carotene compositions whenever high temperatures, i.e., 325° through 550°F are called for, and in particular, such as stable carotene composition is commercially successful in preparing yellow popped corn.

Example 2: This example illustrates the manner of preparing popped corn using the stabilized carotene in a popping oil. A popping oil was first prepared by simple mixing of the following composition:

	Grams
20% carotene (derived from palm oil) by weight of, and partially dissolved in, the unsaponifiable portion of palm oil	240
Butylated hydroxy anisole	30
n-Propyl gallate	12
Citric acid	10
Propylene glycol	28
Total	320

0.3 g of the above composition was then added to 300 g of coconut oil, thus forming a popping oil composition. 20 g of this popping oil were placed in an electric popper which had been preheated to a temperature of 450°F. 65 g of popcorn were then added to the popper. The corn was completely popped within 70 seconds. At the end of the popping cycle, the temperature was 510°F. The popped corn upon visual examination had an attractive butter-like yellow color.

Example 3: This example illustrates the use of the stabilized carotene compositions at a temperature of from 350° to 400°F. A popping oil composition was first prepared by simple mixing of the following:

	Grams
20% carotene (derived from palm oil) by weight of, and partially dissolved in, the unsaponifiable portion of palm oil	60
n-Propyl gallate	3
Butylated hydroxy anisole	12
24.5% citric acid by weight dissolved in propylene glycol	10.4
Total	85.4

42.8 mg of the above composition was added to 39.7 g of coconut oil, thus forming a popping oil composition. The popping oil composition was then placed in an electric popper and the temperature of the popping oil was raised to 350°F within six minutes. When the temperature reached 350°F, 65.7 g of popcorn were added to the oil. After an additional four minutes, the corn was completely popped. The temperature at the conclusion of the popping cycle was 400°F. The popped corn had an extremely attractive butter-like yellow color.

Addition of Alkali Metal Bisulfite

Water dispersible carotenoid preparations can be improved as regards their water dispersibility qualities after storage at room temperature. *B. Borenstein and R.H. Bunnell; U.S. Patent 3,316,101; April 25, 1967; assigned to Hoffmann-La Roche Inc.* report that the life of such water dispersible carotenoid preparations can be considerably enhanced, indeed more than tripled, by the addition of an alkali metal bisulfite in a mol ratio to carotenoid of at least 0.25 mol of alkali metal bisulfite to one mol carotenoid. The carotenoids must have one or more carbonyl groups. For example, apo-carotenal, canthaxanthin, capsanthin, capsorubin, astacene, β-citraurin, as well as all other carotenoids having a free carbonyl group can be used.

A carotenoid is dispersed in an aqueous gelatin medium so as to form an oil in water emulsion. If the carotenoid is first dissolved in a volatile organic carotenoid solvent, particularly a lower polyhalogenated methane, e.g., chloroform, on evaporation of the solvent an emulsion having globules of less than two microns is produced. The advantage of such emulsions is that they approach colloidal state resulting in both high tinctorial potency for the colloid, and when the beadlet is used to color aqueous foodstuffs, the Brownian movement prevents separation of the carotenoid from the foodstuff. When the emulsion prepared by the dispersion of carbonyl carotenoids in gelatin contains particles substantially all of a size less than two microns, it is believed the fine particle size increases the effective concentration of the carotenoid in contact with the gelatin in the exterior phase, and it is due to the contact that water dispersibility decreases upon aging. The alkali metal bisulfite acts so as to prevent the deleterious effect that gelatin would otherwise have on the fine particles of carotenoid.

There are a variety of processes for preparing water dispersible carotenoids where the carotenoid particles in the emulsion will be present, for the most part, in a size of less than two microns in diameter. In the preparation of the emulsion various additives can be used, such as a plasticizer, e.g., sugar and/or sugar alcohols, i.e., sorbitol, mannitol, etc.; an emulsifying agent such as a salt of a higher fatty acid ester of ascorbic acid, particularly ascorbyl palmitate; an edible oil, e.g., peanut oil, vegetable oil, etc.; an antioxidant, e.g., α-tocopherol, butylated hydroxy anisole, butylated hydroxy toluene, etc.; buffering agents, e.g., sodium carbonate; coloring agents other than carotenoids; etc.

A mixture is prepared with a carotenoid having one or more carbonyl groups together with an emulsifying agent, e.g., ascorbyl palmitate, in chloroform into an aqueous solution of gelatin containing sodium bisulfite in a molar amount equal to at least 0.25 times the molar quantity of carotenoid being employed, to form an emulsion containing carotenoid particles of less than two microns in diameter. The chloroform is then removed, preferably by heat. The resulting emulsion can then be dried into particulate form in the method for forming dry beadlets by introducing droplets of the emulsion into a collecting powder such as spraying the emulsion onto a starch bed, which is subsequently dried so as to set up the emulsion droplet into a particle or beadlet. The droplets sprayed onto the collecting powder could be dried prior to making contact with the powder by, for example, spraying the droplets into a heated inert gas or using an extended residence time in the gas. The optimum size of the particles or beadlets is 20 to 200 mesh. It is preferred that the gelatin content be 25 to 70% by weight of the particle or beadlet.

Example 1: 60 g of apo-carotenal, 6 g of d,l-α-tocopherol, 23 g of peanut oil and 35 g of ascorbyl palmitate were dissolved in 450 g chloroform. 100 g of gelatin, 80 g of sucrose, 15 g of sodium bisulfite and 6 g of sodium carbonate were dissolved in 500 ml of water. The chloroform solution was emulsified into the aqueous solution with an Eppenbach homo rod. The solution was heated to 65°C and maintained at that temperature, with stirring, until the chloroform was substantially all removed. The emulsion was sprayed through a revolving nozzle and the droplets collected on a layer of starch. The mixture was dried at 45°C and the excess starch removed by screening.

The resulting beadlets disperse readily in 40°C water after storage at 45°C for 69 days, i.e., the beadlets shrink to the point where there are no obvious solid particles in the solution and no particles remain on filtration with medium speed filter paper, e.g., S & S filter paper No. 604. A similar formula without bisulfite is insoluble in 40°C water after storage at 45°C for 21 days, i.e., a substantial quantity of the beadlets remain visible in the solution and can be filtered out with medium speed filter paper, e.g., S & S filter paper No. 604.

Example 2: 34 g of canthaxanthin, 2.25 g of BHT, 10 g of coconut oil and 22 g of ascorbyl palmitate were dissolved in 480 g of chloroform. 85 g of gelatin, 85 g of sucrose and 15 g of sodium bisulfite were dissolved in 120 ml of water. The chloroform solution was emulsified into the aqueous solution with an Eppenbach homo rod. The procedure was the same as that used in Example 1 above. The resulting beadlets disperse readily in 40°C water after storage at 70°C for 53 days, i.e., the beadlets readily disappear from view and are not recoverable by filtration with medium speed filter paper, e.g., S & S filter paper No. 604. A similar formula without bisulfite is insoluble in 40°C water after storage under the same conditions, i.e., 70°C for 53 days; a substantial number of beadlets remain visible and can be recovered by filtration with medium speed filter paper, e.g., S & S filter paper No. 604.

Soybean Lecithin for Xanthophyll Oil

Xanthophyll oil is obtained as a by-product of the extraction of zinc from corn gluten. It is essentially a mixture of the fatty compounds and the oil-soluble pigments from corn gluten. The total carotenoid pigment content of this oil will usually run from 0.2 to 0.4% with the hydroxy carotenes making up 70 to 90% of the total. The chief pigment is zeaxanthin, 3,3'-dihydroxy-beta-carotene; cryptoxanthin, the 3-hydroxy-beta-carotene, is also present in appreciable quantities as is beta-carotene itself. Alpha-carotene and its corresponding mono and dihydroxy derivatives are present in small quantities.

Xanthophyll oil has found a ready market as a component in chicken feeds. The hydroxy-carotenes in this oil have been shown to cause the fat of the bird to become yellow, a feature which improves the saleability of the chicken considerably. One of the problems in the manufacture and sale of xanthophyll oil is the lack of stability of its pigments. On exposure to air, especially at elevated temperatures, the pigment content of xanthophyll oil drops rapidly.

Soybean lecithin was found by *R.A. Reiners and R.E. Morgan; U.S. Patent 3,081,171; March 12, 1963; assigned to Corn Products Company* to be an extremely effective anti-oxidant for xanthophyll oil. The effectiveness at various concentrations is shown below.

Effect of Various Amounts of Soybean Lecithin on the Stability of Xanthophyll Oil

Amount Lecithin (Percent)	Amount Xantho-phyll Oil (Percent)	Pigment Stability at 50° C. (days)	Protection Factor
None	100	2	1.0
2	98	3	1.5
10	90	9	4.5
25	75	16	8.0
50	50	36	18.0
75	25	28	14.0

The method for determining the oxidative stability of these xanthophyll oil-soybean lecithin blends is as follows. Samples in the amount of 3 ml each of the blend to be tested were put into 50 ml beakers and these beakers placed in a forced air circulation oven held at 50°C. Usually eight samples of each blend were started at one time. At frequent intervals, a beaker was removed from the oven, the entire sample dissolved in hexane, and the total carotenoid pigment content determined. The results from a series of samples taken at different times were plotted to determine the time required for a 50% pigment loss.

Blends of xanthophyll oil and soybean lecithin are easily prepared; the two liquids need merely be mixed together. This procedure, although effective, is slow due to the high viscosity of the lecithin; therefore, it is desirable to heat the mixture mildly to assure better mixing. Common laboratory practice is described in the example given below.

Example: 50 g of soybean lecithin (Gliddol grade N) were added to 50 g of xanthophyll oil and the mixture heated to 60°C under a nitrogen blanket with mild agitation. The mixture was removed from the steam bath and stirred an additional ten minutes to insure good mixing. Samples (3 ml) of this blend were pipetted into clean 50 ml beakers and the beakers placed in a forced circulation air oven held at 50°C. At the desired intervals a beaker was withdrawn from the oven and its content analyzed. The results are given below:

Days at 50°C	0	14	21	36
Total pigment (mg/lb)	1,295	1,100	960	647

The stability of this sample was 36 days, the time required for a 50% loss of pigment.

Soybean lecithin is a commercial product which is available in large quantities. It is made by treating crude soybean oil with water and separating the precipitate from the bulk of the oil. This precipitate is dehydrated under vacuum to yield soybean lecithin, a dark material of grease-like consistency at room temperature. It may be fluidized in a variety of ways to yield a viscous liquid. Both the plastic and fluid grades are equally effective as antioxidants for xanthophyll oil.

The composition of soybean lecithin is quite complex and many of its individual compounds have not been identified. The commercial product contains 35 to 40% triglycerides and 60 to 65% acetone insolubles. The latter are largely phosphatides such as phosphatidylcholine, phosphatidylethanolamine, phosphatidylserine, plasmalogens and the inositol phosphatides. Other compounds which are present are sterols, probably as glycosides and tocopherols. This gross composition is not particularly different from that of corn lecithin, yet the latter is ineffective in stabilizing xanthophyll oil.

The stabilizing effect of soybean lecithin is not due simply to dilution of the pigments. Such diluents as crude corn oil or crude soybean oil either did not appreciably affect the stability of the pigments in xanthophyll oil or improved them by a factor of only 3 to 4. The stabilizing effect of soybean lecithin on the carotenoid pigments of xanthophyll oils is retained even in the presence of diluent oils. This is shown in the data below.

Dilution of Lecithin-Xanthophyll Oil Blends

Composition of Blend (Percent)			Stability at 50° C. (days)	Protection Factor
Xanthophyll Oil	Soybean Lecithin	Diluent		
100	0	0	2	—
50	50	0	22	11.0
50	25	25 Crude soybean oil	16	8.0
50	10	40 Crude soybean oil	7	3.5
50	0	50 Crude soybean oil	4	2.0
50	0	50 Yellow grease	2	1.0
50	25	25 Yellow grease	27	13.5

The unique feature of this process is the effectiveness of large quantities of soybean lecithin in preventing carotenoid oxidation, whereas amounts ordinarily recommended for antioxidants are relatively ineffective. It is characteristic of most antioxidants that dosages above a certain low level are not effective in increasing the stability of the substrate. For example, alpha-tocopherol, a naturally occurring antioxidant, at concentrations above 0.06% did not further extend the stability of lard. Soybean lecithin is unique in that it is the only antioxidant cheap enough to be economically useful at so high a concentration, and its physical characteristics are such that the resulting product is a viscous oil which can be easily handled.

STABILIZED CAROTENOID SUBSTITUTES

There are a number of areas in which colored fatty products have utility. For example, colored oils can be used in salad dressings to produce a more appetizing product. Colored plastic fats likewise present a more appealing appearance to many consumers. Colored cooking fats can be used to impart desirable tints to many fried foods and, particularly, to popcorn.

Coloring materials used in edible fatty compositions generally are limited to oil-soluble dyes such as lycopene, carotene and other carotenoid pigments. Thus the availability and versatility of dyes for use in edible fatty compositions is seriously restricted. A limited number of water-soluble dyes have been used in edible fatty compositions, but only with severe limitations. These compositions have very specific utilities and cannot be adapted to a wide variety of applications. One particularly significant shortcoming of these edible fatty compositions colored with water-soluble dyes is that they are not heat stable; that is, the coloring material does not remain uniformly dispersed when the fatty composition is heated and subsequently cooled.

Heat stable, water-soluble/oil-insoluble dyes are used by *G.W. Brankamp; U.S. Patent 3,489,573; January 13, 1970; assigned to The Procter & Gamble Company* to prepare colored fats utilizing a special emulsifier. The colored fat can be used at normal pan or deep frying temperatures (e.g., 300° to 400°F) without losing any portion of its color due to precipitation of the dye. In spite of repeated temperature changes, the color does not separate from the fatty material. Substitution of other emulsifying agents for the particular esters of polyglycerol does not give the same result. None of the other materials tried, including lecithin, glycerin, monoglycerides of fatty acids, and various fatty acid esters of sorbitan produced heat stable suspensions of the dyes in edible fats.

The edible fats which can be used in the products can be either normally liquid triglyceride compositions or normally plastic triglyceride compositions. Base oils for normally liquid triglyceride compositions can be derived from animal, vegetable or marine sources, and include, for example, such naturally occurring triglyceride oils as cottonseed oil, soybean oil, peanut oil, palm kernel oil, olive oil, corn oil, rapeseed oil, sunflower seed oil, safflower oil and sardine oil. Also suitable oil fractions can be obtained from palm oil, lard and tallow as, for example, by fractional crystallization or directed interesterification, followed by separation of the oil. Oils predominating in glycerides of unsaturated acids may require partial hydrogenation to maintain flavor.

Normally plastic triglyceride compositions are fats which are nonpourable at ambient temperatures (70° to 100°F) but which contain sufficient normally liquid oil and low melting solids that the compositions are converted to a liquid well below frying temperatures. The base oil of such a plastic triglyceride composition generally is a normally liquid triglyceride oil. Uniformly dispersed in the base oil are sufficient high melting, highly hydrogenated triglyceride solids (commonly referred to as hardstock) to give the composition its plastic consistency. Generally, the highly hydrogenated triglyceride hardstock comprises from 5 to 50% by weight of the total composition. By and large, the fats and oils which are suitable for use in plastic shortenings are substantially the same as the base oils for use in normally liquid triglyceride compositions. Preferred triglycerides for use in either the normally liquid or normally plastic triglyceride compositions are soybean oil, cottonseed oil, palm oil, peanut oil, rapeseed oil and mixtures.

The dyes for use in the compositions can be any water-soluble/oil-insoluble dye which is safe for human consumption. Generally this dye is one of those approved by the Food and Drug Commission for use in food, drug and cosmetic products. Typical examples of dyes useful in this process are FD&C Blue #2, FD&C Green #1, FD&C Red #2, FD&C Yellow #5 and FD&C Yellow #6.

The polyglycerol esters which are used in the colored edible fatty compositions to maintain the dye in suspension are commonly prepared from polyglycerol, a polymerization condensation product of glycerol, which has the molecular formula

$$H(OCH_2CHOHCH_2)_nOH$$

where n is the number of glycerol units in the molecule.

The lipophilic-hydrophilic balance in the polyglycerol ester is important for this purpose. It has been found that sufficient lipophilic properties are imparted to the polyglycerol ester by the fatty acid esterification of a single hydroxyl group. However, to maintain sufficient hydrophilic properties in the molecule, the polyglycerol ester cannot contain more than about $n/2$ fatty acid radicals, n being the number of glycerol units in the polyglycerol moiety of the polyglycerol ester. Preferably, the polyglycerol ester will contain between $n/3$ and $n/5$ fatty acid radicals. For example, if a fatty acid ester of decaglycerol, $H(OCH_2CHOHCH_2)_{10}OH$, is used it can contain a maximum of about five fatty acid radicals. Preferably it will contain either two or three fatty acid radicals. Hexaglycerol,

$H(OCH_2CHOHCH_2)_6OH$, can contain a maximum of about three fatty acid radicals and preferably contains either one or two fatty acid radicals.

The minimum number of glycerol units in the polyglycerol esters used in the compositions is two. The maximum number is not material for purposes of this process; however, for reasons of ease of preparation, economics and commercial availability, the practical limit on the number of glycerol units in the polyglycerol esters is about thirty. The preferred number of glycerol units is from two to about ten. Typical examples of polyglycerols which are useful are triglycerol monostearate, tetraglycerol monooleate, hexaglycerol dipalmitate, octaglycerol trioleate, decaglycerol trilinoleate and decaglycerol pentapalmitate.

In preparing the colored edible fatty compositions, the first step is the mixing of the dye and the polyglycerol ester. While not essential, the mixing preferably is accomplished in the presence of water in order to increase the rate of dissolution of the dye. It is desirable to use distilled water, since certain dyes can be precipitated by metallic ions present in most nondistilled water. Dissolution is most easily accomplished by dissolving the dye and the polyglycerol ester in water, preferably accompanied by heating of the mixture to from 160° to 212°F.

Generally a solution suitable for use in coloring an edible fat can be obtained by mixing one part each of dye and polyglycerol ester and three parts of water, if water is used. However, it is usually more convenient to use a greater amount of polyglycerol ester and water. The upper limits on the amounts of polyglycerol ester and water used are determined by considerations of convenience and economics. In the preferred operation, the amount of polyglycerol ester is from 0.5 to 2.0% by weight of the edible fatty composition. Also in the preferred operation, the amount of water, if used, does not exceed 0.5% by weight of the edible fatty composition. To obtain an acceptable final product, water in excess of 0.5% by weight of the edible fatty composition should be removed by boiling or evaporation prior to addition of the dye mixture to the edible fat.

Once the dye-polyglycerol ester mixture is obtained (and, if necessary, its moisture content reduced to the desired level), the mixture is added to and intimately mixed with the edible fat which is to be colored. The mixing step can be performed in any appropriately sized mixing unit, such as a metal mixing tank equipped with a turbine agitator. Naturally, the fat should be in a liquid state to permit proper mixing. To produce a colored plastic fat, the fatty ingredients must be heated to the liquid state prior to mixing with the dye mixture, and subsequently rapidly chilled by any of the methods to obtain the desired plastic consistency.

The amount of dye added to the fatty composition may vary within wide ranges depending upon the particular dye used and the desired color of the fatty product which is being produced. The normal use levels, based on the total weight of the edible fatty composition, is from 1 to 500 parts per million. The colored edible fatty compositions also can contain additives which are commonly used in fatty compositions. The most common of these additives are emulsifiers, such as monoglycerides of fatty acids and various fatty acid ester derivatives of sorbitan, commonly known as Spans and Tweens. These compositions also can contain crystal inhibitors, such as oxystearin, antioxidant agents such as methyl silicone, butylated hydroxyanisole and butylated hydroxytoluene, and other additives commonly used in edible fatty compositions.

Example 1: A colored solution was prepared in a 100 ml beaker by mixing 0.08 g of FD&C Yellow #5 dye and 0.02 g of FD&C Yellow #6 dye in 30 g of 100°F distilled water. To this mixture were added 15 g of decaglycerol trilinoleate. While stirring, the mixture was heated slowly to boiling. The solution had a uniform bright yellow color, and none of the dye particles remained undispersed. Heating of the solution was continued until substantially all of the water was removed by evaporation. 2.5 g of this colored mixture were added to a 1,000 ml beaker containing 400 g of edible oil (refined and bleached soybean oil, partially hydrogenated to an iodine value of 107). After intimate hand stirring, the color was evenly distributed in the oil. After standing for one hour at room temperature, none of the dye precipitated out of the edible oil. The colored solution was heated to 350°F and maintained at that temperature for two hours. After being allowed to cool to room temperature and stand for one hour, the solution was observed. As before, the color remained evenly distributed in the oil, and none of the dye precipitated out of solution. The colored oil then was used for the deep frying (350°F) of French cut potatoes. The potato pieces acquired a slight yellow tint during frying which improved their appearance. The oil again was allowed to cool to room temperature and stand for one hour. As before, the oil retained an even yellow color and no dye precipitated out of solution.

When in the above example the decaglycerol trilinoleate is replaced in whole or in part by triglycerol monostearate, tetraglycerol monooleate, hexaglycerol dipalmitate, octaglycerol trioleate, decaglycerol pentapalmitate and mixtures thereof, substantially similar results are obtained in that heat stable suspensions of the dye in the edible oil are obtained. Further, when in the above example the soybean oil is replaced in whole or in part by cottonseed oil, peanut oil, palm kernel oil, olive oil, corn oil, rapeseed oil, sunflower seed oil, sesame seed oil, safflower oil and sardine oil, substantially similar results are obtained in that heat stable suspensions of the dye in the edible oil are obtained.

Example 2: A colored solution is prepared in a 100 ml beaker by mixing 0.05 g of FD&C Red #2 dye in 10 g of 100°F distilled water and 10 g of tetraglycerol monooleate. The mixture is heated to 212°F and allowed to boil until 8 g of the water is removed. A shortening composition is prepared by heating 500 g of the following to 150°F:

Component	Percent by Weight
Base stock (refined, bleached soybean oil partially hydrogenated to an iodine value of about 107 and having a solids content index of 0.32 at 70°F)	86
Hardstock (substantially completely hydrogenated soybean oil and substantially completely hydrogenated rapeseed oil in a weight ratio of 5.33:1 and having an iodine value of 8)	10
Mono- and diglycerides of partially hydrogenated soybean and cottonseed oils (in a weight ratio of 85:15) having an iodine value of about 80	4

To this composition is added 5 g of the colored mixture prepared above, and hand mixing is continued until the shortening composition is uniformly colored. The colored composition is then rapidly cooled by passing it through a scraped wall heat exchanger to reduce its temperature to 80°F in about 20 seconds. Following the chilling, the colored shortening composition is passed through an agglomerator (picker box) and gently agitated for two minutes. The mixture is then sealed in 12 ounce jars and tempered for 24 hours at 80°F. The tempered product has a uniform pink color and a plastic consistency. This plastic shortening product is useful in all areas where shortening products are commonly used, and when used for frying the color is heat stable.

SALT, BEVERAGES AND FLAVORS

FREE-FLOWING AND LOW PROOXIDANT SALT

EDTA Treatment

Generally the ordinary culinary salt of commerce, while highly refined, contains trace amounts of certain heavy metal salts. Although these heavy metal salts, chiefly copper and iron, are present in relatively small amounts, they have a pronounced effect on the storage stability of certain fats and oils when salt containing them is incorporated therein. The primary problem which has been noted in connection with these trace impurities is that they act in the capacity of catalytic agents or prooxidants when present in certain unsaturated fats, oils or the like, promoting the oxidative degradation which is the chief characteristic of rancidity. It has been the experience of the food industry that when trace amounts of these heavy metal salts are present in a culinary salt which is subsequently incorporated into unsaturated fats, oils, and the like, such as are used for cooking or food purposes, i.e., peanut butter, peanut oil, butter, oleomargarine, lard, corn oil, and the like, oxidative degradation is enhanced.

Thus it has been found that where salts of such metals as copper and iron are present in the salt, the fats and oils have a substantially reduced storage life before rancidity and oxidation render them unpalatable or unfit for use. While it might be deemed desirable to process culinary salts in such a manner as to entirely remove such trace amounts of heavy metal salts or prooxidants as might be present, it has heretofore been the experience that such a processing procedure is so involved as to be economically impractical.

Several alternatives have been proposed to eliminate or reduce the prooxidative effect caused by these impurities, and thereby substantially increase the storage life of unsaturated fats and oils. One such alternative presently in commercial use involves the addition of certain antioxidant materials, such as for example, alkylated phenols which inhibit the oxidative reaction which is promoted by the catalytic activity of these prooxidant heavy metal materials. This expedient, while achieving some degree of success, has not been altogether satisfactory inasmuch as the storage stability is only partially improved; it involves the addition of a foreign material into the salt, which may have deleterious effects when consumed, or impart an off-flavor or color to the food.

Another possibility for the improvement of salts containing trace amounts of copper and iron involves the use of certain organic chemicals classified as chelating agents, which have the characteristic of reacting with heavy metals to produce an organo-metallic complex of such stability that the prooxidant effects previously exerted by the salts or ions of the heavy metal materials, such as copper or iron, are largely eliminated. One such example of this expedient is mixing sodium chloride and a small amount of ethylenediaminetetra-acetic acid, and forming a tablet which is utilized in the canning of various seafoods. While this innovation has been satisfactory to some degree, it is believed that certain drawbacks militate against its use under all conditions. For example, the procedure involves the use of mixture of sodium chloride and ethylenediaminetetraacetic acid introduced into food-stuffs in tablet form, and it is obvious that it would be desirable in many instances to introduce the material as a loose granular or fine powder. This, of course, would involve certain problems of uniformity of mixing to assure that the ethylenediaminetetraacetic acid was uniformly distributed throughout the salt.

A further difficulty resides in the fact that traces of the prooxidant salts are often present within the individual salt crystals as occluded particles, and the use of small amounts of chelating agent as additives to the surface of the salt particles does not assure that all of these occluded salts would be effectively deactivated. Additionally, it is known that ethyl-enediaminetetraacetic acid has chelating powers with respect to salts other than the heavy metals, and that any unreacted chelating agent has the capacity to remove salts, such as calcium, from the body tissues when ingested by mammals, which is obviously detrimental.

A method of preparing a culinary salt with reduced prooxidant tendencies is explained by *F.G. Miller and E.A. Dittmar; U.S. Patent 3,198,608; Aug. 3, 1965; assigned to Morton Salt Company.* The method requires crystallizing sodium chloride by concentrating a saturated brine containing trace amounts of heavy metal prooxidant salts in the presence of an ethylene-diaminetetraacetic acid compound, recovering the crystallized sodium chloride salt from the saturated brine supernate as a slurry, washing the slurry of sodium chloride crystals substantially free of ethylenediaminetetraacetic compound, and drying the sodium chloride.

Tests were carried out on various salts prepared by this process to determine their prooxidant effects. A salt containing 0.95 part per million copper and 0.85 part per million iron was utilized to prepare a saturated brine. To the saturated brine thus prepared was added tetrasodium EDTA in an amount equal to 25.6 parts per million (salt basis). Sodium hydroxide was added to bring the pH of the brine to 10.0. The brine containing the EDTA salt was concentrated by heating and evaporating at elevated temperatures, i.e., 224° to 226°F. The crystals of sodium chloride formed during the evaporation step were recovered as a damp slurry from the supernate brine, washed with fresh brine solution and water to remove all residual EDTA salts.

The rate of washing was carried out so that less than 5% of the salt product was removed. After washing, the product was dried by heating at elevated temperatures, i.e., about 300°F. The salt product so formed was analyzed for the EDTA. No EDTA was found. Analysis for copper and iron indicated 0.3 and 1.8 ppm, respectively.

Elevated temperature oven storage tests were utilized to determine the prooxidant effect of the salt treated by this process. In this experiment 55 grams of salt were added to 19 grams of steam rendered lard, blended and placed in a 100 ml beaker. The salt was completely covered with the fat and no fat layer formed on top of the salt. Each beaker was covered with a filter paper and placed in an oven maintained at 45°C. From time to time a sample was removed and the peroxide value of the fat determined. Curves of peroxide development against time were plotted and the stability of each sample was taken as the time required for the fat to reach a peroxide value of 20 meq/kg. Several other samples of untreated salt were also included to show the effect of various amounts of copper and iron. The results of the test were as follows:

Salts Added to Lard	Stability, hours	Average
(1) Untreated salt containing		
0.95 ppm Cu	45	44
and 0.85 ppm Fe	44	

(continued)

Salts Added to Lard	Stability, hours	Average
(2) Treated salt containing		
0.3 ppm Cu and 1.8 ppm Fe	146	
(3) Untreated salt containing		
0.22 ppm Cu	33	
and 0.6 ppm Fe	37	35
(4) Salt containing		
0.5 ppm Cu	23	
and 0.8 ppm Fe	26	25

Samples 3 and 4 are given in the table for the purpose of showing that the increased stability achieved in the large sample containing salt treated by the process is not due to the decreased copper content. Thus sample 3 containing a salt having 0.22 ppm Cu has an average stability of 35 hours as contrasted to the 146 hours of sample 2. Likewise sample 4 containing a salt with 0.5 ppm Cu has an average stability of 25 hours.

From the foregoing it may be seen that treatment of brines made from salt containing trace quantities of copper and iron by the process produces a culinary salt having greatly reduced prooxidant properties. For example, in the specific test shown above, a stability improvement of about 330% was achieved in fats containing salt treated by the process as contrasted with the untreated control. It is obvious that the process produces a salt which is capable of being used with a broad variety of unsaturated fats and oils to produce a product with a substantially increased storage life.

Potassium Conditioning Agent

Pure sodium chloride, as a salt for food, is deficient in other mineral nutrients, and also has the tendency to agglomerate, especially when exposed to humid weather. It is known that the use of tricalcium phosphate, referred to as TCP or $3Ca_3(PO_4)_2 \cdot Ca(OH)_2$, in salt as a conditioning agent does render it free-flowing even when exposed to humid weather conditions, and although it does give it the added nutritive elements of calcium, phosphorus, hydrogen and oxygen, still such a salt is not nutritively balanced. In addition to all these elements in the conditioning agent, there is a further requirement for potassium in order to achieve nutrition balance.

A potassium-enriched conditioning agent is used by *H.N. Norsen; U.S. Patent 3,306,753; February 28, 1967* in place of TCP. While the addition of a 1% quantity of TCP alone to sodium chloride makes the resulting salt free-flowing so that it will not agglomerate, such a treated salt soon forms a cloudy film or coating of TCP on the inside of a glass salt shaker. On the other hand, if pure sodium chloride did not have the tendency to agglomerate, it would keep the inside of a salt shaker crystal clear. However, when the potassium-enriched conditioning agent is combined with salt, it coats the inside of a glass salt shaker with such a thin transparent film of potassium-enriched conditioning agent as to be hardly detectable, and yet the salt so treated is free-flowing and will not agglomerate.

The potassium-enriched conditioning agent for salt, is especially formulated; the combination of potassium chloride and monopotassium phosphate in a molar ratio of 1 to 6 is contained in a mixture with tricalcium phosphate, as prepared by the wet method such that the mixture exhibits a calcium to phosphorus ratio of 1.15. When a combination of these 3 chemicals is added and thoroughly dry-blended or mixed in such a way as to become a part of sodium chloride to the extent of 0.70%, the resulting conditioned salt has the optimum amount of inorganic buffering ability to make it nutritively-balanced.

The mixture is dry-blended; and then, to insure homogeneousness of composition, is thoroughly wet with water to form a thin smooth paste; uniformly stirred, dried by suitable means, such as using trays for batch or a rotary kiln dryer for continous operations, thus evaporating all the water that had been added; then the dried material is ground to the fineness of the original TCP to yield the potassium-enriched, nutritively-balanced, conditioning agent, which, when added to the salt to the extent of even less than 1%, makes the salt free-flowing so that it will not agglomerate, besides giving it the added characteristic of always keeping the inside of a glass salt shaker nearly crystal clear.

Antioxidant Salt

It has been found convenient to use salt as a carrier for antioxidants with the usual procedure being to dissolve the antioxidants in a solvent such as propylene glycol and then to blend the resultant solution with the salt. In the past the above procedure has been found quite satisfactory with the normal grades of salts found on the market, that is to say, with salt composed of particles substantially between 14 and about 100 mesh Tyler screen size, but it has been found unsatisfactory when the salt is substantially finer than 100 mesh Tyler. It has been felt that a very fine antioxidant salt, namely, one of less than 100 mesh Tyler particle size, if available, could be used for the manufacture of a product such as peanut butter, and it is for this reason that a method for making such a salt product has been considered highly desirable.

One alternate procedure would be to dry blend the finely powdered dry antioxidants with the salt. This is possible with 3 of the ingredients composing the most common antioxidant mixtures. These 3, namely, propyl gallate, butylated hydroxytoluene (BHT), and citric acid are available commercially in a finely powdered grade and present no problems in a dry blending operation. However, the fourth ingredient of many antioxidant formulas, namely BHA, is available in the form of waxy flakes of a rather low melting range, 48° to 55°C. These flakes are too coarse to blend uniformly with a very fine salt since the amount of the antioxidant required is quite small in relation to the total salt composition and, because of their low melting range and waxy constitution the flakes cannot be ground commercially to a fine powder and stored without agglomerating.

Finely powdered antioxidant mixture containing butylated hydroxyanisole is prepared by R. Kolasinski; U.S. Patent 3,502,484; March 24, 1970; assigned to Diamond Crystal Salt Company.

A finely powdered BHA-diluent mixture can satisfactorily be prepared by first melting the BHA and then incorporating a quantity of finely powdered diluent material such as, for example, tricalcium phosphate into the molten BHA, then cooling and powdering the friable mixture before screening it to the desired particle size. This finely powdered BHA-diluent mixture can then be dry blended with very fine salt and other antioxidant ingredients available in powdered form which are desired to be used in the salt composition.

The diluent material used in the antioxidant composition may suitably be from the following group: tricalcium phosphate (TCP), sodium aluminum silicate (Zeolex), finely powdered salt of less than 325 mesh size (i.e., -325 mesh salt), calcium silicate, silicon dioxide, or magnesium carbonate. Other materials could be used as the diluent so long as the material possessed a particle size as stated and a high bulk absorptivity. In addition, the diluent material could be any one of those previously mentioned or a mixture of 2 or more. The particle size of the diluent material should be less than 100 mesh Tyler screen size.

The amount of diluent material required varies depending on its particle size and bulk absorptivity and on the anticipated time during which the BHA-diluent mixture will be stored before incorporating it in the bulk of the fine salt. For example, a satisfactory antioxidant composition has been prepared and stored for six months without caking by using either TCP or Zeolex at the 50% level. However, when microsized salt of approximately 325 mesh size or smaller is used as a diluent a higher amount of the diluent based on its mixture with the BHA is required. For example, the microsized salt was used at a level of about 66⅔% by weight based on the total of diluent and BHA.

Broadly stated, the amount of diluent material used should be at least about 5% by weight based on the total of BHA and diluent in the antioxidant composition. The upper limit of diluent material used is not critical, yet generally it does not exceed approximately 98%. The BHA particle size should be less than 100 mesh Tyler screen size. The amount of the

BHA used in the antioxidant composition will generally make up the balance of the composition over the amount of diluent material used which, as previously stated, is present in an amount of at least 5% by weight or more of the antioxidant composition.

The finely powdered BHA-diluent mixture can be used alone as an antioxidant composition or in conjunction with other powdered antioxidants such as butylated hydroxytoluene, propyl gallate, and citric acid. The finely powdered BHA-diluent mixture can satisfactorily be blended with microsized salt, and other powdered antioxidant ingredients if they are used, to thereby produce a free-flowing fine antioxidant salt composition in particulate form having a size less than 100 mesh Tyler. Generally, the amount of mixture to be used will be less than 2% by weight of the salt and not in excess of about 5%. The total amount of antioxidant composition used with the salt depends on the final application for which the salt product is intended.

Example 1: 50 parts BHA was first melted by heating it to a temperature above its melting range of 48° to 55°C. After the BHA was in a molten state there was incorporated 50 parts of finely powdered tricalcium phosphate (TCP) diluent material having a size of less than 100 mesh. The mxiture was then allowed to cool to approximately room temperature to produce a friable mixture which was then powdered and subjected to a screening operation to produce a finely powdered BHA-tricalcium phosphate mixture having a particle size of less than 100 mesh. It was found that this finely powdered BHA-TCP mixture could be dry blended with very fine salt having a particle size of less than 100 mesh and that other antioxidant ingredients could be readily incorporated in the blending operation with the fine salt.

In order to test this BHA-TCP antioxidant composition it was blended with a microsized salt having a particle size of less than 100 mesh Tyler screen size. In addition, the additional antioxidant ingredients of propyl gallate and citric acid were blended into the salt composition, and all materials were blended together using a twinshell blender. The proportion of ingredients blended together to form the final salt composition was as follows:

	Percent
Microsize salt including 1.5% TCP	99.6700
BHA-TCP mixture (50% BHA, 50% TCP)	0.2550
Propyl gallate	0.0475
Citric acid	0.0275
Total	100.0000

No problems were encountered in blending this mix as set forth in the above formulation.

Example 2: The procedure of Example 1 is repeated except that 50 parts by weight of Zeolex 23 diluent material is used with 50 parts by weight of BHA instead of the tricalcium phosphate diluent material of Example 1.

Example 3: The procedure of Example 1 is repeated except that 66⅔ parts by weight of microsized fine salt is used as the diluent material and having a particle size of less than approximately 325 mesh. This microsized salt is used as the diluent material, in place of the tricalcium phosphate of Example 1, with 33⅓ parts of weight of BHA to form the finely powdered BHA-diluent antioxidant composition.

Example 4: An antioxidant composition as follows was prepared in parts by weight using the procedure of Example 1: BHA, 90; and Cab-O-Sil M5, 10. Cab-O-Sil M5 is a pyrogenic silica (SiO_2).

An alternative method for preparing the BHA-diluent mixture which has been found satisfactory is to incorporate the diluent in a solution of the BHA dissolved in a solvent such as methyl alcohol or ethyl alcohol. After the diluent has been mixed with this solution the solvent is removed for example, by the use of reduced pressure, then the resulting dry mixture of BHA and diluent is screened to give a particle size of less than 100 mesh.

The solvent used as described above may be either methyl alcohol, ethyl alcohol or mixtures thereof, however, the particular solvent chosen is not critical. In a commercial process where the BHA-diluent mixture was to be incorporated in a food product, ethyl alcohol would be a suitable selection as a solvent from the standpoint that it would be nondeleterious to human consumption should any trace amounts of the solvent remain in the mixture.

Propylene Glycol Conditioning

In the past, caking of salt products has long been a problem to those who package, manufacture, store or sell salt. It is desirable that a salt product should be noncaking and free-flowing as this greatly facilitates the use, handling and storage of the salt. The caking of salt is caused by moisture between the crystals of the salt which moisture partially dissolves minute quantities of the sodium chloride to form a salt solution. As the salt redries the salt solution bridges the salt particles at their points of contact, evaporates, and then leaves crystallized sodium chloride causing a structural connection between the salt particles, or in other words, caking of the salt particles and an inhibiting of the free-flowing characteristics of the salt.

For example fine flake salt is subject to severe caking during storage periods, and when the flake salt is placed under pressure in bags that are palletized or stacked one on top of another, the caking is especially severe. If such hot flake salt from its production process is cooled before packaging, the tendency to cake is lessened, however, cooling of the salt before packaging requires expensive extra equipment and does not guarantee caking resistance. Therefore, there is a need for an effective and economical anticaking agent which can be added to flake salt as well as other particulated salt products while still in a hot condition, allowing the bagging of the product immediately without drying or cooling and without experiencing caking of the bagged salt over relatively long storage periods.

Also, the anticaking agent when used with salt should provide an improved salt composition having a capability for absorption to a certain extent of the included moisture in the salt crystals, which moisture comes to the surface of the crystals during drying and long storage.

J.F. Heiss and G.E. Binsley; U.S. Patent 3,374,098; March 19, 1968; assigned to Diamond Crystal Salt Co. use propylene glycol to impart free-flowing and noncaking characteristics to flake salt.

Sorbitol solutions have also been used as anticaking and dust control agents in salt for many years. However, a primary disadvantage of sorbitol is that under low humidity storage conditions, the sorbitol solution crystallizes out, becoming ineffective as an anticaking agent. Such crystallization does not occur with the salt compositions containing propylene glycol as anticaking agent; and also, sorbitol does not have the favorable humectant properties of propylene glycol which remains liquid even at relatively low humidities. A test was conducted to show the improved performances of propylene glycol compared with sorbitol. 900 grams of Alberger fine flake salt was added to each of four 2-liter wide mouth glass vessels. Then 0.067% by weight water was added to the salt in vessel No. 1, and 0.033% water to the salt in vessel No. 2.

A commercial grade of propylene glycol was then diluted with water in the proportions of 1:1 by volume, and this solution was added to the salt in vessel No. 3 in an amount which gave 0.035% by weight anhydrous propylene glycol in the salt. The sorbitol solution was made up by diluting 1 volume of a 70% by weight aqueous solution of sorbitol with 1 volume of water. This 1:1 volume solution of sorbitol was added to the salt in vessel No. 4 so that there was 0.043% by weight of the sorbitol 70% solution in the vessel. Each of the four salt containing vessels contained at least 0.033% by weight water as a control level of moisture.

The salt in each vessel was blended thoroughly so that the added moisture in anticaking additives would be thoroughly distributed. The four salt samples were air dried at room temperature over night and observations were made as follows. The Alberger fine flake salt containing sorbitol was observed to exhibit more caking than the vessel No. 2 flake salt containing 0.033% by weight water. The propylene glycol-flake salt gave least caking and best results, considerably better than the sorbitol sample, even at the low treatment level of 0.035% by weight propylene glycol. The test also demonstrated that the sorbitol dried out as rapidly as the control samples and was ineffective thereafter.

The broad composition limits for the propylene glycol additive in the salt composition expressed on an anhydrous basis, should be such that the propylene glycol is less than 0.05% by weight of the total salt composition, with the lower limit being about 0.005% by weight. Best results have been obtained, however, when the amount of propylene glycol in the salt composition is maintained within a range between about 0.025% and about 0.045% by weight.

The use of relatively small amounts of propylene glycol additive in the salt composition is most advantageous for bagged salt products, especially if the salt is to be preserved in a dry free-flowing condition. However, the propylene glycol additive also can be used for bulk treatment of salt, and at critical humidities below 75% the salt will be relatively dry and free-flowing because of the very low concentration of the propylene glycol additive. Propylene glycol can be used as an anticaking agent for evaporated salts such as grainer, Dendritic, vacuum pan or Alberger, and also, rock, solar, compacted flake and pulverized salts. However, it has been found that the propylene glycol additive gives best results when used with flake salts.

The term flake salt means a salt having an apparent density in the range between about 700 and about 1,000 grams per 1,000 per cubic centimeters. Flake salt contains less than $1/10$ of 1% by weight moisture when the salt is received from its preparation process prior to the addition of propylene glycol. Since the flake salt has a very low moisture content prior to treatment, only a relatively small amount of propylene glycol is required to endow the composition with the desired characteristics, and the salt composition obtained after treating the salt with propylene glycol is subsequently dry, nonckaing and free-flowing, and not moist or sticky.

Example 1: A 1,000 gram sample of fine flake salt taken from production was blended with an additive of USP grade propylene glycol. The amount of additive used was 0.35 ml.

Example 2: A 90% propylene glycol–10% water by volume solution was added to a 1,000 gram sample of fine flake salt in the same amount as in Example 1.

Example 3: An 80% propylene glycol-20% water by volume solution was added to a 1,000 gram sample of flake salt in the same amount as in Example 1.

Example 4: A 60% propylene glycol–40% water by volume solution was added to a 1,000 gram sample of flake salt in the same amount as in Example 1.

Example 5: A 40% propylene glycol-60% water by volume solution was added to a 1,000 gram sample of flake salt in the same amount as in Example 1.

Evaluation tests were conducted with fine flake salt blends to determine the effectiveness of various propylene glycol solutions in comparison with glycerine solutions as an anticaking agent. For the tests, 1,000 gram samples of fine flake salt taken from production were blended with the additives shown in the table below, except for Sample 1, which was plain fine flake salt without any additive; 0.35 ml of additive was used with each 1,000 gram sample of salt. 400 ml tall form beakers were filled with salt after preparation and placed on top of an oven where they were subjected to mild drying conditions for a period of six days.

The beakers were then inverted over a U.S. 12 mesh test screen with salt retained on same being weighed as an indication of the amount of caking taking place. The results are expressed in the following table as percent retained beside the description of the additive used.

Sample No.	Additive Composition	Weight Percent Retained on Screen
1	Plain fine flake (nothing added)	97.0
2	Propylene glycol (USP)	0.7
3	Propylene glycol 90%-water 10% by volume	0.3
4	Propylene glycol 80%-water 20% by volume	1.1
5	Propylene glycol 60%-water 40% by volume	9.9
6	Propylene glycol 40%-water 60% by volume	35.6
7	Glycerine (USP)	93.8
8	Glycerine 80%-water 20% by volume	91.3
9	Glycerine 60%-water 40% by volume	93.5
10	Glycerine 40%-water 60% by volume	95.0

The conclusions from the test data shown in the above table are that, compared with glycerine, propylene glycol is a superior anticaking additive and the same holds true for comparisons of aqueous solutions of either of the two when using aqueous solutions of equal volume percentages.

BEVERAGE STABILIZATION

Glucose Oxidase System

The method of *D. Scott; U.S. Patent 3,193,393; July 6, 1965; assigned to Fermco Laboratories, Inc.* involves contact of water-containing foods and beverages with heat stabilized enzymatic oxidase preparation in which the enzyme components are first stabilized against destruction by heat and oxygen through essentially complete drying and then protected against heat in the presence of moisture by coating the dry stabilized enzymatic oxidase preparation with a moisture permeable barrier layer. The moisture impermeable barrier prevents transfer of enzyme-activating moisture from the aqueous product to the stabilized enzymes in the preparation for a predeterminable minimal interval of time in order to insure that the enzyme activation will take place at an appropriate time for overcoming the deleterious action of oxygen either in a gaseous atmosphere and/or in solution.

Heat-processed foods and beverages are adversely affected by oxygenation of the product. Oxidation can result in the impairment of color and flavor of carbonated or noncarbonated canned or bottled soft drinks, which products must have an appreciable storage life. Not only does the color of such soft drinks tend to fade and the flavor thereof becomes impaired, but there may be iron pickup from the container and eventual perforation of the container because of the direct or indirect chemical and corrosive action of oxygen which is present in solution or in the space above the top surface of the beverage.

The glucose oxidase reaction is specific for the aldose sugar, glucose. The product of reaction of the enzymatic elimination of free oxygen from the canned food stuff is gluconic acid, in amounts determined by the amount of free oxygen present, which product is well recognized as a harmless by-product in the food. While it has been long recognized that the treatment of products with glucose oxidase and catalase enzyme mixtures prior to sealing the cans or packages is an effective way of removing oxygen, the practical application of such a system has of necessity been confined to the incorporation of enzymatic oxidase preparation while the foodstuff is at or near room temperature.

The available enzyme preparations when added to the food product at high temperatures, such as are encountered during hot filling and vacuum packing operation, utilized, for example, in the preparation of baby foods, apple juice, canned dog foods and the like, are rapidly destroyed due to the effect of high temperatures in the presence of the aqueous medium. This destruction of the enzyme is appreciable at elevated temperatures and destruction is accelerated as the temperature is increased. Temperatures above about 60°C, depending upon the characteristics of the aqueous medium, cause such rapid destruction of the enzyme, particularly glucose oxidase enzyme, as to substantially completely impair the capacity of the enzyme to consume the available oxygen in the container of foodstuff.

Enzymatic oxidase preparations may be stabilized in a dormant reactivatable form and that such stabilized enzymatic oxidase compositions, when provided with a moisture barrier coating, can be adapted to withstand destruction at temperatures in excess of 40°C and even at temperatures in excess of 100°C, as are employed during such operations as hot filling, pasteurizing and the like. The enzyme preparation is stabilized by thoroughly drying the same to a substantially constant weight. In drying the aqueous enzyme preparation, the dehydration may be carried out at progressively higher temperatures as the water content is reduced. While the enzyme is in an aqueous medium or in the form of a solid containing an appreciable percentage of water, the temperature thereof should be maintained below about 45°C.

As the enzyme solids become progressively drier, the temperature of dehydration may be progressively raised until at moisture contents of less than 0.5%, the dehydration of enzyme may be carried out at temperatures of the order of 60° to 70°C without deleterious effects upon the activity of the enzyme. In general, the enzymes are in a satisfactory dry condition for coating if they lose no weight when subjected to temperatures of 45°C in a vacuum chamber for 30 minutes.

This heat stabilized enzyme preparation, prior to incorporation in the aqueous product at the high temperatures encountered in packing, is coated with a composition whose rate of dissolution or permeation by water or water vapor will maintain the enzyme dry for any period of time during which the enzyme would be expected to be subject to destructive or inactivating influences in the presence of activating quantities of moisture, thereby providing a predeterminable interval between contact of the coated enzyme with aqueous medium and moisturizing of the enzyme to a water content such that, if glucose and oxygen are present, the enzyme will act thereon.

Coating compositions are film forming materials having a slow rate of dissolution in aqueous media or a slow rate of penetration by moisture. Materials useful in the formation of these films or coatings are the so-called water-insoluble synthetic gums such as ethylcellulose, alkali soluble cellulose and cellulose derivatives, nitrocellulose, and cellulose acetate, water-insoluble, alcohol-soluble natural resinous gums such as shellac, keratin and the like, water-insoluble alcohol-soluble proteins such as zein, formaldehyde treated proteins and the like, and water-insolubilized synthetic polymeric material such as polyvinyl acetate copolymer.

The coating composition serves to effectively separate the stabilized glucose oxidase-catalase enzyme preparation from the aqueous product. After a period of time dependent upon characteristics of the coating, coating thickness, etc., the coating is penetrated by moisture and the enzymatic oxidase composition will be moisturized and subsequently reactivated. The water of the aqueous product may reach the enzyme through any one or more of a variety of means such as slow solution of the coating, distintegration of the coating as evidenced by flaking or by rupture, erosion of the coating, diffusion through the coating and the like.

The stabilized enzyme preparation may be included in the internal surface coating of the can, can top, bottle or jar top or cap, the surface coating containing the thoroughly dry enzyme preparation encased in protective coating. The binder in such coating may be moisture resistant film former in which is incorporated the dry enzyme stabilized against heat. This coating may be subjected, if necessary, to a preliminary drying operation in order to insure that the stabilization of the heat stabilized glucose oxidase-catalase preparation has not been impaired.

In certain instances, such as with tablets or coatings containing the heat-stabilized desiccated enzyme preparation, it may be desirable to introduce the glucose substrate for the oxidase enzyme present in the tablet or coating. If glucose is already present in the product to be protected from the harmful effects of free oxygen, then such addition of glucose in the enzyme preparation is not necessary. The glucose already present in the food is caused to oxidize with the ultimate production of harmless gluconic acid.

Frequently, glucose is not present in the product, but instead there is present some other sugar, such as maltose, sucrose, lactose, etc. In such case, glucose is introduced into the enzyme preparation in an amount which is sufficient to provide the substrate for the glucose oxidase and results in the elimination of the free oxygen by the oxidation of the glucose which has been added.

A further alternative is presented by incorporating in the tablet or coating where glucose is absent and such sugars as maltose, sucrose, lactose or the like are the sugars in the aqueous food, the additional corresponding enzyme, namely, maltase, invertase, or lactase, respectively, which corresponds to the particular sugar substrate and breaks down the substrate in the food to glucose. Thus, in the example of beer which contains no glucose, maltose may be present. By addition of maltase to beer, the maltose is hydrolyzed into glucose which in turn will act as a suitable substrate for the glucose oxidase enzyme.

Similarly, if the condensed sugar is sucrose, the enzyme invertase converts the substrate into glucose. With lactose as the condensed carbohydrate in the food, lactase in the added enzyme preparation converts the lactose into glucose for the oxidase present. In the event that it is desired to convert starch as the condensed product into a glucose substrate for the oxidase present, the enzyme diastase in combination with maltase can be used.

In certain kinds of oxidation reactions such as the oxidation of ascorbic acid or iso-ascorbic acid, the formation of hydrogen peroxide results in the same objectionable effects upon the taste and color of the aqueous food product as does the presence of free oxygen. The improved heat-stable enzyme compositions are useful in the elimination of this hydrogen peroxide because of the presence of the enzyme catalase in the preparation. Catalase consumes hydrogen peroxide even though it is formed by an action other than that which may involve glucose oxidase.

In the foregoing modifications of the heat-stable enzyme preparations wherein glucose is incorporated in the enzyme composition, the oiling of the granulated or tableted desiccated enzyme preparation prior to the coating operation is of particular advantage. In this oiling operation of the thoroughly dried enzyme preparation mineral oil is utilized as a thin coating over the enzyme. After the subsequent application of a coating, for example, of zein about the oiled preparation, the drying of the protected composition is easier than in the case where the oiling is omitted.

The enzyme may be directly combined with a coating substance in solution and thereafter dried to provide a heat stabilized preparation in accordance with the process, or the preparation may be in granule, pellet, or tablet form. Pellets or tablets containing the heat-stabilized preparation of the process, may contain various binders such as gum acacia, gelatin, sucrose, glucose, starch, and the like.

To facilitate disintegration of the pellets or tablets, a disintegrator may be used such as potato starch or corn starch in amounts of about 5 to 10% by weight based on the total weight of the composition. If the tablet is to be used in an acid food product, sodium bicarbonate may be incorporated in the tablet to aid disintegration. In other instances, if desired, a stoichiometrically balanced blend of an alkali metal bicarbonate, such as sodium bicarbonate, together with an organic dry food acid may be used as the disintegrator in dry form in amounts of about 3 to 8% by weight. The dry food acid, such as citric acid, reacts with the sodium bicarbonate upon coming in contact with free water in the food to cause the evolution of carbon dioxide and the disintegration of the tablet which liberates the enzyme into the aqueous food product.

The binders and disintegrators as well as other solids in the tablet or coating serve as diluents which are inert with respect to the enzyme activity. These solid diluents may include other materials such as filter aid, kaolin, salt and the like. Another form of diluent having utility is mannitol. In any coating composition there is always the danger that the coating will be imperfect and that pores will exist extending from the surface to the interior of the coated solids, which would permit premature wetting of the enzyme, for example, with water at a temperature sufficiently high to destroy the enzyme activity. If the wetting is of a limited extent, the effect of contact between enzyme and hot water may be alleviated by the use of mannitol. Mannitol dissolves with an endothermic reaction which will have a cooling effect and will lengthen the time between wetting and appreciable deactivation of the enzymatic composition.

One unit of glucose oxidase may be defined as that amount of enzyme which will cause the uptake of 10 cu ml of oxygen per minute at 30°C under assay conditions described by Scott, *Journal of Agricultural and Food Chemistry,* volume 1, 727-30 (1953).

To assay a standard enzyme solution for catalase activity, 0.04 cc of the enzyme preparation is added to a 250 cc beaker. Then 100 cc of 5 volume hydrogen peroxide (1.5%) buffered to pH 7.0 is added to the beaker and the mixture is allowed to stand for about 1 hour at about 25°C. A 4 cc sample of the resulting solution is withdrawn and mixed with 5 cc of 2 N H_2SO_4 and 2 grams of potassium iodide. The resulting solution is then titrated with 0.25 N thiosulfate solution. Likewise a 4 cc sample of the 5 volume hydrogen peroxide solution (without enzyme addition) is mixed with 5 cc of 2 N H_2SO_4 and 2 grams of potassium iodide and also titrated with the same 0.25 N thiosulfate solution. In each case the disappearance of an iodine color in the titrated solution serves as the end point.

Subtract the difference between the 2 titrations from which may be calculated the equivalent of hydrogen peroxide decomposed by the enzyme in the first solution. A unit of catalase may be defined as that amount of enzyme which, under the above conditions, will decompose 0.0155 equivalent or 0.264 gram of hydrogen peroxide under assay conditions.

In terms of aqueous glucose oxidase enzyme preparations which are readily available in commerce, preparations can be obtained which vary in solid concentration from about 3 to about 100 mg or upwards of solids per cu cm of enzyme solution. This may represent a glucose oxidase activity of from 10 to 1,500 units/cu cm of enzyme solution. On a dry solids weight basis, the activity of such solids may vary between 1,000 and 90,000 units of glucose oxidase activity/gram. Since each unit on a dry solids basis cause the removal of 10 cu mm of oxygen/minute at 30°C under assay conditions, it is obvious that a dosage in excess of that required for the amount of free oxygen which is to be removed may readily be supplied to meet the specific requirements of the aqueous food product in the amount present in the standard size container in which the food is packaged.

For example, tablets or pellets containing the heat-stabilized glucose oxidase-catalase and coating materials when used for the removal of deleterious oxygen in 12 ounce containers of soft drinks, noncarbonated or carbonated, or of apple juice or of citrus fruit drinks in the same size container, provide effective stabilization permitting a shelf life extending to beyond 10 months and up to about a year with a glucose oxidase activity of about 15 to 25 units for the single pellet added to the container. Larger size containers of the food require only slightly larger pellet additions, for example, a 24 oz container may be stabilized to the same extent with from 20 to 40 units of glucose oxidase enzyme. In such glucose oxidase enzyme preparation about 2 to 4 units of catalase/10 units of glucose oxidase is sufficient to catalyze the conversion of hydrogen peroxide produced enzymatically from the glucose substrate by the glucose oxidase.

Example 1: Preparation of Heat-Stabilized Glucose Oxidase-Catalase Enzyme – An aqueous solution of glucose oxidase-catalase enzyme preparation is mixed with 1.1 volumes acetone in the cold. The precipitate is filtered off and then air-dried to remove the solvent and provide a solid pulverulent or powdered product and thereafter dried in a vacuum desiccator over calcium oxide. By checking the weight loss until a constant value is noted, the complete dryness of the preparation is assured. 25 grams of this powdered enzyme assayed about 4,000 units as glucose oxidase/gram and about 2,000 units of catalase/gram.

Drying of the air dry product at normal room humidities is insufficient in the above preparation to provide satisfactory stability in accordance with the heat stability requirements in the presence of aqueous food as dictated by the filling temperature conditions for use in accordance with the process. Rapid inactivation of the air dried product at temperatures of 85° to 90°C was noted in that deterioration to about one-half enzyme strength in about 10 minutes was observed, depending on the room humidity during drying. This inactivation is suppressed and substantially eliminated by the complete and thorough removal of water in a vacuum desiccator or drying in a forced draft laboratory area at about 30°C until constant weight is achieved, this latter checked against the weight loss of the former procedure. Dried enzyme may be pelleted into tablet form by machine.

Example 2: After cleaning away the rough edge of tablets averaging approximately 300 mg per tablet and assaying about 1,000 units of glucose oxidase and 350 units of catalase, 10,000 tablets were placed in a coating pan and the pan rotated. Approximately 100 ml of wax-free shellac in alcohol solution of about 25% solids content was added. The tablets were allowed to roll until they showed signs of solvent evaporation by becoming tacky. At this point a small amount of finely powdered talc was added as a dusting agent. When the tablets were rolling freely in the pan, a blast of room temperature air was directed against the tablets to remove most of the remaining alcohol. Air blasting was continued for about 10 minutes. The tablets were then dried for about ½ hour using air at a temperature of about 85° to 100°F while the pan continued to rotate.

The coating operation was repeated until a coating was formed from a total of 15 coating operations. Tablets prepared according to Example 2 and uncoated tablets from the same source were tested as follows: 1 tablet was added to apple juice heated to a temperature of about 180°F. The container was closed and the apple juice cooled to room temperature. The activity of glucose oxidase introduced in the form of the coated tablets of Example 2 was substantially unimpaired by introduction into the hot apple juice and by retention therein during the cooling operation. Glucose oxidase introduced using uncoated tablets have the enzymatic activity substantially completely eliminated by introduction into the 180°F apple juice by the time the apple juice had cooled to room temperature.

Gallic Acid Esters

Biological stabilization of beer, wine, and cider is accomplished with use of esters of gallic acid according to *M. Loncin and S.P. Leeuw; U.S. Patent 3,490,913; January 20, 1970; assigned to Jos. Schlitz Brewing Company.*

Alcoholic beverages must previously be filtered carefully and possibly pasteurized in bulk, in order to reach a sufficiently low content in living microbial cells, preferably less than 100 cells/ml. Generally the active concentration of ester of gallic acid is between 5 and 15 ppm, e.g., 10 ppm. This concentration depends on the length of the alkyl chain and is lower as the alkyl chain is longer and the alcoholic beverage is less infected.

With the highest concentrations, turbidity may occur, due to the low solubility of these esters or to the presence of reaction products between these esters and the proteins contained in the alcoholic beverage. It is possible to avoid proteic turbidity by stabilizing the beverage, e.g., by adding proteolytic enzymes. This treatment is well known and currently in use, especially in breweries. It was certainly impossible to anticipate the fact that these esters, which are only slightly soluble in water and have at the most a weak fungicidal action, would be so efficient at such weak concentrations in their inhibition of microbial developments in beer, wine and cider.

The weak antimold action of these esters is known. It was however not possible to deduce from this fact that these esters would be active against yeast, lactobacilla, acetic acid producing bacteria and sarcina, because these microorganisms are completely different from molds in biological, morphological and biochemical terms. The use of parahydroxybenzoates of long chain alcohols as stabilizing agents for beer is suggested.

Esters of gallic acid have certain advantages compared with parahydroxybenzoates, in particular better solubility in alcoholic beverages and water. So, for example, the solubility of octyl gallate in water and beer is nearly seven times as great as that of octyl parahydroxybenzoate. Moreover, parahydroxybenzoates of long chain alcohols must be used as alkaline solutions which are not very stable and may influence the pH of the beverage. The esters of gallic acid, the alcohols of which contain 8 to 10 carbon atoms, can be added as such or in solution in a suitable solvent, such as ethanol or propylene glycol. This solution can be injected continuously by means of a metering pump. The dissolution of gallic acid esters in alcoholic beverages when they are in a crystallized state is generally slow and incomplete. Better solubility can be obtained by passing the beverage through a porous inert material which has been previously mixed with a gallic acid ester or impregnated with a solution of this ester in a solvent such as methanol, isopropanol, acetone or propylene glycol, the solvent being eliminated afterwards by heating or water rinsing.

As porous inert material, one can use for instance diatomaceous earth (kieselguhr), asbestos or cellulose. The length of the gallic acid ester chain, the concentration of the ester in the porous material and the rate of flow of the alcoholic beverage through a layer of this material having a given thickness can be selected so as to dissolve an efficient bacteriostatic concentration, i.e., generally one between 5 and 15 ppm.

Example 1: 10 ppm of pure octyl gallate were added to bottled beer. This beer was previously filtered on asbestos plates in accordance with the traditional method of brewers. There was a concentration corresponding to 3.3 mg pure octyl gallate in each bottle having a content of 330 cm^3. Octyl gallate was added as ethanol solution. The bottles were then closed, shaken and stored at a temperature of 25°C. After 25 to 35 days, the control bottles (without any octyl gallate) showed a growth of several microorganisms, i.e., yeast, lactobacilla and sarcina, which gave rise to turbidity and deposits. After 200 days the bottles with octyl gallate did not show any signs of deterioration. With nonyl gallate, the results were about the same. When the length of the alkyl chain is further increased the preservative action was maintained, but the very poor solubility of these products made their utilization much more difficult.

Example 2: Italian white wine from Trebbiano grapes was used which had an alcohol content of 11%, total acidity in terms of tartaric acid, 6.2 g/l and a volatile acidity in terms of acetic acid, 0.57 g/l. First 1% glucose and then a mixture of microorganisms of spoiled wines were added to the wine. Finally, 5, 10, 15 and 20 ppm of octyl gallate were added as alcoholic solution. The wine was preserved in half-filled bottles at room temperature for 28 days. The following results were achieved.

Wine without octyl gallate: Very strong development of microorganisms. Appearance of cloudiness, turbidity and deposit. Volatile acidity, 1.98 g/l in terms of acetic acid.

Wine with 5, 10, and 15 ppm of octyl gallate: No visible deterioration. Volatile acidity, 0.60 to 0.62 g/l in terms of acetic acid.

Wine with 20 ppm of octyl gallate: Occurrence of slight nonmicrobial turbidity. Volatile acidity, 0.58 g/l in terms of acetic acid.

Example 3: 100 parts of diatomaceous earth (kieselguhr) were mixed with 20 parts in weight of a solution of 10% octyl gallate in methanol. The solvent was later eliminated by mixing and gentle heating. From this product 500 g/m^2 were spread as top layer of a suitable support, consisting in a finely meshed metal gauze.

Beer at 2°C was then run through the filtering layer at a rate of 250 to 300 l/hr/m^2. It came out with a content of about 10 ppm of octyl gallate. This beer could be kept in bottles or barrels for 3 months without showing any yeast or mold development. During filtration, diatomaceous earth was added as a slurry at the rate of 50 to 100 grams of earth for every 100 liters of beer. The earth was impregnated with octyl gallate as described above. Even when filtration was interrupted for several hours, the octyl gallate content never exceeded 20 ppm.

PRESERVATION OF BEER

Myristic and Lauric Diethanolamides

It has been the practice to inactivate bacteria and yeast in beer beverages by pasteurization. This process is effective but is troublesome and results in an undesirable change in the flavor of the beverage. It is the object of the process of *T.F. McFadden and T.A. Schueler; U.S. Patent 3,440,057; April 22, 1969,* to preserve the freshness of the flavor of the beverage by adding a chemical which is compatible with the beverage but acts as a nonionic surfactant to inactivate the bacteria and yeast.

For this purpose any one of a number of diethanolamides can be used, these additives having the general formula:

$$RCON \begin{cases} CH_2CH_2OH \\ CH_2CH_2OH \end{cases}$$

wherein R is an aliphatic hydrocarbon radical containing from 7 to 16 carbon atoms. The preferred members of the diethanolamide series are the lauric and the myristic, or a mixture of these two in any proportion, relatively low concentrations of these members being found to be effective in preserving beverages, as illustrated in the following examples showing the results of experiments with samples of a finished fermented malt beverage containing various amounts of preservative. In each case the sample was forced by daily shaking of the bottle to fill the head-space in the bottle completely with foam, the bottles being stored at temperatures between 75° and 85°F. In the following tables the figures in the left-hand column represent the number of grams of preservative/mml of beverage. The figures in the other columns indicate the sediment formed in the time intervals noted, the figures in the columns representing the following conditions: (1) none, (2) trace, (3) very slight, (4) slight, (5) moderate, and (6) heavy.

TABLE 1: MYRISTIC DIETHANOLAMIDE

Parts per Million	1 Month	2 Months	3 Months
1	3	4	5
2	3	4	4
3	3	4	4
4	3	4	4
6	2	3	3
8	2	2	3
10	2	2	2
12	2	2	2
0 - pasteurized control	3	4	5
0 - unpasteurized control	5	6	6

TABLE 2: LAURIC DIETHANOLAMIDE

Parts per Million	1 Month	2 Months	3 Months
3	3	5	6
4	3	4	5
6	3	4	5
8	2	4	5
10	2	4	5
12	2	3	4
14	2	3	4
16	2	3	3
0 - pasteurized control	3	4	5
0 - unpasteurized control	5	6	6

It appears from the foregoing tables that the more carbon atoms a diethanolamide has in its aliphatic chain, the smaller the quantity required to preserve a given quantity of beverage. However, for practical operation, other factors must be considered. For example, when using myristic diethanolamide, the matter of foam stabilization begins to be a problem which is increasingly greater for higher members of the diethanolamide series unless that tendency is counteracted by the addition of some other chemical substance suitable for the purpose. The foregoing tables also indicate that although pasteurization leaves no viable organisms in the beverage, yet considerable sediment was formed by 3 months forcing as a result of · the formation of insoluble proteinaceous matter in the beverage. Superior results in suppressing the formation of sediment were obtained by the use of sufficient myristic and/or lauric diethanolamides.

The use of diethanolamides as preservatives for beverages has the additional advantage of being easily applied. The point of application of such preservatives to a finished malt beverage should be as closely subsequent as practicable to the final filters. The following procedure is preferred. A 10% solution of the diethanolamide in water, propylene glycol, alcohol, or mixtures of 2 or more of these liquids is prepared. This solution is diluted to a precise 2% solution (grams/milliliters). By means of an apportioning pump, the amount of the 2% solution necessary to produce the desired concentration of the preservative in the final volume of beverage is added to the beverage.

For this operation high-precision metering of the preservative solution is not required since the diethanolamides will not produce a difficulty redissolvable precipitate in the beverage in local excesses of up to 100 parts per million (grams/milliliters). Concentrations of this magnitude far exceed what is needed for effectively preserving the beverage. Thus chemical preservatives of the kind described can readily be added to finished fermented malt beverages on a commercial scale irrespective of whether the beverage is to be packaged in cans or bottles or distributed as a draught beverage.

Imidazolines

In another process by *T.F. McFadden and T.A. Schueler; U.S. Patent 3,440,056; April 22, 1969* another class of preservatives are used as surfactants to inactivate bacteria and yeasts in beer beverages. One group of effective additives are the imidazolines and their acid salts, this group being represented by the formula:

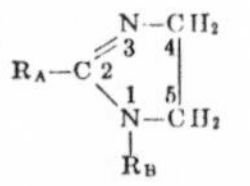

wherein R_A is an aliphatic hydrocarbon chain, the carbon atoms of which range in number from 8 through 18, and R_B is an alkanol group.

Any food-grade acid can be reacted in stoichiometrically equivalent amounts with any of the imidazolines, preferably those wherein R_A is the undecyl radical, the lauryl radical, the myristic radical, the heptadecyl radical or the oleyl radical to produce the acid salts. For the preservation of fermented malt beverages, lactic acid is preferred as a reagent for an imidazoline. If lactic acid is reacted with the imidazoline having the undecyl radical, the resulting cationic surfactant would be:

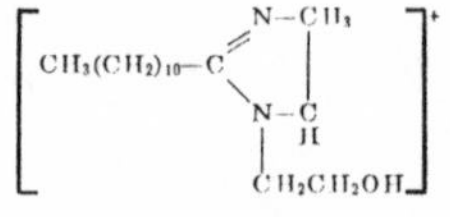

For commercial application of imidazoline to fermented malt beverages, the following procedure is recommended. Prepare a 10% (g/ml) solution of imidazoline. Add 2 stoichiometrically equivalent weights of food-grade lactic acid and sufficient water and/or propylene glycol at about 70°C to make a 2% (weight/volume) solution, agitating the mixture until solution is complete. Introduce this solution into the beverage as closely subsequent as practicable to the final filters, preferably by means of an apportionating pump, the rate of addition being such as to produce the desired concentration of additive in the final volume of beverage, the optimum concentration, up to 50 parts per million, depending on the condition of the beverage as indicated by periodic tests.

An advantageous feature of the use of an imidazoline is that precise metering of the additive is not required since an imidazoline salt in concentrations up to 100 ppm by weight will not produce a permanent or difficultly redissolvable precipitate in the beverage. Slight variations in the action of apportioning equipment are tolerable in view of the almost immediate dispersability of imidazoline solutions in the beverage.

p-Hydroxybenzoic Acid Esters

F.B. Strandskov and J.B. Bockelmann; U.S. Patent 2,175,912; March 30, 1965; assigned to F. & M. Schaefer Brewing Company describe a method to control microbiological growth in finished packaged beer with a synthetic, organic chemical preservative of the general formula:

$$\left(R-O-\overset{\overset{\text{O}}{\|}}{C}-\left\langle\underline{\quad\quad}\right\rangle-O\right)_n -X$$

where R is an aliphatic hydrocarbon radical; X is either a hydrogen atom (H), an alkali metal, e.g., sodium (Na) and potassium (K), or an alkaline earth metal, e.g., calcium (Ca); and n is an integer equal to the valence of X. The preferred preservative is the normal heptyl ester of parahydroxybenzoic acid, which corresponds to the formula:

$$\overset{\overset{\text{O}}{\|}}{C}-O-(CH_2)_6CH_3$$

These synthetic organic chemical preservatives are very effective; they are both fungistatic and bacteriostatic.

The microbiological spoilage of beer is a recognized problem. This is overcome by pasteurization of bottled and canned beer at this time. Draught beer is not pasteurized and often spoils during the warm weather when left unrefrigerated. Spoilage is readily observed by a marked amount of sediment in the beer and the unpleasant taste and odor produced by the microbial growth and metabolism. Pasteurization of beer is a very costly and unpleasant operation. A typical modern pasteurizer for instance occupies approximately 900 square feet of floor space. The volume of steam used is large and the humidity caused by all the steam and hot water results in almost impossible working conditions during the hot summer months.

A further disadvantage of pasteurization is the temperature of the beer when it comes out of the pasteurizer. Although attempts are made to cool the beer, it is difficult to get the temperature below 90°F. Beer at this temperature ages rapidly and develops the undesirable age taste. Preservation with a synthetic organic chemical compound is accomplished without heat, allowing the beer to be packaged and shipped at about 32°F; at this temperature practically no aging takes place. A further advantage of preservation with a chemical compound over that with pasteurization is the elimination of a considerable amount of bottle breakage and resulting loss of beer due to pressure generated in the bottle by high temperature. This loss amounts to as much as 0.5% of the total production.

In the preservation of beer with chemicals, it is appreciated that residual microorganisms contained in the beer may and usually do include both yeasts and bacteria, thus necessitating the use of both a fungistatic chemical and a bacteriostatic chemical. The prime purpose of the process is to eliminate pasteurization and to preserve finished beer chemically.

Esters of para-hydroxybenzoic acid and alkali metal and alkaline earth metal salts containing a hydrocarbon chain can be made up in stock solutions with a suitable solvent, such as ethyl alcohol, but preferably propylene glycol. The alkali metal and alkaline earth metal salts are most conveniently made up in stock aqueous solutions, i.e., water is the solvent. Such a stock solution can then be added to the finished beer to obtain an appropriate concentration of the preservative. It has been found that as little as 10 ppm of the heptyl ester of para-hydroxybenzoic acid in finished beer results in essentially as good preservation over a period of up to 18 weeks as pasteurization of finished beer. Parts per million are actually parts by weight of the heptyl ester per million parts by volume of finished beer. The inhibition of both bacteria and yeast growth in the beer is thus as great as that achieved by pasteurization over the preiod tested.

The effectiveness of the compounds as a beer preservative is evidenced by preparing a series of stock solutions in propylene glycol of the heptyl ester so that with the addition of at most 0.1 ml of any solution to a test bottle, the desired chemical preservative concentration is obtained when the test bottle (a clean, empty standard 12 ounce brown export bottle) is filled with unpasteurized beer and capped on the regular production line. The bottles are incubated at room temperature (about 75°F) and are compared at weekly intervals with a pasteurized control for the development of sediment. The quantity of sediment is determined visually; values from 1 to 9, based on fixed sediment standards, being assigned to the various degrees of sediment formation.

The pasteurized beer which is used is free from all viable microorganisms. The secondary yeast count in the unpasteurized beer is of the order of from 10 to 100 cells per milliliter, and the bacteria count is extremely low (a few per milliliter).

Concentration of Heptyl Ester (p.p.m.) [1]	Sediment Formed After—			
	6 Weeks	8 Weeks	12 Weeks	18 Weeks
0	9+-9+-9+	9+-9+-9+	9+-9+-9+	9+-9+-9+
1.0	9+-9+-9+	9+-9+-9+	9+-9+-9+	9+-9+-9+
2.0	4-9+-9+	9+-9+-9+	9+-9+-9+	9+-9+-9+
5.0	4-4-3	9+-9+-9+	9+-9+-9+	9+-9+-9+
10	3-3-3	2-3-3	3-4-3	5-4-5
Pasteurized Control	3-3-3	4-4-4	4-4-5	4-5-4

[1] Parts by weight of heptyl ester per million parts by volume of finished beer.

The data of the table are obtained using as the preservative the normal heptyl ester of para-hydroxybenzoic acid. These data illustrate the degree of preservation both in relation to the pasteurized control and the unpasteurized sample to which no chemical was added.

Example 1: Into a clean empty standard 12 ounce brown export bottle is placed 0.1 ml of a propylene glycol solution of the normal heptyl ester of para-hydroxybenzoic acid. The concentration of acid (which is the chemical preservative) in the propylene glycol solution is 0.035 part by weight per part by volume. The brown export bottle with the added chemical preservative is then filled with unpasteurized finished beer and capped on the regular production line. In this example, 350 ml of unpasteurized lager beer are placed in the bottly prior to capping. The concentration of preservative in the capped bottle is 10 parts (by weight) per million parts (by volume) of beer. This bottle of lager beer, maintained at room temperature (75°F) for as long as 18 weeks, does not form an excess of sediment.

Example 2: Into a clean standard half-barrel (about fifteen-and-one-half-gallon container) is placed 0.57 ounce of propylene glycol solution of the normal heptyl ester of para-hydroxybenzoic acid. The concentration of ester (the chemical preservative) is 0.035 part by weight per part by volume of propylene glycol solution. The half-barrel containing the propylene glycol preservative solution is then filled with fifteen-and-one-half gallons of unpasteurized lager beer and sealed on the regular production line. The concentration of the chemical preservative in the sealed half-barrel is 10 parts (by weight) of preservative per million parts (by volume) of beer. This half-barrel of beer (containing added preservative), maintained at a temperature of 75°F for as long as 18 weeks, does not form an excessive amount of sediment.

p-Hydroxybenzoic Acid Esters plus Pimaracin

Undesirable microbial growth may be prevented by means of the heptyl or octyl ester of p-hydroxybenzoic acid or an alkali metal or alkaline earth metal salt thereof in conjunction with the antibiotic Pimaracin, which is described in British Patent 846,933 of September 7, 1960, or the alkali metal salts of Pimaracin or the alkaline earth metal salts of Pimaracin.

According to the process of *F.B. Strandskov and J.B. Bockelmann; U.S. Patent 3,232,766; February 1, 1966; assigned to The F. & M. Schaefer Brewing Company* the additives are incorporated into the finished beer in the form of a stock solution wherein the components are dissolved in a solvent which itself has no deleterious effect upon the beer. This may be, for example, propylene glycol, ethanol, etc. The Pimaracin is employed in an amount ranging from 5 to 50 ppm. The amount of the p-hydroxybenzoic acid ester compound used is from 6 to 300 ppm (the greater amount being employed for the lower esters).

Example: 0.5 ml of additive stock solution is placed in 12 oz brown beer bottles which are then filled from the regular production line with cold, unpasteurized beer. The beer is foamed up to expel headspace air and then capped so that each bottle contains 350 ml of beer together with the stock solution. The bottles are stored at room temperature and examined weekly for microbiological spoilage which may be readily observed by sediment, unpleasant taste, and odor. The table shown below sets forth the results obtained with different additive combinations.

	ppm	Pimaracin, ppm	Microbiologically sound after—	Microbiologically spoiled after—
n–Heptyl p–hydroxybenzoate	6	0	5 weeks	7 weeks
	0	50	3 weeks	4 weeks
	6	10	>20 weeks	
n–Octyl p–hydroxybenzoate	6	0	5 weeks	7 weeks
	6	10	>20 weeks	
n–Hexyl p–hydroxybenzoate	10	0	6 weeks	8 weeks
	10	50	>20 weeks	
n–Pentyl p–hydroxybenzoate	20	0	6 weeks	8 weeks
	20	50	>20 weeks	
n–Butyl p–hydroxybenzoate	50	0	8 weeks	10 weeks
	50	50	>20 weeks	
n–Propyl p–hydroxybenzoate	100	0	6 weeks	8 weeks
	100	50	>20 weeks	
n–Methyl p–hydroxybenzoate	200	0	6 weeks	8 weeks
	200	50	>20 weeks	
Unpasteurized[1] control	–	–	2 weeks	3 weeks
Pasteurized[2] control	–	–	>20 weeks	

[1] The control product is beer to which no additive had been introduced.
[2] The pasteurized control is beer to which no additive has been introduced and which has been subjected to standard pasteurization treatment.

The results show that the beer preserved with the respective benzoates alone spoils between the fifth and eighth weeks of storage while the beer preserved with the Pimaracin alone spoils between the third and fourth weeks. But the beer to which both the components have been added is not spoiled even after more than 20 weeks at which time the tests were discontinued. This establishes the synergistic aciiton obtained when using the combination of the additives. In producing beer preserved against microbial growth in accordance with this process in commercial production quantities, the procedure is show below.

A stock solution is prepared containing the additives in such amount that 2 gal of the stock solution equally distributed throughout 100 barrels of beer gives the desired concentration of the additives in the beer. The stock solution is injected into the beer pipeline at the desired stage of production of the beer as the beer flows through the line. The rate of injection into the line is correlated to the flow rate of the beer through the line. This correlation may be achieved through methods known in the art, as for example, flow meters. This method insures a thorough admixture of the stock solution with the beer. The beer containing the additive intimately admixed therewith is then filled into bottles, cans or kegs, as desired.

p-Hydroxybenzoate-Octyl Gallate

In another process by *F.B. Strandskov and H.L. Ziliotto; U.S. Patent 3,764,342; October 9, 1973; assigned to The F. & M. Schaefer Brewing Co.* octyl gallate (n-octyl-3,4,5-trihydroxybenzoate) is used with para-hydroxybenzoic acid. The preservation of finished beer with octyl gallate alone or with the para-hydroxybenzoic acid ester compounds alone, is sometimes not feasible because, inter alia, the comparatively large amount of the materials necessary to achieve preservation for a satisfactory length of time can have a tendency to create adverse effects upon the quality of the finished beer product.

It has been found that when utilized in combination with the para-hydroxybenzoic acid ester compounds, not only is the amount of the octyl gallate necessary for preservation reduced to an acceptable level, but also the amount of the para-hydroxybenzoic acid ester compound is more than correspondingly reduced. That is, the compounds when used in conjunction with each other display a synergistic action. It has additionally been found that the mixture of the compounds results in a finished beer product having extremely good foam retention properties notwithstanding the fact that it has been previously recognized that the para-hydroxybenzoic acid ester compounds alone have a tendency to adversely affect the properties of the finished beer product in which they are used. In the examples the following procedure is used.

Various stock solutions are prepared for use in the examples. The stock solution numbers conform to the corresponding example number.

Stock Solution No. 1 is prepared as follows. Dissolve 420 mg of n-heptyl para-hydroxybenzoate acid is a sufficient amount of ethanol to make 100 ml of solution. The addition of 0.5 ml of this solution, containing 2.1 mg of the n-heptyl para-hydroxybenzoate to 350 ml of beer yields a solution containing 6 ppm of the benzoate.

Stock Solution No. 5 is prepared as follows. Dissolve 420 mg of octyl gallate in a sufficient amount of ethanol to make 100 ml of solution. The addition of 0.5 ml of this solution, containing 2.1 mg of the octyl gallate to 350 ml of beer yields a solution containing 6 ppm of the octyl gallate.

In the same manner as for Stock Solutions 1 and 5, additional solutions are made up such that 0.5 ml of the solution added to 350 ml of beer yields a solution containing the amounts of materials indicated below where the additive is in ppm.

Stock Solution	Additive	Amount
2	n-Heptyl p-hydroxybenzoate	8
3	n-Heptyl p-hydroxybenzoate	10
4	n-Heptyl p-hydroxybenzoate	12
6	Octyl gallate	9
7	Octyl gallate	12
8	Octyl gallate	15
9	Octyl gallate	18
10	Octyl gallate	21
11	Octyl gallate	24
12	n-Heptyl p-hydroxybenzoate	6
	Octyl gallate	6
13	n-Heptyl p-hydroxybenzoate	6
	Octyl gallate	12
14	n-Heptyl p-hydroxybenzoate	6
	Octyl gallate	18

Example 1: Into each of several 12 ounce brown beer bottles place 0.5 ml of Stock Solution No. 1. Fill these bottles from the regular production line with cold, unpasteurized beer. Foam up the beer to expel headspace air, and cap. Each of these bottles contains 350 ml of the beer together with the stock solution. Store these bottles at room temperature (about 15° to 30°C), periodically (weekly) examining them for microbial spoilage. Such spoilage may be readily observed by a marked amount of sediment in the beer and by the unpleasant taste and odor produced by microbial growth and metabolism.

Example 2: Into each of several 12 ounce brown beer bottles place 0.5 ml of Stock Solution No. 2. Fill these bottles from the regular production line with cold, unpasteurized beer. Foam up the beer to expel headspace air, and cap. Each of these bottles contains 350 ml of the beer together with the stock solution. Store these bottles at room temperature (about 15° to 30°C), periodically (weekly) examining them for microbiological spoilage.

Example 3: Into each of several 12 ounce brown beer bottles place 0.5 ml of Stock Solution No. 3. Fill these bottles from the regular production line with cold, unpasteurized beer. Foam up the beer to expel headspace air, and cap. Each of these bottles contains 350 ml of the beer together with the stock solution. Store these bottles at room temperature (about 15° to 30°C), periodically (weekly) examining them for microbiological spoilage.

Examples 4 to 14: The same procedure is followed as in the preceding examples with Stock Solutions 4 through 14. The following table sets forth the results obtained in the examples.

Example	(1) n-Heptyl p-hydroxy-benzoate (p.p.m.)	(2) Octyl gallate (p.p.m.)	Sediment [1] reading after—			
			4 wks.	6 wks.	10 wks.	18 wks.
Control [2]	0	0	9+-9+	9+-9+	9+-9+	9+-9+
I	6	0	6-7	9+-9+	9+-9+	9+-9+
II	8	0	6-7	9+-9+	9+-9+	9+-9+
III	10	0	3-3	4-3	4-4	5-6
IV	12	0	3-3	4-4	4-4	5-6
V	0	6	9+-9+	9+-9+	9+-9+	9+-9+
VI	0	9	3-5	6-7	9+-9+	9+-9+
VII	0	12	3-4	5-5	9+-9+	9+-9+
VIII	0	15	4-4	4-5	9+-9+	9+-9+
IX	0	18	3-4	3-4	9+-9+	9+-9+
X	0	21	4-4	4-4	4-5	5-5
XI	0	24	3-3	3-4	4-5	5-7
XII	6	6	3-3	4-4	4-4	4-5
XIII	6	12	4-4	4-5	4-5	5-5
XIV	6	18	4-5	4-4	5-5	5-6
Pasteurized control [3]	0	0	3-3	4-5	5-6	5-6

[1] A sediment reading of 9 or above indicates microbial spoilage.
[2] The control product is beer prepared as in Example I to which no preservative had been introduced.
[3] The pasteurized control is beer prepared as in Example I to which no preservative had been introduced and which had been subjected to standard pasteurization treatment.

The results set forth in the table show that 10 ppm of the benzoate alone and 12 ppm of the benzoate alone and 21 ppm of the gallate alone are required to achieve preservation. The beer to which as low as 6 ppm of both of the components have been added is not spoiled even after more than 18 weeks at which time the tests were discontinued. This establishes the synergistic action obtained when using the combination of the additives. In producing beer preserved against microbial growth in commercial production quantities, the following procedure is followed:

A stock solution is prepared containing the additives in such amount that 8 gal of the stock solution equally distributed throughout 100 barrels of beer gives the desired concentration of the additives in the beer. The stock solution is injected into the beer pipeline at the desired stage of production of the beer as the beer flows through the line. The rate of injection into the line is correlated to the flow rate of the beer through the line. This proportioning may be achieved through methods known in the art, as for example flow meters. This method insures a thorough admixing of the stock solution with the beer. The beer containing the additive intimately mixed therewith is then filled into bottles, cans or kegs, etc.

The chill stability of the beer can be improved by the addition of propylene glycol aliginate. Any sufficiently soluble propylene glycol alginate, i.e., sufficiently soluble in the beer medium may be used for this purpose. As an example of commercially available propylene glycol alginates which are satisfactory are Kelcoloid-O and Kelcoloid-L (Kelco Company), Kelcoloid-L being preferred. As to the amount of propylene glycol alginate to be added, a preferred range is from about 40 to about 120 ppm. The time of addition of the propylene glycol alginate is likewise not particularly critical and may be at any time after the fermentation is completed and before or after the addition of the benzoate and octyl gallate preservatives. In a preferred process it is added after the first filtration but before the final filtration of the beer.

According to this aspect of the process (1) individual stock solutions may be prepared containing the predetermined amounts of heptyl para-hydroxybenzoate or salt, octyl gallate or salt, and propylene glycol alginate; or, (2) stock solutions may be prepared containing the predetermined amounts of any of these elements in any combination. If a production quantity of beer is to be treated, the proper amount of stock solution(s) is added to the beer stream at the desired stage in the plant operation. If laboratory quantities of beer are to be treated, the proper amount of stock solution of heptyl para-hydroxybenzoate or salt and octyl gallate or salt is added to the empty bottle, the unpasteurized finished beer is placed in the package, and the package sealed.

After the beer has come to rest, the package is unsealed, the proper amount of the stock solution containing the propylene glycol alginate is added and the package is resealed and mixed. There is thus provided a beer composition which is preserved against microbial growth and need not be pasteurized or refrigerated and which in addition possesses commercially acceptable chill stability as well as the other properties which are indicative of a commercially acceptable beer.

p-Hydroxybenzoate-Pyrocarbonate

Diethyl pyrocarbonate (DEPC) is used with octyl or heptyl p-hydroxybenzoate in the process of *F.B. Strandskov; U.S. Patent 3,751,264; August 7, 1973; assigned to The F. & M. Schaefer Brewing Company.*

Stock Solution No. 1 is prepared as follows. Dissolve 420 mg of n-heptyl para-hydroxybenzoate acid in a sufficient amount of ethanol to make 100 ml of solution. The addition of 0.5 ml of this solution, containing 2.1 mg of the n-heptyl para-hydroxybenzoate, to 350 ml of beer yields a solution containing 6 ppm of the benzoate.

Stock Solution No. 5 is prepared as follows. Dissolve 700 mg of DEPC in a sufficient amount of absolute ethanol to make 100 ml of solution. The addition of 0.5 ml of this solution, containing 3.5 mg of the DEPC, to 350 ml of beer yields a solution containing 10 ppm of the DEPC.

In the same manner as for Stock Solutions 1 and 5, additional solutions are made up such that 0.5 ml of the solution added to 350 ml of beer yields a solution containing the amounts of materials indicated below. All amounts are ppm.

Stock Solution	Additive	Amount
2	n-Heptyl p-hydroxybenzoate	8
3	n-Heptyl p-hydroxybenzoate	10
4	n-Heptyl p-hydroxybenzoate	12
6	DEPC	20
7	DEPC	50
8	DEPC	100
9	DEPC	200

Example 1: Into each of 3, 12 ounce brown beer bottles place 0.5 ml of Stock Solution No. 1. Fill these bottles from the regular production line with cold, unpasteurized beer containing no chemical preservative. Foam up the beer to expel headspace air, and cap. Each of these bottles contains 350 ml of the beer together with the stock solution. Store these bottles at room temperature (about 15° to 30°C). At the end of 3, 16 and 20 weeks, respectively, the bottles are examined for the development of sediment. Spoilage may be readily observed by a marked amount of sediment in the beer.

Example 2: Into each of 3, 12 ounce brown beer bottles, place 0.5 ml of Stock Solution No. 2. Fill these bottles from the regular production line with cold, unpasteurized beer containing no chemical preservative. Foam up the beer to expel headspace air, and cap. Each of these bottles contains 350 ml of the beer together with the stock solution. Store these bottles at room temperature (about 15° to 30°C). At the end of 3, 16 and 20 weeks, respectively, the bottles are examined for the development of sediment.

Example 3: Into each of 3, 12 ounce brown beer bottles place 0.5 ml of Stock Solution No. 3. Fill these bottles from the regular production line with cold, unpasteurized beer containing no chemical preservative. Foam up the beer to expel headspace air, and cap. Each of these bottles contains 350 ml of the beer together with the stock solution. Store these bottles at room temperature (about 15° to 30°C). At the end of 3, 16 and 20 weeks, respectively, the bottles are examined for the development of sediment.

Examples 4 to 18: The same procedure is followed as in the preceding Examples using Stock Solutions 4 through 9. In the Examples were a combination of the preservatives is used, these are added serially by introducing the benzoate-containing stock solution into the bottle, filling the bottle with beer and subsequently adding the DEPC-containing solution and mixing. The following table sets forth the results obtained in the examples.

The data of the examples reveal that a concentration of DEPC as high as 200 ppm is not effective as a beer preservative (Examples 5 to 7). The beer containing the high level of DEPC spoils as rapidly as the control beer. The benzoate ester when used alone preserves the test beer at a concentration of 12 ppm. However, preservation was obtained with 10 ppm of the benzoate when as little as 10 ppm of DEPC was added and with 8 ppm of benzoate when 20 ppm of DEPC was used. This is seen to represent substantial synergism when one considers that 200 ppm DEPC is clearly inactive.

It is not entirely clear why the mixture is effective in achieving preservation when the high level of DEPC is ineffective. It can be theorized that DEPC has little or no effect on the growth of bacteria in beer. It is, however, extremely effective against yeast. The benzoate esters are much more effective against bacteria than against yeast. Thus, while it requires 12 ppm of the benzoate to completely inhibit yeast growth, the bacterial growth may be inhibited with 8 ppm. DEPC inhibits yeast growth completely at a level of 20 ppm and so a combination of 8 ppm of the benzoate and 20 ppm of DEPC gives complete protection. The data indicate that the 8 ppm of benzoate also has a considerable effect on yeast growth and may exert an effect on the 20 ppm of DEPC. Regardless of the accuracy of the theory, it is clear that synergism is demonstrated when considering the complete preservation.

Example	n-Heptyl p-Hydroxy-benzoate (ppm)	DEPC (ppm)	Sediment Reading[1] After 3 Weeks				16 Weeks		20 Weeks		
Control	0	0	9+	9+	9+	9+	9+	9+	9+	9+	9+
1	6	0	6	6	7	9+	9+	9+	9+	9+	9+
2	8	0	4	4	4	9+	9+	9+	9+	9+	9+
3	10	0	2	2	2	3	4	9+	9+	9+	9+
4	12	0	1	1	2	2	3	4	3	3	4
5	0	50	9+	9+	9+	9+	9+	9+	9+	9+	9+
6	0	100	9+	9+	9+	9+	9+	9+	9+	9+	9+
7	0	200	9+	9+	9+	9+	9+	9+	9+	9+	9+
8	6	10	5	6	6	9+	9+	9+	9+	9+	9+
9	8	10	2	3	3	5	6	9+	5	9+	9+
10	10	10	1	2	2	4	3	3	3	4	5
11	6	20	6	7	7	9+	9+	9+	9+	9+	9+
12	8	20	2	3	5	6	4	5	5	3	5
13	10	20	1	2	2	3	3	5	3	4	5
14	12	20	1	1	2	2	3	4	3	3	4
15	6	50	4	5	9+	9+	9+	9+	9+	9+	9+
16	8	50	1	1	2	3	3	4	4	3	3
17	10	50	1	2	2	2	4	3	2	4	3
18	12	50	1	1	2	3	4	3	3	3	4

[1]A sediment reading of 9 or above indicates microbial spoilage.

FLAVOR STABILIZATION

Coffee Bean Extract

G. Lehmann, O. Neunhoeffer, W. Roselius, and O. Vitzhum; U.S. Patent 3,663,581; May 16, 1972; assinged to Hag AG, Germany claim that an effective antioxidant for foods can be extracted from green coffee beans. This extract is obtained by:

(a) dissolving in hot water the caffeine extract obtained from green coffee beans,

(b) cooling the hot aqueous phase whereby coffee wax is separated on the surface and the caffeine is deposited as a sediment,

(c) removing the caffeine, making the aqueous solution containing the coffee wax strongly alkaline and separating the aqueous alkaline phase,

(d) extracting the separated aqueous alkaline phase with an aliphatic halohydro-carbon and separating the aqueous alkaline phase from the organic phase,

(e) acidifying this aqueous alkaline phase with a mineral acid,

(f) extracting the acid aqueous solution with a water-immiscible solvent, drying the solvent phase and drawing off the solvent.

The residue remaining after the solvent is drawn off is a brown substance of oily consistency having an exceedingly pronounced antioxidative action. Substances susceptible of undergoing autooxidation, that are protected from such oxidation are fats, oils, aromatic substances and powdered milk, but also plastics for the preparation of consumer goods and packaging materials. It is particularly effective to add the antioxidant to roasted coffee to increase the keeping quality. The quantity of added antioxidant lies, in all cases, in the range between 0.1 to 2% by weight.

In the preparation of the antioxidative-active substance the green coffee beans used as starting material (which may or may not be comminuted) are extracted in a known manner with a caffeine extraction solvent (aliphatic halohydrocarbons, such as $CHCl_3$, CH_2Cl_3, CCl_4, dichloroethylene, trichloroethylene, benzene, ethyl acetate, under certain conditions also ether, etc.) The extract is treated for the removal of the solvent and then dissolved in hot water. Beneficial effects are obtained, if the extract, at first, is not completely thickened by evaporation and if the residual caffeine extraction solvent is removed by a jet of steam. The steam that condenses at the surface of the extract serves at the same time to make up the caffeine extraction residue.

At this point, the hot aqueous phase is allowed to cool off (to about room temperature) whereby coffee wax is separated on the surface and the caffeine is deposited as a sediment, the latter being recovered for example by decanting or with the aid of a discharge device at the bottom of the reaction vessel, leaving behind only an aqueous solution and the coffee wax. This aqueous solution, if necessary after modifying the volume of liquid, is made strongly alkaline by means of a basic substance, most efficiently with NaOH or KOH, so as to provide a preferred pH value in the range 12 to 14. The solution is then heated if necessary to about 50° to 70°C and vigorously agitated, by shaking or stirring, together with the coffee wax, the latter being then separated suitably after cooling to room temperature.

The aqueous alkaline phase is now extracted with an aliphatic halohydrocarbon, preferably with CCl_4, $CHCl_3$, CH_2Cl_2, di- or trichloroethylene, and the aqueous phase is separated from the organic phase. The strongly alkaline aqueous extract is acidified with a mineral acid, preferably with 5 N HCl or H_2SO_4. The aqueous solution, which is now acid, is extracted with a water-immiscible solvent, for example with ether (diethyl ether, diisopropyl ether). The solvent phase is then separated, dried in the usual manner and the solvent drawn off, leaving behind a brown oil. From 100 grams finely ground green coffee beans it is possible to obtain about 10 mg of a brown oil having an antioxidative action. This brown oil, according to several orientating chromatographic tests, does not represent a unitary substance but rather a mixture that includes even crystalline bodies.

The very high degree of antioxidative activity of the oils is shown by the following auto-oxidation tests of benzaldehyde carried out with air-oxygen under the same conditions at room temperature in the presence of known antioxidants and the special antioxidant oil.

Auto-oxidation of—	Antioxidant	O_2-absorption in microliters	After—
0.1 ml. C_6H_5CHO	4.4 mg. D,L-tocopherol	150	7 hours.
	5.6 mg. n-propyl-gallate.	150	5 hours.
	6.1 mg. Nor-dihydro-guaiaretic acid.	150	3 hours.
	7.3 mg. oil according to the invention	140	30 hours.

According to the test results listed in the above table, the oil obtained from green coffee beans has an antioxidative action that surpasses even that of tocopherol which is known to be one of the most potent organic antioxidants. The oil obtained is odorless and tasteless and obviously also nonpoisonous and harmless. It is therefore suitable for use in food articles; since the green coffee beans used as starting materials for the oil are known to be unobjectionable from the standpoint of palatableness, and since the preparation of the oils does not involve the use of conditions that fundamentally alter these substances.

Needless to say, the oils prepared in accordance with the process can also be utilized admixed with other antioxidants for the protection of autooxidizable substances. The addition can be made either to the particular end product, in which case it is recommended to mix thoroughly, or during a suitable step of its manufacturing process, or partly during this step and partly to the end product.

Example 1: 50 liters of the aqueous phase, obtained as a result of the caffeine extraction in accordance with step (b) above, upon removal of the caffeine precipitated in the course of cooling, is made alkaline with 1 liter of 5 N NaOH. Residual caffeine is removed by consecutive agitations carried out with chloroform. The aqueous alkaline phase is acidified with 3.5 liters 5 N HCl and the acid solution is agitated by shaking with 35 liters, 25 liters, 15 liters, 15 liters of ether. The combined ether extracts are dried over sodium sulfate and the ether is distilled off. This leaves behind 140 grams of a brown oil having high antioxidative activity.

Example 2: To a roasted coffee extract containing 18% by weight of solid matter, there is added 0.1% by weight, based upon the total coffee extract, of the oil. The extract is then concentrated in a vacuum to a solids content of 30%, whereby the volatile aromatic substances are recovered separately by condensation into a cooling device. These are combined with the same amount by weight of the antioxidant that was used for the coffee extract and this mixture is applied as a spray onto the dry powder of the roasted coffee extract. There is obtained a highly aromatic stable product.

Example 3: Roasted coffee beans are treated with a spray of 0.05% by weight of the oil suspended in an aqueous phase, and are then finely ground in the presence of an inert gas. The ground roasted coffee powder is distinguished by its outstanding flavor stability.

Example 4: Aniseeds are treated by spraying on 0.5% by weight of the oil which is suspended in 10 parts by weight of an 0.3 weight percent aqueous ethanol solution of sodium citrate, whereupon they are dried and ground in the usual way. The resulting anise powder has a stable flavor.

Example 5: Linseed oil to be used for dietary purposes is mixed with 0.1% by weight of the oil. The resulting product is stable even in containers that were opened.

Spice Antioxidant

It is known that certain spices have antioxidant properties, whereby they extend the stability of certain foods and prevent fat oxidation in certain meats and meat products. Spices are generally dried plant products which have distinctive aroma and flavor characteristics. A majority of spices are grown and harvested in tropical countries and imported in whole form into the United States. In harvesting a spice crop, little or no effort is made to remove the insects that are normally associated with the spice; accordingly, it is not unusual to find that spice lots are adulterated with insects and insect fragments when they are imported into this country. Other impurities and adulterants may also be found in the imported spices.

While spice processors exercise care in handling and storing spices, it is not unusual to find that the spices are further contaminated in storage. It is extremely difficult to remove impurities and adulteration from spice products, especially such impurities as insect fragments and animal hairs which find their way into the spices as the result of the usual rodent population found on ships and in warehouses. Consequently, when the food processor utilizes spices in the production of his products, he runs the risk of incorporating adulterated or impure agents into his product.

Consider, for instance, a manufacture of processed meat products, such as sausage, luncheon meats, etc., products which traditionally have high spice contents. Such a manufacturer has in the past incorporated spices essentially in the form they are imported into this country, save for comminution to acceptable ground size, into his product to enhance the flavor and aroma. Recognizing that the moist meat product is a fertile media for the growth and multiplication of dormat bacteria and spores found in the spices, elaborate and costly procedures have been developed for sterilizing the spices prior to incorporation into meat products. Despite these procedures, certain contaminants such as animal hairs and insect fragments are not eliminated from the spices, and consequently it is possible to have an entire batch of a food product rendered unsalable due to the presence of even minute quantities of contaminant.

Another approach has been to extract spice principles such as those responsible for flavor, aroma, and antioxidant characteristics of the spice and incorporate the extracts into the product being prepared. Extraction procedures generally provide a product free of contaminants. However, due to the complex nature of the spice components responsible for such characteristics as flavor and aroma, extraction procedures are usually complex and difficult to design so that they produce a uniform product each time the extraction procedure is utilized. Moreover, previous attempts to isolate the antioxidant principle from spices by various extraction procedures have not met with success.

When the food manufacturer desires to utilize spices for their antioxidant properties he is, of course, faced with the problems noted above. Synthetic antioxidants are available commercially; however, only those recognized as being safe for consumption by the FDA may be utilized. Additionally, synthetic antioxidants are volatile at relatively low temperatures, (circa 100°C), thus, food processing above these temperatures results in a loss of the antioxidant activity.

D.L. Berner, G.A. Jacobson and C.D. Trombold; U.S. Patent 3,732,111; May 8, 1973; assigned to Campbell Soup Company describe a process for an efficient extraction of the antioxidant principle from spices. In their process ground spice is extracted with heated animal or vegetable oil. The oil soluble extract containing volatile and nonvolatile components is separated from the spice solids. The extract is heated under vacuum conditions and simultaneously sparged with steam to obtain a deodorized oil extract containing the oil soluble nonvolatile spice antioxidant principle.

Members of the Labiatae family, such as sage, rosemary, basil, peppermint, spearmint, and blends of these spices may be utilized to provide the desired spice antioxidant principle. The antioxidant principle extracted from these spices has been found to exhibit greater antioxidant activity in comparison with an equivalent amount of the dry spice. While whole spice may be utilized in the process, it is preferred that the spice or mixture of spices should be coarsely ground. The oil selected for extraction may be a liquid vegetable oil such as cottonseed oil or peanut oil or it may be a solid animal oil such as beef or mutton tallow. Hydrogenated or saturated oils may also be utilized. The preferred oils are cottonseed oil and peanut oil. The ground spice is combined with the oil in a proportion of from about 15 to about 20%, by weight based on the weight of the oil, for the extraction process.

It has been found that the oil-soluble spice components are extracted most efficiently when the oil is heated to a temperature of from about 120° to about 125°C during the extraction step. In practice, it is preferred to heat the vegetable oil prior to the addition of the ground spice. The extraction takes place in about 2 hours, with continuous agitation of the oil-spice mixture; however, the agitation should not cause aeration of the oil. Following the oil extraction, the extract containing the oil soluble volatile and nonvolatile spice components is separated from the spent spice solids. Separation may be accomplished by any suitable means, one convenient means involving centrifuging the mixture followed by isolation of the oil phase from the spent spice solids.

After separating the oil extract from the spent spice solids, the oil extract is deodorized to recover the desired spice antioxidant principle. The goal of the deodorization step is to remove the more volatile spice components from the less volatile spice components of the oil extract. Deodorization is accomplished by heating the oil extract under vacuum conditions while simultaneously sparging with steam. In the deodorization step, heating temperature and vacuum conditions go hand-in-hand; higher temperatures requiring less vacuum and vice-versa to remove the unwanted volatile components. Steam sparging is essentially utilized to effect more efficient removal of the unwanted volatile material.

The extraction procedure outlined above removes about 85% of the antioxidant principle from the treated spices, the extracts having greater antioxidant activity as compared with an equivalent amount of dry spice. Furthermore, the antioxidant principle is recovered from the spice essentially free of adulterants and impurities and the process permits recovery of the antioxidant principle in a form which is uniform from lot to lot.

Example: 225 grams of cottonseed oil is heated in an open container to about 125°C. 45 grams of sage (20 to 60 mesh) is added to the oil and the mixture is continuously agitated without significant aeration of the oil for a period of 3 hours while maintaining the mixture at 125°C. The mixture is transferred to a centrifuge, and the oil extract phase is separated from the spent spice solids. The oil extract is filtered to remove fine space particles, transferred to a closed deodorized vessel and heated to 175°C with steam sparging for ½ hour under 2 to 4 mm mercury. 180 grams of oil, having a bland taste and little or no aroma is recovered.

Varying amounts of the spice antioxidant principle are added to pork fat and the treated pork fat is evaluated for the antioxidant activity of the added extract. Comparisons are made with a control sample containing no antioxidant principle and with pork fat containing synthetic antioxidants. The results are set forth in the table below which gives the active oxygen method (AOM) evaluations of antioxidant activity of deodorized cottonseed oil-sage antioxidant principle.

Percent antioxidant principle in pork fat	AOM time,[1] min.	Anti-oxidant index[2]
Control (no antioxidant)	20	1.0
1.25% extract	480	24.0
2.50% extract	1,050	32.5
3.75% extract	1,290	64.5
5.00% extract	1,110	55.5
7.50% extract	1,620	81.0
0.05% mixture[3]	1,725	86.3
0.02% BHA (butylated hydroxyanisole)	570	28.4
0.02% PG (propyl gallate)	940	47.0

[1] Time to reach a peroxide value (PV) of 20 at 110° C. and 2.33 cc. air per sec.

[2] Antioxidant index = $\dfrac{\text{Sample time required to reach PV 20}}{\text{Control time required to reach PV 20}}$

[3] The mixture consisted of 20% BHA, 6% propyl gallate, 4% citric acid, and 70% propylene glycol.

Propylene Oxide Treatment of Spices

As directed by legislation, the FDA has reevaluated its approval of the use of ethylene oxide as a fumigant or sterilizing agent for food materials. At one time the FDA was very concerned about the formation of ethylene glycol, which occurs during the sterilization process using ethylene oxide as the sterilizing agent. Ethylene glycol is toxic and gave cause for concern. However, propylene glycol, which is formed during the analogous sterilizing process using propylene oxide, is considered quite safe. Research showed that in addition to the glycols, chlorohydrins are sometimes formed during sterilizing processes using the alkylene oxides. Ethylene chlorohydrin has been found to be toxic, while propylene chlorohydrin has a much lower toxicity value.

Propylene oxide is not as effective as ethylene oxide as a sterilizing agent or fumigant when used on spices. The FDA has permitted the use of propylene oxide as a fumigant or sterilizing media for ground spices, but has not approved the use of ethylene oxide in ground spices. Although the FDA continues to refuse to permit the use of ethylene oxide generally, they do permit the use of ethylene oxide to sterilize whole, unground spices.

It was found by L. Sair; U.S. Patent 3,647,487; March 7, 1972; assigned to The Griffith Laboratories, Inc. that whole unground spices can be substantially sterilized by a treatment of the whole spice using ethylene oxide. The spices thus sterilized may then be ground and given a final sterilization treatment with propylene oxide. While the propylene oxide is not as effective as the ethylene oxide as a sterilizing agent, it has been found that during the grinding process, the spices do not pick up a great deal of bacteria or other contaminants, and that the final sterilizing step is adequate to give an effective kill. It has been found that this process does not cause the formation of undesirable quantities of ethylene chlorohydrin.

Effective bacteria kills on spices can be achieved when ethylene oxide is applied to whole unground spices and the resulting ethylene chlorohydrin level will be much lower than that resulting from a corresponding sterilization process on ground spice. To demonstrate this, a series of whole spices were treated with ethylene oxide using the standard 0.75 oz of ethylene oxide per cubic foot in the retort for a period of 4 hours. A duplicate group of spices were then ground and samples were treated in a similar fashion in the retort using ethylene oxide. The samples were then analyzed for ethylene chlorohydrin residues.

In general, the chlorohydrin residue on a spice when treated whole is approximately 50%
or lower than the ethylene chlorohydrin level when the spice is treated in the ground state.
It has been discovered that the bulk of bacteria associated with spices are found on the
surface of the whole spice. This discovery is easily demonstrated by the following test.
A series of whole unground spices were treated for 4 hours at 125°F with 0.75 oz of ethylene oxide per cubic foot of retort space. The whole spices were then ground under commercial conditions, and the bacteria count taken. The results are shown in the table below.

	Total Bacterial Count/gm.	
	Before Treatment	After Treatment
Whole Black Pepper	16,000,000	500
Whole Fennel Seed	320,000	40
Whole Allspice	1,600,000	1,500
Whole Muntok White Pepper	193,000	Less than 100

The first step of this process, namely, the ethylene oxide sterilization of the whole, unground
spices, may be carried out under conventional conditions using conventional equipment.
For instance, the ethylene oxide concentration may vary from about 0.25 oz to 1.50 oz
per cubic foot of retort, while the retention time may vary from about 30 minutes to 6 hours
or more. This process requires only that a fairly effective bacteria kill be accomplished by
the ethylene oxide sterilization of the whole spice. The precise conditions under which
this is accomplished are not critical.

The next step is to grind the sterilized whole spices. The grinding may immediately follow
the first sterilization step or the spices may be first stored and ground later. If the storage
conditions are sufficiently clean and sanitary, no further sterilization before grinding should
be required. The grinding is accomplished under the most sanitary conditions possible in
order to avoid unnecessary bacteria pickup. However, it has been found that, under normal
plant conditions, the spices will pickup bacteria during the handling, storage, or grinding.
While it is possible, under laboratory conditions, to nearly obviate such bacteria pickup,
such precautions or conditions are not practicable in commercial operations.

The final step, the propylene oxide sterilization can be accomplished under a wide variety
of conditions. For instance, it has been found that 2.25 oz of propylene oxide per cubic
foot of retort for 3 hours will give a very good bacteria kill on the ground spices, if the
bacteria count is not too high to begin with. However, from as little as 1 oz to about 4
oz of propylene oxide for from 30 minutes to 6 hours may be successfully used. Likewise,
the temperatures and pressures may be varied over wide limits.

Example: A whole unground black pepper sample having a bacteria count of 18,000,000
per gram, is divided into 3 portions. The first portion of the pepper sample is ground under
conditions simulating commercial spice grinding, and was then placed in a retort. A vacuum
is pulled on the retort and 0.75 oz of ethylene oxide per cubic foot of retort is admitted to
the retort. After 4 hours, a vacuum is again pulled, and then broken with air and the first
portion of the sample tested. The first portion of the sample shows a bacteria count of
3,000 per gram and an ethylene chlorohydrin content of 1,520 ppm.

The second portion of the sample is similarly ground and sterilized, but using 2.25 oz of
propylene oxide per cubic foot of retort rather than ethylene oxide. The thus treated second
portion of the sample contains no ethylene chlorohydrin, but has a bacteria count of
65,000 per gram.

For the third portion of the sample the whole unground pepper is sterilized using ethylene
oxide at 0.75 oz per cubic foot of retort volume for 4 hours. The treated pepper has a
bacteria count of 2,000 per gram and an ethylene chlorohydrin content of 425 ppm. Following a grinding operation under simulated commercial conditions, the bacteria count is
3,500 per gram with the ethylene chlorohydrin unchanged at 425 ppm. The ground pepper

is sterilized as described above, using 2.25 oz of propylene oxide per cubic foot of retort for 4 hours. The final testing shows a bacteria count of 500, with the ethylene chlorohydrin still unchanged at 425 ppm.

The process may be applied to any spices which are to be ground. For instance, the process has been successfully used on allspice, basil, celery, korintji, cinnamon, coriander, ginger, mace, nutmeg, oregano, black pepper, cayenne pepper, white pepper, thyme and turmeric.

Oleoresinous Spice Extract

The addition of antioxidants has often proven effective in improving the stability of the color, flavor, and odor, and to deter the development of rancidity in essential oils and in oleoresinous spice extractives themselves, that is, when they are stored as concentrates without incorporation on carrier materials. However, when the extract concentrates are deposited on solid carrier materials, and particularly when soluble carriers such as salt and sugar are used, the effectiveness of antioxidants to prevent deteriorative changes in the spice extractives is greatly reduced. And spice extractives, when combined with soluble carriers, are susceptible to such rapid color loss, taste loss and general deterioration that the commercial practicability and usefulness of such products has been seriously limited.

It is postulated that the increase in the rate of deterioration of the spice extractives is due to the action of contaminants comprising metallic salts or trace metals such as copper, iron, and nickel, etc. These trace metals, which have a known pronounced effect in catalyzing the oxidative deterioration of other materials such as oils and edible fats, are believed to have a similar deleterious effect on spice concentrates or extractives.

A special procedure is used according to *M.R. Peat; U.S. Patent 3,095,306; June 25, 1963; assigned to Wm. J. Stange Co.* as follows: A carrier material, such as salt or sugar, is treated with a special metal chelating or sequestering agent. The spice extractive (oleoresin or essential oil, etc.) is treated separately with an antioxidant preparation. And then the spice extractive and the carrier material are combined to form a stabilized and protected soluble seasoning.

The resulting product is resistant to color loss and to oxidative deterioration caused by metal-catalyzed reactions. The metal complexing or metal sequestering agent is added directly to the carrier material itself rather than to the spice extractive or to the final combined product. This procedure ensures the most effective and complete association between the complexing agent and the trace metal contaminants in the carrier.

A representative example of the process relates to a method of stabilizing the oleoresin of paprika against oxidative deterioration and color loss catalyzed by a metal or metal salt. The method comprises treating the salt carrier, on which the oleoresin is to be incorporated, with from about 0.001% to about 0.05% by weight (based on the weight of carrier) of the calcium chelate of the disodium salt of ethylenediaminetetraacetic acid, the oleoresin itself having an antioxidant in an amount ranging from about 0.001 to about 2.0% by weight, based on the weight of oleoresin. Based on the weight of the salt-carried oleoresin product, the antioxidant concentration would be from about 0.1 to about 1,000 ppm, depending upon the concentration of oleoresin in the seasoning and upon the concentration of antioxidant in the oleoresin.

The chelating agents are also useful in preventing loss of color, and impairment of taste and aroma in spice extractives not combined with carriers, but which do themselves contain trace metals such as iron, copper, and nickel. Many known metal complexing agents or chelating agents were investigated. These were tested at various concentrations, both alone and in combination with antioxidants. In some instances, where solubility characteristics permitted, the chelating agents were added directly to the spice extractives or combined with the antioxidants; in other experiments, the chelating or sequestering agents were added directly to the carrier salt before combining the extractive and the carrier. The effectiveness of any particular combination of variables was measured in a series of experiments in which 3 to 5 parts of oleoresin of paprika were blended with about 95 parts of salt. It having been established that a direct relationship exists between the loss of color of the paprika oleoresin, on the one hand, and the development of oxidative rancidity and the decline of general acceptability of the spice product, on the other hand, the extent of deterioration of the oleoresin carried on the salt was followed and measured as a function of the loss of color of a standardized preparation.

Various antioxidants were added to the oleoresin and various metal complexing or chelating agents were added to the salt. Control experiments in which neither antioxidants nor chelating agents were used were conducted along with the general experimental program. In other control or reference experiments, antioxidants were used without chelating agents; and, in still other controls, chelating agents were used without antioxidants. It was found that, in general, better results were obtained when the sequestering, chelating, or metal complexing agent was added directly to the salt rather than to the oleoresin, and, with one exception, the sequestering agent was blended into the salt prior to combining the salt and the oleoresin. The sequestering or chelating agent is the disodium salt of the calcium chelate of ethylenediaminetetraacetic acid dihydrate ($C_{10}H_{12}O_8N_2CaNa_2 \cdot 2H_2O$) or calcium disodium EDTA, CaNa$_2$ EDTA.

It has been found that a concentration of about 30 ppm (based on the soluble carrier) of the chelating compound is ordinarily effective to complex the trace metals normally present in food grade salt; and in the work recorded about 25 to 100 ppm has proved a practical range. The metal deactivators or metal chelating agents may be distinguished from other additives generally added to fatty or fat-containing materials for various purposes. For example, antioxidants are added to edible fats, oils, and similar substances to retard the development of rancidity therein. These antioxidants fulfill the role of controlling the formation of free radicals which are the active agents in instigating the rancidity chain reaction. The antioxidants will not suppress the catalytic effect of metals in accelerating the deterioration of edible fats, oils, and related substances; and the metal deactivators will not substantially suppress rancidity development except to the extent that such rancidity may result from the catalytic effect of trace metals.

Example 1: Antioxidant No. 1 (AO-1) contains 4% of nordihydroguaiaretic acid (NDGA), 40% of butylated hydroxy anisole (BHA), 2% of citric acid, 20% of Atmos 300 (mono- and diglycerides), and 34% of propylene glycol.

Example 2: Antioxidant No. 2 (AO-2) contains 10% of nordihydroguaiaretic acid, 20% of butylated hydroxy anisole, 6% of citric acid, 44% of Atmos 300 and 20% of propylene glycol.

It will be observed that antioxidant compositions No. 1 and No. 2 above each includes, as a component, citric acid which is a known metal chelating agent. In addition to the mixtures of antioxidants, specified above, solutions of nordihydroguaiaretic acid and of butylated hydroxy anisole in mixtures of mono- and diglycerides were investigated. Typical formulations are antioxidant preparations No. 3 and No. 4 below.

Example 3: Antioxidant No. 3 (AO-3) contains 16% of nordihydroguaiaretic acid and 84% of Atmos 300.

Example 4: Antioxidant No. 4 (AO-4) contains 40% of butylated hydroxy anisole and 60% of Atmos 300.

In the experiments carried out and summarized in Table 1, the antioxidants were used by adding 1% to 3% of the antioxidants solution to the oleoresin. The total antioxidant content of the prepared antioxidant solutions (see Examples 1 through 4) varies from about 15 to about 50%. As indicated above, and as shown in Table 1, the use of antioxidants alone (without a metal chelating agent) has some retarding effect on the fading of the color of oleoresins, such as paprika and capsicum carried on a salt base. Whereas the control, having no antioxidant and no chelating or sequestering agent, suffers complete loss of color and becomes rancid within 2 days, the incorporation of either the antioxidant No. 1 composition or the antioxidant No. 2 composition gives definite improvements, as shown in Table 1, although more than two-thirds of the original color is lost at the end of about 2 weeks. In the following tables, the suffix P after the number of a test or of an experiment indicates an oleoresin paprika experiment; the suffix C indicates an oleoresin capsicum experiment. The letters AO refer to antioxidant, and the letters SA referred to sequestering agent. The oleoresin used was oleoresin of paprika.

TABLE 1

Test No.	Antioxidant Composition	Sequestering or Chelating Agent	After 5 Days	After 10 Days	After 15 Days
			- - - - - - Color Loss (%) - - - - - -		
1 P (Control)	None	None	100*		
2 P	AO-1	None	11	30	65
19 P	AO-1	None	7	43	68
23 P	AO-1	None	6	40	73
3 P	AO-2	None	12	33	69

*Completely faded and rancid within 2 days.

The ability of calcium disodium EDTA to deter the deterioration of color and of aroma and taste in salt-carried oleoresin of paprika was first indicated in exploratory experiments carried out as a part of a general research program. This program was aimed at improving the overall quality and the shelf life of soluble seasonings and of commercial products in which spice extractives play a major role. In order to establish a basis of comparison by which to evaluate the calcium disodium EDTA properly, a series of experiments was carried out in which the effect of that chemical agent could be compared with other chelating agents in the system: spice extractive-salt carrier. Among the reference chelating agents or sequestering agents used were citric acid (at several concentrations), stearyl acid phosphate, disodium acid phosphate, and the disodium salt of ethylenediaminetetraacetic acid. The various formulations used are described below.

Sequestering agent solution C-3 (SA-C3) consists of 3 g/100 cc of citric acid. Sequestering agent solution C-10 (SA-C10) consists of 10 g/100 cc of citric acid. Sequestering agent solution C-33 (SA-C33) consists of 33 g/100 cc of citric acid. The effectiveness of citric acid as a chelating agent in oleoresins in the presence of antioxidants was studied in a series of experiments in which three concentrations of citric acid were used with one antioxidant preparation, and in which a single concentration of citric acid was used with three different antioxidant preparations. The citric acid concentrations were 0.008%, 0.025%, and 0.075% based on the salt used; and the oleoresin was 5%. In each case, 0.25 gram of the sequestering agent solution was blended with 95 grams of salt prior to the blending of the oleoresin into the salt.

The antioxidant was incorporated into the oleoresin before blending the oleoresin with the salt carrier. The oleoresin, containing the antioxidant, was blended with the salt in a ratio of 5 grams of oleoresin per 95 grams of salt. The oleoresin was the oleoresin of paprika.

It will be observed that the low 3% solution (test 11 P) of citric acid produced a final system somewhat less stable than the other systems. Increase from a 10% solution (test 12 P) to a 33% solution of citric acid (test 8 P) yielded no further apparent benefits, that is, no appreciable improvement over the 10% solution. Attention is directed to the fact that the antioxidant AO-1 includes 2% of citric acid. Since the antioxidant is used at a concentration of about 1% to 3% of a 10% solution, based on the oleoresin, this additional citric acid amounts to an additional 0.01% based on the salt carrier used.

A comparison of 15 day color loss, in Tables 1 and 2, indicates that the addition of citric acid has little, if any, beneficial effect on color retention in the spice extractives system under investigation. This may be because the citric acid merely duplicates the effect of the antioxidant compositions, or, on the other hand, it may be that citric acid is actually inactive as a chelating agent in the particular experimental environment involved. There are many unknown factors which affect the rate of reaction of natural products, such as spice extractives, with chemical agents, and it is, therefore, not possible to be certain that all variables are controlled completely at all times. Thus it becomes impractical to compare one series of experiments directly with another run conducted at a different time, even though presumably under the same controlled conditions. As a result, one must, in general, rely upon base reference or control experiments carried out with each phase of the overall project.

TABLE 2

| Test No. | Antioxidant Composition | Sequestering or Chelating Agent | - - - - - - Color Loss (%) - - - - - - | | |
			After 5 Days	After 10 Days	After 15 Days
11 P	AO-1	SA-C 3	19	74	100
12 P	AO-1	SA-C 10	17	40	65
8 P	AO-1	SA-C 33	31	50	66
6 P	AO-3	SA-C 33	29	48	72
9 P	AO-4	SA-C 33	37	67	70

Sugar Polymer Fixation

The process of *H. Sugisawa; U.S. Patent 3,695,896; October 3, 1972; assigned to Canadian Patents and Development Limited, Canada* provides for preservation of volatile flavors and aromas in a stable dry form. Briefly, the process includes heating a sugars-containing mixture having a sucrose content between about 50% and about 80%, by weight to a temperature and for a period of time sufficient to cause formation of oligosaccharides in such an amount as to ensure a resultant sugar polymers base. The sugar polymers base is diluted with water to bring the total sugars content to between about 70% and about 80% by weight and also is brought to a temperature less than approximately 50°C.

The desired flavor and aroma essences are mixed with a binding agent, the binding agent being in an amount sufficient to constitute from about 1% to about 5% by weight, of the final product, and the mixture then blended with the sugar polymers base. The product then is dried at a temperature not exceeding about 50°C to a moisture content of less than about 5%. Optionally, the product may be ground to give a powdered, amorphous product. The product is a stable, dry food product having oligosaccharide sugar polymers as a major constituent, with the volatile flavor and aroma essences of the food product being essentially entrapped in the complex stereostructure of the sugar polymers.

The required oligosaccharides may be prepared from several different sugars and sugar-containing products such as sucrose, corn syrup, maltose, dextrose, malto-dextrin, dextrin, the hydrol by-product of starch hydrolysate preparation, hydrogenated hydrol products, and mixtures thereof. The sucrose content preferably should be from about 50% to about 75% in the initial mixture so as to provide a resultant sugar polymer product in an amorphous form which is amenable to subsequent treatment.

Starch hydrolyzates such as corn syrup, malto-dextrin and dextrin may be utilized in the starting materials. The oligosaccharides utilized for the flavor-locking are prepared by heating a mixture of sucrose and other sugar-containing products to a temperature and for a period of time sufficient to cause formation of the necessary sugar-polymer stereostructures. This is preferably accomplished by adding a small amount of water to the mixture, bringing the mixture to a boil, increasing the temperature gradually to about 140° to 145°C and holding it there for approximately 5 minutes.

Following formation of the oligosaccharide stereopolymers, water is added to the mixture to bring the total sugars content of the mixture to between 70% and about 80%. Preferably this is accomplished by cooling the mixture to approximately 90° to 110°C and adding the water. The addition of water to the sugars syrup at this higher temperature is preferred since it has been found that it tends to drive off any undesirable flavors from the sugars resulting from any minor caramelization, although the tendency for the sugars to caramelize has been reduced by avoiding prolonged heating at higher temperatures. It has also been found easier to adjust the sugars content of the molten sugar mixture at this temperature since the syrup becomes quite viscous upon cooling.

The sugars syrup containing the oligosaccharides is brought to about room temperature (about 20° to 30°C) for the addition of the volatile food essences. The essences are first dissolved in a binding agent, preferably glycerin, of an amount sufficient to become about 1% to 5% of the final product. The binding agent may also be propylene glycol, polyethylene glycol, or mixtures with glycerin. The binding agent serves to minimize the loss of any low-boiling water-soluble flavor essences during drying, and acts as a solvent for both the sugars and the flavor compounds. The binding agent containing the oligosaccharide sugar polymers, and is stirred for sufficient time to achieve proper blending. It is necessary to mix the flavor essences with the glycerin or other binding agents at about room temperature or at as low a temperature as possible to minimize the loss of easily volatilized compounds. For example, temperatures in the neighborhood of 50°C will result in significant loss in the case of delicate fruit flavors.

The sugar polymers base containing the flavor volatiles and binding agent is dried and ground to form a powdered product having a moisture content of between 2% and 4%. It is desirable that drying of the product takes place at or below room temperature to minimize loss of flavor and aroma with the water evaporation. If drying is undertaken at atmospheric pressure a temperature of up to 50°C may be tolerated in some instances; under reduced pressure, for example 25 to 29 inches Hg, the temperature should not exceed 30°C. Drying methods such as vacuum belt drying, vacuum drum drying, dessicated air drying, or freeze-drying may be utilized.

It has been found advantageous, although not necessary, to add to the resulting dry, powdered mixture from about 0.1% to about 5% of an anticaking agent to reduce any tendency of the powder to cake during storage. Suitable anticaking agents which may be used for this purpose are microsilicagel, calcium silicate, tricalcium phosphate, or one of the well-known cereal-base anticaking agents.

Example: 300 grams (3 parts) of sucrose, 100 grams (1 part) of corn syrup (43° Baumé, 20% water, total carbohydrates 80%) having a low dextrose equivalent (17.5%), and 15 ml of water were mixed in a one liter stainless steel vessel provided with a stirrer and thermometer. The contents of the vessel were brought to a gentle boil over a hot plate and water was boiled off until the temperature of the sucrose-corn syrup mixture reached a temperature of about 145°C. The mixture was held at 145°C for 5 minutes while slowly stirring to permit the sugars to undergo a themal polymerization. The mixture was then cooled to 100°C and 15 ml of water was added to adjust the sugar concentration of the mixture to about 70%. The mixture was then cooled to room temperature (20°C), the stirrer speed was increased, and 4 ml of 100 fold strength apple essence concentrate dissolved in 4 ml of glycerin was added slowly over a period of about one-half a minute.

Stirring was continued and 4 grams of a foaming agent (methylcellulose 2 grams and acetylated monoglyceride 2 grams) were added. After additional stirring for about 2 minutes, the foamed syrup was transferred to a drying dish and dried under vacuum at room temperature. After dehydration the product was ground to a powder (moisture content 3%) in a mortar. In the solid dried state the product did not have detectable aroma, was not hygroscopic, but was readily water-soluble. When dissolved in water the resultant solution possessed a satisfactory apple aroma. When the product was blended with an equal weight of dry unsweetened applesauce flakes, and sufficient water to reconstitute the mixture to a sauce, a sweetened applesauce having a very satisfying aroma and flavor was produced.

One part of the sugar polymer containing 100 fold apple flavor concentrate and 3 parts of dried unsweetened applesauce flakes were blended and stored at room temperature (20°C) under a nitrogen pack for a period of 3 months. When reconstituted with sufficient water to form a sauce, the applesauce had the aroma and flavor of fresh applesauce.

Stabilized Halogenated Oils

Halogenated oils, specifically brominated oils such as brominated sesame oil, brominated apricot kernel oil, brominated corn oil, brominated soybean oil and other brominated oils of a similar nature, have been used in the preparation of citrus-flavored beverages and have been suggested for use in the confectionery and baked foods industries. Since citrus oils have a specific gravity of less than 1 and are insoluble in water, one of the most desirable means for suspending these flavoring oils in water is by combining the citrus oil with high density miscible liquids such as brominated oils. By this means it is possible to suspend the citrus oil-brominated oil mixtures in water and produce a cloudy citrus-flavored liquid which can then be further formulated into an attractive, typical citrus-flavored beverage.

The brominated oils which have been employed in the past for this purpose possess a high specific gravity in the range about 1.2 to 1.3 and reputedly provide a good stable cloud in citrus-flavored beverages in which they are incorporated. Moreover, these brominated oils are said to enhance the stability of the oil-water emulsion and offer no interference with the citrus flavor.

J.M. Becktel, F.E. Kuester, and E. Fritz; U.S. Patent 3,008,833; November 14, 1961; assigned to Swift & Company provide a stabilizing agent for increased shelf-life of products formulated with brominated oils. Brominated oils include, generally, animal, vegetable and marine glycerides which contain sufficient bromine to provide a specific gravity in the range desired. Glyceride oils containing fatty acid radicals having unsaturated bonds which may be reacted with bromine are desirable as the source of the brominated oils. In addition to the unsaturated higher aliphatic fatty acid glyceride esters, other materials such as unsaturated higher aliphatic alcohols and unsaturated higher aliphatic acid esters of mono- and dihydric alcohols may also be brominated to provide the high density oil. A wide range of specific gravity oils is possible because of the great variation in the degree of unsaturation contained in naturally occurring fatty materials.

The stabilizing agent which is incorporated in the brominated oil to provide desirable color properties may be characterized as a fatty material containing one or more oxirane groups in the molecule. Epoxidized fatty materials are particularly satisfactory for this purpose. Such epoxidized oils are epoxidized soybean oil, epoxidized cottonseed oils, epoxidized linseed oil, epoxidized rapeseed oil, epoxidized menhaden oil, epoxidized peanut oil, epoxidized lard oil, epoxidized tallow and epoxidized safflower oil. Other epoxy-containing esters such as methyl epoxystearate, ethyl epoxystearate and triepoxystearin are suitable as the oxirane-containing component.

It is possible to add substantial quantities of the epoxy fatty material to the brominated oil and control the specific gravity of the mixture while also insuring that color deterioration is inhibited. The addition of the stabilizers to the high density oil can be accomplished readily, mere mixing being sufficient to prepare a homogeneous product. The amount of a particular epoxidized fatty material to be added depends to some extent upon the oxirane content of the material being added. Thus, smaller amounts of polyepoxidized fatty materials such as epoxidized soybean oil are equal in effectiveness to greater quantities of such materials as methyl epoxystearate.

In order to demonstrate the stabilizing effect which the additives impart to brominated oils to which they are added, a mixture of 1 part soybean oil and 5 parts cottonseed oil, which had been brominated with bromine at low temperature, was treated with varying amounts of epoxidized fatty materials. The oil had a specific gravity of 1.33 and a desirable light color (11 on the FAC scale). This brominated oil was divided into several samples, one of which was retained as a control. To the other samples of this oil, varying amounts of epoxidized soybean oil (oxirane content 6.0%), epoxidized linseed oil (oxirane content 7.7%), and methyl epoxy stearate (oxirane content 3.6%) were added. All of the samples were subjected to accelerated aging tests, i.e., holding at 150°F for several days. The following table shows the color stabilization afforded by the epoxy-containing material:

	Oxirane Oxygen Content, %	FAC Color — Days at 150°F						
		0	3	5	7	10	14	21
Control (fresh brominated oil)	0	11	19	21	21	21	21	21
Control + 0.1% epoxidized soybean oil	0.006	11	17	21	21	21	21	21
Control + 0.5% epoxidized soybean oil	0.03	11	11	15	17	19	21	21
Control + 1.0% epoxidized soybean oil	0.06	11	11	13	13	13	15	21
Control + 2.0% epoxidized soybean oil	0.12	11	11	11	11	11	11	13
Control + 0.08% epoxidized linseed oil	0.006	11	17	21	21	21	21	21
Control + 0.39% epoxidized linseed oil	0.03	11	11	15	17	19	21	21
Control + 0.79% epoxidized linseed oil	0.06	11	11	13	13	13	17	21
Control + 1.58% epoxidized linseed oil	0.12	11	11	11	11	11	11	13
Control + 0.16% methyl epoxystearate	0.006	11	17	21	21	21	21	21
Control + 0.79% methyl epoxystearate	0.03	11	11	15	15	19	21	21
Control + 1.58% methyl epoxystearate	0.06	11	11	13	13	13	15	21
Control + 3.16% methyl epoxystearate	0.12	11	11	11	11	11	11	13

The oxirane oxygen content, which is expressed in terms of percentage, is the product of the weight percent of the epoxidized material added and the oxirane oxygen content of that sample. It should be mentioned that after 21 days at 150°F striking differences in color are apparent, but this difference is not noted in the color values reported above because of the poor sensitivity of the FAC color standards in this range. An additional advantage provided by the stabilizers lies in the fact that these materials also appear to provide a degree of bleaching or color improvement to darkened oils to which they are added.

A sample of brominated sesame oil which had been standing at room temperature for a period of about 18 months and which had become very dark in color and appeared to have undergone substantial degradation was divided into 2 parts. One part was retained as the control and about 1% based on the weight of the brominated oil of epoxidized soybean oil having an oxirane content of about 6% was added to the other sample. The samples were held for about a week at room temperature to allow time for the bleaching action. The curves run on the spectrophotometer clearly illustrate the degree of stabilization imparted by the epoxidized oil. Curves drawn from the data show the striking advantage that the epoxidized material offers in bleaching or improving color of these brominated oils.

A further advantage in the use of the stabilizers with brominated oils lies in the resistance to heat deterioration of the oil provided by the stabilizers. For example, unstabilized brominated oils which have developed an unpleasant flavor or odor can be deodorized to improve the flavor, provided a stabilizer is first added to the oil. In the absence of the stabilizer, an oil subjected to deodorization will darken in color, and flavor improvement is minimal. Any of the deodorization procedures such as steam deodorization and inert gas deodorization may be employed to improve flavor and odor characteristics of the oil. All of these procedures involve heating the oil however, thus accelerating degradation. The stabilizers provide protection against this heat degradation.

ALL-PURPOSE PRESERVATIVES AND ANTIOXIDANTS

POTENTIATING AGENT IN COMBINATION WITH PRESERVATIVE

A food preserving agent inhibits or retards the formation or development of various groups of microorganisms including yeasts and molds. Known preserving agents that are suitable for use in food products are generally effective against some microorganisms at relatively low concentrations. However, substantially higher concentrations of these agents may be required before a satisfactory growth inhibition of other microorganisms occurs.

For the safety of the consumer, food products must be protected against various groups of microorganisms including yeasts and molds to insure the suitability of the food for consumption after a period of storage. Protection of food against deterioration by microorganisms has been generally provided by one or more of the common preserving agents.

Of these agents, sorbic acid, benzoic acid and its lower alkyl esters, propionic acid, and their edible alkali metal and alkaline earth metal salts have been most widely used in food products. However, it is not always possible to utilize a sufficient amount of these agents to insure adequate and lasting protection. An effective amount may either exceed the approved level for food preserving agents or affect the flavor of the food adversely.

Even in the few areas of use where the common preserving agents are fully effective, as regards protection against microorganisms, this effectiveness is quite limited in time. That is, protection is not adequate because the preserving agent loses its potency before the food or other product is used by the consumer.

It is desirable to provide a method for potentiating, enhancing, and substantially extending the life-span of the antimicrobial activity of the common food preserving agents. Enhancing the protective power of these agents would increase the effectiveness of the preserving agents in food products. It would also permit the effective use in food products of unusually low levels of the food preserving agents without jeopardizing their protection against microbial attack. In addition, it would make it possible to protect fully those food products in which ineffective levels of preserving agents are now used because of the undesired detectable taste of these agents when used in larger quantities. Further, extending ing the life-span of the preservative would render a longer shelf life to any product thus greatly reducing the incidence of spoilage.

The use of auxiliary compounds to potentiate and enhance the effectiveness of food preserving agents is described by *J.A. Kooistra and J.A. Troller; U.S. Patent 3,404,987;*

October 8, 1968; assigned to The Procter & Gamble Company.

Salts which can be used as the potentiating agent are the phosphates, carbonates, chlorides, nitrates, sulfates, pyrophosphates, and hydroxides of iron, manganese, zinc, tin and silver. Salts which give especially favorable results are manganese chloride, manganese phosphate, ferric chloride and ferrous sulfate. Chlorides, sulfates, and phosphates are preferred anions for use in food compositions because they exhibit very little taste.

The preserving agent used in conjunction with the potentiating agent can be propionic acid, sorbic acid, benzoic acid and edible salts of these acids. Suitable salts include alkali metal and alkaline earth metal salts such as sodium, potassium, magnesium and calcium; methyl and ethyl para-hydroxy benzoate are also suitable. These compounds all possess a recognizable degree of antimicrobial activity. Sorbic acid, potassium sorbate, and calcium propionate are particularly suitable compounds because of their high degree of antimicrobial activity and their comparatively less detectable taste.

The cooperative effect between the preserving agent and the potentiating agent is not clearly understood except that it is apparent that the cation in the potentiating agent is responsible for the enhanced and sustained antimicrobial activity of the preserving agent. Although the use of potentiating agents and the preserving agents, individually, does give some degree of antimicrobial protection, the combinations of these agents give outstanding activity against microorganisms.

This combination of potentiating agents and preserving agents is especially useful in the following food products in which food preserving agents have been shown to have utility: meats; fish; cheese (particularly cottage cheese); milk; ice cream; fruit juices such as apple juice, orange juice and tomato juice; corn syrup, maple syrup; chocolate syrup and candy; fruits including dried, fresh, and citrus fruits; vegetables; beer, wine; farinaceous-containing products such as bread (when the preserving agent is introduced in a manner which does not interfere with the proofing of the dough) and cakes; butter; oleomargine and butter substitutes; vegetables and animal oils and fats; candies; icings and toppings.

When the food product is prepared in a completely sterile manner, the product can be protected by coating the surface of the prepared product. This coating can be applied by dipping the food product in a solution containing the preserving and potentiating agents. Alternatively, a solution of these agents can be washed, sprayed, or otherwise applied to the surface of the food product.

A solid dusting compound can be prepared by using a dry mixture of the preserving agent and the potentiating agent either alone or in mixture with another ingredient such as flour or milk solids. These mixtures can be either dusted on the surface of a sterile food product or, if mixed with an ingredient of the food product, incorporated in the product itself. The preserving agent and the potentiating agent can also be dispersed in other materials, particularly vegetable oils and fats, intended for use in food products. In addition, the mixtures can be effectively incorporated in or coated onto the surface of wrapping materials used to intimately surround food products which are to be protected.

The following methods are most convenient. The preserving agent and the potentiating agent are dissolved in fruit juice, beer, wine or other substantially liquid products. In the instance of milk, it may be more desirable to utilize the protection on the wall of the paper milk carton by coating the carton with a solution of the preservative and potentiating agent.

A product such as bread can be protected by including the food additive compositions in the dough in such a manner that they do not interfere with the biological activity of the yeast or by applying the protective agents to the bread wrapper. Articles such as fruits and vegetables can be washed in, dipped in, or sprayed with a solution containing the preservative agent and potentiating agent, or these articles may be incorporated onto the material utilized for wrapping such products.

To insure the required protection of food products it is necessary to use from about 0.02% to about 1.0% by weight of the food product. This is a generally preferred amount of the composition since it provides extremely effective antimicrobial activity for a long period of time without adding any detectable taste to the food product. In many food products, however, adequate protection can be achieved by using from about 0.02 to 0.5% by weight of the food product of the compositions of this process.

Example: The effectiveness of this combination of preserving agent and potentiating agent in controlling the growth and development of microorganisms responsible for the deterioration of food products is illustrated by the results obtained in the following in vitro tests with a strain of *Aspergillus niger*, one of the most common food spoilage molds. Substantially similar results are observed with other molds such as *Penicillium citrinum, Aspergillus sydowi,* and *Aspergillus repens*; and, a yeast, *Saccharomyces cereviseae.* In conducting the test with *Aspergillus niger,* 1 liter of an agar medium was first prepared having the following composition:

Substance	Grams/Liter of Water
Asparagine monohydrate	5.0
Glucose	10.0
$MgSO_4 \cdot 7H_2O$	0.25
KH_2PO_4	0.25
$FeSO_4 \cdot 7H_2O$	0.001
Agar	20.0

Each preserving agent, potentiating agent, and each combination thereof, as shown in the following table, was added to each of three 20 cc samples of the agar medium.

The pH of the samples was adjusted to 5.0 by adding a small amount of 0.1 N NaOH. The samples were then sterilized with steam at 121°C for 15 minutes. Each 20 cc sample was then poured into a Petri dish and allowed to cool to room temperature before being stab inoculated with a bacteriological inoculating needle previously dipped in a saline suspension of *Aspergillus niger* spores. The length of time required for the commencement of mold growth was visually determined and a measurement of the rate of enlargement of the mold colony was recorded. An averaged value of these mold growth data for the three samples of each preserving agent, potentiating agent, and each combination thereof, is shown in the following table.

Aspergillus niger Colony Diameter in Agar Medium (Millimeters)

Days of Storage of Agar Medium at 30°C	A 0.02%* Potassium Sorbate	B 0.076%* $SnCl_2$	A + B	C 0.15%* Potassium Sorbate	C + B	D 0.055%* $ZnCl_2$	C + D
2	10	10	0	5	0	5	0
4	25	35	0	15	0	15	0
6	40	55	0	28	0	25	1
8	55	80	0	37	0	38	5
10	70	100	5	48	0	50	8

Days of Storage of Agar Medium at 30°C	E 0.20%* Calcium Propionate	F 0.055%* $ZnCl_2$	E + F	G 0.10%* Sodium Benzoate	H 0.05%* $ZnCl_2$	G + H
2	15	30	5	5	5	0
4	22	38	5	15	8	5
6	35	55	10	25	18	8
8	50	75	17	35	40	10
10	60	90	15	45	50	18

*Percent by weight of the agar medium.

The table vividly illustrates how the potentiating agent not only substantially extends the life-span of the preserving agent but also greatly enhances the antimicrobial efficacy of the preserving agent. The enhancing effect is demonstrated, for instance, by referring to the above table, columns A and B. After 10 days' storage at 30°C the *Aspergillus niger* colony diameter had increased to 70 and 100 mm in the presence of 0.02% potassium sorbate and 0.076% $SnCl_2$ respectively. However, when these same amounts of potassium sorbate and $SnCl_2$ were combined, as shown in column A + B of the table, the colony diameter had increased to only 5 mm after 10 days' storage at 30°C.

These results also can be readily translated to food products. For instance, the combination of calcium propionate and zinc chloride, as shown in column E + F of the table, is effective against mold growth in an ordinary loaf of white bread.

HEPTYL HYDROXYBENZOATE FOR CANNED FOOD

In the food canning industry, it is necessary to conduct the processing operation in such a manner as to protect the health of the consumer as well as to prevent the spoilage of the food in order that the taste, color, texture and nutritional value of the food may be as high as possible. In protecting the health of the consumer the microbiological factor to be considered is toxin produced by spore-forming bacteria, as for example *Clostridium botulinum*. This Clostridium is toxic through the agency of an exotoxin. In other words, when the organism grows in a food, it elaborates a toxin. All of the other known food poisoning organisms, such as Salmonella, the Streptococci and the Staphylococci are non-spore formers. Although these others will grow in low and medium acid foods when inoculated into such substrates, their heat resistance is so low that they are not considered to be an important problem as far as canned food processes are concerned.

In respect to protecting against spoilage, it is noted that there are generally three groups of spore-forming spoilage organisms: a thermophilic (high temperature loving) group; a mesophilic (medium temperature loving) group; and a psychrophilic (low-temperature loving) group. There are various types of organisms falling within each of these groups causing various types of spoilage. The importance of each of these types will vary with the type of foodstuffs employed.

Considerable research has been done to uncover the most acceptable processing method for achieving both the protection of the public as well as the prevention of spoilage of the food. Experimenters have endeavored to achieve sterilization by steam injection, high temperature processing for very short periods of time, sonics, microwaves, radiation and other mechanical means. The addition of certain chemical preservatives including antibiotics, has also been considered from time to time.

Presently, however, the only acceptable method commercially employed in the canning industry is that of heat processing at temperatures sufficiently high and for periods sufficiently long to achieve the desired level of sterilization.

Among the problems encountered in the use of the above methods has been the unavoidable deterioration of flavor, color, texture, and the destruction of vitamin content arising out of the high processing or retorting temperatures required to achieve sterilization. These relatively high temperatures which are detrimental to the organoleptic and nutritional qualities of the foodstuff are required if the sterilization of the contents is to be achieved with the necessary degree of certainty. If a method could be found in which lower temperatures and/or shorter processing times could be employed while maintaining the same degree of certainty of sterilization, the appearance and value of the food could be improved.

G.R. Di Marco, J.F. Hogan, Jr., H.L. Schulman and F.B. Standskov; U.S. Patent 3,443,972; May 13, 1969; assigned to F & M Schaefer Brewing Company and Washine Chemical Corp. recommend the use of n-heptyl ester of p-hydroxybenzoic acid as an additive to canned foods to reduce the risk of spoilage due to resistant types and spores of *Clostridium botu-*

linum. Also, this process enables the use of lower degrees of heat and time to sterilize canned foods.

Although it is impossible to define limits of degrees of heat and time, the preferred aspect of this process relates to a strong, heat treatment of about 250°F for about five minutes. To illustrate this aspect, reference may be made to mushrooms which constitute an especially complex problem to the canning industry. Despite extremely thorough washing techniques and the employment of sanitary canning conditions, in order to prevent spoilage of the pack during shelf storage, it is necessary to process the canned mushrooms at a temperature of 250°C for 18 to 30 minutes depending upon the size of the container normally used. Such extreme conditions are detrimental to the taste, color, texture, and nutritional value of the canned product. By adding to the pack, prior to sealing, a microbial growth inhibiting amount of one of the esters the desired level of sterilization may be achieved by processing the sealed pack in a manner that the quantum of heat applied is below that in which the taste, color, texture and/or nutritional value are normally markedly adversely affected. It is thus possible to place in the hands of the consumer a better product from the standpoint of taste, color, texture and nutritional value.

To further illustrate the applicability of the process, attention is directed to the recent rise in the use of automatic vending machines dispensing hot canned foods such as soups and sauces. This has created a problem from the standpoint of spoilage of the food stored in the machine. While in the machine, the canned product is stored at a temperature ranging from 110° to 130°F and is heated just prior to dispensing to a temperature of about 150°F. Certain thermophilic bacteria which cause spoilage of foods show optimum growth at about the storage temperatures in these vending machines. Thus, unless special precautions are taken for the processing of products to be stored in these machines a great amount of spoilage is likely to occur. These precautions are not normally necessary for foods to be transported and stored under the standard conditions prevailing in the United States, since storage temperatures are much lower than 130°F and the thermophilic bacterial growth is not encouraged as it is in the vending machines. The special precautions necessitated by these conditions can decrease the quality of the product, can be very expensive, and may involve special ingredients, equipment, process control and other measures.

By adding to the canned product a microbial growth inhibitor and processing under conditions similar to or even less extreme than those commonly in use, a canned product is achieved which may be stored in the automatic vending machines at temperatures approximating the optimum growth conditions for thermophilic bacteria without spoilage resulting and which at the same time possesses superior taste, color, texture and nutritional value.

The amount of the ester or mixture of esters which is used will of course vary depending upon the nature of the foodstuff in question. The degree of heat utilized with the specific foodstuff will also produce a variation in the amount used. Some foods are affected less by a high degree of heat than are others. It is possible with these foods to utilize higher temperatures and less additive to obtain the desired level of sterilization.

Also affecting the amount of the additive to be employed is the type of spoilage to which the food is susceptible. High, medium and low acid foods for example, such as tomatoes, peas, corn and milk may exhibit flat-sour spoilage wherein surviving spores germinate and produce acid but no gas during growth. In other cases putrefactive anaerobic spore-forming bacteria lead to swelling of the container and putrid odors. Acid foods on the other hand show somewhat different spoilage. In this instance, acid-tolerant spore-formers may produce abnormally high amounts of acid without swelling the container. Nonspore formers, such as yeasts, molds and lactic acid bacteria, may spoil container contents with or without container swelling.

Also affecting the amount of the additive to be employed is the composition, i.e., the chemical makeup of the foodstuff employed. For example, it has been found that if a foodstuff has a high fat content, it may be necessary to utilize a greater amount of the additive. This is thought to be due to the fact that the additive is withdrawn from the liquid medium surrounding the foodstuff and taken into the fat content of the foodstuff. Experimentation with vaseline overlay of broth medium in glass tubes in the presence of the propyl and heptyl esters of p-hydroxybenzoic acid indicated that the vaseline reduced the inhibitory power of the compound. These esters are known to be very soluble in oils. A concentration of 100 ppm heptyl ester gave a cloudy suspension in media because it would not dissolve completely. However, tubes containing this much heptyl ester in T-BEST medium sealed with vaseline were found to become clear after 72 hours incubation at 98°F, while tubes without vaseline remained cloudy. This seemed to indicate absorption of the heptyl ester into the vaseline. The heptyl ester was soluble in the medium up to about 16 ppm at room temperature.

Also effecting the amount of ester to be employed is the type and size of container in which the foodstuff is to be placed. Chemical analysis of the heptyl ester concentration in various types of containers gave the results shown in the following table. The heptyl ester (5 and 10 ppm) was prepared in distilled water and filled into glass stopped bottles, 211 x 400 mm plain cans, and thermal-death-time (208 x 006) de-enameled cans. After sealing the cans under 25 inches of vacuum, all containers were held approximately 7 days before analysis. The results in the following table show that the heptyl ester tends to retain its initial concentration when stored in glass. Retention in plain cans is apparently influenced by the ratio of exposed can surface area to the volume of the container. A larger can surface area per unit volume of can contents contributed to the reduction of recoverable ester. Effect of container type and size on the retention of the heptyl ester of p-hydroxybenzoic acid as determined by chemical analysis

Original Concentration	Recovery*, ppm
10 ppm:	
Glass stopped bottle	9.8
211 x 400 plain can	9.8
TDT plain can	7.6
5 ppm:	
Glass stopped bottle	5.0
211 x 400 plain can	5.0
TDT plain can	3.3

*Following approximately 7 days storage.

The additive may be added to the pack at any time prior to sealing. It is also possible, for example, to coat the inside of the container with the desired additive and then place the food contents in the container and seal. In a preferred aspect, the additive is intimately mixed with the foodstuff immediately prior to the canning, the foodstuff containing the additive is placed into the container and the container is sealed.

The form in which the additive is added to the container or to the foodstuff is not critical and may be in any form in which good contact between the additive and the food may be obtained, e.g., as a solution or dispersion. One method which is employed is to dissolve or disperse the desired amount of additive in a solvent or carrier such as will not affect the quality of the foodstuff, e.g., propylene glycol, and then add the solution or dispersion in the most acceptable manner. The process is applicable to all types of canned foodstuffs, e.g., fruits and vegetables, eggs, fish, dairy products, poultry and meats.

Example 1: A dispersion of n-heptyl para-hydroxybenzoate in propylene glycol is prepared. Varying amounts of this dispersion are added to cans of peas which have been inoculated with from 1,000 to 2,000,000 spores per 011-400 can of various types and strains of *Clostridium botulinum*. These varying amounts of the solution are chosen to achieve a concentration of from 2 to 100 parts of the ester per million parts of the final product (ppm).

The cans are hermetically sealed and heated to a temperature of 200°F and held at this temperature for 15 minutes. The cans are removed from the heat, cooled and placed in storage for up to two years. During this period of storage it will be found that spoilage of some of the cans will be indicated by swelling. When this occurs, these cans are removed, opened and the contents examined. At the end of this period the remaining cans are opened and examined for microbial content upon which a determination of spoilage or lack of spoilage is based. The results of the tests are recorded below. Note that in all cases some microbiological control is obtained. In the cases where a maximum level of inocula and a low level of ester pertained, spoilage eventually occurred.

Spoilage in Botulinum Inoculated Canned Peas

Spores/Can	Ester, ppm	Percent Spoilage
1,000	0	100
1,000	2	0
1,000	10	0
1,000	100	0
1 to 2 million	0	100
1 to 2 million	2	20
1 to 2 million	10	6
1 to 2 million	20	4
1 to 2 million	100	0

Example 2: A stock solution (1,000 ppm) of n-heptyl para-hydroxybenzoate is made by dissolving 0.1 gram in 100 ml of 95% alcohol. Next a ten TDT-tube (208 x 006 mm) series containing canned beef bouillon is sterilized by autoclaving, and then cooled. The pH of each tube is about 7.0 after autoclaving. To each of the ten TDT tubes is added a quantity of stock solution to achieve a 15 ppm concentration of n-heptyl para-hydroxybenzoate. Ten TDT tubes containing canned beef bouillon but not heptyl ester are sterilized in the same manner and used as controls. Each of the TDT tubes are then inoculated with Putrefactive Anaerobe No. 3679 spores to achieve spore concentration of 20,000/ml Heating is carried out at 250°F in an oil bath for 0, 3, 6, 8, 10 and 12 minute intervals. After heating, the tubes are water cooled and incubated at 85°F under a 25 inch vacuum.

Results show that although the controls at the zero time were positive and 3 minute control tubes were positive, neither the zero time tubes nor any other were positive with 15 ppm of n-heptyl para-hydroxybenzoate. No change took place even after 6 months of incubation.

EVERNINOMICIN ANTIBIOTIC

The standard industrial heat sterilization of canned goods is designed to destroy bacterial cells, some of which resist destruction up to 6 hours at atmospheric boiling temperature. However, the susceptibility of some food products to degradation upon heating precludes the degree of heat treatment necessary to kill the spores of *Clostridium botulinum*. Therefore, new and better methods of food preservation are constantly being sought. Recently, a considerable amount of research has been conducted in the use of antibiotics to prevent bacterial spoilage of foods.

An antibiotic, Everninomicin, has been discovered which is effective in inhibiting the activity of some species of harmful bacteria. The antibiotic activity of Everninomicin is highly active against gram-positive bacteria, including strains resistant to other antibiotics. Five active components of this agent have been isolated and identified as Everninomicin A, B, C, D and E. The major chemical component is Everninomicin D.

The use of this antibiotic is described by *G.J. Haas and N. Insalata; U.S. Patent 3,607,311; September 21, 1971*. The antibiotic may be introduced into foodstuffs by any standard method. For example, it may be infused directly into the food or it may be added to a diluent and blended with the food. Among the many diluents suitable for this purpose are liquids such as water, alcohol and propylene glycol or dry powdered substances such as lactose and starch. The diluent permits more accurate measurement of the antibiotic and more uniform distribution of it throughout the food.

The exact method of operation of Everninomicin in preventing food spoilage is not completely known. Examination of foods treated with Everninomicin have shown viable *Clostridium botulinum* spores indicating that the antibiotic is not sporicidal. It is believed that the antibiotic prevents the complete development of dividing vegetative cells and renders the bacteria incapable of producing exotoxin. It is possible that it also destroys the exotoxin as it is produced. Whatever the action, foodstuffs treated with Everninomicin have not produced the exotoxin of *Clostridium botulinum* in quantities measurable by accepted standard test methods.

Everninomicin has no harmful effect on the human body. Tests have shown that, administered orally, Everninomicin is not absorbed into the blood stream in measurable amounts. Tests have shown that Everninomicin is destroyed by the acid in the stomach of humans and is passed harmlessly through the intestines. (Black, et al, "Pharmacological Properties of Everninomicin D," *Antimicrobial Agents and Chemotherapy*, 1964 pp 38 through 46.)

In addition to treatment with Everninomicin foods may be subjected to other sterilization methods. Thus, the food may be subjected to a heat treatment for more thorough sterilization. Since the Everninomicin contributes to the prevention of bacterial activity it may be possible to use shorter time periods and lower temperatures when heat sterilizing Everninomicin-treated foods, thereby preserving much of the texture, flavor and color of the food that would otherwise be lost.

Example 1: A quantity of Everninomicin-D, rated at 1,600 units per milligram was dissolved in 95% ethyl alcohol and added to a sufficient amount of distilled water to yield a solution having an Everninomicin concentration of 5 parts per million. Test samples were prepared by adding 2 ml of the Everninomicin solution to each of duplicate test tubes containing 7 ml of Duff's enrichment medium and 1 ml of spore suspension containing 1,000 exotoxin-free spores.

A first control tube containing 9 ml of Duff's enrichment medium and 1 ml of the spore suspension, and a second control tube containing 8 ml of Duff's enrichment medium, one ml of the spore suspension and 1 ml of 10% ethyl alcohol were also prepared. Exact replicates of the tubes described above were prepared and heated at 100°F for 15 minutes. All tubes were incubated at 85°F under a 10% CO_2-90% N_2 atmosphere.

After 5 days incubation all tubes were analyzed for *Clostridium botulinum* exotoxin by standard mouse inoculation procedures. Tests showed that no exotoxin was produced in any of the tubes containing Everninomicin but it was produced in all of the control tubes. This experiment shows that Everninomicin interferes with the normal production of *Clostridium botulinum* exotoxin.

Example 2: In this example ground beef which has been autoclaved at 15 psig for 15 min to destroy indigenous microflora was used as the test material. Control samples were prepared by blending 200 grams of the sterile ground meat for 15 minutes in a five quart Hobart mixer with 30 ml of an aqueous suspension containing 2.4×10^6 spores of *Clostridium botulinum* Type A. The mixture was divided into six equal portions and sealed in a nitrogen atmosphere in polymylar bags.

Test samples were prepared as follows: 200 grams of sterile ground beef was blended for 15 minutes with 15 ml of water containing 18.2 mg of the water-soluble sodium salt of Everninomicin B containing 1,011 units of Everninomicin per mg. 15 ml of spore suspension containing 2.4×10^6 spores were then mixed with the meat for 15 minutes. The resulting mixture was divided into six equal portions and sealed in a nitrogen atmosphere in polymylar bags. The samples were stored at room temperature and analyzed after one week. The control samples contained exotoxin but the Everninomicin-protected samples did not. After 12 days the samples were again analyzed with the same result. Standard mouse inoculations of the samples demonstrated that the Everninomicin prevented intoxication. The spores in the Everminomicin samples were found to be still viable and toxigenic when removed from the effects of the Everninomicin in inhibiting the production of botulinum exotoxin in meat.

GLUTATHIONE-PRESERVATIVE COMBINATIONS

J.A. Troller; U.S. Patent 3,276,881; October 4, 1966; assigned to The Procter & Gamble Company reports that a sulfhydryl-containing compound like glutathione is effective as a potentiating agent in combination with a regular preserving agent used for cheese. Any nontoxic sulfhydryl-containing compound can be used. Illustrative examples of sulfhydryl-containing compounds include cysteine and its various acid salts such as, for example, cysteine hydrochloride and di-L-cysteine sulfate. Other preferred compounds include thioglycolic acid and its sodium and potassium salts. Glutathione is a particularly preferred compound because of its relatively bland flavor.

Other specifically useful potentiating agents include: thioctic acid, homocysteine, sodium thioglycolate, potassium thioglycolate, thioglycerol, thiolactic acid, thiomalic acid and thiosorbitol. Thiol enriched materials such as Thiol-gel, a thiol enriched protein substance, are also useful potentiating agents.

The preserving agent used with the auxiliary potentiating agent can be sorbic acid, or benzoic acid. Sorbic acid and potassium sorbate are preferred because of their high degree of antimicrobial activity in an acid medium and their comparatively less detectable taste.

Saturated lower aliphatic carboxylic acids, such as propionic acid and diacetic acid and their esters and salts, are not potentiated in the same manner as other preserving agents. The cooperative effect between the preserving agent and the potentiating agent is not clearly understood except that it is apparent that the sulfhydryl group in the potentiating agent is responsible for the enhanced antimicrobial activity of the preserving agent.

The effectiveness of the combinations of this process is illustrated by the results obtained in the following in vitro tests with a strain of *Aspergillus niger,* one of the most common food spoilage molds. Substantially similar results have been observed with other molds such as *Penicillium cirinium, Aspergillus sydowi,* and *Aspergillus repens;* a bacterium, *Micrococcus candidus;* and, a yeast, *Saccharomyces cerevisiae.*

In conducting the test with *Aspergillus niger,* an agar medium was first prepared having the following composition.

Substance	Grams/Liter of Water
Asparagine monohydrate	5.0
Glucose	10.0
$MgSO_4 \cdot 7H_2O$	0.25
KH_2PO_4	0.25
$FeSO_4 \cdot 7H_2O$	0.001
Difco agar	20.0

The pH of the agar medium was adjusted to 5.0 with 0.1 N NaOH and sterilized with steam at 121°C for 15 minutes. The medium was then poured into Petri dishes and allowed to cool to room temperature before being stab inoculated with a bacteriological inoculating needle previously dipped in a saline suspension of *Aspergillus niger* spores. The length of time required for the commencement of growth was visually determined and a measurement of the rate of enlargement of the mold colony was also noted. These observations were made on Petri dishes containing the following perserving agents and potentiating agents and on Petri dishes containing the same percentage of preserving agent alone. The results were compared and are reported below as the percent increase in relative inhibition of mold growth.

Preserving Agent + Potentiating Agent	Percent Increase in Relative Inhibition
0.1% sorbic acid + 0.05% cysteine	94
0.1% potassium sorbate + 0.1% potassium thioglycolate	100*
0.1% sorbic acid + 0.05% glutathione	79
0.1% potassium sorbate + 0.1% glutathione	100*
0.5% potassium sorbate + 0.05% cysteine hydrochloride	25
0.5% potassium sorbate + 0.05% potassium thioglycolate	15
0.05% sorbic acid + 0.1% sodium thioglycolate	100*
0.075% sorbic acid + 0.05% di-L-cysteine sulfate	100*
0.01% 2,3-decenoic acid + 0.05% cysteine	40
0.01% nonanoic acid + 0.05% cysteine hydrochloride	50

*Denotes equal to or greater than 100.

The food additive compositions of this process can be used in the following food products in which food preserving agents have been shown to have utility: meats; fish; cheese (particularly cottage cheese); milk; ice cream; fruit juices such as apple juice, orange juice and tomato juice; corn syrup; maple syrup; fruits including dried, fresh, and citrus fruits; vegetables; beer; wine; farinaceous-containing products such as bread (when the preserving agent is introduced in a manner which does not interfere with the proofing of the dough) and cake; butter; oleomargine and butter substitutes; vegetable and animal oils and fats; candies; icings and toppings.

The preserving agent and the potentiating agent can be incorporated directly in the food product. When the food product is prepared in a completely sterile manner, the product can be protected by coating the surface of the prepared product. This coating can be applied by dipping the food product in a solution containing the preserving and potentiating agents. Alternatively, a solution of these agents can be washed, sprayed, or otherwise applied to the surface of the food product.

A solid dusting compound can be composed by using a dry mixture of the preserving agent and the potentiating agent either alone or in mixture with another ingredient such as flour or milk solids. These mixtures can be either dusted on the surface of a sterile food product or, if mixed with an ingredient of the food product, incorporated in the product itself. The preserving agent and the potentiating agent can also be dispersed in other materials, particularly vegetable oils and fats, intended for use in food products. In addition, the food additive compositions can be effectively incorporated in or coated onto the surface of wrapping materials used to intimately surround food products which are to be protected.

The following illustrative methods are given: the preserving agent and the potentiating agent are dissolved in fruit juice, beer, wine or other substantially liquid products. In the instance of milk, it may be more desirable to utilize the protection on the wall of the paper milk carton by coating the carton with a solution of the preservative and potentiating agent. A product such as bread can be protected by including the food additive compositions in the dough in such a manner that they do not interfere with the biological activity of the yeast or by applying the protective agents to the bread wrapper. Articles such as fruits and vegetables can be washed in, dipped in or sprayed with a solution containing the preserving agent and potentiating agent, or these articles may be incorporated onto the material utilized for wrapping these products.

It is generally desirable to insure the required protection of the food product to use the maximum possible level of the preserving agent and from about 0.05 to 0.15% by weight of the food product of the sulfhydryl-containing potentiating agent. This is a preferred amount of potentiating agent since it provides the desired enhancement in the antimicrobial activity of the preserving agent without adding any detectable taste to the food product.

GENERAL ANTIOXIDANTS

Bisphenol Antioxidants

Bis(3,5-dialkyl-4-hydroxyphenyl)methanes for stabilizing foodstuffs subject to oxidative deterioration are described by *G.R. Ferrante and R.C. Morris; U.S. Patent 3,041,183; June 26, 1962; assigned to Shell Oil Company.* Preferably, each of the alkyl substituents on the phenyl ring is selected from secondary and tertiary alkyl radicals having from 3 to 8 carbon atoms. Also representative of such bisphenols are those bisphenols substituted with both secondary and tertiary alkyl radicals.

Of these, the most preferred antioxidant, which combines the most desirable properties of low toxicity and superior antioxidant capability, is bis(3,5-di-tert-butyl-4-hydroxyphenyl)-methane.

Edible oils which are stabilized with the tetraalkyl bisphenols to yield the oxidation-resistant compositions include linseed oil, menhaden oil, cod liver oil, safflower oil, castor oil, olive oil, rapeseed oil, coconut oil, sesame oil, peanut oil, babassu oil, palm oil, and corn oil. Edible fats which are stabilized in the same manner are represented by oleomargarine, lard, beef tallow, animal fat and hydrogenated vegetable shortening products. Other oils and fats which have been specially treated by air-blowing and heating, may also be stabilized with the bisphenols.

Meat, particularly fatty meats, such as bacon, sausage, ham and hamburger; and poultry, including turkey, squab and duck, may be stabilized against rancidity by treatment with these tetraalkyl bisphenols. Bakery products, including bread, cookies, pretzels, pastries, pies and cakes, and candies, particularly those made of chocolate, cream filling, butter, and nuts, which tend to become rancid on storage, are also afforded considerably extended shelf life by incorporation of the bisphenols. For example, a package of walnut meats which contains a few crystals of bis(3,5-di-tert-butyl-4-hydroxyphenyl)methane has a shelf life many times that of a similar package not so protected.

When oils or fats containing the tetraalkyl bisphenols are used for cooking, frying or baking, it will be observed that sufficient oil or fat is incorporated in the cooked product so as to render the food stable and retard development of rancidity. Thus, potato chips fried in an oil consisting of 50% vegetable shortening and 50% vegetable oil, containing 1 wt %, based on the total oil, of bis(3,5-diisopropyl-4-hydroxyphenyl)methane, have greater stability than potato chips fried in oil not containing the bisphenol.

The method by which the tetraalkyl bisphenols are brought into intimate contact with the foodstuff described will, of course, depend on the nature of the foodstuff. In many cases, the bisphenol may be physically incorporated into the foodstuff. When the substrate is a liquid, such as an oil, milk, cream or fruit juice, such as citrus fruit juice, the bisphenol may be dissolved in the liquid.

In solid foods, the bisphenol may be incorporated by dissolving it in one of the ingredients, such as the shortening in pie crusts, or by dispersing it on the surface in solid, liquid or emulsion form. In some cases, the bisphenol may be incorporated in the packaging which is used in direct contact with the surface of the food; under these circumstances the antioxidant serves to prevent oxidation at the surface of the packaged item.

It is well known that traces of chemicals present in paper products actually catalyze the oxidation of fatty organic materials in contact with the paper. By incorporating a stabilizing amount of the tetraalkyl bisphenol in the paper, the oxidation catalysis is suppressed and the surface of the food is also preserved against normal oxidation resulting from prolonged exposure to air. The bisphenol may be combined with the paper by conventional papermaking methods as, for example, by applying it to paper leaving the Fourdrinier machine as a solution or oil-in-water emulsion. It may also be used in such other cellulosic wrapping materials as paperboard, cellophane, tissue paper, vegetable parchment paper, and the like. Papers treated in this manner are found to be especially effective to stabilize packaged foods such as butter, bacon, fish, candy, nuts and oleomargarine.

For example, in the manufacture of chocolate board, the tetraalkyl bisphenol may be incorporated in the dye solution which is then used to treat the paperboard. On the other hand, in waxed papers, the bisphenol is conveniently added to the paraffin wax when the wax is at its lowest viscosity, around 140° to 160°F, and stirred in until complete dispersion is achieved. The paper is then impregnated with the wax-bisphenol dispersion in the conventional manner.

Meats and seafoods, such as fish, crab, lobster and shrimp may also be preserved by spraying a dispersion of the tetraalkyl bisphenol onto their surface prior to storage or transportation of the foods.

One important feature of the compositions is that the bisphenol serves to prevent the degradation of vitamins, including those naturally contained in oils and fats, those prepared synthetically, and those recovered from other sources such as ascorbic acid and vitamin A. The bisphenols may be readily dispersed in vegetable or fruit juices for this purpose. Compositions of particular utility are citrus fruit juices such as lemon juice, orange juice and grapefruit juice, and other juices including pineapple juice, tomato juice, apple juice, and grape juice, containing a stabilizing amount of the bis(3,5-dialkyl-4-hydroxyphenyl)methanes described above. Animal feeds, such as alfalfa, silage, chicken feed, dog and cat food, and fish meal may also be stabilized against the development of rancidity, off-taste and degradation of vitamins through oxidation by incorporation therein of a stabilizing amount of the bis(3,5-dialkyl-4-hydroxyphenyl)methanes of the process. The amount of stabilizer required will in general be that corresponding to the amount needed for other foods.

Thus, the nutritional values offered to poultry feeds by menhaden fish meals are reduced when the feed is stored at temperatures on the order of 150°F or above through denaturation of protein and oxidative degradation of vitamins, particularly those of the vitamin B complex. Inclusion in the fish meal of about 0.2% by weight of bis(3,5-di-tert-butyl-4-hydroxyphenyl)methane materially increases the useful storage life of poultry feeds based on such meal without impairing the taste or nutritional value of the feed in any way.

Another important advantage of the compositions is that they retain their stabilized properties under conditions of heat and moisture as, for example, during processing, when many other food compositions lose their stabilizer through volatilization. This is particularly important when the stabilizer is incorporated in the paper wrapping during paper manufacture where the moist paper is subsequently dried. It has been found that at advanced temperatures and humidities the bis(3,5-dialkyl-4-hydroxyphenyl)methane remains in the food substrate and thus preserves it after compositions stabilized with other preservatives have begun to deteriorate.

In some cases the antioxidant composition may include additional components, such as acids including gallic acid, citric acid, ascorbic acid, tartaric acid and phosphoric acid. Esters of these acids, such as isopropyl citrate or propyl gallate, may also be included.

Example 1: To evaluate the effectiveness of several antioxidants in a typical edible fatty oil, induction period measurements were made using the gravimetric procedure of Olcott and Einsett, *J. Am. Oil Chemists' Soc.,* 35, 161 (1958). Samples of various substrates containing different concentrations of the antioxidants were stored in an oven at 50°C and weighed daily until a rapid increase in weight was obtained. The data showing the relative efficiencies of the candidate antioxidants are presented in the following tables. Data presented are days to rancidity. In Table 1, the effectiveness of various concentrations of several antioxidants in methyl linoleate is set forth in weight percent.

TABLE 1

Additive	0.01%	0.02%	0.04%
None	4	4	4
2,2',6,6'-tetra-tert-butyl biphenol	4¾	12	20½
Bis(3,5-di-tert-butyl-4-hydroxyphenyl)methane	11¼	14½	23½

In Table 2 is presented the relative performance of several antioxidants in several other food media. Where antioxidant performance in paperboard was evaluated, the antioxidant was added to the paperboard in methanolic solution and the paperboard was dried before testing. Antioxidant concentrations of 0.02% by weight, based on the oil, were employed except in the paperboard test, wherein the concentration was 0.2% of weight, based on the paperboard.

TABLE 2

Antioxidant	- - Fatty Product- -		- - - - Paperboard- - - -	
	Lard	Soybean Oil	Methyl Linoleate	Lard
None	8½	20½	3¼	7
2,6-di-tert-butyl-4-methylphenol	98	23¼	5½	52½
2,2',6,6'-tetra-tert-butyl biphenol	140	40¼	4	34
Bis(3,5-di-tert-butyl-4-hydroxyphenyl)-methane	110	68¼	8½	51

From these data, it will be seen that compositions containing bis(3,5-di-tert-butyl-4-hydroxyphenyl)methane were generally more stable than those containing the other antioxidants tested.

Example 2: Solutions of two antioxidants in a water-methanol mixture (40% water-60% methanol) were distilled. Each solution contained 0.02 gram of antioxidant. Colorimetric determinations of the quantities of antioxidant present in the overhead and bottoms fraction were made after 75 ml of water had been collected in the overhead fraction. 100% of the 2,6-di-tert-butyl-4-methylphenol was found in the overhead fraction. 8% by weight of bis(3,5-di-tert-butyl-4-hydroxyphenyl)methane was found in the overhead fraction, and 92% in the bottoms fraction. The data show that under the conditions of the test 2,6-di-tert-butyl-4-methylphenol was readily steam-distilled while the bis-phenol was relatively nonvolatile.

Example 3: When 0.25 pound blocks of butter are stored at room temperature in bleached sulfite pulp handsheets containing 0.25% by weight bis(3,5-di-tert-butyl-4-hydroxyphenyl)-methane, the odor of rancidity from the resulting package requires at least twice as long to become apparent as that from blocks stored under identical conditions in samples of the same paper containing no antioxidants.

Example 4: A slab of bacon is stabilized against rancidity by soaking it in a 10% vegetable oil solution of bis(3,5-di-tert-butyl-4-hydroxyphenyl)methane until a stabilizing amount of the inhibitor is incorporated in the bacon. Development of discoloration and odor of rancidity in the slab of bacon, when stored at room temperature, requires considerably more time than that in nonstabilized bacon.

Isoascorbic Acid Phosphates

The use of ascorbic and isoascorbic acids as antioxidants can effectively retard the oxidative deterioration with the resultant occurrence of off flavors, off odors, and discoloration, in a wide variety of food materials processed and packed in the conventional ways. Experience has shown that isoascorbic and ascorbic acids possess inherent properties that detract sometimes from their usefulness as antioxidants. For example, although in the dry form both ascorbic and isoascorbic acid and their salts are stable for long periods when stored under cool, dry conditions, discoloration may occur during prolonged storage and this decomposition is accelerated by the presence of moisture or elevated temperatures and exposure to air. Likewise, solutions of these acids will rapidly undergo oxidation if exposed to air, alkaline conditions or high temperatures, even for relatively short periods.

Furthermore, certain metals such as copper and iron greatly accelerate the destruction of ascorbic acid, either in dry form or solution. Thus, in order to effectively use isoascorbic or ascorbic acid as antioxidants in the preservation of food and beverages, it is often necessary to use an excess over the amount actually needed or effect addition at a late stage in the process, such as immediately prior to canning or packing the food to be processed.

In certain cases, even the use of excesses or the late addition of the antioxidant may still be inadequate to afford extended shelf life to some products due to what may be termed enzymatic, residual, or autoxidative changes occurring in the food or beverage after processing and packaging.

The instability of both isoascorbic and ascorbic acids is known to be due to the high sensitivity of the enolic hydroxyl groups of the lactone ring to oxidative influences. In order to prevent this oxidation, one or both of the enolic hydroxyl groups may be blocked; for example, by formation of an ester or ether. Unfortunately, the esterification or etherification of these hydroxyl groups often renders isoascorbic or ascorbic acid unsatisfactory or useless as an antioxidant for different reasons; namely, the resistance to in vitro or in vivo cleavage of the ether or ester group.

Thus, it is desirable to protect the enolic hydroxyls on the lactone ring by means of a group which is stable to aerial or other rapid oxidation influences, yet is slowly hydrolyzed under in vitro or in vivo conditions by enzymatic or other controlled mechanisms. It has been found that the phosphate esters of isoascorbic acid meet the desired criterion of stability and availability.

Details on the preparation of isoascorbic acid phosphates and their use as antioxidants in foods are given by *D.F. Hinkley; U.S. Patent 3,718,482; February 27, 1973; assigned to Merck & Co., Inc.*

Phosphate esters of isoascorbic acid are derivatives of isoascorbic acid formed by the phosphorylation of one or both of the hydroxyl groups contained on the 2 and 3 position of the isoascorbic molecule. Thus, the compounds are namely isoascorbic acid-2-phosphate, isoascorbic acid-3-phosphate, isoascorbic acid-2,3-diphosphate, and isoascorbic acid-2,3-cyclic phosphate, and the mono-, di-, or tri-alkali and their alkaline earth metal salts.

These compounds function as antioxidants via the slow liberation of isoascorbic acid by medium hydrolysis or enzymatic cleavage. By virtue of this ability to slowly hydrolyze to isoascorbic acid, the compounds offer many advantages over the free ascorbic and isoascorbic acids when used as antioxidants in food and beverage processing since, for the most part, they are stable for longer periods towards oxidative influences that ordinarily lead to the premature decomposition of isoascorbic acid.

For example, the 3-phosphate ester of isoascorbic acid is stable in aqueous solutions for prolonged periods even at elevated temperatures and in the presence of air, and metals do not appear to have a deleterious effect on this form of isoascorbic acid. Similarly, the 2-phosphate ester, although not quite as stable as the 3-ester, is significantly more stable than isoascorbic ester under the same conditions. The 2,3-diphosphate ester and the 2,3-cyclic esters, however, appear to be quite labile, offering only a slight resistance to cleavage of the ester moiety.

The advantages which may be realized by the use of the compounds as antioxidants are quite significant, for it is apparent that they can be made up in solution for longer periods of time prior to use, are more economical since excesses are not necessary, effectively act over a more prolonged period of time and thus extend shelf life, and can be added to any stage of the food processing procedure.

The relative amounts of the various phosphate esters of isoascorbic acid and the free isoascorbic contained in these compositions may vary widely. For example, a suitable antioxidant composition may contain anywhere from 5 to 95% of the 2- and/or 3-phosphate ester of isoascorbic acid, 10 to 70% of the 2,3-diphosphate ester, 5 to 60% of the 2,3-cyclic phosphate ester of isoascorbic acid, and 5 to 95% of isoascorbic acid. A preferred composition is one that contains 3-phosphate esters of isoascorbic acid combined with isoascorbic acid or the sodium salts, each ingredient comprising either a major or minor proportion of the composition.

In utilizing the phosphate esters, an antioxidative amount of one of the compositions of the process is added to the food or beverage prior to packaging. By the term prior to packaging is meant at any step of the processing procedure including just prior to sealing in the unit container. By the term antioxidative amount is meant the minimum amount of antioxidant needed to consume the dissolved oxygen, the headspace oxygen, or any residual oxidative elements which might form during the processing or upon storage of the processed food or beverage. Thus, the exact amount and the relative proportions of the active ingredients in the compositions utilized will vary depending on the type of food to be preserved, the size and headspace in the packing container, the processing procedure utilized, and the point in the processing step at which the antioxidant composition is added.

For example, if the oxidative influences are greatest during processing, i.e., temperature and other conditions, then the composition used will predominantly contain the more stable antioxidants such as the 2- and/or 3-phosphate esters of isoascorbic acid and relatively smaller proportions of the less stable esters and/or the free acid itself. On the other hand, if it is desired to have a prolonged residual antioxidant effect and where there is also present a need for immediate consumption of available oxygen, it is desirable to have the free acid in excess with minor amounts of the more stable 2- and 3-ester.

The compounds may be prepared by contacting a 5,6-O-loweralkylidene isoascorbic acid dissolved in an anhydrous organic solvent, such as acetone, with a phosphorylating agent, suitably phosphorus oxychloride, in the presence of a tertiary amine such as pyridine. The phosphorylation is effected at room temperature, but preferably at a lower temperature of from about -5° to about 15°C, and continued until a sample when titrated with iodine using a starch indicator becomes negative for free isoascorbic acid.

The reaction mixture is then neutralized with a base such as an alkali or alkaline metal hydroxide or bicarbonate such as sodium bicarbonate, aged, and filtered, and the filtrate containing a mixture of the 5,6-loweralkylidene isoascorbic acid-2- and 3-phosphates, and minor amounts of -2,3-diphosphate and -2,3-cyclic phosphate esters of 5,6-loweralkylidene isoascorbic acid. The mixture of esters can be isolated from the reaction mixture by recrystallization from a methanol solution and the 5,6-O-isopropylidene protecting group hydrolyzed with, for example, 0.1 N HCl. The respective phosphate esters of isoascorbic acid may be separated from the reaction mixture by column chromatography using as a solvent, for example, isopropanol:water:acetic acid:concentrated ammonium hydroxide (65:30:30:30), the esters fractionating in the following order, the 2,3-cyclic ester of isoascorbic acid, the 2-phosphate ester of isoascorbic acid, the 3-phosphate ester of isoascorbic acid and finally the 2,3-diphosphate ester of isoascorbic acid. The respective esters can be obtained in pure form by further purification on ion exchange resins.

Alternatively, the 3-phosphate ester of isoascorbic acid may be obtained by treating the mixture of esters with concentrated hydrochloric acid. This treatment serves to selectively hydrolyze the 2-phosphate, the 2,3-diphosphate and 2,3-cyclic phosphate ester moieties such that there remains in solution the 3-phosphate ester of isoascorbic acid and free isoascorbic acid. The isolation of the desired 3-phosphate ester may be accomplished by standard methods.

The alkali and alkaline earth metal salts of the phosphate esters of isoascorbic acid may be prepared by treating the respective ester with the desired amount of alkali metal or alkaline earth metal base such as sodium hydroxide, potassium hydroxide, lithium hydroxide, calcium hydroxide, and magnesium hydroxide. Thus, the trisodium salt of isoascorbic acid-2- or 3-phosphate can be obtained by adding sufficient base until the pH is greater than 9. The mono- and di-metal salts may be prepared by the partial acidification of the tri- metal phosphate ester of isoascorbic acid; for example, the trialkali or trialkaline earth metal salts of the 2- or 3-phosphate ester of isoascorbic acid, suitably the trisodium, tri-potassium or trimagnesium salt is dissolved in water and one or two equivalents of acid added per ion mol of phosphate ester moiety.

In a similar manner, the tetra salt of the 2,3-diphosphate ester and the monosalt of the 2,3-cyclic ester of isoascorbic acid may be prepared, that is, by adding sufficient of the desired base until a pH of 10 is obtained. A mixture containing the tri-alkali metal salt of the 2- and 3-phosphate ester, the tetra alkali metal salt of the 2,3-diphosphate ester and the mono-alkali metal salt of the 2,3-cyclic ester of isoascorbic acid is obtained by treating the mixture of esters with one of the aforementioned alkali metal hydroxides, preferably 50% sodium hydroxide, until the pH is greater than 9.

The 5,6-O-loweralkylidene derivatives of isoascorbic acid may be prepared by suspending isoascorbic acid in a suitable ketone or aldehyde solvent in the presence of an acid catalyst such as p-toluene-sulfonic acid. Thus, 5,6-O-isopropylidene isoascorbic acid may be prepared using acetone as the solvent and 5,6-O-formylidene isoascorbic acid may be prepared using formaldehyde. Other loweralkylidene derivatives of isoascorbic acid may be prepared using other appropriate loweralkylketones or aldehydes.

Example 1: Stabilization of Ale — The ale is fermented in the usual manner. At the end of the fermentation period there is added 1½ lb of a mixture containing 60% of trisodium isoascorbic acid-3-phosphate, 10% of trisodium isoascorbic acid-2-phosphate, and 30% sodium isoascorbate per 100 barrels of ale. The ale is then pasteurized, filtered, and bottled or canned immediately.

Preserved Heat Processed Mushrooms — Mushrooms are prepared in the usual manner. Prior to heat processing there is added to the menstruum 250 mg of the trisodium salt of isoascorbic acid-3-phosphate per lb of mushrooms (drained weight). Immediately prior to packing there is added an additional 100 mg per lb of mushrooms of a mixture containing 40% trisodium isoascorbic acid-2-phosphate and 60% sodium isoascorbate.

Preservation of Frozen Clams — Clams (2.5 gal shucked) are washed and minced, then rinsed and agitated for 5 minutes in an aqueous solution containing 5% sodium isoascorbate and 30% trisodium isoascorbate-3-phosphate. The clams are then drained and cold water added during packaging to obtain the desired consistency. In accordance with the above procedure, the following salts may be employed in place of trisodium isoascorbic acid-3-phosphate: monomagnesium salt of isoascorbic acid-3-phosphate and monocalcium salt of isoascorbic acid-3-phosphate.

Example 2: The preparation and separation of the phosphate ester of isoascorbic acid is accomplished in four steps which are described below.

Phosphorylation of 5,6-O-Isopropylidene Isoascorbic Acid to Form the Phosphate Esters — 195 grams (0.9030 mols) of 5,6-isopropylidene isoascorbic acid is dissolved in 6,700 ml of dry acetone at 40°C under nitrogen and the solution is subsequently cooled to –5°C. To this solution is then added, via a dropping funnel over a 45 minute period, a freshly pre-pared solution containing 97.5 ml (1.005 mols) of phosphorus oxychloride, and 273 ml (3.46 mols) of pyridine which has been cooled to 10° to 15°C. The batch is then aged for another 45 minutes during which time the temperature is maintained at –5° to 2°C. To the aged batch is then added 341 grams (4.06 mols) of sodium bicarbonate followed by a careful addition of 560 ml of water. The resulting slurry is then aged for one hour, filtered, and the resulting salt cake washed twice with 600 ml of acetone. The combined filtrates contain a mixture primarily of the 2- and 3-enol phosphate esters of 5,6-O-isopropylidene isoascorbic acid and the 2,3-diphosphate and 2,3-cyclic phosphate esters of 5,6-O-isopropyl-idene isoascorbic acid and the 2,3-diphosphate and 2,3-cyclic phosphate esters of 5,6-O-isopropylidene isoascorbic acid in lesser amounts. The combined filtrates are then concentrated on a bath under vacuum to a heavy oily syrup containing a mixture of the 2- and 3-enol phosphate esters of 5,6-O-isopropylidene isoascorbic acid, and the 2,3-diphosphate and 2,3-cyclic phosphate esters of isoascorbic acid.

Cleavage of 5,6-O-Isopropylidene Protecting Group — The mixture containing the phosphate esters of 5,6-isopropylidene isoascorbic acid is dissolved rapidly in 1,200 ml of 0.1 N hydrochloric acid. The batch is then aged at room temperature for one-half hour after which it is diluted with 4,160 ml of methanol (precooled in an acetone wet ice bath) and the pH slowly adjusted to 5.5 with about 825 ml of 50% aqueous sodium hydroxide, while the temperature is maintained at about 10° to 15°C. The mixture is then aged for one hour, filtered, and the precipitate containing sodium chloride is washed with 700 ml of methanol. The filtrate and washings contain the crude phosphate esters of isoascorbic acid, traces of free isoascorbic acid and inorganic phosphate salts.

Isolation of Crude Isoascorbic Acid Phosphates — The pH of the filtrate obtained in the step above is slowly adjusted from 5.5 to 6.5 to 7.0 with 50% sodium hydroxide solution, and the pH is slowly raised to 10.0 using 50% sodium hydroxide. The resulting slurry is then aged for a period of 12 to 20 hours at room temperature, after which it is filtered (filtrate contains traces of isoascorbic acid in the form of its sodium salt) and the solid obtained thereby washed twice with 500 ml of methanol followed by 500 ml of ether and sucked dry on the funnel under nitrogen, the precipitate obtained contains a mixture of crude 2- and 3-enol phosphate ester of isoascorbic acid trisodium salt, 2,3-diphosphate isoascorbic acid tetrasodium salt and 2,3-cyclic phosphate isoascorbic acid monosodium salt.

Separation of Phosphate Esters of Isoascorbic Acid — The mixture obtained in the above step is placed on a 1¼ x 20 inch silica gel column prepared in the manner described by B. Love and M. Goodman, (*Chem. and Ind.,* Dec 2, 1967, "Dry Column Chromatography") and eluted with a solvent comprising isopropanol:water:acetic acid:ammonium hydroxide (65:30:30:30). The elute is collected in 20 ml fractions whose composition is evaluated via thin layer chromatography on silica gel using a similar solvent system. The fractions containing each isomer are combined and evaporated to small volume and diluted with 50% sodium hydroxide to pH 10. The solutions are then mixed with 5 volumes of methanol and allowed to stand overnight. The resulting precipitates of sodium salts are filtered and washed with a little aqueous methanol to give, after drying, the several isomeric phosphates, respectively.

Bisulfite Oxygen Scavenger

F. Bloch; U.S. Patent 3,169,068; February 9, 1965; assigned to the U.S. Secretary of Agriculture in this process speaks of the use of oxygen scavengers to protect foods from the effects of free oxygen. The particular scavenger developed as a preservative is a bisulfite which absorbs oxygen according to the following reaction:

$$HSO_3^- + \tfrac{1}{2}O_2 \longrightarrow HSO_4^-$$

Thus a molecule of the bisulfite takes up an atom of oxygen, forming a bisulfite. This relationship furnishes a convenient basis on which one may determine how much of the scavenger is required in any particular situation. Thus, the minimum amount of scavenger is that which will furnish one mol of HSO_3^- per each half-mol of oxygen in the container to be rendered oxygen-free. Generally, to ensure complete and accelerated oxygen removal, the scavenger is used in an amount to furnish the bisulfite ion in excess, for example, at least 1.5 mols of HSO_3^- per half mol of oxygen.

Ordinarily, the bisulfite ingredient of the scavenger is provided directly by use of such compound. However, the equivalent effect may be produced by providing a sulfite plus an acid, typically a mixture of sodium sulfite or calcium sulfite plus sulfuric, phosphoric, lactic, or other acid. Also, the so-called meta-bisulfites or anhydrous bisulfites, such as $Na_2S_2O_5$, may be used instead of regular bisulfites.

It is to be noted that a bisulfite by itself will not absorb oxygen to any practical extent and it is necessary to supplement it to form a composition which will enable the absorption of oxygen at a useful rate. Thus, the bisulfite is increased in surface area by incorporating with it a carrier having an extended surface area. Typical of the carriers which may be used are charcoal, activated carbon, alumina, silica gel, pumice, or other conventional inert material which provides a large surface area.

In addition, to further increase the effectiveness of the composition there is preferably provided an activator. This activator may comprise one or more of the following types of materials: (a) a heavy metal, as for example, iron, manganese, copper, nickel, vanadium, molybdenum, or the like. The heavy metal may be used in elemental form or in the form of an oxide or a salt. Thus typically one may use the metals themselves or their oxides, sulfates, chlorides, nitrates, or phosphates. Preferred agents because of their effectiveness and low cost are the salts of iron, typically iron chlorides or sulfates. In a special preparation the iron salt is used in a mixture with an iron oxide. Such mixtures are readily prepared by applying a solution of an iron salt such as ferric sulfate, ferrous sulfate, ferric chloride, or ferrous chloride to a carrier having extended surface and then drying the material. During the drying step, the iron salt is partially decomposed to the oxide.

(b) A peroxide, generally hydrogen peroxide solution because it is inexpensive and effective. However, other peroxides may be used as, for example, sodium peroxide, calcium peroxide, benzoyl peroxide, acetyl peroxide, t-butyl hydroperoxide, urea peroxide, or the like.

(c) In the event that the various components of the scavenger are essentially anhydrous, it is desirable to moisten the composition with a small proportion of water. In the event that the components contain water of crystallization or other water content, it is usually not necessary to add additional moisture.

An especially efficient scavenger which provides a high rate of oxygen absorption contains, in addition to sodium bisulfite, the following ingredients:

    (1) Activated carbon in a proportion of 0.5 to 10 parts, preferably 2.5 parts, per part of sodium bisulfite.
    (2) A mixture of ferric chloride and ferric oxide in a proportion of 0.01 to 1 part per part of sodium bisulfite.
    (3) Hydrogen peroxide in a proportion of 0.02 to 0.2 part per part of sodium bisulfite.
    (4) Water (including that present in the other ingredients) in a proportion of 0.3 to 5 parts per part of sodium bisulfite.

These scavengers may be utilized in preserving all kinds of materials which are normally subject to being adversely affected by contact with free oxygen. Typical examples of such materials, given merely by way of illustration, are dried fruits, dried vegetables, dried eggs, dried milk, dried fruit or vegetable juices, nuts, cereals, edible fats and oils, butter, margarine, bacon, ham, smoked or dried fish products, dried meats, bread, crackers, and other bakery products, cheeses, etc.

Generally, it is preferred that the scavenger be out of physical contact with the material to be preserved but in oxygen-absorptive relationship with the material and the atmosphere within the container. This goal is readily achieved by enclosing the scavenger in a receptacle through which gases can diffuse. Typical receptacles for such purposes are bags made of porous paper or cloth or metal receptacles provided with a window of paper, cloth or perforated metal. The material to be preserved is filled into a container, the packet of scavenger is added and the container is then sealed. If the material to be preserved is a dehydrated product which requires further dehydration during storage, a conventional desiccant packet such as an envelope of gas-permeable material containing calcium oxide may be added with the other items prior to sealing. When the packages so prepared are stored, oxygen is removed by chemical combination with the activated bisulfite in the scavenger. The removal of oxygen does not occur instantaneously, the rate of oxygen removal being rapid at first and diminishing as the residual amount of oxygen is decreased. Indeed by using the scavenger, the oxygen is removed at a rate faster than that involved in the deterioration of the material to be preserved.

It is obvious that when the scavengers are used in a sealed container, the pressure in the container will be reduced as oxygen is removed from the atmosphere. In some cases it may be preferred to avoid this pressure reduction. For example, where the package is made up of plastic sheet material or metal foil, it may be preferred to retain the original size and plumpness of the package. In such event one may incorporate with the scavenger composition a carbonate such as sodium carbonate or preferably sodium bicarbonate. As oxygen is taken up by the bisulfite forming the bisulfite ion, this latter stronger anion will liberate carbon dioxide from the added carbonate, replacing the absorbed oxygen by released carbon dioxide and maintaining the pressure in the container approximately constant.

Another aspect of the process concerns the utilization of the scavenger compositions containing a carbonate in order to provide atmospheres of a controlled composition. For example, the amount of bisulfite in the scavenger may be less than that required to remove all the oxygen, resulting in an atmosphere containing a reduced proportion of oxygen plus carbon dioxide released from the scavenger composition. Such atmospheres are often desirable in storage of fresh vegetable materials such as lettuce, cabbage, or other fresh leafy produce.

Example 1: (A) 540 mg of $FeCl_3 \cdot 6H_2O$, 20 grams of activated carbon (12 x 20 mesh), and 40 ml of water were mixed. The mixture was then dried in an oven at 106° to 107°C. The product contained a mixture of ferric chloride and ferric oxide on activated carbon.

(B) One gram of sodium bisulfite, 2.56 g of the material prepared in part (A), and 1.25 ml of 3% hydrogen peroxide solution were mixed, thus forming a scavenger composition.

(C) The scavenger of part (B) was tested in the following manner: the scavenger was placed in a glass bottle connected to a manometer. The total contents of the bottle were the scavenger plus the air in the system which was calculated to have a volume of 300 cc. The system was sealed and held at room temperature (about 24°C). From time to time the degree of vacuum created within the system was measured and from these figures the proportion of oxygen removed was determined. The results are tabulated below:

Time, minutes	Vacuum, mm of Hg	Oxygen Removed, percent
0	- -	none
10	46	29
20	68	43
30	81	51
60	103	64
90	116	73
120	125	78
180	137	86
240	145	91
300	150	94
900	160	100

Example 2: Two and one-half grams of the ferric chloride-ferric oxide-carbon composition of Example 1, part (A), was mixed with 1 gram of sodium bisulfite, 0.1 gram of calcium peroxide, and 1 ml of lactic acid diluted 1 to 1 with water. The resulting scavenger was tested as described in Example 1, part (C). It was found that 50% of the oxygen

was absorbed in 3 hours and 100% of the oxygen absorbed in four days.

Example 3: Five grams of the ferric chloride-ferric oxide-carbon composition of Example 1, part (A), was mixed with 2 grams of sodium bisulfite and 2.5 ml of 3% hydrogen peroxide solution. The resulting scavenger was tested as described in Example 1, part (C). It was found that 50% of the oxygen was absorbed in less than 10 minutes and 100% of the oxygen was absorbed in 90 minutes.

Example 4: Two and five-tenths grams of the ferric chloride-ferric oxide-carbon composition of Example 1, part (A), was mixed with 1 gram of sodium bisulfite and 1.2 ml of 7.5% hydrogen peroxide. The resulting scavenger was tested as described in Example 1, part (C), with the following change: the system was kept in a refrigerator at 36°F. It was found that 50% of the oxygen was absorbed in 90 minutes and 100% of the oxygen was absorbed in 36 hours.

Example 5: Five and six-tenths grams of the ferric chloride-ferric oxide-carbon composition of Example 1, part (A), was mixed with 2.2 grams of sodium bisulfite and 2.75 ml of 3% hydrogen peroxide solution. The resulting scavenger was tested as described in Example 1, part (C), but with these changes: (1) the test system contained 650 ml of air; (2) the test system also contained a conventional desiccant packet containing 20 grams of calcium oxide. The desiccant was added to test the efficacy of the scavenger in the presence of a desiccating agent. It was found that 50% of the oxygen was absorbed in 35 min and 100% of the oxygen was removed in 20 hours.

Example 6: A mixture was prepared containing the following ingredients: 5 grams of activated carbon; 1 gram of sodium metabisulfite; 0.84 gram of sodium bicarbonate; and 0.5 gram of H_2O. The above mixture was sealed in a glass vessel containing 320 ml of air. After standing 24 hours at room temperature, the atmosphere in the vessel was analyzed. The results tabulated in percent by volume are as follows: 13.7%, CO_2; 6.0%, O_2; 80.3%, N_2.

Nitrous Oxide Antioxidant

The use of nitrous oxide for the preservation of foods has been employed for several years. The researchers in this field appear to have taken a wrong approach, that is to say that the searchers have considered the use of nitrous oxide as bactericide. Nitrous oxide was in fact used in Germany during the World War of 1914 to 1918 as bactericide under pressures varying between 30 and 40 atmospheres.

The use of nitrous oxide as a general antioxidant is discussed by *C. Balestra; U.S. Patent 3,335,014; August 8, 1967; assigned to Nitrox SA, Switzerland.*

An explanation of the protective action of nitrous oxide may be taken from the consideration of the apolar character of its molecule, due to which it shows a greater solubility in the apolar components (as greases, alcohols, aldehydes, ketones). It may be that due to the saturation of a foodstuff with this gas there are created molecular attachments of nitrous oxide onto the apolar components contained in the food, which form a kind of barrier against the oxygen, just with respect to the components which are the most easily oxidizable. The barrier may be overcome by the oxygen only under specific conditions, as will be described in the following.

A further important feature of the nitrous oxide is that its protective action against oxidation and chemical reactions of the foodstuffs is independent of the temperature, for which it may be used very well at room temperature (20° to 40°C), and also at a lower temperature, of course. The protective action is further not at all dependent on the pressure. The inhibiting action is exerted also when the nitrous oxide saturated room contains a percentage of oxygen under partial pressure or in volume less than 20% in respect of the nitrous oxide of course, this percentage of the oxygen may vary between large limits, but without exceeding the maximum of 20%, this depending on the nature and quality of the foodstuff to be preserved, and on the greater or lesser capacity of being oxidized. This permissible proportion of oxygen is of great importance because such a tolerance for the oxygen permits a much more extended industrialization of the process due to the simplification and acceleration of all its steps, in particular the substitution of the air in the containers, whereby any indispensability of utilizing an extremely high degree of vacuum is advantageously avoided.

A further property of the nitrous oxide is its protective action which is achieved also in presence of other inert gas or gases. In this case, when oxygen is also present, the partial pressure of it will equal but not exceed 20% in respect of the nitrous oxide. From the above information it appears quite clear that foodstuffs can very well be preserved at room temperature, thus also avoiding the cost of the otherwise required freezing or refrigerating. The fact that the action of the nitrous oxide is not at all dependent on any absolute pressure whatever allows to realize the process with whatever type and capacity of receptacle from the smallest bag line to the greatest and largest silo, with the only requirement that the same be impervious to the passage of the nitrous oxide. Thus the container may be made not only of metal, but also of other laminated materials, plastic and/or derivatives, or paper impregnated with synthetic solutions.

The process can be carried out in an economical manner inasmuch as the proportion of nitrous oxide per liter or cubic decimeter of the product to be preserved is not high at all. As a matter of fact, the nitrous oxide has a solubility in water, at 15°C of 0.75 volumes per volume of grease; consequently, also supposing that the total mass of the products should dissolve nitrous oxide, and that the mass should consist entirely of grease, and further supposing a consumption of gas of 33%, the maximum amounts of gas will be about 4 grams (2 liters at normal pressure) per liter or cubic decimeter of foodstuff to be preserved. Furthermore, excess gas, also when mixed with other gases, as air or nitrogen, or still another may be easily recuperated by liquifying, and recycling, thus ensuring an accordingly high yield of the process.

The formation of the nitrous oxide atmosphere can be carried out either during the introduction of the foodstuff into its container, or at the instant of stocking into silos or types of warehousing cells, which will have to be hermetically closed when the foodstuff has not to be submitted to sterilization, as it is the case with butter or roasted coffee, or for a provisional preservation, as for fresh fruits and vegetables. In this last case, the stuff may be preserved under a slight pressure of nitrous oxide (a few of tenths of one atmosphere only) and conveyed directly to the packing machines working in a nitrous oxide atmosphere.

The introduction of the nitrous oxide can be arranged either by effecting a vacuum for the evacuation of the air, e.g., till a residual pressure of 110 mm of Hg, and by re-establishing normal pressure or a slight over pressure by means of the nitrous oxide itself as well, or by means of repeated washing operations and the compression of the gas deprived of oxygen (e.g., nitrogen, sulfur dioxide), and successive exhausting, successively eliminating the quantity of gas fed by sucking and reintegrating to the atmospheric or the pre-established pressure, under addition of nitrous oxide (which procedure is especially adapted for the preservation of coffee), or still by means of successive washing operations under pressure with nitrous oxide and successive exhausting as well, until such an amount of air has been exhausted that the partial pressure of the oxygen will be brought within the allowed limits, as specified.

Example 1: Four bags of pork packed under vacuum were purchased, each weighing about 150 grams, two of sliced salami and two of sliced salted fresh bacon. One bag of sliced salami and one bag of sliced bacon were maintained in their original package at room temperature, while the other two bags were opened and their contents distributed separately slice by slice into a glass receptacle. The latter was hermetically closed, and saturated with nitrous oxide introduced through a suitable cock. The internal pressure reached was 1.5 absolute atmospheres; then the gas was exhausted until ambient pressure was reached. This operation was repeated three consecutive times and at the end a superpressure of 25 cm water column was retained within the receptacle. All samples were left near one another in the same room, the temperature of which was at the beginning 38°C, and was successively lowered to 25°C. After a three month period all containers were opened, and the results were as follows.

The sliced pork meats of the original bags showed a yellowish color of the fatty parts, while the meat itself had acquired a greyish color; further within the bags drops of a muddy liquor were found. In contrast, the nitrous oxide treated samples did not show any change whatever either in color, or in their consistency and savor. While the first samples on opening their bags showed a heavy odor of rancidity and a background of putrefaction odor, the nitrous oxide treated samples had maintained their good odor of fresh sliced pork meat, and the characteristic one of salami and bacon, so that they could be eaten without any impairment of the health of the consumers each of whom consumed 100 g of these products.

Example 2: From the trade there were purchased two packages of fresh butter; the covering paper was taken away and both pieces of butter were divided into two halves each; two portions of the two different packages were submitted to saturation with nitrous oxide until a pressure of 1.05 absolute atmosphere was reached, after preceding evacuation to a residual pressure of 110 mm of Hg while the other two samples were hermetically enclosed in a glass vessel. All butter halves were put in the same room at a temperature varying between 25° and 28°C.

After a period of five months all butter samples were examined with the following result: the butter treated with nitrous oxide was intact, that is without any change in color, consistency and savor, while the two untreated samples were no longer edible, due to their heavy rancidity.

Antioxidant from Rootlets

Antioxidant substances are extracted from plant seeds and sprouts in the process of *D.L. Baker and W.B. Dockstader; U.S. Patent 2,925,345; February 16, 1960; assigned to Basic Products Corporation.* Suitable starting materials are such plant seeds as commercial grain, for example, wheat, rice, maize or barley, and in some instances rye although the seeds of nearly all seed propagated plants in some measure provide a source of antioxidant substance. The starting seed, for example, barley, may be first matured, cleaned, graded and then steeped in several changes of water. It may then be spread in layers and held ventilated at growing temperature, say form 55° to 80°F for one to six days. At the end of the growing or sprouting period rootlet development is evident and the acrospire will have extended to a substantial degree.

During the initial growth period the plant metabolism appears to be such as to develop a substantial content of antioxidant substance in the several plant parts. In the case of sprouted barley or malt little or none of the antioxidant material appears in the hull but other parts of the growing plant carry a substantial concentration and this is particularly marked in the rootlets. As growth of the plant continues the concentration of antioxidant substances in any given part of the plant begins to drop rapidly after the early sprouting stage. It is, therefore, necessary to terminate the plant growth or life process at an effective stage by means which are not destructive of the antioxidant content.

The drying of the sprouted seed can be best accomplished by passing warmed or heated air rapidly over the seed under conditions where the wet bulb temperature remains within the limit above noted until the readily available water content of the plant begins to drop markedly. Thereafter the temperature of the drying air must be regulated more carefully in order to avoid destructive temperatures. Optimum results seem to occur where temperature is held at or below 120°F at this stage.

At the conclusion of the drying process the sprouted seed or malt may be first treated for removal of rootlets and then for separation of the hull from the berry. Fatty food or feed may be prepared by incorporating either the ground berry or ground rootlets or both, directly with the fatty material to be protected, since adequate antioxidant substance will be present in these plant parts to protect substantial quantities of fatty material.

In instances where the presence of added plant material is objectionable, a concentrate or extract of the antioxidant may be used which may be obtained through solvent extraction. As an instance, the sprouted seed material may be subjected to extraction with a lower alcohol, such as methyl or ethyl alcohol, to remove an extract from which a concentrate may then be recovered by evaporation and further purified by extraction and recrystallization. The material, thus recovered, will be found effective in concentrations as low as one part per 10,000 when added to animal or vegetable fats of a type susceptible to rancidification.

To demonstrate the effectiveness of such a concentrate an accelerated oxidation test was conducted upon purified corn oil free of added protective material. A control portion of this oil was maintained at constant temperature at 100°C by immersion of the vessel holding the same in a controlled temperature bath. Air was then bubbled through the heated oil. After five hours, marked deterioration, due to rancidity, became apparent.

As compared with the control portion, another portion of the same oil was treated under the same conditions, except that there was added to this portion 1 part in 10,000 of an antioxidant concentrate prepared in accordance with this process. The oil thus treated withstood the accelerated oxidation treatment for more than eight hours before deterioration due to rancidity reached the level which occurred in the untreated oil for five hours.

The protective capacity of the antioxidant concentrate equals or exceeds the best results obtainable from other commercially available antioxidants.

2,3-Enediols of 3-Ketoglycosides

The prevention of oxidative deterioration is afforded by an inhibitor selected from the group consisting of 2,3-enediols of 3-ketomaltose, 3-ketosucrose, 3-ketolactose, 3-keto-maltobionic acid and 3-ketolactobionic acid and their fatty acid esters. These antioxidants are described by *R.W. Eltz; U.S. Patent 3,372,036; March 5, 1968; assigned to Sun Oil Company.*

The enediols of 3-ketoglycosides are readily obtainable by the oxidation and tautomerization of certain common sugars or sugar acids, as for example, the disaccharides such as maltose, sucrose or lactose, or the bionic acids such as maltobionic acid or lactobionic acid. While chemical methods for oxidizing sugars to their corresponding 3-ketoglycosides are known, these methods are generally characterized by their low yields. However, means by which the oxidation may be achieved microbially have recently been reported whereby nearly stoichiometric conversions are obtained. Thus, for example M.J. Bernaerts et al, have described in *J. Gen. Microbiol.*, 22, 129–136 (1960); *J. Micro. and Serology*, 27, 247–256 (1961); and *Nature*, 197, 406–407 (1963), that 3-ketoglycosides are produced in high yield by the action of *Agrobacterium* sp. on disaccharides and bionic acids.

The 3-ketoglycosides which may be prepared by this method include such compounds as 4-(3-keto-α-D-glucosido)-D-glucose (hereinafter referred to as 3-ketomaltose), derived from maltose; α-3-keto-D-glucopyranosyl-β-D-fructofuranoside (hereinafter referred to as 3-ketosucrose), derived from sucrose; 4-(3-keto-β-D-galactosido)-D-glucose hereinafter referred to as 3-ketolactose), derived from lactose; 4-(3-keto-α-D-glucosido)-D-gluconic acid (hereinafter referred to as 3-ketomaltobionic acid), derived from maltobionic acid; 4-(3-keto-β-D-galactosido)-D-gluconic acid (hereinafter referred to as 3-ketolactobionic acid), derived from lactobionic acid.

These 3-keto compounds are then conveniently tautomerized to their corresponding 2,3-enediol form under alkaline conditions at room temperature, or by boiling with strong acids such as sulfuric acid. Thus, for example, 3-ketolactose converts to its enediol form, when treated with dilute sodium hydroxide, as follows.

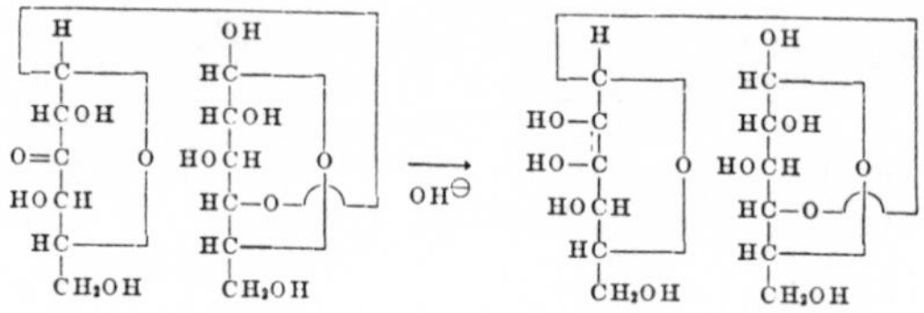

Similarly each of the other 3-ketoglycosides may thus be treated to form a compound having the partial structure:

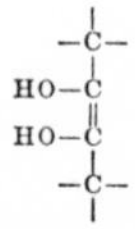

which structure possesses very strong reducing powers necessary to effectively stabilize foodstuffs. These compounds also include compounds having a 3,4-enediol arrangement since the structure of these highly complex molecules has not as yet been fully characterized.

These compounds, are suitable for the stabilization and/or prevention of discoloration of such aqueous or dry foodstuffs as fruits, fruit juices, carbonated beverages, fish, meat, beer, milk and other dairy products, known by the food industry to require the presence of antioxidants. Use of the enediols of 3-ketoglycosides is especially effective in conjunction with the use of preparations containing vitamin C, since these enediols, when present in amounts in excess of the vitamin C, serve to spare the oxidation of this vitamin, which itself is a known reducing agent, and thus permit this material to retain its vitamin activity.

The 2,3-enediols of 3-ketoglycosides may be used to stabilize fatty, nonaqueous material such as cooking oils, shortening, mayonnaise and butter by converting the enediols to their corresponding fat-soluble esters. This may be achieved by a conventional esterification of the enediol compounds with higher fatty acids, their salts, or acyl halides, having from 12 to 20 carbon atoms, such as lauric acid, palmityl chloride, myristyl chloride, stearic acid or oleic acid, to form fat-soluble esters which are particularly useful in preventing rancidity and discoloration in fatty or oily foods, especially during long periods of storage.

These enediol antioxidants and their esters may be added to foodstuffs in amounts varying from about 0.1 to 1.0% by weight of the food and may be used alone or in combination with other food additives and/or suitable inert carriers.

Example 1: To the vegetable oil phase of 1 lb of a commercial mayonnaise is introduced 2 g of the stearate ester of the 2,3-enediol of 3-ketolactose. The stabilized mixture is stored for 30 days at room temperature. At the end of that period, no rancid odor or taste is discernible. An equal amount of untreated mayonnaise stored for the same period of time under the same conditions has a markedly rancid flavor and undesirable odor.

Example 2: Ten pounds of steam-peeled peaches are sliced very thin and 3 g of the 2,3-enediol of 3-ketosucrose are added and mixed in thoroughly with the slices. The mixture is then frozen in sugar syrup; after ten days the peaches are examined and found to have retained their natural flavor and color when thawed. A corresponding amount of untreated sliced peaches, by comparison, is found to have turned brown and is unsuitable for consumption after three days.

Example 3: Fifty pounds of chopped meat preparation used in making frankfurters are treated with ½ ounce of the 2,3-enediol of 3-ketolactose. This meat, together with a separate, untreated ten pound portion are used to make up separate groups of cooked, cured frankfurters. These frankfurters are placed in an open, refrigerated display case lighted by artificial light. At the end of five days, the two groups are examined and it is found that the treated meat has retained a reddish, commercially acceptable color, while the untreated meat has a mottled, greyish-green appearance.

PROPYLENE OXIDE

The technology on the use of ethylene oxide and propylene oxide as food preservatives is given by *W.O. Tundermann and S.D. Friedman; U.S. Patent 3,346,398; October 10, 1967; assigned to Colgate-Palmolive Company.*

One general approach in the field of food preservation, has been to package the particular food commodity in an evacuated container. By the removal of air and moisture, the processes of deterioration are greatly retarded. Other methods involve heating the material to sterilizing temperatures prior to packing or storage. Another approach to the problem, sometimes used in conjunction with the vacuum packing or heat sterilizing methods, has been to flood the container with a sterilizing agent prior to closing of the package, so that a sterilizing atmosphere is maintained until the package is opened.

In processes of the latter type, considerable attention has been directed to the use of epoxides, such as ethylene oxide and propylene oxide, as sterilizing agents. Although the epoxides have been found to exhibit excellent bactericidal and fungicidal activity, certain serious disadvantages have sharply limited their usefulness. Among the more serious drawbacks involved in the use of epoxides as sterilizing agents is the explosive character of epoxide-air mixtures.

Another disadvantage in the use of the epoxides has been their tendency to quickly volatilize and escape from the storage container. When sealed containers into which epoxide sterilizing agents have been introduced are opened, the epoxide vapors escape almost instantaneously. The epoxide vapors also escape very rapidly from containers which are not hermetically sealed and they may even escape rapidly from many types of sealed containers, since a number of conventional packaging materials are readily permeated by the epoxide vapors.

In order to overcome this problem, attempts have been made to utilize the epoxides in the form of solutions in solvents, such as carbon dioxide, isopropyl formate and ethylene dichloride which tend to lower the vapor pressure of the epoxides. However, such epoxide solutions have not been entirely successful due to cost, adulteration of the flavor, color or texture of foods contacted by the solutions, and the production of undesirable residues. In addition, the lowering of the vapor pressure in solutions of this type is only sufficient to enable the epoxide to be handled as a liquid and introduced into the package. Thereafter, the epoxide vaporizes rapidly and is subject to the same drawbacks noted in the previous paragraph.

As for the difference ethylene oxide and propylene oxide, it has been found that the latter is generally more satisfactory as a sterilizing agent, for a number of reasons. For example, upon reaction with water or moisture in the presence of food, propylene oxide hydrolyzes to form propylene glycol which is substantially nontoxic when consumed in amounts of the magnitude that would be found on or in food packaged in an atmosphere of propylene oxide. On the other hand, ethylene oxide hydrolyzes to produce ethylene glycol which is considerably more toxic. Also, while both propylene oxide and ethylene oxide are explosive when mixed with air, propylene oxide-air mixtures are explosive over a much narrower or more restricted composition range than are mixtures of air and ethylene oxide. Another important advantage of propylene oxide is that it is a liquid under normal conditions of temperature and pressure and therefore is easier to handle than ethylene oxide which is normally a gas.

It is now possible to use either ethylene oxide and/or propylene oxide without the disadvantages discussed earlier. In this process compositions are provided which are capable of slowly releasing or diffusing sterilizing ethylene oxide and/or propylene oxide vapors over extended periods of time so that the sterilizing atmosphere may be maintained and continuously regenerated in circumstances where the food or other materials is temporarily exposed to the atmosphere, where the container is not air tight or where the container is constructed at least in part from materials pervious to air, moisture or propylene oxide vapors. Molecular sieve material with absorbed ethylene oxide and/or propylene oxide is used for this purpose.

It has been found that molecular sieves absorb moisture in preference to alkylene oxides so that as moisture is absorbed from the food or from the atmosphere surrounding the food or other organic matter in the container, alkylene oxide is gradually released from the molecular sieve into the atmosphere to maintain the contents in a sterilizing condition. The moisture, when absorbed in the sieve, results in desorption of the ethylene oxide or propylene oxide. The moisture may be taken from the contents of the container, such as a foodstuff, or from the atmosphere present when the container is closed or admitted when the container is temporarily opened. Thus, not only does the molecular sieve function as a reservoir for the epoxide sterilizing agent, but it also simultaneously provides a dehydrating function which further contributes to the inhibition of mold growth and other processes of spoilage.

The slow diffusion or exudation of ethylene oxide or propylene oxide vapors from the molecular sieves has the added important advantage that diffusion takes place more rapidly in high moisture containing atmospheres where the danger of spoilage is most acute.

The use of molecular sieves containing absorbed propylene oxide to sterilize foodstuffs is illustrated in Figure 11.1a and Figure 11.1b. As will be seen by reference to these figures, container 10 is filled with a material 11 which is to be preserved against spoilage. One or more pellets of molecular sieve 12 containing absorbed ethylene oxide or propylene oxide are introduced into the container to furnish the desired sterilizing atmosphere.

FIGURE 11.1: METHOD OF PRESERVING PERISHABLE MATERIAL

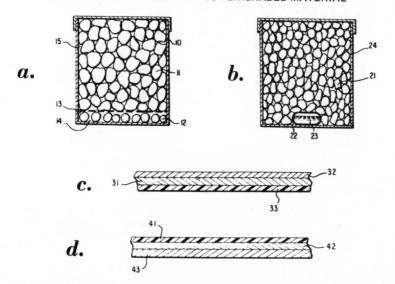

Source: W.O. Tunderman and S.D. Friedman; U.S. Patent 3,346,398; October 10, 1967

The container 10 may be of any size and construction. For example, it may be the familiar small volume metal can or jar employed in sales to the consumer and having a volume of from a few ounces up to one or two quarts. On the other hand, the container may be a drum of considerable capacity used in commercial operations and having a volume of many gallons. Similarly, the container may be constructed of metal, plastic, paper and combinations of these and other conventional packaging materials.

If desired, the molecular sieve pellets 12 may be segregated from the contents 11 of the container by a perforated diaphragm 13 which freely permits transfer of moisture and the ethylene oxide or propylene oxide vapors between the molecular sieve compartment 14 and the other compartment 15 in which material subject to spoilage is located.

Where the contents of the container to be preserved are materials for internal human or animal consumption, propylene oxide is used as the sterilizing agent since its residue is substantially nontoxic.

The amount of molecular sieve and alkylene oxide composition may be varied dependant on the size of the container and its intended or usual method of use. For example, if the container is the type which is sealed air tight and then is ordinarily opened, emptied, and disposed of, then only enough of the sterilizing composition need be included to maintain sterility within the container prior to opening. However, if the container is of a type which is ordinarily opened, partially emptied and reclosed, more of the molecular sieve-propylene oxide composition may be desirable, so that the sterilizing atmosphere may be regenerated a number of times as the container is opened and reclosed.

Molecular sieves of the type used in this process are crystalline metal aluminosilicate materials of the following general formula: $M_{2/n}O \cdot SiO_2 \cdot YAl_2O_3 \cdot ZH_2O$ in the salt form, where n is the valence of the metal cation M, M ordinarily is Na or K but may be other cations substituted by exchange, Y is the number of mols of alumina and Z is the number of mols of water of hydration.

Upon removal of the water of hydration by heating, the crystalline metal aluminosilicates become highly porous and are characterized by a series of surface cavities and internal pores which form an interconnecting network of passageways within the crystal. Due to the crystalline nature of such materials, the diameters of the surface cavities and of the internal pores is substantially constant and is of molecular magnitude. For this reason, the crystalline metal aluminosilicates have found wide use in the separation of materials according to molecular size or configuration, hence the name molecular sieves.

Molecular sieves or crystalline aluminosilicates are also sometimes referred to as crystalline zeolites and are of both natural and synthetic origin. Synthetic materials include, for example, synthetic faujasite. Natural crystalline aluminosilicates exhibiting molecular sieve activity include for example, analcite, paulingite, ptilolite, clinoptilolite, ferrierite, chabazite, gmelinite, levynite, erionite and mordenite.

Since not all of the natural crystalline aluminosilicates are available in abundance, considerable attention has been directed to the production of synthetic equivalents. A number of these are available on a commercial scale and are designated as X molecular sieve, Y molecular sieve, and A molecular sieve. Other molecular sieves which have been synthesized include B, F, G, H, K-G, J, L, M, K-M, Q, R, S, T, U and Z zeolites.

The numerical prefix in the designation or identification of various molecular sieves indicates the approximate pore size in angstroms. For example, 13X molecular sieve is a sieve of the X type which is believed to have pores having diameters of about 13 angstroms. On the other hand, a 5A molecular sieve is a synthetic crystalline aluminosilicate of the A type having pores the diameters of which are approximately 5 angstroms.

Sterilizing compositions are prepared by heating a molecular sieve material, preferably in the form of compacted pellets, or spheres to remove moisture and then causing the dried sieve to be impregnated with ethylene oxide or propylene oxide. The impregnation may be accomplished by immersing crystals or pelleted crystals of the molecular sieve in liquid ethylene oxide or propylene oxide or by placing the pellets or crystals in a closed vessel containing the oxide vapors.

As an example of the later method, 123 tablets of 13X molecular sieve having a thickness of ¼" and a diameter of ¾" were baked overnight in an oven at a temperature of 500°F to drive off absorbed moisture. The dried tablets were then quickly transferred to a porous bag and the bag was suspended in a closed glass container above the level of liquid propylene oxide placed in the bottom of the container. The molecular sieve was thus exposed for a period of 5 days during which time the temperature was maintained at about 45°F. Upon weighing, it was found that the molecular sieve had adsorbed about 20% by weight of propylene oxide into its pores.

Example 1: Sixty-nine gram portions of pumpkin meat were placed in each of three polyethylene bags. The first bag, used as a control, was folded closed and the fold was secured with adhesive tape.

The following amounts of molecular sieve saturated with from about 20 to 40% by weight of propylene oxide were then inserted in the other two polyethylene bags, and they were closed in the same manner as Bag 1: 1.6 grams in Bag 2 and 3.2 grams in Bag 3.

After four days, the pumpkin meat in Bag 1 was observed to be covered with heavy black fungus. The pumpkin meat in Bag 2 showed some softening and the presence of a considerable amount of liquid, but the fungus growth was not as extensive as in Bag 1. The pumpkin meat in Bag 3 was in good condition.

Example 2: A slice of bread and two drops of water were placed in a polyethylene bag and the bag was folded closed. A second slice of bread, two drops of water and a tube of polyethylene containing 10 cc of propylene oxide were then placed in a second polyethylene bag which was likewise folded closed.

After twelve days, it was observed that fungus growth had begun on the slice placed in the first bag, but that no fungus growth had commenced on the slice of bread placed in the bag with the polyethylene tube containing propylene oxide. Upon observation after several months, no fungus growth was found to have commenced on the slice of bread in the bag containing the tube of propylene oxide. Another approach is illustrated in Figure 11.1b.

Food 21 and a tube 22 containing ethylene oxide and/or propylene oxide 23 are placed into an outer container 24 and the container is then closed. The tube is polyethylene or other plastic which is pervious to the alkylene oxide vapors and which enables the vapors to diffuse through the wall of the tube into the atmosphere surrounding the food 21.

Tube 22 may be dropped into the container with the food or may be adhesively attached to the bottom of container 24 or to its side walls or may be placed in a pocket, clamp or recess in a wall of the container or in the closure.

As one alternative to this process, the plastic used to form the tube may be of a water-soluble or disintegrable nature and may be edible as well, so that if the food placed in the container contains substantial amounts of water, the tube will disintegrate after release of the propylene oxide and will leave no undesirable residue.

As illustrated in Figure 11.1c, a specific example of material may be a conventional paper-foil laminate of the type widely used for packaging foods. The laminate comprises a layer of paper 31 adhesively bonded to a foil of aluminum 32 or other impermeable material. The other surface of the paper is coated with a film 33 of a fixative containing propylene oxide. The film 33 may, for example, comprise a concentrated solution of ethylene oxide and/or propylene oxide in a polyethylene glycol polymer having a molecular weight of 200.

The film-forming fixative solution containing ethylene oxide and/or propylene oxide may also be applied as an interlayer between an impermeable foil and a vapor permeable layer.

The laminated sheet material shown in Figure 11.1d is a vapor permeable layer **41**, such as polyethylene, cellulose acetate or the like, a vapor impermeable layer **43**, such as aluminum foil or heavy waxed paper, and an interlayer **42** which is a film of fixative containing ethylene or propylene oxide. As the sterilizing agent is gradually released from the fixative film, such as cellulose nitrate, it diffuses through the permeable layer **41**.

BENZOHYDROXAMIC ACID

According to *J.J. Beereboom; U.S. Patent 3,446,630; May 27, 1969; assigned to Chas Pfizer & Co., Inc.* small amounts of benzohydroxamic acid preserves and protects various foodstuffs from spoilage by microorganisms. Foodstuffs particularly suited for the process are those which it is not economically feasible to freeze or to subject to cold storage and those foods whose flavor is lost on freezing. Examples of foods in the former category are dog food, cake mixes, apple cider and bakery products such as bread, cakes, cookies and pies. Foods whose flavor is lost or deteriorates on freezing and cold storage are fruit and vegetable juices, cheeses, fruit and vegetables and, in particular, sliced peaches and bananas. By the term processed foods is meant those foods which are mixed or treated in some manner prior to being sold. Such foods are, for example, cakes, pies, breads, dog foods, canned peaches and fruit juices.

Fresh fruits including berries, and vegetables are dipped in an aqueous solution of benzohydroxamic acid or one of its nontoxic salts. A solution of 1% benzohydroxamic acid or one of its salts has been found to be convenient for this purpose. The previously weighed foodstuff is reweighed to determine the wet pick-up of solution and is then dried, either at room temperature or at a higher temperature, with the aid of a compressed air stream. Concentrations of from 0.1 to 1.0 weight percent of benzohydroxamic acid and its salts are effectively applied in this manner. The foods so treated can then be stored at room temperature, frozen or put in cold storage areas.

Another convenient method of applying these preservatives is by spraying an aqueous solution of benzohydroxamic acid or one of its salts on the food. For convenience and economy, a 1% solution can be effectively applied to large batches of fruit or vegetables in open containers with a hand sprayer. Compressed air or an electric sprayer would also be useful. Where the application must be performed in the shortest possible time, a 10% aqueous solution of the sodium or potassium salt of the acid is most conveniently used. In this manner, a 0.5% treatment on the weight of foodstuff can be applied in a single spraying.

Benzohydroxamic acid and its nontoxic salts are most conveniently added to processed foods by simply mixing in a weighed amount of the acid or by dissolving the compound or its salt in water or ethanol and adding the solution to the food. For example, in mixing large commercial quantities of cake mix, the benzohydroxamic acid may be added to the shortening in any mixer of adequate size to ensure proper blending before the cake mix is prepared. In this manner, the benzohydroxamic acid is used to preserve the shortening in storage, the cake mix on the grocer's shelves and the final finished cake in the home or bakery.

For the preservation of fresh bananas an amount of the sodium salt of benzohydroxamic acid in sufficient water to equal 0.01% of the weight of the bananas is sprayed on the bananas either as a hanging stalk or in bunches. The means of spraying the solution can be any type of rust resistant commercial sprayer. An alternate procedure involves dipping 50 lb stalks or bananas in large tanks of an aqueous solution of benzohydroxamic acid in water. A more concentrated solution would, of course, be necessary since only a small volume of the solution is deposited on the bananas.

Another method which has been found effective in applying benzohydroxamic acid to edibles, involves applying solutions of benzohydroxamic acid or one of its salts, such as the ammonium salt, to wrapping papers, paper cartons or carton liners used for storing food. For this type of application, the packing papers or paper cartons are weighed and dipped in a 10% aqueous solution of a salt of benzohydroxamic acid. The wet pick-up is determined and the paper is dried. The papers are then inserted into the cartons so as to give an effective concentration of benzohydroxamic acid based on the weight of the foodstuff. Similarly, the food may be placed in paper boxes treated with either benzohydroxamic acid or one of its salts.

Example 1: To one liter of fresh unpasteurized apple cider was added sufficient dry benzohydroxamic acid to give a total of 0.1% by weight. The cider did not develop microbial growth after two weeks storage at room temperature of about 25°C. Untreated cider from the same pressing spoiled within three days when stored under the same conditions.

The process is repeated with the sodium and potassium salts of benzohydroxamic acid at levels of 0.1% by weight. After two weeks storage at room temperature (25°C), no microbial growth develops.

Example 2: Oranges were inoculated by scratching the surfaces of the fruit and dipping in a suspension of *Penicillium digitatum* spores. The fruit was stored at room temperature (about 25°C) for 24 hours and then dipped in a 0.5% aqueous solution of the potassium salt of benzohydroxamic acid for one minute. The oranges were stored in cardboard cartons at 20°C for eleven days. Examination showed 76% less decay than inoculated control samples which were not treated with the benzohydroxamic acid and which were stored under the same conditions.

Example 3: To samples of melted American cheese were added sufficient benzohydroxamic acid to give 0.05% based on the weight of the cheese. The melted cheese was inoculated with *Penicillium digitatum* spores and did not develop mold on storage for 20 days at room temperature. Inoculated, untreated control samples developed mold after 3 days storage at room temperature.

On one pound wedges of Muenster cheese was sprayed a 5% aqueous solution of ammonium salt of benzohydroxamic acid. The samples were stored at room temperature for 20 days and did not develop mold growth. Untreated slices developed mold growth within 5 days when stored at room temperature.

ANTIOXIDANTS AND ANTIMICROBIALS IN LIQUID FREEZANTS

More modern means of food freezing comprise directly contacting the food with liquid chillants and ebullient freezants. These liquids permit the rapid chilling or freezing of foods of all types due to excellent heat transfer from the food to the freezing medium. Such processes are considerably faster than the tunnel freezing process.

It is often desirable to add to the foods which are to be frozen various food adjuvants which will contribute to improved flavor, color and storage stability of the frozen foods. Art processes teach that such adjuvants can be added to the foods before freezing by the tunnel process or after freezing.

It has been found by *K.D. Dastur; U.S. Patent 3,792,180; February 12, 1974; assigned to E.I. du Pont de Nemours and Company* that food adjuvants which are normally essentially insoluble in ebullient dichlorodifluoromethane freezant, such as flavors, odorants, preservatives, surfactants and fats can be dispersed in dichlorodifluoromethane freezant and transferred to food by direct contact. By regulation of the mode of contact between the food and the freezant, the amount of adjuvant distributed to the food surfaces can be readily controlled.

The specific dichlorodifluoromethane direct contact freezing device in which the process is carried out is not critical, but it is preferably carried out either by dropping the food particles into a pan of the liquid freezant, passing the food particles under cascading sprays of the liquid freezant or both.

The manner of deposition of the dispersed materials on food surfaces is different from what one would expect from a true solution. It appears that particles once deposited on food tend largely not to be entirely removed on further contact with fresh freezant as would be the case were the dispersed material truly soluble in the freezant. Since deposition occurs for the most part on evaporation to dryness or nearly to dryness of the freezant on the food surfaces there is given a means for controlling the amount of deposition on the surface of the food.

The mechanism of deposition is not clearly understood and is thought to be related to a sorption process, and it is known that when one freezes food by contact with a dispersant-freezant mixture by regulating the flow of freezant so that the freezant evaporates from the food surfaces at about the rate at which it is added, maximum deposition of dispersed materials results. On the other hand, should one immediately immerse or float the food in a large amount of freezant and in this manner extract from the food all the heat that is to be extracted, then a minimum amount of deposition will occur. It follows therefore that by adjustment of the freezing conditions, a wide range of adjuvant deposition can be obtained. The amount of deposition will vary with the concentration of the dispersed material. It is, however, generally observed that the amount of deposition is not linearly proportional to the concentration under all conditions due to the role played by the manner of freezant-food direct contact.

As applied to freezing devices, when both a pan and sprays are used for direct food-freezant contact, it follows that the ratio of residence time in the liquid freezant-containing pan to the residence time under the freezant sprays controls to a great extent the amount of adjuvant deposition. Maximum deposition is obtained when the residence time in the pan is very short or even zero, i.e., there is no pan at all, and most of the heat is extracted from the food under slow-flowing sprays. The amount of deposition could thus be readily adjusted simply by varying the residence time of the food as it contacts the adjuvant containing freezant in either the pan or spray or both.

The preparation of the adjuvant dispersions in dichlorodifluoromethane will vary, with the nature of the adjuvant and the desired effect. Fats, fat-like and other easily meltable materials described below such as the mono- and di-glyceride surfactants are generally melted and sprayed into agitated dichlorodifluoromethane. The resultant dispersion is sometimes not permanent and the fatty materials may accumulate in the upper areas of the quiescent freezant. For this reason it is often preferred to maintain the freezant in the turbulent state. Often, in the usual freezing process, ebullience is adequate to maintain the materials in dispersion.

Normally fatty materials are added at a concentration of from about 1 to 5% by weight. Oleomargarine is preferably added at a concentration of about 2.5% by weight. Such a concentration transfers to vegetables about the amount of fat usually found, for example, on frozen buttered peas.

Coloring and flavoring materials, in part because of the very small amount of their active ingredient required, are conveniently mixed in an edible solvent such as ethanol and the mixture is introduced into the freezant in a turbulent zone. Other useful materials such as antioxidants and antimicrobial agents may be placed in dispersion in the same manner. Such antioxidants are BHA (butyl hydroxyanisole), BHT (butyl hydroxytoluene), propyl gallate, ethoxyquin and NDGA (nordihydroquaiaretic acid). Useful antimicrobial agents are for example propyl-p-hydroxybenzoate, benzoic acid, sorbic acid, sodium and calcium propionate, etc.

Powdered materials can often be dispersed in dichlorodifluoromethane by adding the finely divided powder in a thin stream to agitated dichlorodifluoromethane.

Some of the adjuvants used in the process, e.g., mono- and di-glycerides, are soluble in dichlorodifluoromethane at room temperature but are essentially insoluble at the normal boiling point. In the case of adjuvants having practical solubility in room temperature dichlorodifluoromethane, such as mono- and di-glycerides, it is preferred to prepare a solution within a pressure cylinder and to add measured amounts of the liquid phase to the process dichlorodifluoromethane freezant.

The adjuvants, being essentially insoluble in the cold ebullient dichlorodifluoromethane immediately come out of solution and disperse themselves surprisingly uniformly in the cold ebullient dichlorodifluoromethane.

Although the process, when using dispersed fats and oils alone as adjuvants, is operable to deposit these materials on some foods, it is preferred to use them in the presence of edible surfactants. The evenness of deposition on foods of the congealed fats and oils as tiny droplets is thereby improved.

Even distribution of fats and oils is especially important in the case of breaded foods such as chicken parts, seafood and vegetables such as onion rings. It has been observed that the fat normally present in breaded chicken parts is disturbed when frozen according to some direct contact processes and as a result small areas of the chicken parts are sometimes left, at the end of the process, with little or no fat. As is known, the fat deficient areas tend to become more quickly dehydrated on cold storage with the result that they appear white on cooking. By adding a fat and a surfactant to the dichlorodifluoromethane freezant this undesirable result is largely overcome. The improvement is most striking when small amounts of carboxymethylcellulose or of a polysaccharide are incorporated into the breading.

Example 1: This example demonstrates the dispersion of an antioxidant in ebullient dichlorodifluoromethane and its transfer to a sample of meat during direct contact freezing. The freezant mixture consisting of 89% dichlorodifluoromethane, 10% ethanol, and 1% butylated hydroxyanisole (Tenox) by weight was prepared by dissolving the anisole in the ethanol and pouring the resultant solution into the ebullient dichlorodifluoromethane. A pork chop was frozen by floating it in the ebullient freezant mixture for 8 minutes.

On removal from the freezant and after the dichlorodifluoromethane had evaporated from the surface, the pork chop was washed with 200 ml ethanol. The ethanol solution gave a strong positive test for the antioxidant on addition of Ehrlich's Reagent, according to Test No. 26.107 (c) of the *Official Methods of Analysis of the Association of Official Agricultural Chemists,* p. 444 (1965).

Example 2: This example demonstrates the transfer to food of propyl p-hydroxybenzoate dispersed in freezant dichlorodifluoromethane. Propyl p-hydroxybenzoate is a well-known antimicrobial agent often used with foods.

39 grams of propyl p-hydroxybenzoate were dissolved in 150 ml ethanol and the resultant solution was poured into 3 liters of agitated ebullient dichlorodifluoromethane. 190 grams of fresh green beans were frozen by floating them in the dispersion for 35 seconds. Thereafter they were suspended in the vapors over the freezant for 60 seconds, experience having shown that this procedure reduces the average temperature of the beans to 0°F (-18°C), the usual storage temperature.

The frozen beans were transferred to 100 ml of ethanol and were allowed to stand with occasional stirring for five minutes. Thereafter the extract was hydrolyzed by refluxing for two hours with 5 ml of 0.5 N potassium hydroxide in ethanol. After concentration to less than 25 ml total volume, the solution was diluted with deionized water, acidified with N sulfuric acid and saturated with ammonium sulfate. The ether solution resulting from three-fold ether extraction was tested for p-hydroxybenzoic acid using Millon's Reagent.

The test solution turned to a deep purplish red color in less than 5 minutes whereas a test solution from beans frozen in pure dichlorodifluoromethane gave no red color at all, indicating the presence of p-hydroxybenzoic acid in the former case and its absence in the latter case.

PYROCARBONIC ACID DIESTER

In the process of *D. Eolkin and R.J. Bouthilet; U.S. Patent 3,341,335; September 12, 1967; assigned to Norda Essential Oil and Chemical Co.* dry particulate foods are sterilized by heating a diester of pyrocarbonic acid. This material may be represented by the general formula:

$$R_1 - O - \underset{\underset{O}{\|}}{C} - O - \underset{\underset{O}{\|}}{C} - O - R_2$$

where R_1 and R_2 may be alkyl radicals such as methyl, ethyl, propyl, butyl, isopropyl, amyl, or isoamyl radicals.

Pyrocarbonic acid esters have been used in the preservation of foodstuffs by combining a suitable amount of the ester with the food to be preserved. In all cases the ester was added to the food in a fluid or slurry medium and usually in the presence of a considerable amount of water. In order to obtain a thorough dispersion of the preservative in the food in the relatively small quantities used, a liquid or slurry medium was necessary. The above type esters are used because they react in the presence of water and are converted to end products including CO_2 and alcohols. A process has been developed to accomplish the preservation without the use of large amounts of water.

Consequently, it permits the preservation of dry particulate materials such as foods and including items such as nuts, spices, cereal grains, cereal flours, prepared food mixes, starches, dry vegetables, dry juices, cake mixes, dry egg and protein products.

It has been found that if the selected ester is vaporized and contacted with the food in gaseous form in combination with a suitable carrier, the desired sterility will be obtained without introducing the problems present where liquid media such as water are employed. It has been found that the decomposition of the ester is achieved as in previous procedures by small amounts of moisture that are present in commercially dry materials. Materials that are normally referred to as commercially dry usually contain several percent moisture, i.e., on the order of about 5% but variable depending on the material and other conditions. It has been found that this moisture is sufficient to react with the ester to convert it to the nontoxic end products.

Figure 11.2 illustrates schematically the apparatus for the process. A dry particulate substance 10 to be sterilized is placed in a treatment chamber 11. Treatment chamber 11 is adapted for fluidizing materials 10 and contains a porous plate 12 upon which materials 10 rest. The treatment chamber includes a removable cover 13 for filling and emptying the treatment chamber.

Associated with the treatment chamber are inlet conduit 14 and outlet conduit 15. The inlet conduit links the area below the plate into fluid communication with vaporizing chamber 16. A selected hydrocarbon ester of pyrocarbonic acid 17 is placed in the vaporizing chamber. A heater 18 is associated with the vaporizing chamber for raising the temperature of ester 17 as desired. A conduit 19 and pump 20 communicate interiorly with the vaporizing chamber adjacent the bottom thereof. The carrier gas enters the chamber through the conduit under impetus from the pump.

As an example of operation, black pepper may be used as material 10. Ester 17 is suitably the diethyl ester of pyrocarbonic acid. It is a liquid at room temperature and pressure. A suitable carrier gas is air (although another gas could equally well have been used such as nitrogen). The air is forced into the chamber beneath the surface of ester 17 by pump 20 and is allowed to bubble upwardly through ester 17. Suitably, the pump is actuated so as to create a generally very slow rate of bubbling. At the same time the heater is actuated

FIGURE 11.2: METHOD OF STERILIZING DRY PARTICULATE MATERIAL

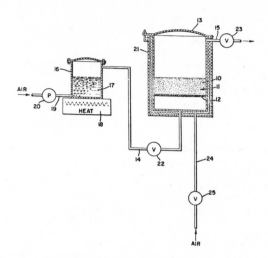

Source: D. Eolkin and R.J. Bouthilet; U.S. Patent 3,341,335; September 12, 1967

sufficiently to raise the temperature of ester **17** to about 30° to 50°C. While ester **17** at
room temperature and pressure has a very low vapor pressure under the influence of the
elevated temperature and the passage of carrier gas therethrough, a sufficient amount of
ester **17** is vaporized and carried through conduit **14** and upwardly through plate **12** into
contact with the pepper **10** thereabove. The process is continued for a time sufficient to
sterilize the black pepper. While the time will vary depending upon the temperature created
by heater **18** and the rate of air flow, sterilization may be obtained in times on the order
of about 15 minutes. (A control sample of black pepper that was not sterilized in this
manner had a microorganism count of 160.)

Sterilization is obtained by the absorption of the ester vapor by material **10** as the gas flow
proceeds through chamber **11**. During this phase of the process, the material **10** is not
moistened by the gas and the particulate nature of the material is retained. To ensure drying and absence of liquid condensation of the ester vapor, it is advantageous to maintain
chamber **11** at an elevated temperature with heater jacket **21**. If the jacket is utilized, it
is suitably adjusted to bring the temperature in chamber **11** near that of the vaporization
chamber **16**.

Vaporization of the ester **17** in chamber **16** may in some cases be accelerated by applying
a negative pressure to the system. In this event instead of utilizing a pump **20** for flowing
the carrier gas and vaporized ester through the component parts, a source of vacuum (not
shown) can be applied at conduit **15**. This will serve to create the desired flow of carrier
gas and vapor through the lines and will also accelerate the vaporization process. In this
alternative heat may be beneficially used as before.

The precise amount of ester needed with respect to the materials **10** under treatment is
variable over a considerable range. It has been discovered, however, that when the present
dry vapor sterilization treatment is employed in contrast with the liquid phase sterilization
previously used, the amount of ester required for sterilization is significantly reduced. Thus
it has been discovered that where the ester vapor (by weight) absorbed by the material is
between 0.000001 to 0.001%, sterilization is obtained.

After sterilization has been accomplished, the residual absorbed ester may be disposed of
by a number of suitable alternatives as desired. As before, the ester (in the absorbed vapor

phase) may be permitted to remain in the material sterilized to react with the moisture present even in dry material to produce carbon dioxide and alcohol. As one alternative, or in addition to such degradation of the ester, the residual ester may be removed by applying a negative pressure to chamber 11 through conduit 15. To accomplish this a valve 23 in conduit 15 is suitably opened while conduit 15 is connected to a vacuum source (not shown). Another alternative would be to close valve 22, open valve 23 and inject a flow of a pure gas without ester and flow it through chamber 11 and thereby carry away the residual ester. To accomplish this a conduit 24 with a valve 25 may be suitably employed. The valve 25 is opened and a flow of air, for example, moved into chamber 11 through materials 10 and out through conduit 15.

The process is also used utilizing conventional gas sterilization techniques that do not use fluidized bed concepts as above described. Thus, if valves 22 and 25 are closed and valve 23 opened, conduit 15 may suitably be linked to a source of vacuum and a vacuum created in chamber 11. At that time valve 23 may be closed and valve 22 opened while a carrier gas such as air is permitted to pass through ester 17 and with the vapor obtained therefrom, flow through conduit 14 into chamber 11 to release the vacuum. (Pump 20 need not be operated for this purpose.) Valve 22 is then closed and the gas permitted to remain in chamber 11 until the sterilization is complete. Valves 23 and 25 may then be opened and the residual ester vapor flushed out with a pure flow of air.

Pyrocarbonic acid esters are used by *W. V. White and C. Dame, Jr.; U.S. Patent 3,337,348; August 22, 1967; assigned to General Foods Corporation* to prevent browning in foods containing carbohydrate and protein.

The effective compounds are the dimethyl and diethyl esters of pyrocarbonic acid. These esters hydrolyze into carbon dioxide and methyl alcohol in the case of the dimethyl ester and carbon dioxide and ethyl alcohol in the case of the diethyl ester. When using animal gelatin as the proteinaceous material these compounds have a maximum inhibitory effect when present at a level of about 2.5 to 5.0% by weight of the gelatin present in the food material to be treated. Levels of above 5% by weight of the pyrocarbonate do not add to the inhibitory effect while presenting collateral problems, e.g., imparting a mild fruity-wine taste to the gelatin formulations. The inhibitory effect of the pyrocarbonate begins at levels as low as 0.5% by weight of the gelatin and reaches its peak at levels above 2.5%.

It is theorized that the inhibitory effect of the pyrocarbonates, involves at least a partial reaction with the free amino groups present in the proteinaceous material. In this manner the amino groups are rendered incapable of reaction with free carbonyl groups in the carbohydrate material. Therefore, it is understood that the level of pyrocarbonate, or other antibrowning organic compounds, will vary somewhat with the protein material employed and the relative number of free amino groups available for reaction. However, in almost all cases the selected compound will be present at a range of 0.5 to 5.0% by weight of the total nitrogen in the proteinaceous material employed.

Although the antibrowning agents in this process are principally directed at prevention of chemical browning due to the carbonyl-amino reaction, or more specially the reactions of reducing saccharides and their precursors and amine-containing compounds such as amino acids, peptides and so-called naturally occurring proteins, it should be understood that browning due to pyrolysis and browning due to enzymatic action may also be avoided incident or preparatory to this treatment. In cases where it is desired to inhibit the browning of materials containing enzymes care should be taken to inhibit enzyme activity by blanching or heat treating the foodstuff being protected from chemical browning, fresh fruits and vegetables preferably, prior to treatment with the anitbrowning compounds although collateral treatment in the blanch water may be used.

Example 1: Fresh bread was prepared in the following manner. 3.5 g of diethyl pyrocarbonate was mixed with 120 ml of water and the resulting mixture folded into 90 g of a commercial bread mix. No yeast was necessary in the bread formulation since the diethyl pyrocarbonate provided all of the leavening action required. The batter was poured into a greased pan and baked at 375°F for 25 minutes. The baked bread had a white crust.

Example 2: Two 100 g samples of fresh, shredded coconut were prepared and to one of them was added 1 ml of diethyl pyrocarbonate with thorough mixing. Both samples were stored at 90°F in sealed glass jars. After eight days the samples were examined. The sample containing diethyl pyrocarbonate remained white, and free of mold while the control sample had turned brown and was moldy.

ENZYME INACTIVATION ADDITIVE

Freeze dried and spray dried products are treated with ethyl alcohol or other enzyme de-activators in the process of *A. Starke; U.S. Patent 3,615,727; October 26, 1971.* The capillaries and cavities inside the dried particles communicate with each other and, owing to the absence of hardening, also with the outside of the particles so that, with the same weight, the dried products produced by means of the freeze drying method, have a sub-stantially larger surface area than those produced in accordance with other methods, and offer therefore a larger attacking surface to the enzyme poison.

For avoiding hardening or incrustation, the process also includes the use of the known spray drying method. In this method, the products, formed into particles by spray drying, are subjected to accumulation, to a vacuum and to the treatment. Preferably, the mass is introduced into containers, such as bags, permitting the diffusion, and then treated under a vacuum.

With freeze drying, prior to the reduction of the vacuum, i.e., after the completion of the drying, the alcohol is introduced into the chamber as a gas or liquid; in the latter case, the evaporation must take place in the chamber. With the introduction of the alcohol into the vacuum chamber there occurs a pressure rise by about 10 to 30 mm Hg, according to the amount of alcohol applied, so that another gas, e.g., nitrogen, is then introduced to reduce the vacuum. Thus, an additional protective layer is provided on the surface of the dried material, against atmospheric oxidation.

An example, of the use in connection with spray drying involves the manufacture of cream powder. This cream powder is placed into fabric bags, e.g., of linen, and is then subjected to a vacuum, and more particularly to a high vacuum of the order of 0.6 mm Hg. During a first reduction step, in which the enzyme poison is added, the vacuum is reduced to 12 mm Hg. This is followed by flooding with a known inert gas, such as nitrogen. How-ever, the reduction of the vacuum may also be effected in one step under addition of the enzyme poison.

With this method of enzyme deactivation, frequently very small amounts of alcohol are sufficient, of the order of magnitude of about 0.3% by weight, relative to the foodstuff, which are substantially below the quantities which must be added according to other meth-ods. The effects of a deactivating agent are intensified, there is the advantageous possibility of using rejected substances in amounts in which they have been regarded as ineffective. Accordingly, also sulfur dioxide may be used as enzyme deactivator, in addition to alcohol. The efficiency of this substance in an amount of 0.05% has been regarded as questionable and in an amount of 0.03% as negative.

It has already been stated that, for raising the pressure, another gas, e.g., nitrogen, is intro-duced. The packaging is effected in a gastight manner under a nitrogen cover or under a vacuum, after they have been exposed to the vapor or gas. In both methods, also with the gastight packaging, a drop of alcohol is added to the package; this improves the effects yet further. The process is described further by Figure 11.3.

Figure 11.3 shows diagrammatically a freeze drying installation 1. At the end of a tunnel-shaped housing which may contain transportation means for the foodstuffs to be frozen, there is a gate having end walls 2 and 3. These end walls are equipped with passages 4 and 5.

FIGURE 11.3: FREEZE DRYING INSTALLATION

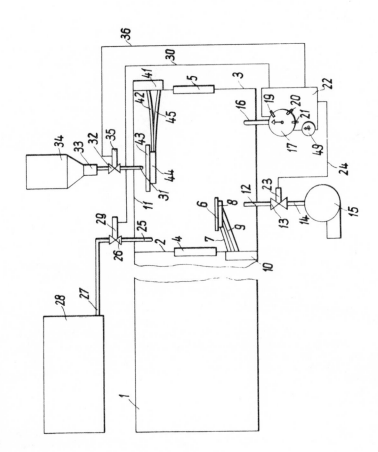

Source: A. Starke; U.S. Patent 3,615,727; October 26, 1971

These passages are adapted to be closed by doors in a gastight manner, as shown in the drawing. It should be mentioned that such an entire installation may also consist only of such a chamber between end walls **2**, **3**, and having, for example, a cylindrical shell **11**, which connects these end walls.

In the latter case, the gate is equipped with heating plates, one of which is indicated at **6**. This plate is held on a bracket **7** on the end wall **2** and is equipped with a heating device **8** connected by a lead **9** with a terminal unit **10**. Into the cylindrical shell **11** leads a tube socket **12** which is terminated by a blocking valve **13**. This valve **13** communicated via a tube socket **14** with a vacuum pump **15**, the housing of which also houses an electric motor.

In addition, into the housing also leads a connecting socket **16** for a pressure gauge **17**, the pointer of which is shown at **18**. This pointer may be associated, for example, with three stop contacts **19**, **20**, **21**, adapted to be adjusted along the scale and which carry out a control function upon contact with the pointer or with a member attached thereto.

Control members in which the impulse generated by the pointer may be transformed and/or amplified and/or maintained for a given period of time, the termination of which depends on the position of the pointer, are housed in the associated control device **22**. Obviously, the construction of the contacts may be such that they become effective only when actuated by the pointer moving in a certain direction.

The blocking valve **13** is equipped with an electromechanical actuating device **23**, connected to the control device **22** by connecting means **24**. This electromechanical actuating device **23** serves for adjusting the valve **13** as a function of signals, applied through the connecting means **24**.

The upper portion of the housing **11** is also equipped with a further tube socket **25**, terminating inside the housing shell and in an adjustable metering valve **26**. This valve is connected by way of a connecting conduit **27** to a storage tank or source **28** for nitrogen. The second valve **26** is also equipped with an electromechanical actuating device, **29**, whereby the valve can be adjusted, and more particularly opened and closed, as a function of signals, applied by the connecting means **30**. These connecting means connect the actuating device with the control device **22**.

A third connecting socket **31** terminates also in the upper part of the housing shell **11** and leads to a blocking or metering valve **32**. This valve is connected by a conduit with a connecting fitting **33**. This fitting communicates with a storage tank **23** for a so-called enzyme deactivator or enzyme poison, and more particularly alcohol. The valve **32** is equipped with a further electromechanical-actuating mechanism **35** for opening and closing the valve. This action is triggered off by signals applied through connecting means **36**, extending between the actuating mechanism **35** and the control device **22**.

The connecting device **41** is provided, if the container **34** contains, for example, alcohol, as enzyme deactivator. In this case, a heating plate **43**, associated with a heating device **44**, is arranged under the tube socket **31** and is supported on a wall **3** by a bracket **42**. It is connected to the connecting device **41** through connecting means **45**. If liquid alcohol is applied to the hot plate **42**, it is thus caused to evaporate. This hot plate **43** is not necessary if the alcohol is supplied as vapor. The bottle may contain, for example, SO_2, acting as enzyme poison.

The apparatus is made ready for operation by closing the main switch **38**. In this case, freeze dried material or material to be dried, is located in the chamber between the end walls **2**, **3**. The vacuum pump **15** evacuates the device to a certain pressure. The heating plates **6** and **43** are heated. When the vacuum has been held for a certain period, which may be measured by means of the stop contact **21** and a timing clock **49** in the control device **22**, the vacuum pump is stopped through the connection **24**, and the timing member may be triggered at the same time. After the rundown of this timing member, the valve **32** is actuated via **36** and alcohol is applied to the heating plate **43** where it evaporates.

This causes the vacuum in the chamber to drop. When, for example, the pointer **18** now reaches the stop contact **20**, a signal is applied via the connecting means **36** for closing the valve **32**, and via the connecting means **30** for triggering off the actuating mechanism **29** for opening the valve **26**. Nitrogen flows into the chamber, whereby the vacuum is further reduced. When the pressure is reached, for example, at which the pointer of the pressure gauge **17** contacts the stop contact **19**, another signal is applied via the means **30** for closing the valve **26**. Now the pressure in the chamber is such that the doors can be opened and the material can be extracted for further packaging.

cis-PENTOSES AND HEXOSES

A special additive, classified as a 2,3-cis-hydroxyl compound, is reported to prevent inter-action between lipids and proteins during storage by *F.A. Andrews; U.S. Patent 3,429,716; February 25, 1969*. Examples of such substances are, for example, the tetrose d-erythrose, the pentose d-ribose, the hexoses d-allose and d-gulose.

Example 1: A mixture containing 80% by weight of citronella oil in 20% by weight of sulfated castor oil was prepared and divided into two equal parts. To one 100 g aliquot was added 50 mg of d-ribose with stirring. To the other aliquot was added 50 mg of d-arabinose with stirring. These samples were then irradiated with ultraviolet light of moderate intensity for one hour. The samples were then left at room temperature and tested for smell periodically. The oil containing ribose continued to have a fresher odor with less terpene inversion than the sample containing the inactive arabinose. The conclusion reached was that the ribose functioned as a primary antioxidant rather than as a pro-oxidant or chelator for trace metals which may be present in the system.

Example 2: A fresh potato was cut in two halves. To one-half was added 0.05% by weight of d-erythrose, and to the other half was added 0.05% by weight of d-xylose. The raw potato was chopped in a blender, dried by being passed through a freeze dry cycle, and thereafter permitted to be exposed to ordinary ambient air. After 24 hours of exposure, it was observed that there was significantly less darkening on the sample treated with the erythrose than there was on the half containing xylose.

Example 3: A slice of U.S. Choice prime beef was dipped in a one percent solution of d-erythrose, and was then freeze dried. Another slice of the same meat was similarly freeze dried without such a treatment. On reconstitution after one week standing in air, the meat that had been freeze dried after the treatment was d-erythrose showed vastly better flavor retention.

Example 4: Coffee was brewed from 35 grams of vacuum packed Maxwell House coffee using 400 cc of tap water in a commercial type percolator with automatic cut-off set for medium. The coffee was brewed and divided into two equal portions. To one of these portions was added 40 mg of d-ribose, to the other was added 40 mg of d-arabinose. Both samples were freeze dried and stored in ambient air for 30 days. In the sample containing the d-arabinose, a substantially complete loss of fragrance occurred, while in the sample containing the d-ribose, desirable fresh aroma was retained.

ACID ADDITIVES FOR CANNED FOODS

A particular problem in heat sterilization is that the degree of heating required to destroy the infesting microbial life, especially spores, often causes undesirable changes in the in-strinsic properties of the substances in question. Depending on the composition of the substance being subjected to the heat sterilization, such deleterious changes may occur as for example: denaturation of proteins; degradation of starch or other high polymers into smaller fragments; hydrolysis of ester, peptide, and other structures susceptible to hydrol-ysis; decomposition of labile compounds such as vitamins, flavor components, etc. The common canning method of preserving perishable foods involves placing the food in a sealed container, usually a can, and then subjecting the container and its contents to heat for an extended period of time. Modern processors use temperatures well above 212°F and long periods of heating (as high as 60 to 200 minutes, or more) to insure adequate destruction of spores. As a result of such drastic heat treatment, the color, flavor, and texture of the product are necessarily adversely affected. Thus the majority of conventional canned foods are inferior in color, texture, and flavor to the fresh cooked products. Another undesirable feature of the widely used canning process is the fact that live spores of thermophilic bacteria remain in the heat-processed product. If the product is then accidentally or, as in food vending machines, purposely held at thermophilic temperatures (50° to 80°C), troublesome spoilage results sometimes from the germination and growth of these thermophiles.

It is known that an acid condition during heat sterilization is beneficial as decreasing the thermal resistance of spores. Thus, as an acid pH, the spores are killed by a lesser degree of heating than would otherwise be the case. This phenomenon has been advantageously applied in canning foods. For example, an edible acid such as acetic acid is added to vegetable products, typically artichoke hearts, when they are canned so that sterility may be attained with a lesser degree of heating than would be the case with the vegetable at its natural pH. One important point is that the flavor of the food product is changed; it has a sour taste because of the added acid.

Another factor is that chemical changes in the substance being treated are encouraged by the combination of acid conditions plus elevated temperature. Thus many of the deleterious changes discussed above in connection with heat sterilization are magnified by the presence of the acid in combination with the heat of sterilization. Typical among these changes are those involving hydrolysis, for example: splitting of the peptide linkages in protein components; splitting of glycosidic linkages in starches, sugar, and other polysaccharides; splitting of ester groups in flavor or other components; decomposition of complex structures in dyes, pigments, vitamins, flavors, etc.

G. Alderton; U.S. Patent 3,328,178; June 27, 1967; assigned to U.S. Secretary of Agriculture has developed a procedure which substantially lessens the problems outlined above. Basically, the process involves these steps: (1) The material in question is treated with an acid. (2) The acid-treated material is restored to its original pH, as by neutralization with a base. (3) The material is then subjected to a heat sterilization treatment.

(1) In the first step the material to be sterilized is contacted with an acid. The nature of the acid is not material since its only function is to supply hydrogen ions. Typical acids are for example nitric, hydrochloric, sulfuric, phosphoric, hydrobromic, trichloroacetic, acetic, chloroacetic, fumaric, maleic, salicylic, tartaric, etc. Acid salts such as sodium bisulfate may also be used. Hydrochloric acid is generally preferred as on neutralization with sodium carbonate or hydroxide (in the second step) there is formed ordinary salt, a common ingredient of foods. The acid is applied in aqueous solution and at a concentration to provide an acidic pH. Low pH's are desirable to expedite the treatment and generally a pH of about 1.5 to 5 is preferred. The formation of the hydrogen form of the spores, or stripping as it is termed herein takes place relatively slowly and sufficient time must be allowed for the desorption of metal ions naturally present in the spores. This rate is further dependent on temperature so that, for example, where extensive stripping will be obtained in about 1 to 5 hours at 50°C, the same degree of stripping at 25°C will require anywhere from 1 to 10 days.

Also, the concentration and valence of metal cations which are present in the material are factors influencing the rate of stripping. For example, polyvalent metal cations such as Ca and Mg take a longer time and/or a lower pH to strip than monovalent cations such as Na and K. A higher concentration of metal cations will also require a longer time of stripping. In any particular case, thermal death trials may be run from time to time during the acid stripping operation to determine when an effective decrease in heat resistance has been achieved. Taking into account the various factors mentioned above, the acid stripping is continued for a period long enough to obtain such a degree of stripping that the thermal resistance is markedly decreased, this time being anywhere from 1 hour to 10 days, depending on the circumstances. The acid stripping may be conducted at room temperature (25°C) or above. Since the rate of stripping increases with increasing temperature, it is preferred to use as high a temperature short of damaging the material being treated.

The acid treatment by itself exhibits little if any ability to destroy spores on contact. The function of the acid is to put the spores into a condition where their thermal resistance is lowered. However, it is obvious that if the acid stripping is conducted at a high enough temperature there will be spore destruction at this stage. Such spore destruction is, however, merely an incident to the main function of the acid treatment and the process is operative even when the acid treatment is conducted at sublethal temperatures.

(2) After the material has been contacted with acid to obtain the desired stripping effect, is is treated to restore it to substantially its original pH. The neutralization is generally accomplished by adding a base. The particular base to be used is not critical but preferably a base of a monovalent metal such as sodium or potassium is employed. Typical of the bases which are used are sodium hydroxide, calcium hydroxide, potassium hydroxide, sodium or potassium carbonate or bicarbonate, disodium phosphate, etc. The amount of base used is that required to restore the material to substantially its original pH. In the case of solid materials, particularly those with relatively impervious surfaces, the acid removal can be effected by simply withdrawing the material from the acid solution and washing it with water to eliminate residual acid.

(3) In the next step the neutralized material is subjected to heat treatment to destroy the spores which are in a heat labile condition (plus, of course, to destroy whatever vegetative microbial forms are present on the material). Ordinarily, the material is sealed in a container prior to heat treatment thus to prevent reinfection of the sterilized product with microbial forms from the environment. Thus, for example in preserving foods, the food is treated with acid and neutralized as above described and then sealed in a can or other container and the container and contents subjected to a heat treatment. The temperature and time for heating the material will vary, depending on such factors as the effectiveness of heat transfer attained, the nature and previous history of the material, the types of microorganisms with which it is infected and the amount of adventitious microflora.

For example, in the preservation of foods, such items as low pH of the food and good sanitary condition of the food make for a lesser degree of heat treatment. Also as in conventional canning, one must take into account the size of the container since with larger containers one must allow more time for penetration of heat into the interior than with a smaller container. For best results, it is preferred to use a system where the material is brought up to high temperature in as short a time as possible to minimize any possibility of the spores taking up metal cations during the period that the material is heating up.

For this reason, programs which involve high temperature and short time are preferred to those of relatively lower temperatures for longer periods. In any event, the degree of heat processing will be substantially less than with conventional sterilization. Thus, by rendering the spores sensitive to heat, as described, the temperature or heating time or both will be substantially less than with conventional sterilization.

The process is of wide versatility and can be applied to materials of every type. A typical application of the process is in the preservation of foodstuffs, for example, fruits, vegetables, milk, eggs, meat, spices, fish, cereal products and cheeses. Liquid foods such as juices, purees, concentrates, sauces, soups, extracts, and beverages of every type are included.

Example: Fresh frozen peas were thawed, pureed, heated at 100°C for 1 hour and centrifuged. The separated juice was filtered and autoclaved for 20 minutes under a steam pressure of 15 psi to provide a sterile, clear pea juice. One portion of the pea juice was acidified to pH 2.5 by addition of hydrochloric acid (136 ml juice to 9 ml of 1.02N HCl). Also, spores of *Bacillus stearothermophilus* were added to provide a level of one million spores per milliliter. The mixture was allowed to stand overnight (about 16 hours) at room temperature. The next day the acid treated, inoculated pea juice was treated with sufficient 1N sodium hydroxide to restore it to its original pH of 6.0. Another portion of the pea juice was inoculated with the same amount of the spores but without acidification, thus to provide a control. This was kept cool to prevent germination of spores. To provide comparative conditions, an equivalent amount of water and salt (NaCl) was added to the control.

Two-ml portions of the treated juice and control juice were sealed into thermal death time tubes and these were heated for various times in an oil bath maintained at 120.6°C. There were 10 tubes for each of the juices and for each time of heating. After heating, the tubes were cooled quickly and each was placed on glucose-tryptone agar. After incubation at 55°C for 6 days the colonies on each plate were counted. The results are tabulated below:

Sample	Heating Time, min	Results
Acid treated and neutralized	10	2 plates had 2 colonies each
		8 plates had no colonies
Acid treated and neutralized	15	No colonies on any of the plates
Control	20	All 10 plates had 20 to 50 colonies each
Control	25	3 plates had one colony each
		7 plates had no colonies
Control	30	No colonies on any of the plates

APPENDIX I

THE USE OF CHEMICALS IN FOOD PRODUCTION, PROCESSING, STORAGE, AND DISTRIBUTION

National Academy of Sciences, 1973

INTRODUCTION

Nutritional status studies continue to reveal nutritional deficiency diseases in many parts of the world (World Health Organization, 1971). Although some evidence of malnutrition has been discovered in the United States in recent years (Center for Disease Control, U.S. Department of Health, Education, and Welfare, 1972), frank nutritional deficiency diseases, such as pellagra and rickets, were practically eliminated from the United States from 1920 to 1950, as a result of the application of science and the understanding of the chemistry of foods and nutrition.

Through the utilization of science in food production and food technology, the American people today have access to the most abundant and varied food supply in history. Although 11% less land is being used for crops than in 1950, the overall output of farm products has expanded by 40%, and food consumption by the average person in the United States has increased about 6%. This larger output is the result of improved utilization of land and labor, which has been achieved by such new and improved technologies as more efficient farm organization, more effective use of agricultural chemicals, improved crop and livestock species, more irrigation, and the shift to better land. Food production and utilization must be increased, however, in order to feed an estimated 100 million additional people in the United States by the end of this century (Krause, 1971).

As farming patterns change and as each farmer has become able to feed more people, dramatic shifts in populations to cities and towns have occurred. The result, a more complex food distribution system, has made it necessary to preserve food longer without loss of freshness, palatability, and nutritional quality. Centralized processing of many foods into service-ready forms before they reach the consumer has become increasingly necessary in order to minimize shipment of inedible parts of raw commodities. The application of many new discoveries in biology, chemistry, and engineering; the invention and use of a great variety of high-production machines; and the development of improved storage facilities, packaging materials, and means of distribution have made it possible to meet the continually increasing need for food.

Although the engineering and mechanical aids to this task are clearly evident to most people, many are unaware of the contribution of chemistry and of the chemical nature of food itself. They are especially ill informed as to the use of chemicals in food — an area that has been greatly misunderstood. This report concerns the technologic rationale for the use of chemicals in food production, processing, storage, and distribution; the problems associated therewith; the research necessary to solve these problems; and the legislative measures that have been enacted to ensure protection of the public.

THE CHEMICAL NATURE OF FOODS

All components of foods are chemicals. The great bulk of foods is comprised of chemicals classified as carbohydrates, fats, proteins, minerals, and water. Foods also contain small amounts of such other naturally occurring chemicals as vitamins, antioxidants, antimycotics, buffers, thickeners, emulsifiers, chelating agents, colors, and flavors.

The carbohydrates consist of various sugars, starches, dextrins, celluloses, and gums. Most of these are oxidized in the body, thereby releasing energy for muscular activity and other body functions. The fats and oils supply energy and essential fatty acids and facilitate absorption of fat-soluble vitamins. The proteins are composed of amino acids that, when made available by digestion, are used as building materials for the proteins of soft tissues of the body, and for the bone matrix, or as a source of energy. The minerals have many functions. Some are used primarily in building skeletal structures and teeth, while others are essential aids in metabolism. The various vitamins are essential cell components; they form a part of many coenzymes and participate in metabolic reactions. The colors and flavor-imparting compounds may have nutrient value but most frequently are primarily of aesthetic value.

Such factors as the variety of crop, soil fertility, and nutrient content, duration and intensity of sunlight during growth, rainfall, disease, and methods of harvest, handling and storage may affect the chemical composition and hence the flavor, quality, and nutritional value of each plant product. Chemical composition of animal products is also influenced, to some extent, by environmental factors and, most importantly, by genetics.

Each foodstuff consists of chemicals that are characteristic of it, but because of the natural variation, it is frequently necessary to adjust the composition in order to provide a product of constant quality. Milk, for example, contains variable amounts of butterfat, milk sugar (lactose), proteins (principally casein and lactalbumin), minerals (notably calcium and phosphorus), various vitamins, and other constituents. The composition of milk varies somewhat according to the breed, the individual cow, the period of lactation, and the nature of the feed provided to the cow. Milk from Holstein cows is somewhat lower in fat and carotenoid pigments than is that of either Jersey or Guernsey cows. Milk destined for general distribution comes from various sources and is generally blended or standardized to a uniform level of butterfat. Milk received at evaporating plants at different seasons and from different herds varies in its stability to heat. This property is dependent upon the balance in the milk of the natural mineral salts, particularly the proportions of calcium, phosphate, and citrate. It is often necessary, therefore, to add one or another of these chemicals in order to make it sufficiently heat-stable for an acceptable product. This is a case of adding a constituent normal to a food to help standardize its composition.

Standardization procedures comparable to those described for milk are carried out with other foods as, for example, in the blending of wheat varieties to secure a flour of uniform baking quality.

Natural foodstuffs contain substances that are not now known to have nutritive value or are harmful if taken in amounts substantially larger than those encountered in normal usage of the food. (For further information, see *Toxicants Occurring Naturally in Foods*, Committee on Food Protection, National Academy of Sciences — National Research Council Publ. 1354, Washington, D.C., 1966.) Coffee and tea, for example, contain caffeine and theobromine, which have well-known physiological effects; cocoa contains a related compound. Small amounts of arsenic and other toxic metals are found in most foodstuffs. Substances that interfere with thyroid function, resulting in development of goiter, have been identified in rutabagas and other species of the genus *Brassica*, and oxalates are present in rhubarb, spinach, and chard. Many foodstuffs contain traces of toxic alkaloids and cyanide-generating compounds that are toxic in high concentrations. These foods are considered acceptable, however, since long usage has indicated that the substances in question are apparently of no harmful consequence when consumed in the amounts likely to appear in a balanced diet.

WHAT IS A FOOD ADDITIVE?

Throughout this report, unless otherwise indicated, the term "food additive" is used as defined by the Committee on Food Protection and as it is generally used, except in the legal sense in the United States: A food additive is a substance or a mixture of substances, other than a basic foodstuff, that is present in a food as a result of any aspect of production, processing, storage, or packaging. The term does not include chance contaminants. This definition differs from the legal definition of food additive which appears on page 365.

In addition to those chemicals that constitute foodstuffs per se, chemicals may be incorporated — either directly or indirectly — during the growing, storage or processing of foods. These chemicals may be described for convenience as food additives. When they are purposely introduced to aid in processing or to preserve or improve the quality of the product, they are called intentional additives. Such materials as vitamins and minerals for enrichment, mold inhibitors, bactericides, antioxidants, colors, flavors, sweeteners, and emulsifiers are intentional additives. They are added to the food product in carefully controlled amounts during processing, and the amounts necessary to achieve the desired effect are usually very small. In addition to the intentional additives, certain other chemicals may find their way into foods as a result of their use in some phase of the production, handling or processing of food products. They are known as "incidental additives." Under provisions of the federal food and drug laws, such additives are permitted in foods only if they cannot be avoided by invoking good production and processing practices and, then, only if the amounts that occur under these conditions are known to be safe.

SITUATIONS IN WHICH FOOD ADDITIVES ARE ACCEPTABLE

Several circumstances justify the use of food additives to the advantage of the consumer. Every chemical used in food processing should serve one or more of these purposes:

- improve or maintain nutritional value
- enhance quality
- reduce wastage
- enhance consumer acceptability
- improve keeping quality
- make the food more readily available
- facilitate preparation of the food

SITUATIONS IN WHICH FOOD ADDITIVES SHOULD NOT BE USED

Apart from the question of safety, the use of food additives in some situations is not in the best interests of the consumer and should not be employed when:

- used to disguise faulty or inferior processes
- used to conceal damage, spoilage, or other inferiority
- used to deceive the consumer
- otherwise desirable results entail substantial reduction in important nutrients
- the desired effects can be obtained by economical, good manufacturing practices
- used in amounts greater than the minimum necessary to achieve the desired effects

COMMON TYPES OF INTENTIONAL FOOD ADDITIVES

The purpose of intentional additives and the technologic benefits of their use can be best illustrated by considering the more important classes of these substances.

Acids, Alkalies, Buffers and Neutralizing Agents

The degree of acidity or alkalinity is a very important property of many foods. Chemical leavening agents are used in the baking industry as well as in the home to produce carbon dioxide, which makes the batter light and porous, thereby providing a finished product of good volume, crumb texture, and palatability. This reaction requires an ingredient that acts as an acid in the presence of moisture or heat. The acid ingredients used are such compounds as potassium acid tartrate, sodium aluminum phosphate, tartaric acid, mono-calcium phosphate and sodium acid pyrophosphate. Sodium bicarbonate is the gas-producing substance normally used, although ammonium bicarbonate is employed in the commercial production of some cookies and crackers. The carbon dioxide gas produced by leavening agents, such as baking powder, is the same as that produced by the growth of yeast in some types of bread. Chemical leavening is preferred in products such as cakes and cookies, when the yeasty flavor is not desired.

The tart taste of soft drinks — other than cola-type beverages — is imparted by the addition of organic acids from either natural or synthetic sources. Citric acid (a component of citrus fruits), malic acid (a component of apples), and tartaric acid (a component of grapes) are the major organic acids employed. Buffering agents, generally the sodium salts of these acids, are frequently used to control the degree of acidity in soft drinks. The concentrations of acids and buffers employed are essentially the same as the levels at which these substances occur naturally in fruits. In cola-type beverages, the most commonly used acidulant is phosphoric acid.

Adjustment of acidity is necessary in the production and use of several dairy products; for example, emulsification and a desired tartness in processed cheese and cheese spreads are obtained by the addition of such acids as citric, lactic, malic, tartaric and phosphoric. Excessive acidity that may develop in cream must be neutralized for satisfactory churning and to produce a butter of acceptable flavor and keeping quality. Acids are also used as flavoring agents in confections, and alkalies may be employed in the processing of chocolate.

Bleaching and Maturing Agents: Bread Improvers

Wheat flour in its natural, freshly milled state has a pale-yellow tint. When such flour is stored, it slowly becomes white and undergoes an aging process that improves its baking qualities. About fifty years ago, it was discovered that certain oxidizing agents added to the flour in small amounts would markedly accelerate this natural process, thus reducing storage costs and the hazards of spoilage and insect and rodent infestation. Some of the permitted compounds, e.g., benzoyl peroxide, exert only a bleaching action and are without influence on baking properties. Others, such as the oxides of nitrogen, chlorine dioxide, nitrosyl chloride and chlorine, have both bleaching and maturing or improving properties.

Bread improvers are used by the baking industry to provide assurance that the dough will ferment vigorously and evenly and have uniform quality. They contain small amounts of such oxidizing substances as potassium bromate, potassium iodate, and calcium peroxide. Improvers also contain limited amounts of other inorganic compounds, e.g., ammonium chloride, ammonium sulfate, calcium sulfate and ammonium and calcium phosphates, that serve as yeast food and dough conditioners. The quantity of oxidizing substances required is small, and, since excessive treatment results in an inferior product, their use is self-limiting. Bleaching agents may also be used in the preparation of certain cheeses in order to improve the appearance of the finished product.

Emulsifying, Stabilizing and Thickening Agents

Emulsifying agents are used in baked goods, cake mixes, ice cream, frozen desserts, and confectionery products. Examples of those used are lecithin and mono- and diglycerides. In bakery products, these substances improve volume, uniformity, and fineness of grain. They facilitate machining in bread doughs, and the resulting bread has a softer crumb and a somewhat slower firming rate than do breads prepared without their use. The whipping properties and physical nature of frozen desserts are improved by the use of small amounts of an emulsifier. In candies, they are employed to maintain homogeneity and improve keeping quality. Butter oil and, occasionally, sorbitan derivatives are also used to retard "bloom" — the whitish deposits of high-melting components of cocoa butter that sometimes appear on the surface of chocolate candies.

The texture of ice cream and other frozen desserts is dependent, in part, on the size of the ice crystals in the product, which can be controlled by the addition of small amounts of stabilizing agents. Alginates, gelatin, cellulose gum, and other vegetable gums are among the substances used. Certain of these compounds are also used in chocolate milk to increase the viscosity of the product and thus prevent the settling of cocoa particles to the bottom of the container. Gelatin, pectin and starch are used in confectionery products to give a specific texture. Agar-agar, alginates, gum arabic and gum tragacanth are used as stabilizers or thickeners in certain types of hard gums.

Sugar-sweetened beverages ordinarily possess a certain amount of "body." Since beverages that are sweetened with nonnutritive sweeteners do not have this property, so-called "bodying agents" are used in their production. These include such natural gums as sodium alginate and pectins, cellulose gum derivatives and sorbitol. The foaming properties of brewed beer can also be improved by the addition of certain of these stabilizing agents.

Flavoring Materials

A wide variety of spices, natural extractives, oleoresins, and essential oils are used in processed foods. In addition, today's flavor chemist has produced many synthetic flavors, most of which are similar to the naturally occurring flavor compounds. Both types of products are used extensively in soft drinks, baked goods, ice cream, and confectionery. They are usually employed in small amounts ranging from a few to 300 parts per million. Amyl acetate, benzaldehyde, carvone, ethyl acetate, ethyl butyrate, and methyl salicylate are representative of compounds that are employed in the preparation of flavoring materials. Many of the compounds used in flavoring preparations are also found in natural products or are esters of natural acids. Many spices and spice extractives are used in sausages and prepared meats. Protein hydrolysates are also employed to enhance the flavor of some foods.

Food Colors

For a comprehensive discussion on the use of food coloring materials, see *Food Colors*, Committee on Food Protection, National Academy of Sciences-National Research Council Publ. 1930, Washington, D.C., 1971.

Food colors of both natural and synthetic origin are used extensively in processed foods, and they play a major role in increasing the acceptability and attractiveness of these products. The indiscriminate use of color, however, can conceal damage or inferiority, thus making the product appear to be better than it actually is. In view of these factors, food colors must be used with discretion. Classes of foods that are frequently colored include confectionery; bakery goods; soft drinks; and some dairy products such as butter, cheese, and ice cream. Natural colors used in foods include annatto, alkanet, carotene, cochineal (carmine), chlorophyll, saffron, paprika, and turmeric.

Nutrient Supplements

Vitamins and minerals frequently are added to processed foods to improve their nutritive value. It is recognized, for example, that the processing of cereal grains to produce many familiar foods, such as breakfast cereals and bread products, removes a large portion of the vitamins originally present. It is important to restore these lost vitamins and minerals. Definitions and standards of identity have therefore been established by the Food and Drug Administration (FDA) for the enrichment of wheat flour, farina, cornmeal, corn grits, macaroni, noodle products, and rice. These standards define the minimum and maximum levels of thiamin, riboflavin, niacin, and iron permitted to be added and, in some cases, provide for the optional addition of sources of calcium and vitamin D. Many manufacturers of ready-to-eat breakfast foods add thiamin, riboflavin, niacin, and other supplements on a voluntary basis to provide products that contain amounts of these nutrients equal to or greater than those present in the cereals from which the foods are made.

Vitamin A and, in some cases, vitamin D are added to margarine. Vitamin D is added to both fluid and evaporated milk, and both vitamins may also be added to skim milk. Vitamin A may be added to Blue cheese and Gorgonzola cheese to replace that lost in the bleaching process and to low-fat milk to compensate for that removed with the separated butterfat. Iodized salt contains a small amount of potassium iodide to furnish the iodine necessary to prevent simple goiter.

Preservatives and Antioxidants

Preservatives are substances added to foods to prevent or inhibit microbial growth. Certain molds have been found to produce toxins, and the food-poisoning potential of many bacteria is well established. These circumstances require that every practicable measure be taken to impede the growth of microorganisms in food. A number of different types of preservatives are used, depending on the food product and the organism involved.

Although baking destroys the spores of most fungi and bacteria present in flour and other ingredients, baked goods are constantly exposed to spores present in the air and on baking equipment. Under summer conditions, the organisms become active and produce a condition in the bread called "rope" that renders the product inedible. Sodium diacetate, the propionates of sodium and calcium and such acidic substances as acetic acid, lactic acid, and monocalcium phosphate are effective in retarding the growth of fungi and "rope" bacteria. Sorbic acid or its potassium salt are used as antimycotic agents in cheeses, syrups, and pie fillings. Benzoic acid and sodium benzoate are employed in oleomargarine, certain fruit juices, pickles, and confections to inhibit bacterial or fungal growth. Sulfur dioxide is widely used for the preservation of dried fruits. Sugar, salt, and vinegar are also effective in preventing microbial growth.

Fatty foods are susceptible to oxidative changes that take place in the fat molecule, with the consequent production of off-flavors, odors, and potentially toxic materials. The substances used to prevent this type of spoilage are known as antioxidants. The compounds most widely employed for this purpose are butylated hydroxyanisole, butylated hydroxytoluene, and propyl gallate. They are used in such foods as lard, shortening, crackers, soup bases, and potato chips. It has been found that certain acidic substances, e.g., citric acid, ascorbic acid, and phosphoric acid, enhance the properties of the antioxidant, and these substances frequently are added in combination with the antioxidant. Ascorbic acid has been found effective in preventing the oxidative discoloration (browning) of such frozen fruits as sliced peaches.

Miscellaneous Intentional Additives

A number of additional substances are employed for various purposes. Clarifying agents, e.g., tannin, gelatin and albumin, are used to remove small particles and minute traces of copper and iron in the production of vinegar and certain beverages. Failure to remove these trace minerals leads to flocculation and undesired color changes in the product. Such changes are usually harmless, but they make it difficult to distinguish between normal products and potentially harmful products that may be unsafe because of other types of contamination.

Such sequestering agents as ethylenediaminetetraacetic acid and its salts prevent the adverse effects of metallic ions in certain food products by forming chemically inactive complexes with the metals. To prevent their drying out, humectants are necessary in the production of some types of confections and candy. Without a humectant, shredded coconut, for example, would not remain soft and pliable. Substances used for this purpose include glycerin, propylene glycol and sorbitol. Glazes and polishes, such as beeswax and natural resins, are used on coated confections to give luster to an otherwise dull surface. Magnesium carbonate, tricalcium phosphate, and various silicates are employed as anticaking agents in products such as table salt that must remain free-flowing; calcium stearate is used for a similar purpose in garlic salt.

Chemicals sometimes are added to processed fruit and vegetable products in order to improve their texture. Because canned tomatoes, potatoes and apple slices, for example, tend to become soft and fall apart, a small amount of calcium chloride or other calcium salt is added to the product to serve as a firming agent. Sodium nitrate and nitrite are used in the curing of meats to help prevent botulism and to develop and stabilize the pink color commonly associated with these products. Nitrogen, carbon dioxide, and nitrous oxide are employed in pressure-packed containers of certain foods to act as whipping agents or to serve merely as propellants.

SOME POTENTIAL SOURCES OF INCIDENTAL ADDITIVES

Although incidental additives are not deliberately added to foods, they sometimes appear there as a result of operations inherent in production, storage, processing, packaging, or marketing. Some of the chemicals that can occur as incidental additives are reviewed below.

Pesticides

According to a 1969 report (Committee on Plant and Animal Pests), the average annual loss from insect pests and the cost of control in the United States, from 1951 to 1960, was approximately $6.8 billion. When most of the U.S. population was rural, insect damage was taken for granted as a part of the normal hazard of crop and animal production. Today, however, with a large urban population, increased total demand for food, and consumer demand for top-quality produce, the grower must continually seek increased productivity if his operation is to remain viable. In order to meet these requirements, with decreasing arable land, there will be a continuing need for pest control chemicals in the foreseeable future. The Report of the Health, Education, and Welfare (HEW) Secretary's Commission on Pesticides and Their Relationship to Environmental Health (1969) predicts that the production and use of pesticides in the United States will continue to grow at an annual rate of about 15% and that their use will more than double between 1969 and 1975.

Although it is generally recognized that the use of chemical pesticides has conferred substantial benefits upon society, increasing concern has developed over the undesirable side effects. Concern for human health hazards, environmental pollution and damage to such nontarget organisms as beneficial insects, birds and fish has been voiced by many (HEW Secretary's Commission on Pesticides and Their Relationship to Environmental Health, 1969; U.S. Office of Science and Technology, 1971; Newsom, 1967). Reports have been issued describing populations of pest species that have developed resistance to commonly used chemical insecticides (Oppenoorth, 1965; Newsom, 1970). According to Pickett (1959), the adverse effect of broad-spectrum pesticides on predators and parasites has released certain species of insects from the natural control of these agents. These several concerns have stimulated the imposition of limitations on the use of pesticides, as illustrated by the recent restrictions on use of DDT for control of crop pests in the United States (Federal Register, 1972a). Threatened curtailment of other chemical pesticides has engendered augmented research activity in the search for alternative means for pest control.

Various biological control techniques (Burges and Hussey, 1971), genetic manipulation (Smith and von Borstel, 1972), and such chemical agents as chemosterilants and sex attractants ("Pheromones") are under investigation at the present time. The safety of these chemical agents for use on agricultural products will have to be demonstrated in the same way as those chemicals currently in use.

Despite much imaginative research now in progress, the HEW Secretary's Commission expressed the opinion in 1969 that noninsecticidal control techniques are not likely to have a significant impact on the need for chemical insecticides in the foreseeable future. The committee stated further that:

> There are many factors that are influencing the changing use patterns of pesticides. In addition to new pest infestations, resistance to selected pesticides, alterations in economics of crop production and changing agricultural and social patterns, the impact of public opinion is having a growing influence on the use of pesticides. The increased concern for new legislation and regulation of the manufacture, sale and use of pesticides must not be so structured as to destroy the incentive for development of new pesticides more compatible with other environmental qualities.

Fertilizers

Very large amounts of commercial fertilizer are used each year in the production of crops.

Hundreds of different chemical and physical combinations of the mineral elements essential for plant growth are used; many of these are mined from the earth in areas in which they occur in great abundance. The plant nutrient elements applied as fertilizers are taken up and utilized by plants in precisely the same manner as are the elements occurring naturally in soils.

Feed Adjuvants and Drugs

Research has shown that physiologic responses of animals are influenced by many chemicals. Some of these responses can be translated into improved efficiency of production of commercially important animal products. Antibiotics, hormones, tranquilizers, and enzymes have been used to increase the efficiency of feed utilization and thus of animal production. Drugs also are added to animal feeds for prophylactic or for therapeutic purposes. Traces of some of these chemicals may appear in animal products used for food and thus become incidental food additives. Because questions have arisen as to the wisdom of those practices, existing regulations concerning the use of these materials are being reviewed, and research to reestablish safe conditions of use is under way.

Packaging Materials

Present-day food packages are designed to protect their contents during storage — both before sale and in the home — from contamination by dirt and other foreign material; infestation by insects, rodents and microorganisms; and loss or gain of moisture, odors or flavors. They frequently are designed to protect the food from deterioration resulting from contact with air, light, heat, and contaminating gases. They may serve as containers in which the food is processed as well as receptacles in which the food is heated for serving. Highly processed or "convenience" foods, such as heat-and-serve products, prepared dinners, and mixes, may have specific and exacting packaging requirements.

A large number and variety of materials are required to meet all the needs of food packaging. They range from materials of simple composition as, for example, a metal foil, to chemically complex synthetic films. The primary packaging materials, e.g., metals, glass, wood, fabric, paper and synthetic films, are modified in many ways for particular purposes. They may be treated to withstand the abrasive action of foods and the solvent action of acidic, basic, neutral, alcoholic, or fatty foods they will contain. They may be specially formulated to exclude or permit passage of gases and moisture, made shrinkable by heat or relatively heat-stable, cold-resistant, flexible or rigid, colored, and transparent or opaque.

The basic packaging materials are simple or complex chemicals, and they are modified, as indicated, usually by chemical means. Because they are closely associated with foods, they may contribute incidental additives to the foods they contain. This source of additives is recognized and is closely regulated by those officials in industry and government responsible for the safety of the food supply.

THE SAFE USE OF FOOD ADDITIVES

Safety of a food chemical should be appraised in relation to the extent of its proposed use in foods, the amount that would be consumed under all probable dietary patterns of the populations concerned, the nature of the biologic response it may induce, and the minimal intake that might provoke response. No use of a chemical in food should be permitted if there is a reasonable basis on which to expect that its intake from the permitted use will be so high as to produce adverse effects. These and other principles have been developed in detail by the Committee on Food Protection (1954; 1959a,b; 1965; 1967; 1970), the Joint FAO/WHO Expert Committee on Food Additives (1957; 1958; 1961), the World Health Organization (1967), and the Food and Drug Administration (Division of Pharmacology, 1959).

Hazard Versus Toxicity

A distinction must be made between the hazard associated with use of an additive and the toxicity of the material. Toxicity is the capacity of a substance to produce injury; hazard is the probability that injury will result from use of a substance in a proposed quantity and manner. Safety is the practical certainty that injury will not result from such use of the substance.

The fact that an additive may give rise to adverse effects when taken in large amounts does not necessarily mean that its proper use will entail a hazard to man. As already indicated, additives can be used without hazard in food on the basis of information gained from investigations carried out to determine the toxicity of the materials to laboratory animals. Through consideration of the toxicity of the additive, the amount proposed for inclusion in foods, and the variety and types of foods to which it is proposed to be added, it is possible to reach a reasoned judgment of the hazard involved. Legal measures are designed to protect the public from the presence, in the food supply, of any additives that have not been demonstrated to be safe under the recommended conditions of use.

Risk-Benefit Relation

Any decision to use a chemical in such a way that it may remain in food as consumed should be based on the assurance that the use will be safe (i.e., that there is virtual certainty that no injury will result), and that — directly or indirectly — the use will, in some way, benefit consumers. Factors to be considered include the following:

- hazard to the consumer
- consumer needs and wishes
- requirements of food supply and public health
- needs of the food producer and processor
- economic factors
- availability of methods and mechanisms for regulatory control
- threat to the adequacy, wholesomeness, or availability of the food supply in the absence of protection afforded by the additive

No single formula can relate all these factors to one another under all conditions, but they should be taken into account whenever a decision is being made concerning whether to permit or deny use of a chemical. The risk-benefit relation is affected by the circumstances that pertain to a given situation. For example, if the supplies of a major food staple in a food-deficient area are threatened by pests, use of a chemical to control the pests might well be justified, even though levels of residues greater than those usually permitted remain in the food.

Chemical and Physical Properties

It should be possible to characterize the substance in question in chemical and physical terms. Constancy of composition, stability, and freedom from harmful amounts of impurities should be assured. The *Food Chemicals Codex* (Committee on Food Protection, 1966, 1972) is designed to provide national standards of identity, purity, and composition for food chemicals, as has been done for drugs in the United States and elsewhere. Standards for food-grade chemicals are described in the *Codex* by chemical and physical specifications that can be attained through good manufacturing practice and that, at the same time, provide reasonable standards of identity and assurance of safety. Obviously, such a compendium is never complete, nor are its listed specifications necessarily permanent; it must be continuously reviewed and revised. Patterns of use of foods and new knowledge on toxicology may be expected to determine the chemicals to be included as well as their specifications.

The persistence of the chemical in the foods in which it is to be used, its reactivity with the components of these foods, and the identity of the substances to which it may be converted in foods should be known. Methods for estimation of chemicals in food must be sufficiently accurate and sensitive to permit determination of quantities that would be hazardous to health. When conversion products are likely to be hazardous, appropriately accurate and sensitive assay methods for their detection should be available. When, however, the proposed use of a chemical cannot reasonably be expected to result in hazardous amounts being present in food, the development of analytical methods practicable for routine application to foods in commerce is not as essential.

Biological Considerations

Results of critically designed animal studies of the physiologic, pharmacologic, and biochemical behavior of a proposed food chemical provide the crucial information needed to evaluate safety at a specified level of intake by man. Because of species variation, however, judgment must be exercised in arriving at a safe level of use by man on the basis of data derived from studies with various species of animals. No method is at hand — and none is in sight — for establishing, with absolute certainty, the safety of a food chemical under all conditions of use. Experience has shown, however, that properly conducted and interpreted animal experiments can provide that degree of assurance of safety reasonably expected in the evaluation of chemicals for use in human food. In rare instances, adverse responses in animals, discovered well after the material in question had come into regular use, have led to discontinuance of that use, even though no hazard to man had been demonstrated. Nitrogen trichloride, certain synthetic colors, and a few naturally occurring flavoring components, e.g., safrole and coumarin, are examples.

The appropriate study of responses in human beings is a valuable phase of the safety evaluation of chemicals used, proposed for use, or present in food. Occupationally or accidently exposed subjects provide some data; another possibility is the use of volunteers in controlled studies. The latter should be carried out only after careful investigation of biochemical and toxicologic properties in animals has clearly indicated that the risk to which the subjects may be exposed will be trivial in relation to the benefit of the knowledge to be expected. Data from human studies are especially appropriate in assessing the safety of a material of exceptionally widespread potential use.

Once a food chemical is approved for use, there should be continual observation of the exposed population for deleterious effects that may emerge under prolonged and varying conditions of use, and safety of the use should be reappraised whenever warranted by experience in use or advances in knowledge.

It has been customary to permit use of a new chemical on the basis of long-term feeding studies in at least two species of animals. Chronic toxicologic studies in animals are designed to find a dosage level that will produce a deleterious effect in the animal as a whole or in one or more of its organs. It is not always possible, however, to find a deleterious effect unequivocally related to toxic properties of the substance being studied. In addition to a toxic level, a "no-adverse effect" level is sought. This level, adjusted by incorporating an adequate margin of safety, provides an estimated safe level for use in the human diet.

Responsibilities and Role of Industry in Assuring Safety

The initial evaluation of toxicity of chemicals that might be used in foods is made by the companies that create them. This is necessary to protect the employees who handle the materials in laboratories and factories. Toxicologic investigations are started as soon as a new material shows promise of being useful. These initial experiments may be made in the manufacturers' own laboratories or in private toxicologic laboratories. In addition, some companies have developed fellowship and grant-in-aid programs with colleges, universities, and hospitals where they can secure research services and advice from competent scientists.

Most pesticide manufacturers will have conducted acute toxicity measurements (responses to one or a few large doses) on at least one animal species by the time they have completed laboratory studies on the usefulness of chemicals under consideration. Chronic toxicity studies (responses to repeated smaller doses over an extended period of time) may be made while the material is being tested in experimental field plots. Governmental agencies cooperate with industry so that the necessary field experiments can be conducted on a limited commercial scale while toxicologic research is still in progress. Pesticidal chemicals are rarely accepted for use with less than 3 to 5 years of experimentation.

Many food processors also have established their own laboratories, and much of the progress in detecting residues of pesticides and in developing methods of removing traces of these chemicals has come from these sources. In addition, they have sponsored extensive research in colleges and universities.

Although responsible chemical manufacturers and food processors conduct extensive toxicity studies on the food additives they are recommending for use, there may be honest differences of opinion in assessing the hazards involved. These differences of opinion can usually be resolved by further study and consultation with appropriate governmental or scientific agencies.

Functions of Federal and State Agencies in Assuring Safety

Several federal agencies participate in assuring the safety of the food supply. They function by the administration of different laws and usually have both regulatory and research responsibilities. In addition to federal statutes, most states also have laws pertaining to the safety of foods. These laws are administered by either the department of public health or department of agriculture of the respective state. Some of these laws are modeled on a uniform pattern, while others vary from the pattern in important respects. Efforts are being made to encourage the standardization of the state laws, so that all foods in all states will meet uniform standards of safety under both state and federal regulations.

The Department of Health, Education, and Welfare [and its various agencies, including the Food and Drug Administration (FDA), the National Institutes of Health (NIH), and the Center for Disease Control (CDC)]is responsible for public health in its broadest aspects. Investigations are made into the use of chemicals to control diseases and insects affecting man, the effects of exposure of man to these chemicals during manufacturing and application in the field, and the effect of the ingested chemicals on man. In addition to using small animals for research, NIH also performs epidemiological studies and carries out clinical investigations on human subjects. The Food and Drug Administration also acts in an advisory capacity to state and local health departments in devising and enforcing ordinances regulating the handling of food and milk. Finally, it sponsors an extensive program in support of research in toxicology, nutrition, and environmental health.

The meat and poultry inspection program of the U.S. Department of Agriculture (USDA) controls the use of chemical additives in meat and poultry and their products under the authority of the Federal Meat Inspection Act (1907), as amended by the Wholesome Meat Act and the Wholesome Poultry Act. Under these acts, no chemical additives may be used in meat or meat products without the approval of the USDA.

The Pesticides Regulation Division of the Environmental Protection Agency is responsible for the registration of all pesticides under the Insecticide, Fungicide, and Rodenticide Act of 1947. Before any pesticide can be sold interstate, data must be submitted on its ability to control pests, its safety for use by spray operators and on crops, its effects on quality and safety of food, and the potential damage to other forms of life and to soil. This division does not itself conduct research of this type.

In order to provide effective control over the use of pesticide chemicals and to minimize the potential hazard from their misuse, Congress, in 1954, amended the Federal Food, Drug and Cosmetic Act. The amendment provides a means of establishing safe tolerances for residues of pesticide chemicals in or on raw agricultural commodities. The term "pesticide chemical," as used in this legislation, covers insecticides and other agricultural chemicals used to control a wide variety of pests. This law provides that a food shall not be marketed if it bears residues of a pesticide chemical unless:

(1) the pesticide chemical is generally recognized by experts as safe; or
(2) upon consideration of an adequate amount of scientific evidence, the government has established a safe tolerance for residues of the chemical or has exempted it from the requirements of a tolerance; and
(3) if a tolerance has been established, the residues remaining on the food are within the safe tolerance.

This law requires that before a safe tolerance for a pesticide can be established, the petitioner must submit to the government the following information:

(1) the name, chemical identity, and composition of the pesticide chemical;
(2) the amount, frequency, and time of application of the pesticide chemical;
(3) full reports of investigations made with respect to the safety of the pesticide chemical;
(4) the results of tests on the amount of residue remaining, including description of the analytical methods used; and
(5) practicable methods for removing residues if excessive.

The petition should also give the proposed tolerance, if any, and reasonable grounds in support of this request. Since 1970, the Pesticides Tolerances Division of the Environmental Protection Agency has had the responsibility of establishing tolerances.

The Food and Drug Administration has the responsibility for enforcing established pesticide tolerances. In general, actual residues are found to be well within the limits established, but from time to time it has been necessary to remove a shipment of food from interstate commerce when the residues exceed the established tolerances. Most of these instances have resulted from the misuse of the pesticide.

The Federal Food, Drug, and Cosmetic Act of 1938 prohibited the addition of any poisonous or deleterious substance to a food. However, it did not provide adequate protection against those chemicals that might be used before being tested sufficiently to determine whether or not they were harmful. The law was amended in 1958, therefore, to provide such additional protection.

The term "food additive" under this legislation meant:

any substance the intended use of which results or may reasonably be expected to result, directly or indirectly, in its becoming a component or otherwise affecting the characteristics of any food (including any substance intended for use in producing, manufacturing, packing, processing, preparing, treating, packaging, transporting, or holding food; and including any source of radiation intended for any such use), if such substance is not generally recognized, among experts qualified by scientific training and experience to evaluate its safety, as having been adequately shown through scientific procedures (or, in the case of a substance used in food prior to January 1, 1958, through either scientific procedures or experience based on common use in food) to be safe under the conditions of its intended use.

This legal definition differs from that given on page 355, which defines "food additive" as used throughout this statement. For the purposes of this amendment, pesticide chemicals in or on raw agricultural commodities (since the use of these materials is governed by other provisions of the law) and substances used in accordance with a sanction or approval granted prior to the enactment of this legislation are not classified as additives.

Food and chemical manufacturers are now required to file a petition with respect to any intended use of a food additive, as defined in this legislation, proposing the issuance of a regulation prescribing the conditions under which the additive may be safely used. Such a petition must contain the following information:

(1) the name and all pertinent information, including — when available — the chemical identity and composition;
(2) a statement of the conditions of the proposed use of such an additive;
(3) all relevant data on the effect the additive is intended to produce and the quantity required;
(4) a description of a practicable method for the determination of the additive in or on food; and
(5) full reports of investigations made with respect to the safety for use of the additive.

If, after consideration of these data, the scientists of the FDA believe that the additive will be safe, a regulation is issued permitting its use. Such a regulation may place a limit or tolerance level on the amount that may be used. However, according to the 1958 amendment,

> no additive shall be deemed to be safe if it is found to induce cancer when ingested by man or animal, or if it is found, after tests which are appropriate for the evaluation of the safety of food additives, to induce cancer in man or animal.

If the evidence regarding the safety for use of the substance is not considered adequate, the additive will not be permitted. The law applies to any residues that may carry over into meat, milk, or eggs as a result of the use of the substance in animal feeds. It further specifies that the additive shall not be used in an amount greater than that required to produce the desired effect and shall not result in deception of the consumer. The Food and Drug Administration inspectors are continuously on the lookout for foods containing any additives not cleared by FDA under the new law.

The Federal Food, Drug, and Cosmetic Act was later amended to bring the sections dealing with color additives into accord with those pertaining to other additives.

In the development of the Food Additives Amendment, not all substances added to foods are subjected to the requirements of the premarketing clearances mentioned above. Congress provided for certain exceptions, the principal one being to exclude those substances that scientifically qualified experts generally recognize as safe (GRAS) under the conditions of intended use on the basis of either scientific procedures or experience based on common use in food. The effect of this exemption is to use scientific judgment informally in deciding whether it is necessary to require submission of a formal petition, along with supporting evidence of safety, for a regulation to permit and limit the use of a food chemical.

Such common food ingredients as salt, pepper, sugar, vinegar, and baking powder, together with hundreds of other substances, have been regarded as GRAS when used in accordance with "good manufacturing practice." The latter term implies that the substances in question are of suitable food grade and that they are used and remain in food in amounts no higher than necessary to accomplish their intended effects. Judgment that a substance is GRAS also takes into consideration the history of the occurrence or use of the substance in food, experience through its use in medicine or in industry, and biochemical, metabolic, and toxicologic evidence relating to the substance itself or to chemically or pharmacologically comparable substances.

Present Situation

In recent years, there has been considerable discussion of the possibility of hazard from the use of additives in the production and processing of foods. Although many statements are exaggerated, many new chemicals have indeed been and are being introduced. Since the original GRAS list was formulated in the early 1960's, toxicity test procedures have become more sophisticated, and the demands of modern technology have increased the uses of certain substances well beyond the exposure patterns then anticipated. As a result, several substances have been removed from the GRAS list in accordance with the provisions of the 1958 amendment, even though their use had no observable effects in man. The cyclamates, for example, are no longer approved for use in either foods or drugs. Some other substances that were removed from the GRAS list are not permitted to be used in foods only under certain specified conditions and at limited levels of use.

Such actions caused increased concern among consumers regarding the safety of other substances on the GRAS list. President Nixon took cognizance of this situation when, in his consumer message of 1969, he directed the FDA to undertake a full-scale reevaluation of the safety of all substances on the GRAS list. The review of the GRAS list was begun in 1970 (Spiher, 1970).

In connection with the GRAS list review, the FDA in 1971 issued new criteria concerning the eligibility of substances for classification of GRAS (Federal Register, 1971). Once the GRAS review is completed, it is intended that the substances will be placed, under the new criteria, in one of four categories (Grand, 1971).

> An item formerly on the GRAS list becomes a regulated food additive when the substance does not meet the new criteria for GRAS eligibility, and there is sufficient toxicity information to support a food additive petition.
>
> Continuation, in a new GRAS status, where the substance meets the new GRAS criteria.
>
> Use in food to be discontinued until such time as experiments establishing the conditions of safe use have been conducted.
>
> Use in food to be continued during a testing period, where this can be done without undue risk to the public health, under an interim regulation, while additional toxicity testing is performed to complete the required safety information.

The Food and Drug Administration (Federal Register, 1972b) has also established criteria under which the GRAS status of certain substances not previously considered might be affirmed. This regulation applies to those substances (including packaging materials) that industry has deemed to be generally recognized as safe but which FDA had not previously included on the GRAS list or in the GRAS review. After review of the petition requesting GRAS affirmation, the FDA would either: (a) affirm the GRAS status of the substance; (b) issue an interim food additive order, permitting temporary use of the substance until additional data were obtained; (c) classify the substance as a regulated "food additive," rather than as an exempt GRAS substance; or (d) require that the substance be eliminated from the food supply.

The preponderance of data concerning food additives confirms that they have been used safely, effectively, and to the benefit of consumers. The occasional actions taken to remove previously used additives, in the light of new data that limit their margin of safety, confirm the general effectiveness and stringency of the regulatory systems. There is no evidence that consumption of foods in which these substances and regulated additives were properly used has endangered human health.

LOOKING AHEAD

As our population grows and becomes increasingly urbanized, less and less agricultural land per capita, and proportionately less of the population, can be devoted to the production of food. To provide a constant, wholesome, and adequate supply of food for this population, increased production per acre and per man and increased reliance on protection of the food from deterioration during storage and distribution will be required. New chemical aids to production, processing, packaging, and distribution will be needed.

Experience has amply demonstrated that chemical aids of this kind can be used safely and beneficially, but it is imperative that competent, strong regulatory agencies and a public-spirited, ethical industry apply the measures necessary to ensure that they will be so used in the future. It should be pointed out that methods of detecting toxicity, especially long-term effects, have steadily improved and are likely to become more sensitive and reliable, and that health safeguards will continue to require revision according to gains in knowledge.

REFERENCES

Burges, H.D., and N.W. Hussey. 1971. Microbial control of insects and mites, p. 861. Academic Press, New York.

Center for Disease Control. 1972. Ten-State Nutrition Survey 1968-1970. DHEW Publ. No. (HSM) 72-8134. Health Services and Mental Health Administration, U.S. Department of Health, Education, and Welfare, Washington, D.C.

Committee on Food Protection, Food and Nutrition Board. 1954. Principles and procedures for estimating the safety of intentional chemical additives in foods. National Academy of Sciences, Washington, D.C.

Committee on Food Protection, Food and Nutrition Board. 1959a. Principles and procedures for evaluating the safety of food additives. NAS-NRC Publ. 750. National Academy of Sciences, Washington, D.C.

Committee on Food Protection, Food and Nutrition Board. 1959b. Problems in evaluating the carcinogenic hazard from use of food additives. NAS-NRC Publ. 749. National Academy of Sciences, Washington, D.C.

Committee on Food Protection, Food and Nutrition Board. 1965. Some considerations in the use of human subjects in safety evaluation of pesticides and food chemicals. NAS-NRC Publ. 1270. National Academy of Sciences, Washington, D.C.

Committee on Food Protection, Food and Nutrition Board. 1966. Food Chemicals Codex, First ed. NAS-NRC Publ. 1406. National Academy of Sciences, Washington, D.C.

Committee on Food Protection, Food and Nutrition Board. 1967. Use of human subjects in safety evaluation of food chemicals: Proceedings of a conference. NAS-NRC Publ. 1491. National Academy of Sciences, Washington, D.C.

Committee on Food Protection, Food and Nutrition Board. 1970. Evaluating the Safety of Food Chemicals. NAS-NRC Publ. 1859. National Academy of Sciences, Washington, D.C.

Committee on Food Protection, Food and Nutrition Board. 1972. Food Chemicals Codex, Second ed. NAS-NRC Publ. 1949. National Academy of Sciences, Washington, D.C.

Committee on Plant and Animal Pests, Agricultural Board. 1969. Insect-pest management and control. NAS-NRC Publ. 1695. National Academy of Sciences, Washington, D.C.

Division of Pharmacology, Food and Drug Administration. 1959. Appraisal of the safety of chemicals in foods, drugs and cosmetics. The Association of Food and Drug Officials in the U.S., Baltimore.

Federal Register 37, No. 131, 1972a.

Federal Register 37, No. 233, 1972b.

Federal Register 36, No. 123, 1971.

Grant, J.D. 1971. The GRAS review, an overall perspective. Presented at Industry Briefing on the GRAS Questionnaire, New York, May 27, 1971.

Joint FAO/WHO Expert Committee on Food Additives. 1957. General principles governing the use of food additives: First report. FAO Nutrition Meetings Report Series No. 15. WHO technical report.

Joint FAO/WHO Expert Committee on Food Additives. 1958. Procedures for the testing of intentional food additives to establish their safety for use. Second report. FAO Nutrition Meetings Report Series No. 17. WHO technical report.

Joint FAO/WHO Expert Committee on Food Additives. 1961. Evaluation of the carcinogenic hazards of food additives: Fifth report. FAO Nutrition Meetings Report Series No. 29. WHO technical report.

Krause, O.E. 1971. Farm production capacity can meet our needs, pp. 278-284. In The yearbook of agriculture 1971. U.S. Department of Agriculture, Washington, D.C.

Newsom, L.D. 1967. Consequence of insecticide use on new target organisms. Annu. Rev. Entomol. 12:257-286.

Newsom, L.D. 1970. The end of an era and future prospects for insect control. Proceedings Tall Timbers Conf. 2:117-136.

Oppenoorth, F.T. 1965. Biochemical genetics of insecticide resistances. Annu. Rev. Entomol. 10:185-206.

Pickett, A.D. 1959. Utilization of native parasites and predators. J. Econ. Entomol. 52:1103-1105.

Secretary's Commission on Pesticides and Their Relationship to Environmental Health. 1969. I and II. U.S. Department of Health, Education, and Welfare, U.S. Government Printing Office, Washington, D.C.

Spiher, A.T. 1970. The GRAS list review. FDA Papers 4:12-14.

Smith, Roger H., and R.C. von Borstel. 1972. Genetic control of insect populations. Science 178 (4066):1164-1174.

U.S. Office of Science and Technology. 1971. Ecological effects of pesticides on non-target species, p. 200. U.S. Government Printing Office, Washington, D.C.

World Health Organization Scientific Group. 1967. Procedures for investigating intentional and unintentional food additives. WHO Technical Report Series No. 348.

World Health Organization. 1971. Fourth report on the world health situation, 1965-1968. Off. Rec. WHO, No. 192.

APPENDIX II

FEDERAL FOOD, DRUG, AND COSMETIC ACT
Selected Abstracts

CHAPTER II — DEFINITIONS

Sec. 201 (s): The term "food additive" means any substance the intended use of which results or may reasonably be expected to result, directly or indirectly, in its becoming a component or otherwise affecting the characteristics of any food (including any substance intended for use in producing, manufacturing, packing, processing, preparing, treating, packaging, transporting, or holding food; and including any source of radiation intended for any such use), if such substance is not generally recognized, among experts qualified by scientific training and experience to evaluate its safety, as having been adequately shown through scientific procedures (or, in the case of a substance used in food prior to January 1, 1958, through either scientific procedures or experience based on common use in food) to be safe under the conditions of its intended use; except that such term does not include—

(1) a pesticide chemical in or on a raw agricultural commodity; or
(2) a pesticide chemical to the extent that it is intended for use or is used in the production, storage, or transportation of any raw agricultural commodity; or
(3) a color additive; or
(4) any substance used in accordance with a sanction or approval granted prior to the enactment of this paragraph pursuant to this Act, the Poultry Products Inspection Act (21 U.S.C. 451 and the following) or the Meat Inspection Act of March 4, 1907 (34 Stat. 1260), as amended and extended (21 U.S.C. 71 and the following).

CHAPTER IV — FOOD

Adulterated Food

Sec. 402: A food shall be deemed to be adulterated—
(a) (1) If it bears or contains any poisonous or deleterious substance which may render it injurious to health; but in case the substance is not an added substance such food shall not be considered adulterated under this clause if the quantity of such substance in such food does not ordinarily render it injurious to health; or

(2) (A) if it bears or contains any added poisonous or added deleterious substance (other than one which is (i) a pesticide chemical in or on a raw agricultural commodity; (ii) a food additive; or (iii) a color additive) which is unsafe within the meaning of section 406, or (B) if it is a raw agricultural commodity and it bears or contains a pesticide chemical which is unsafe within the meaning of section 408(a), or (C) if it is, or it bears or contains, any food additive which is unsafe within the meaning of section 409: *Provided,* That where a pesticide chemical has been used in or on a raw agricultural commodity in conformity with an exemption granted or a tolerance prescribed under section 408 and such raw agricultural commodity has been subjected to processing such as canning, cooking, freezing, dehydrating, or milling, the residue of such pesticide chemical remaining in or on such processed food shall, notwithstanding the provisions of sections 406 and 409, not be deemed unsafe if such residue in or on the raw agricultural commodity has been removed to the extent possible in good manufacturing practice and the concentration of such residue in the processed food when ready to eat is not greater than the tolerance prescribed for the raw agricultural commodity;

(3) if it consists in whole or in part of any filthy, putrid, or decomposed substance, or if it is otherwise unfit for food; or

(4) if it has been prepared, packed, or held under insanitary conditions whereby it may have become contaminated with filth, or whereby it may have been rendered injurious to health; or

(5) if it is, in whole or in part, the product of a diseased animal or of an animal which has died otherwise than by slaughter; or

(6) if its container is composed, in whole or in part, of any poisonous or deleterious substance which may render the contents injurious to health; or

(7) if it has been intentionally subjected to radiation, unless the use of the radiation was in conformity with a regulation or exemption in effect pursuant to section 409.

(b) (1) If any valuable constituent has been in whole or in part omitted or abstracted therefrom; or

(2) if any substance has been substituted wholly or in part therefor; or

(3) if damage or inferiority has been concealed in any manner; or

(4) if any substance has been added thereto or mixed or packed therewith so as to increase its bulk or weight, or reduce its quality or strength, or make it appear better or of greater value than it is.

[Sec. 403: A food shall be deemed to be misbranded—]

(k) If it bears or contains any artificial flavoring, artificial coloring, or chemical preservative, unless it bears labeling stating that fact: *Provided,* That to the extent that compliance with the requirements of this paragraph is impracticable, exemptions shall be established by regulations promulgated by the Secretary. The provisions of this paragraph and paragraphs (g) and (i) with respect to artificial coloring shall not apply in the case of butter, cheese, or ice cream. The provisions of this paragraph with respect to chemical preservatives shall not apply to a pesticide chemical when used in or on a raw agricultural commodity which is the produce of the soil.

[Regulation] **Food; labeling; artificial flavoring or coloring, chemical preservatives**

§1.12 (a) (1) The term "artificial flavoring" means a flavoring containing any sapid or aromatic constituent, which constituent was manufactured by a process of synthesis or other similar artifice.

(2) The term "artificial coloring" means a coloring containing any dye or pigment, which dye or pigment was manufactured by a process of synthesis or other similar artifice, or a coloring which was manufactured by extracting a natural dye or natural pigment from a plant or other material in which such dye or pigment was naturally produced.

(3) The term "chemical preservative" means any chemical that, when added to food, tends to prevent or retard deterioration thereof, but does not include common salt, sugars, vinegars, spices or oils extracted from spices, substances added to food by direct exposure thereof to wood smoke, or chemicals applied for their insecticidal or herbicidal properties.

(b) A food which is subject to the requirements of section 403(k) of the act shall bear labeling, even though such food is not in package form.

(c) A statement of artificial flavoring, artificial coloring, or chemical preservative shall be placed on the food, or on its container or wrapper, or on any two or all of these, as may be necessary to render such statement likely to be read by the ordinary individual under customary conditions of purchase and use of such food.

Tolerances for Poisonous Ingredients in Food

Sec. 406: Any poisonous or deleterious substance added to any food, except where such substance is required in the production thereof or cannot be avoided by good manufacturing practice shall be deemed to be unsafe for purposes of the application of clause (2) (A) of section 402(a); but when such substance is so required or cannot be so avoided, the Secretary shall promulgate regulations limiting the quantity therein or thereon to such extent as he finds necessary for the protection of public health, and any quantity exceeding the limits so fixed shall also be deemed to be unsafe for purposes of the application of clause (2) (A) of section 402(a). While such a regulation is in effect limiting the quantity of any such substance in the case of any food, such food shall not, by reason of bearing or containing any added amount of such substance, be considered to be adulterated within the meaning of clause (1) of section 402(a). In determining the quantity of such added substance to be tolerated in or on different articles of food the Secretary shall take into account the extent to which the use of such substance is required or cannot be avoided in the production of each such article and the other ways in which the consumer may be affected by the same or other poisonous or deleterious substances.

Food Additives

Unsafe Food Additives

Sec. 409: (a) A food additive shall, with respect to any particular use or intended use of such additives, be deemed to be unsafe for the purposes of the application of clause (2) (C) of section 402(a), unless—

    (1) it and its use or intended use conform to the terms of an exemption which is in effect pursuant to subsection (i) of this section; or

    (2) there is in effect, and it and its use or intended use are in conformity with, a regulation issued under this section prescribing the conditions under which such additive may be safely used.

While such a regulation relating to a food additive is in effect, a food shall not, by reason of bearing or containing such an additive in accordance with the regulation, be considered adulterated within the meaning of clause (1) of section 402(a).

Petition to Establish Safety

(b) (1) Any person may, with respect to any intended use of a food additive, file with the Secretary a petition proposing the issuance of a regulation prescribing the conditions under which such additive may be safely used.

    (2) Such petition shall, in addition to any explanatory or supporting data, contain—

        (A) the name and all pertinent information concerning such food additive, including, where available, its chemical identity and composition;

        (B) a statement of the conditions of the proposed use of such additive, including all directions, recommendations, and suggestions proposed for the use of such additive, and including specimens of its proposed labeling;

        (C) all relevant data bearing on the physical or other technical effect such additive is intended to produce, and the quantity of such additive required to produce such effect;

        (D) a description of practicable methods for determining the quantity of such additive in or on food, and any substance formed in or on food, because of its use; and

        (E) full reports of investigations made with respect to the safety for use of such additive, including full information as to the methods and controls used in conducting such investigations.

    (3) Upon request of the Secretary, the petitioner shall furnish (or, if the petitioner is not the manufacturer of such additive, the petitioner shall have the manufacturer of such additive furnish, without disclosure to the petitioner) a full description of the methods used in, and the facilities and controls used for, the production of such additive.

(4)   Upon request of the Secretary, the petitioner shall furnish samples of the food additive involved, or articles used as components thereof, and of the food in or on which the additive is proposed to be used.

(5)   Notice of the regulation proposed by the petitioner shall be published in general terms by the Secretary within thirty days after filing.

Action on the Petition

(c) (1) The Secretary shall—

(A) by order establish a regulation (whether or not in accord with that proposed by the petitioner) prescribing, with respect to one or more proposed uses of the food additive involved, the conditions under which such additive may be safely used (including, but not limited to, specifications as to the particular food or classes of food in or on which such additive may be used, the maximum quantity which may be used or permitted to remain in or on such food, the manner in which such additive may be added to or used in or on such food, and any directions or other labeling or packaging requirements for such additive deemed necessary by him to assure the safety of such use), and shall notify the petitioner of such order and the reasons for such action; or

(B) by order deny the petition, and shall notify the petitioner of such order and of the reasons for such action.

(2) The order required by paragraph (1) (A) or (B) of this subsection shall be issued within ninety days after the date of filing of the petition, except that the Secretary may (prior to such ninetieth day), by written notice to the petitioner, extend such ninety day period to such time (not more than one hundred and eighty days after the date of filing of the petition) as the Secretary deems necessary to enable him to study and investigate the petition.

(3) No such regulation shall issue if a fair evaluation of the data before the Secretary—

(A) fails to establish that the proposed use of the food additive, under the conditions of use to be specified in the regulation, will be safe: *Provided,* That no additive shall be deemed to be safe if it is found to induce cancer when ingested by man or animal, or if it is found, after tests which are appropriate for the evaluation of the safety of food additives, to induce cancer in man or animal, except that this proviso shall not apply with respect to the use of a substance as an ingredient of feed for animals which are raised for food production, if the Secretary finds (i) that, under the conditions of use and feeding specified in proposed labeling and reasonably certain to be followed in practice, such additive will not adversely affect the animals for which such feed is intended, and (ii) that no residue of the additive will be found (by methods of examination prescribed or approved by the Secretary by regulations, which regulations shall not be subject to subsections (f) and (g) in any edible portion of such animal after slaughter or in any food yielded by or derived from the living animal; or

(B) shows that the proposed use of the additive would promote deception of the consumer in violation of this Act or would otherwise result in adulteration or in misbranding of food within the meaning of this Act.

(4) If, in the judgement of the Secretary, based upon a fair evaluation of the data before him, a tolerance limitation is required in order to assure

that the proposed use of an additive will be safe, the Secretary—

   (A) shall not fix such tolerance limitation at a level higher
than he finds to be reasonably required to accomplish
the physical or other technical effect for which such
additive is intended; and

   (B) shall not establish a regulation for such proposed use
if he finds upon a fair evaluation of the data before
him that such data do not establish that such use
would accomplish the intended physical or other
technical effect.

(5) In determining, for the purposes of this section, whether a proposed use
of a food additive is safe, the Secretary shall consider among other relevant factors—

   (A) the probable consumption of the additive and of any
substance formed in or on food because of the use
of the additive;

   (B) the cumulative effect of such additive in the diet of
man or animals, taking into account any chemically
or pharmacologically related substance or substances
in such diet; and

   (C) safety factors which in the opinion of experts qualified
by scientific training and experience to evaluate the
safety of food additives are generally recognized as appropriate for the use of animal experimentation data.

Regulation Issued on Secretary's Initiative

(d) The Secretary may at any time, upon his own initiative, propose the issuance of a regulation prescribing, with respect to any particular use of a food additive, the conditions under which such additive may be safely used, and the reasons therefor. After the thirtieth
day following publication of such a proposal, the Secretary may by order establish a regulation based upon the proposal.

Publication and Effective Date of Orders

(e) Any order, including any regulation established by such order, issued under subsection
(c) or (d) of this section, shall be published and shall be effective upon publication, but
the Secretary may stay such effectiveness if, after issuance of such order, a hearing is sought
with respect to such order pursuant to subsection (f).

Objections and Public Hearing

(f) (1) Within thirty days after publication of an order made pursuant to subsection (c) or (d) of this section, any person adversely affected by such
an order may file objections thereto with the Secretary, specifying with
particularity the provisions of the order deemed objectionable, stating
reasonable grounds therefor, and requesting a public hearing upon such
objections. The Secretary shall, after due notice, as promptly as possible hold such public hearing for the purpose of receiving evidence relevant and material to the issues raised by such objections. As soon as
practicable after completion of the hearing, the Secretary shall by order
act upon such objections and make such order public.

   (2) Such order shall be based upon a fair evaluation of the entire record
at such hearing, and shall include a statement setting forth in detail
the findings and conclusions upon which the order is based.

   (3) The Secretary shall specify in the order the date on which it shall take
effect, except that it shall not be made to take effect prior to the
ninetieth day after its publication, unless the Secretary finds that
emergency conditions exist necessitating an earlier effective date, in

which event the Secretary shall specify in the order his findings as to such conditions.

Judicial Review

(g) (1) In a case of actual controversy as to the validity of any order issued under subsection (f), including any order thereunder with respect to amendment or repeal of a regulation issued under this section, any person who will be adversely affected by such order may obtain judicial review by filing in the United States Court of Appeals for the circuit wherein such person resides or has his principal place of business, or in the United States Court of Appeals for the District of Columbia Circuit, within sixty days after the entry of such order, a petition praying that the order be set aside in whole or in part.

(2) A copy of such petition shall be forthwith transmitted by the clerk of the court to the Secretary, or any officer designated by him for that purpose, and thereupon the Secretary shall file in the court the record of the proceedings on which he based his order, as provided in section 2112 of title 28, United States Code. Upon the filing of such petition the court shall have jurisdiction, which upon the filing of the record with it shall be exclusive, to affirm or set aside the order complained of in whole or in part. Until the filing of the record the Secretary may modify or set aside his order. The findings of the Secretary with respect to questions of fact shall be sustained if based upon a fair evaluation of the entire record at such hearing. The court shall advance on the docket and expedite the disposition of all causes filed therein pursuant to this section.

(3) The court, on such judicial review, shall not sustain the order of the Secretary if he failed to comply with any requirement imposed on him by subsection (f)(2) of this section.

(4) If application is made to the court for leave to adduce additional evidence, the court may order such additional evidence to be taken before the Secretary and to be adduced upon the hearing in such manner and upon such terms and conditions as to the court may seem proper, if such evidence is material and there were reasonable grounds for failure to adduce such evidence in the proceedings below. The Secretary may modify his findings as to the facts and order by reason of the additional evidence so taken, and shall file with the court such modified findings and order.

(5) The judgment of the court affirming or setting aside, in whole or in part, any order under this section shall be final, subject to review by the Supreme Court of the United States upon certiorari or certification as provided in section 1254 of title 28 of the United States Code. The commencement of proceedings under this section shall not, unless specifically ordered by the court to the contrary, operate as a stay of an order.

Amendment or Repeal of Regulations

(h) The Secretary shall by regulation prescribe the procedure by which regulations under the foregoing provisions of this section may be amended or repealed, and such procedure shall conform to the procedure provided in this section for the promulgation of said regulations.

Exemptions for Investigational Use

(i) Without regard to subsections (b) to (h), inclusive, of this section, the Secretary shall by regulation provide for exempting from the requirements of this section any food additive, and any food bearing or containing such additive, intended solely for investigational use by qualified experts when in his opinion such exemption is consistent with the public health.

APPENDIX III

FEDERAL REGISTER
Selected Abstracts

FEDERAL REGISTER, VOL. 38, NO. 148, AUGUST 2, 1973

Department of Health, Education, and Welfare, Food and Drug Administration
[21 CFR Part 1]

LABELING OF CHEMICAL PRESERVATIVES IN FOOD

Notice of Proposed Rule Making

Section 403(i) of the Federal Food, Drug, and Cosmetic Act requires the common or usual name of each ingredient used in a fabricated food to be listed on the label, and section 403(k) of the act requires that any food containing a chemical preservative bear labeling stating that fact. The Food and Drug Administration has consistently interpreted these requirements to mean that a preservative ingredient contained in a food must be designated in the statement of ingredients by both the common or usual name of the ingredient and a separate designation of that ingredient as a preservative (e.g., BHA, a preservative, or calcium propionate, to retard spoilage).

At one time, the United States Department of Agriculture permitted the label designation of preservatives in meat and poultry products by the use of such general designations as "oxygen interceptor" and "freshness preserver," without the specific name of the ingredient. The USDA regulations have now been changed, and require that the label designation of preservatives in meat and poultry products identify the substance by both its common name and its purpose (9 CFR 317.2(f)(1), (j)(10), (12) published in the *Federal Register* on October 3, 1970 (35 FR 15551, 15581, 15583).

The Food and Drug Administration is therefore aware of no exception to the rule that every preservative in food must be labeled both by its common or usual name and its function. Because some food labels presently do not comply with these requirements, and because new food labels are now being prepared for compliance with food labeling regulations published in the *Federal Registers* of January 19 and March 14, 1973, and elsewhere in this issue of the *Federal Register*, the Commissioner of Food and Drugs has concluded that it is appropriate to propose a specific regulation covering this matter.

Therefore, pursuant to provisions of the Federal Food, Drug, and Cosmetic Act [secs. 403, 701(a), 52 Stat. 1047–1048, 1055; 21 U.S.C. 343, 371(a)] and under authority delegated to the Commissioner (21 CFR 2.120), it is proposed that Part 1 be amended in §1.12 by adding thereto a new paragraph (j) to read as follows:

§1.12 *Food; labeling; spices, flavorings, colorings and chemical preservatives.*

(j) A food to which a chemical preservative(s) is added shall, except when exempt pursuant to §1.10a, bear a label declaration stating both the common or usual name of the ingredient(s) and a separate description of its function, e.g., "preservative," "to retard spoilage," "a mold inhibitor," "to help protect flavor," or "to promote color retention."

FEDERAL REGISTER, VOL. 38, NO. 143, JULY 26, 1973

Department of Health, Education, and Welfare, Food and Drug Administration
[21 CFR Parts 3, 121]

SUBSTANCES PROHIBITED FROM USE IN FOOD

Notice of Proposed Rulemaking

The Food and Drug Administration has prohibited the use of various substances in food on the basis of toxicological data showing a potential hazard to public health or because inadequate data exist to conclude that they are safe for use in food. Some of these actions were taken prior to enactment of the Food Additives Amendment of 1958, and others have been taken pursuant to that Amendment.

Because information on these actions is presently scattered throughout existing regulations, *Federal Register* notices not codified in the Code of Federal Regulations, old trade correspondence (TC), and unpublished correspondence, and thus are either difficult to find or are not generally available to the public, the Commissioner of Food and Drugs has concluded that they should be consolidated in one regulation. All of the substances presently proposed for inclusion in this regulation were the subject of action previously taken. The Commissioner is not now proposing such action against any additional substances. Should the current review of the safety of direct human food ingredients classified as generally recognized as safe (GRAS) or subject to a prior sanction justify additional action of this type, this proposed new section will also be used for that purpose.

Some of these food additives were prohibited from use in food on the conclusion that the available evidence did not establish safety, and not on the basis of a determination that the ingredient was in fact unsafe. Section 409 of the act places the burden on the manufacturer or distributor of a food additive to prove its safety prior to use. Accordingly, the Commissioner recognizes that, as additional scientific information becomes available, it may well be possible to approve one or more of these ingredients for food use and thus to delete it from this section. The proposed regulation provides for such transfers to and from this section on the Commissioner's initiative or on the petition of any interested person.

The fact that a substance does not appear on this list of prohibited substances does not mean that it may lawfully be used in food. This proposed new section includes only a partial list of prohibited substances, for easy reference purposes, and is not a complete list of substances that may not lawfully be used in food. Before any substance may be used in food, it must meet all of the applicable requirements of section 401 and 409 of the act.

Accordingly, the Commissioner of Food and Drugs concludes that it is in the public interest and will promote efficient enforcement of the act to provide a section in the food additive regulations to contain a listing of food ingredients for which use in food has been prohibited.

Therefore, pursuant to provisions of the Federal Food, Drug, and Cosmetic Act [secs. 201(s), 409, 701(a), 52 Stat. 1055, 72 Stat. 1785-1787, as amended; 21 U.S.C. 321(s), 348, 371 (a)] and under authority delegated to him (21 CFR 2.120) the Commissioner proposes that Trade Correspondence No. 377 (December 29, 1941) be revoked, and that Title 21 of the Code of Federal Regulations be amended.

Substances Prohibited from Use in Food

(a) The food ingredients listed in this section have been prohibited from use in food by the Food and Drug Administration because of a determination that they present a potential risk to the public health or have not been shown by adequate scientific data to be safe for use in food. Use of any of these substances in violation of this section causes the food involved to be adulterated in violation of the act.

(b) This section includes only a partial list of substances prohibited from use in food, for easy reference purposes, and is not a complete list of substances that may not lawfully be used in food. No substance may be used in food unless it meets all applicable requirements of the act.

(c) The Commissioner of Food and Drugs, either on his own initiative or on behalf of any interested person who has submitted a petition, may publish a proposal to establish, amend, or repeal a regulation under this section on the basis of new scientific evaluation or information. Any such petition shall include an adequate scientific basis to support the petition, shall be in the form set forth in §2.65 of this chapter, and will be published for comment if it contains reasonable grounds.

(d) Substances prohibited from direct addition to food:

(1) *Calamus, oil of calamus, extract of calamus.* (i) Calamus is the dried rhizome of Acorus calamus L. It has been used as a flavoring compound, especially as the oil or extract. (ii) Food containing any added calamus, oil of calamus, or extract of calamus is deemed to be adulterated in violation of the act based upon an order published in the *Federal Register* of May 9, 1968 (33 FR 6967).

(2) *Dulcin.* (i) Dulcin is the chemical 4-ethoxyphenylurea. It is a synthetic chemical having a sweet taste about 250 times that of sucrose, is not found in natural products at levels detectable by the official methodology, and has been proposed for use as an artificial sweetener. (ii) Food containing any added or detectable level of dulcin is deemed to be adulterated in violation of the act based upon an order published in the *Federal Register* of January 19, 1950 (15 FR 321).

(3) *P-4000.* (i) P-4000 is the chemical 5-nitro-2-n-propoxyaniline. It is a synthetic chemical having a sweet taste about 4,000 times that of sucrose, is not found in natural products at levels detectable by the official methodology, and has been proposed for use as an artificial sweetener. (ii) Food containing any added or detectable level of P-4000 is deemed to be adulterated in violation of the act based upon an order published in the *Federal Register* of January 19, 1950 (15 FR 321).

(4) *Coumarin.* Coumarin is the chemical 1,2-benzopyrone. It is found in tonka beans, among other natural sources, and is also synthesized. It has been used as a flavoring compound. (ii) Food containing any added coumarin as such or as a constituent of tonka beans or tonka extract is deemed to be adulterated under the act based upon an order published in the *Federal Register* of March 5, 1953 (19 FR 1239).

(5) *Cyclamate; calcium, sodium, magnesium and potassium.* (i) Calcium, sodium, magnesium and potassium salts of cyclohexane sulfamic acid. Cyclamates are synthetic chemicals having a sweet taste 30 to 40 times that of sucrose, are not found in natural products at levels detectable by the official methodology, and have been used as artificial sweeteners. (ii) Food containing any added or detectable level of cyclamate is deemed adulterated in violation of the act based upon an order published in the *Federal Register* of October 21, 1969 (34 FR 17063).

(6) *Safrole.* (i) Safrole is the chemical 4-allyl-1,2-methylene-dioxybenzene. It is a natural constituent of the sassafras plant. Oil of sassafras is about 80% safrole. Isosafrole and dihydrosafrole are derivatives of safrole, and have been used as flavors. (ii) Food containing any added safrole, oil of sassafras, dihydrosafrole, or safrole or as a constituent of any food or extract is deemed to be adulterated in violation of the act based upon an order published in the *Federal Register* of December 3, 1960 (25 FR 12412).

(7) *Monochloroacetic acid.* (i) Monochloroacetic acid is the chemical chloroacetic acid. It is a synthetic chemical not found in natural products,
and has been proposed as a preservative in alcoholic and nonalcoholic
beverages. Monochloroacetic acid is permitted in food package adhesives with an accepted migration level up to 10 ppb under § 121.2520.
The official methods do not detect monochloroacetic acid at the 10
parts per billion level. (ii) Food containing any added or detectable
level of monochloroacetic acid is deemed adulterated in violation of
the act based upon trade correspondence dated December 29, 1941
(TC-377).

(8) *Thiourea.* (i) Thiourea is the chemical thiocarbamide. It is a synthetic
chemical, is not found in natural products at levels detectable by the
official methodology, and has been proposed as an antimycotic for
use in dipping citrus. (ii) Food containing any added or detectable
level of thiourea is deemed to be adulterated under the act.

(9) *Cobaltous Salts; acetate, chloride and sulfate.* (i) Cobaltous salts have
been used in fermented malt beverages as a foam stabilizer and to
prevent gushing. (ii) Food containing any added cobaltous salts is
deemed to be adulterated in violation of the act based upon an order
published in the *Federal Register* of August 12, 1966 (31 FR 8788).

(10) *NDGA (Nordihydroguaiaretic acid).* (i) Nordihydroguaiaretic acid is
the chemical 4,4'-(2,3-dimethyltetramethylene)dipyrocatechol. It
occurs naturally in the resinous exudates of certain plants. The commercial product, which is synthesized, has been used as an antioxidant
in foods. (ii) Food containing any added NDGA is deemed to be
adulterated in violation of the act based upon an order published in
the *Federal Register* of April 11, 1968 (33 FR 5619).

(11) *DEPC (Diethylpyrocarbonate).* (i) Diethylpyrocarbonate is the chemical pyrocarbonic acid diethyl ester. It is a synthetic chemical not
found in natural products at levels detectable by available methodology and has been used as a ferment inhibitor in alcoholic and nonalcoholic beverages. (ii) Food containing any added or detectable
level of DEPC is deemed adulterated in violation of the act based
upon an order published in the *Federal Register* of August 2, 1972
(37 FR 15426).

(e) Substances prohibited from indirect addition to food through use in food contact surfaces:

(1) *Flectol H.* (i) Flectol H is the chemical polymerized 1,2-dihydro-
2,2,4-trimethylquinoline. It is a synthetic chemical not found in
natural products, and has been used as a component of food packaging adhesives. (ii) Food containing any added or detectable level
of this substance is deemed adulterated in violation of the act based
upon an order published in the *Federal Register* of April 7, 1967
(32 FR 5675).

(2) *4,4'-Methylenebis(2-chloroaniline).* (i) 4,4'-Methylenebis(2-chloroanaline)
is a synthetic chemical not found in natural products and has been
used as a polyurethane curing agent and as a component of food
packaging adhesive and polyurethane resins. (ii) Food containing
any added or detectable level of this substance is deemed adulterated
in violation of the act based upon an order published in the *Federal
Register* of December 2, 1969 (34 FR 19073).

FEDERAL REGISTER, VOL. 38, NO. 143, JULY 26, 1973

Department of Health, Education, and Welfare, Food and Drug Administration
[21 CFR Part 121]

FOOD CATEGORIES AND FOOD INGREDIENT FUNCTIONS

Proposed Designation

The Food and Drug Administration is conducting a study of the direct human food ingredients classified as generally recognized as safe (GRAS) or subject to a prior sanction. As this study progresses, the Commissioner of Food and Drugs will publish in the *Federal Register,* appropriate proposals to (1) affirm GRAS status, (2) publish a prior sanction, (3) establish an interim food additive regulation, (4) establish a permanent food additive regulation, or (5) eliminate food use of the ingredient under review. The Commissioner is proposing regulations in this issue of the *Federal Register* with respect to the first ingredients subject to this review.

In regulations published since 1958 under the Food Additives Amendment, it has frequently been appropriate to designate broad food categories in which an ingredient may properly be used, and to state the functional purpose for which the ingredient may be used. To date, no standardized definitions of the food categories or functional descriptions have been adopted.

In conducting the industry survey of production and use of GRAS and prior-sanctioned food substances, under contract with the Food and Drug Administration, for use in the review of the safety of these ingredients, the National Academy of Sciences developed standardized food categories.

Food categories of a similar type were also used by the United States Department of Agriculture and the Market Research Corporation of America (MRCA), in their respective surveys, to determine the sizes of servings used by consumers and the frequency of consumption of specific foods.

The Commissioner has concluded that the food categories adopted by the National Academy of Sciences (NAS) represent a valid and useful method of dividing food products into general classes of related products. Where tolerances or limitations are established for the use of direct human food ingredients, and there is significant variation with respect to appropriate tolerances or limitations for one or more specific food categories, this method of classification will permit designation of the foods to be covered without requiring a detailed list of each of the individual products included.

In many instances, tolerances or limitations may be imposed uniformly for all foods. It may also be necessary to impose, with relatively few exceptions, tolerances or limitations for specific food categories at levels higher or lower than the general rule. It is the Commissioner's intent to utilize the broadest possible approach, in the interests of simplification, wherever justified by the available safety data and information.

It is appropriate that the same food classification system developed by the NAS for its production and consumption survey should also be utilized by the Food and Drug Administration in imposing tolerances and limitations for use of specific ingredients. NAS Survey data was accumulated using these food categories, and they are consequently of great assistance in determining whatever tolerances and limitations are justified.

The same classification system, already cross-indexed to the MRCA and USDA consumption data, also provides an immediate reference to consumption patterns on which those tolerances and limitations are in part based.

The proposal set out below contains a general description of each food category, without attempting to list in detail all the products within it. Wherever any question arises with respect to the proper classification of a specific food product which might reasonably fall within two or more categories, proper classification will be determined by referring to the more detailed and specific classification lists established by the MRCA and cross-indexed to NAS food categories, as contained in the final NAS report to the Food and Drug Administration. The Final Report of the NAS is now available from the National Technical Information Service (NTIS), in accordance with the notice on this matter published in this issue of the *Federal Register*. Accordingly, the Commissioner is incorporating this specific classification list, by reference, into this proposed regulation, for purposes of resolving close questions with respect to proper classification.

The NAS also found it necessary to establish a similar classification system with respect to the technical functions performed by the various specific ingredients directly added to human food. These functional effects are contained in the final NAS report to the Food and Drug Administration and are the subject of production, use, and consumption data on the technical functions of numerous food ingredients, added to the various NAS food categories. Thus, these tables describe the specific technical purposes for which GRAS and prior-sanctioned ingredients are added to NAS food categories, and they consequently serve as an excellent reference to consumption patterns on which ingredient tolerances and limitations are in part based. Accordingly, the Commissioner is proposing to standardize the technical functional descriptions submitted by the NAS, so that regulations permitting the use of ingredients in food will accurately describe their purpose. A standardized system of classification will also assist consumers in understanding the functions performed by these ingredients in the foods they consume.

Therefore, pursuant to provisions of the Federal Food, Drug, and Cosmetic Act [secs. 201(s), 409, 701(a), 52 Stat. 1055, 72 Stat. 1784–1788 as amended; 21 U.S.C. 321(s), 348, 371(a)] and under authority delegated to him (21 CFR 2.120), the Commissioner proposes to amend Part 121 by adding to §121.1 the following two new paragraphs:

§121.1 *Definitions and interpretations.*

(l) The following general food categories are established to group specific related foods together for the purpose of establishing tolerances or limitations for the use of direct human food ingredients. Individual food products will be classified within these categories according to the detailed classification lists contained in Exhibit 33B of the report of the National Academy of Sciences on "A Comprehensive Survey of Industry on the use of Food Chemicals Generally Recognized as Safe" (September 1972).

- (1) Baked goods and baking mixes (includes ready-to-eat or ready-to-bake products and all dry flour, meal, and multipurpose mixes).
- (2) Beverages, alcoholic (includes malt beverages, wines, and distilled liquors).
- (3) Beverages and beverage bases, nonalcoholic (includes dry and liquid imitation concentrates and ready-to-drink products, except for coffee or tea).
- (4) Breakfast cereals (includes cold and hot breakfast cereals).
- (5) Cheeses (includes standardized, nonstandardized, snack, and other miscellaneous cheeses).
- (6) Chewing gum (includes all flavored gums).
- (7) Coffee and tea (includes regular and instant products).
- (8) Condiments and relishes (includes plain seasoning sauces and spreads, olives, pickles, and relishes).
- (9) Confections and frostings (includes candy and flavored frostings and frosting sugars).
- (10) Dairy product analogs (includes nondairy derived products such as toppings, mixes, and coffee whiteners).
- (11) Egg products (includes liquid, frozen, or dried eggs, and egg products).
- (12) Fats and oils (includes salad dressings, margarines, butter, and cooking oils).
- (13) Fish products (includes all prepared or frozen products containing fish, shellfish, and other aquatic animals, except fresh fish).

(14) Fresh eggs (includes only whole fresh eggs).

(15) Fresh fish (includes only fresh and home frozen fish, shellfish, and other aquatic animals).

(16) Fresh fruits and fruit juices (includes only raw fruits and fruit juices and fruit blends).

(17) Fresh meats (includes only fresh and home frozen beef, pork, lamb, and game animals).

(18) Fresh poultry (includes only fresh and home frozen poultry).

(19) Fresh vegetables and potatoes (includes only fresh, home canned, and home frozen vegetables and potatoes).

(20) Frozen dairy desserts and mixes (includes ice cream, ice milk, sherbets, and frozen novelties).

(21) Fruit and water ices (includes all frozen fruit and water ices).

(22) Gelatins, puddings, and fillings (includes flavored gelatins, puddings, custards, parfaits, and pie fillings).

(23) Grain products and pastas (includes macaroni, noodle, and rice dishes, without meat or vegetables).

(24) Gravies and Sauces (includes flavored meat sources, gravies, and marinades).

(25) Hard candy and cough drops (includes all hard sucker type candies).

(26) Herbs, seeds, spices, seasonings, blends, extracts, and flavorings (includes all natural spices and blends and artificial flavors).

(27) Jams and jellies, homemade (includes fruit butters and preserves).

(28) Jams, jellies, and sweet spreads (includes fruit butters and preserves).

(29) Meat products (includes all prepared and frozen products containing beef, pork, or lamb).

(30) Milk, whole and skim (includes only whole and skim milks).

(31) Milk products (includes dried and fluid milk products such as concentrated, evaporated, and flavored milk, and cream products).

(32) Nuts and nut products (includes whole or shelled nuts, coconut, and nut spreads).

(33) Poultry products (includes all prepared and frozen products containing poultry).

(34) Processed fruits and juices (includes canned or frozen fruits and fruit juices, concentrates, dilutions, ades, and drink substitutes).

(35) Processed vegetables and juices (includes canned or frozen vegetables, vegetable juices, and blends).

(36) Reconstituted vegetable proteins (includes only meat substitute products and dishes).

(37) Snack foods (includes chips, pretzels, and other novelty snacks).

(38) Soft candy (includes all soft and nougat candies).

(39) Soups, homemade (includes all homemade soups).

(40) Soups and soup mixes (includes all meat and vegetable soups).

(41) Sugar, white, granulated (includes only white granulated sugar).

(42) Sugar substitutes (includes all forms of sugar substitutes).

(43) Sweet sauces, toppings, and syrups (includes all fruit, berry, or other flavored products).

(m) The following terms describe the physical or technical functional effects for which direct human food ingredients may be added to foods. They are adopted from the National Academy of Sciences national survey of food industries, reported to the Food and Drug Administration under the contract title, "A Comprehensive Survey of Industry on the Use of Food Chemicals Generally Recognized as Safe" (September 1972).

(1) "Anticaking agents and free-flow agents:" substances added to finely powdered or crystalline food products to prevent caking, lumping, or agglomeration.

(2) "Antioxidants:" substances used to retard deterioration, rancidity, or discoloration due to oxidation.

(3) "Colors and coloring adjuncts" (including color stabilizers, color fixatives, color-retention agents, etc.): substances used to impart, preserve, or enhance the color or shading of a food.

(4) "Curing and pickling agents:" substances imparting a unique flavor and/or color to a foodstuff, usually producing an increase in shelf life stability.

(5) "Dough conditioners" (including yeast foods): substances used to modify the gluten and enhance the property of making an elastic and stable dough.

(6) "Drying agents:" substances with moisture-absorbing ability, used to maintain an atmosphere of low moisture.

(7) "Emulsifiers and emulsifier salts:" substances which modify surface tension in the component phase of an emulsion to establish a uniform dispersion or emulsion.

(8) "Enzymes:" enzymes used to improve food processing.

(9) "Firming agents:" substances added to precipitate residual pectin, thus strengthening the supporting tissue and preventing its collapse during processing.

(10) "Flavor enhancers:" substances added to supplement, enhance, or modify the original taste and/or aroma of a food without imparting a characteristic taste or aroma of its own.

(11) "Flavoring agents and adjuvants:" substances added to impart or help impart a taste or aroma.

(12) "Flour-treating agents" (including bleaching and maturing agents): substances added to milled flour to improve its color and baking qualities.

(13) "Formulation aids" (including carriers, binders, fillers, plasticizers, film-formers, and tableting aids, etc.): substances used to promote or produce a physical state or texture in food.

(14) "Fumigants:" volatile substances used for controlling insects or pests.

(15) "Humectants" (including moisture-retention agents and antidusting agents): hygroscopic substances incorporated in food to promote retention of moisture.

(16) "Leavening agents:" substances used to produce carbon dioxide in baked goods to impart a light texture.

(17) "Lubricants and release agents:" substances added to food contact surfaces to prevent confections and baked goods from sticking to their containers.

(18) "Nonnutritive sweetners:" substances used as a substitute for sugar when intake of sugar or its bulk is undesirable.

(19) "Nutrient supplements:" food components, or their synthetic substitutes, which are necessary for the body's nutritional and metabolic processes.

(20) "pH control agents" (including buffers, acids, alkalies, and neutralizing agents): substances added to change or maintain active acidity or basicity.

(21) "Preservatives" (including antimicrobial agents, fungistats, and mold and rope inhibitors, etc.): substances added to prevent growth of contaminating microorganisms and subsequent spoilage.

(22) "Processing aids" (including clarifying agents, clouding agents, catalysts, flocculents, and filter aids, etc.): substances used as manufacturing aids to enhance the appeal or utility of a food or food component.

(23) "Propellants, aerating agents, and gasses:" chemically inert gasses used to supply force to expel a product or used to reduce the amount of oxygen in contact with the food in packaging processes.

(24) "Sequestrants:" substances which combine with polyvalent metal ions to form a soluble metal complex, to improve the quality and stability of products.

(25) "Solvents and vehicles:" substances used to extract or dissolve another substance.

(26) "Stabilizers and thickeners" (including suspending and bodying agents, setting agents, jelling agents, and bulking agents, etc.): substances used to produce viscous solutions or dispersions, to impart body, improve consistency, or stabilize emulsification.

(27) "Surface-active agents" (other than emulsifiers, but including solubilizing agents, dispersants, detergents, wetting agents, rehydration enhancers, whipping agents, foaming agents, and defoaming agents, etc.): substances used to modify surface properties of food components for a variety of effects.

(28) "Surface-finishing agents" (including glazes, polishes, waxes, and protective coatings): substances used to increase palatability, preserve gloss, and inhibit discoloration of foods.

(29) "Sweeteners:" substances used to sweeten the taste of food.

(30) "Synergists:" substances used to act or react with another food ingredient to produce a total effect different or greater than the sum of the individual effects.

(31) "Texturizers:" substances which affect the appearance or feel of the composition of a food.

ADDITIVE INDEX

387

COMPANY INDEX

The company names listed below are given exactly as they appear in the patents, despite name changes, mergers and acquisitions which have, at times, resulted in the revision of a company name.

INVENTOR INDEX

U.S. PATENT NUMBER INDEX

3,585,223 - 21	3,622,351 - 115	3,695,899 - 225
3,595,681 - 97	3,623,884 - 103	3,709,696 - 207
3,597,235 - 232	3,627,543 - 145	3,716,381 - 69
3,597,236 - 57	3,628,971 - 19	3,718,482 - 324
3,600,198 - 121	3,634,104 - 100	3,732,111 - 301
3,600,200 - 59	3,637,772 - 257	3,732,112 - 93
3,607,304 - 139	3,647,487 - 302	3,734,748 - 188
3,607,311 - 317	3,663,581 - 298	3,751,264 - 296
3,615,691 - 55	3,667,970 - 76	3,753,734 - 211
3,615,703 - 222	3,674,510 - 246	3,754,938 - 255
3,615,717 - 151	3,690,893 - 209	3,759,719 - 249
3,615,727 - 346	3,692,534 - 75	3,762,933 - 49
3,617,312 - 72	3,694,224 - 210	3,764,342 - 294
3,620,773 - 236	3,695,896 - 307	3,792,180 - 340

NOTICE

Nothing contained in this Review shall be construed
to constitute a permission or recommendation to
practice any invention covered by any patent without
a license from the patent owners. Further, neither the
author nor the publisher assumes any liability with
respect to the use of, or for damages resulting from
the use of, any information, apparatus, method or
process described in this Review.

SUGAR SUBSTITUTES
AND ENHANCERS 1973

by Roger Daniels

Food Technology Review No. 5

Because of recent restrictions on the use of synthetic sweeteners, there has developed considerable interest and research activity concerning the use of naturally occurring materials to induce or expand the sweetness of natural sugars, so that lower sugar levels can be used for lower caloric content of foods and beverages.

All of these approaches are covered in this volume, as are new synthetic sweeteners with very low toxicity figures, such as chemical modifications of natural amino acids.

The appendix contains important directives by the U.S. Food and Drug Administration on the status, use, and labeling requirements of effective combinations of such sugar substitutes and enhancers.

Except for the appendix, the book is based on 105 recent U.S. patents. Numbers in ( ) indicate the number of processes per topic. Chapter headings are given here, followed by examples of important subtitles.

1. **MIRACULIN, GLYCYRRHIZIN, AND JERUSALEM ARTICHOKE SWEETENERS (8)**
   Stable, Solid Product
   Solubilizing Method
   Chewing Gum Coating
   Combination with Sucrose
   Citrus Juice Sweetener
   Basic Extraction Methods
   Artichoke Flour for Bread
   Nigerian Berries

2. **DIPEPTIDES, CHALCONES, MALTOLS (13)**
   Aspartic Acid Lower Alkyl Esters
   L-Aspartyl-L-phenylalanine Esters
   L-Aspartyl-L-(β-cyclohexyl)alanine
   Dihydrochalcone Derivatives
   Preparation from Flavanone Glycosides
   Hesperetin Dihydrochalcone
   Saccharin and Dihydrochalcone
   Maltol Derivatives
   Isomaltol Derivatives

3. **STRUCTURAL VARIATIONS AND SYNTHETICS (11)**
   Diacetone Glucose
   5-(3-Hydroxyphenoxy)-1H-tetrazole

Kynurenine Derivatives
8,9-Epoxyperillartine Sweeteners
6-(Trifluoromethyl)tryptophan

4. **COMBINATIONS WITH SACCHARIN (15)**
   Saccharin and Dipeptides
   Saccharin + Glucono-delta-lactone
   Sodium Gluconate Buffer
   Saccharin + Lactose
   Saccharin and Pectin
   Saccharin + Maltol

5. **INCREASING BULK OF MIXES (16)**
   Pillsbury Low Calorie Drink Mix
   Mix with Fumaric Acid
   Peebles Process
   McKesson Process
   Malto-Dextrin Agglomeration

6. **REDUCED CALORIE PRODUCTS (11)**
   Maltol Sweetness Potentiator
   Sugars and Saccharin
   Arabinogalactan
   Arabinogalactan and Corn Starch
   Lactitol Sweeteners
   Maltitol Sweeteners

7. **PRODUCTS: DRINKS, JELLIES, FRUITS (16)**
   Dry Cola Beverage Mix
   Basic Effervescent Process
   Glycerol Sweetener
   Carragheenan-Pectin Base
   Honey-Malt Flavoring
   Freezedried Fruit Sweeteners

8. **FROZEN DESSERTS, DAIRY PRODUCTS, BAKED GOODS, CONFECTIONS (15)**
   Use of Polyoses
   Dietary Dry Cake Mix
   Insoluble Protein Approach
   Gum Acacia + Corn Syrup
   Sugarless Hard Candy

9. **APPENDIX**
   Exemption of Certain Food Additives from the Requirement of Tolerances.
   Combinations of Nutritive and Nonnutritive Sweeteners in Canned Fruits.
   Food Additives Permitted in Food for Human Consumption, or in Contact with Food, for Limited Periods of Time.

ISBN 0-8155-0492-6

275 pages

DEHYDRATION OF NATURAL AND SIMULATED DAIRY PRODUCTS 1974

by M. T. Gillies

Food Technology Review No. 15

Commercial production of dried dairy products began early in this century and was expanded considerably during World War II, when the U.S. government bought huge quantities of dried milk and eggs for the armed services and for lend-lease programs.

This book begins with an overview of patented dehydration processes which are suitable for a broad range of dairy products. It then deals with the special problems of dehydrating whole milk and combinations of fats plus milk proteins. Many processes on the manufacture of nonfat dry milk also discuss agglomeration techniques to improve this product's wettability and dispersion characteristics upon rehydration.

Numerous miscellaneous dairy foods capable of dehydration are given extensive treatment and the last section discusses the drying of simulated dairy products for many purposes, including coffee whiteners, dessert mixes, whipped toppings, etc. These products are commonly made from edible caseinates plus fats and carbohydrates.

A partial and condensed table of contents follows. Numbers in parentheses indicate the number of processes per topic. Chapter headings are given, followed by examples of important subtitles. 173 distinctly different processes are described.

ISBN 0-8155-0539-6

328 pages

SHORTENINGS, MARGARINES, AND FOOD OILS 1974

by M. T. Gillies

Food Technology Review No. 10

Margarines and prepared shortenings are now widely accepted by the public due to their lower cost and absence of saturated fat.

Polyunsaturated fats are believed to be essential for growth and the maintenance of normal skin conditions, also to reduce blood cholesterol levels. Since the newer margarines contain polyunsaturated fats in liberal amounts, even better public acceptance and increased sales can be predicted with certainty.

All margarines and many shortenings may be considered as emulsions. Without proper additives or interesterification of the constituent oils their texture can break down from excessive freezing or heating, making them unsalable because of granular, lumpy appearance.

The processing of oils and fats, plus antispattering agents, crystallization inhibitors, flavors and colors, has grown into a highly developed technology, mostly because of the physico-chemical aspects of the raw materials and of the finished products.

How to prepare and apply these essential food products on a commercial scale is the subject of this book, illustrated by processes from 174 late U.S. patents.

A partial and condensed table of contents follows. Numbers in parentheses indicate the number of processes per topic. Chapter headings are given, followed by examples of important subtitles.

ISBN 0-8155-0517-5